Receptors and Recognition

General Editors: P. Cuatrecasas and M.F. Greaves

About the series

Cellular recognition — the process by which cells interact with, and respond to, molecular signals in their envrionment — plays a crucial role in virtually all important biological functions. These encompass fertilization, infectious interactions, embryonic development, the activity of the nervous system, the regulation of growth and metabolism by hormones and the immune response to foreign antigens. Although our knowledge of these systems has grown rapidly in recent years, it is clear that a full understanding of cellular recognition phenomena will require an integrated and multidisciplinary approach.

This series aims to expedite such an understanding by bringing together accounts by leading researchers of all biochemical, cellular and evolutionary aspects of recognition systems. The series will contain volumes of two types. First, there will be volumes containing about five reviews from different areas of the general subject written at a level suitable for all biologically oriented scientists (Receptors and Recognition, series A). Secondly, there will be more specialized volumes, (Receptors and Recognition, series B), each of which will be devoted to just one particularly important area.

Advisory Editorial Board

Receptors and Recognition

Series A

Published

Volume 1 (1976)
M.F. Greaves (London), Cell Surface Receptors: A Biological Perspective
F. Macfarlane Burnet (Melbourne), The Evolution of Receptors and Recognition in the Immune System
K. Resch (Heidelberg), Membrane Associated-Events in Lymphocyte Activation
K.N. Brown (London), Specificity in Host-Parasite Interaction

Volume 2 (1976)
D. Givol (Jerusalem), A Structural Basis for Molecular Recognition: The Antibody Case
B.D. Gomperts (London), Calcium and Cell Activation
M.A.B. de Sousa (New York), Cell Traffic
D. Lewis (London), Incompatibility in Flowering Plants
A. Levitski (Jerusalem), Catecholamine Receptors

Volume 3 (1977)
J. Lindstrom (Salk, California), Antibodies to Receptors for Acetylcholine and other Hormones
M. Crandall (Kentucky), Mating-Type Interaction in Micro-organisms
H. Furthmayr (New Haven), Erythrocyte Membrane Proteins
M. Silverman (Toronto), Specificity of Membrane Transport

Volume 4 (1977)
M. Sonenberg and A.S. Schneider (New York), Hormone Action at the Plasma Membrane: Biophysical Approaches
H. Metzger (NIH, Bethesda), The Cellular Receptor for IgE
T.P. Stossel (Boston), Endocytosis
A. Meager (Warwick) and R.C. Hughes (London), Virus Receptors
M.E. Eldefrawi and A.T. Eldefrawi (Baltimore), Acetylcholine Receptors

Volume 5 (1978)
P.A. Lehmann (Mexico), Stereoselective Molecular Recognition in Biology
A.G. Lee (Southampton, U.K.), Fluorescence and NMR Studies of Membranes
L.D. Kohn (NIH, Bethesda), Relationships in the Structure and Function of Receptors for Glycoprotein Hormones, Bacterial Toxins and Interferon

Volume 6 (1978)
J.N. Fain (Providence, Rhode Island), Cyclic Nucleotides
G.D. Eytan (Haida) and B.I. Kanner (Jerusalem), Reconstitution of Biological Membranes
P.J. O'Brien (NIH, Bethesda), Rhodopsin: A Light-sensitive Membrane Glycoprotein
Index to Series A, Volumes 1–6

Series B

Published

The Specificity and Action of Animal, Bacterial and Plant Toxins (B1)
edited by P. Cuatrecasas (Burroughs Wellcome, North Carolina)

Intercellular Junctions and Synapses (B2)
edited by J. Feldman (London), N.B. Gilula (Rockefeller University, New York)
and J.D. Pitts (University of Glasgow)

Microbial Interactions (B3)
edited by J.L. Reissig (Long Island University, New York)

Specificity of Embryological Interactions (B4)
edited by D. Garrod (University of Southampton)

Taxis and Behavior (B5)
edited by G.L. Hazelbauer (University of Uppsala)

In preparation

Virus Receptors Part 1 *Bacterial Viruses* (B7)
edited by L. Randall and L. Philipson (University of Uppsala)

Virus Receptors Part 2 *Animal Viruses* (B8)
edited by K. Lonberg-Holm (Du Pont, Delaware) and L. Philipson (University of
    Uppsala)

Membrane Receptors: Methods for Purification and Characterization (B9)
edited by P. Cuatrecasas and S. Jacobs

Neurotransmitter Receptors
edited by S. Enna and H. Yamamura

Purinergic Receptors
edited by G. Burnstock

Receptor Regulation
edited by R. Lefkowitz

Transplantation Antigens
edited by P. Parham and J. Strominger

Receptors and
Recognition

Series B Volume 6

Bacterial Adherence

Edited by
E. H. Beachey

*Veteran's Administration Hospital,
Memphis, Tennessee*

1980
LONDON AND NEW YORK

CHAPMAN AND HALL
150th Anniversary

First published 1980
by Chapman and Hall Ltd.,
11 New Fetter Lane, London EC4P 4EE

Published in the U.S.A. by
Chapman and Hall
in association with Methuen Inc.,
733 Third Avenue, New York, NY 10017

© *1980 Chapman and Hall*
Softcover reprint of the hardcover 1st edition 1980

ISBN-13:978-94-009-5865-4

British Library Cataloging in Publication Data

Receptors and recognition.
   Series B. Vol. 6: Bacterial adherence
   1. Cell interaction
   I. Cuatrecasas, Pedro   II. Greaves, Melvyn
   Francis   III. Beachey, E.H.
   574.8'76   QH604.2   80-40093

   ISBN-13:978-94-009-5865-4   e-ISBN-13:978-94-009-5863-0
   DOI: 10.1007/978-94-009-5863-0

Contents

Contributors *page* ix

Preface x

1 **General Concepts and Principles of Bacterial Adherence in Animals and Man** 1
Itzhak Ofek and Edwin H. Beachey

2 **Adherence of Normal Flora to Mucosal Surfaces** 31
Dwayne C. Savage

3 **Bacterial Adherence and the Formation of Dental Plaques** 61
Ronald J. Gibbons and Johannes van Houte

4 **Mechanisms of Adherence of *Streptococcus mutans* to Smooth Surfaces *in vitro*** 105
Shigeyuki Hamada and Hutton D. Slade

5 **Structure and Cell Membrane-Binding Properties of Bacterial Lipoteichoic Acids and their Possible Role in Adhesion of Streptococci to Eukaryotic Cells** 137
Anthony J. Wicken

6 **Attachment of *Mycoplasma pneumoniae* to Respiratory Epithelium** 159
Albert M. Collier

7 **Adhesive Properties of Enterobacteriaceae** 185
J.P. Duguid and D.C. Old

8 **The Adhesive Properties of *Vibrio cholerae* and other *Vibrio* Species** 219
Garth W. Jones

9 **Adherence of *Neisseria gonorrhoeae* and other *Neisseria* Species to Mammalian Cells** 251
Peter J. Watt and Michael E. Ward

10 **Structure and Cell Membrane-Binding Properties of Bacterial Fimbriae** 289
William A. Pearce and Thomas M. Buchanan

vii

11 **Adherence of Marine Micro-organisms to Smooth Surfaces** *page* 345
Madilyn Fletcher

12 **Microbial Adherence in Plants** 375
James A. Lippincott and Barbara Lippincott

13 **Cell Recognition Systems in Eukaryotic Cells** 399
David R. Phillips and T. Kent Gartner

14 **Prospects for Preventing the Association of Harmful
Bacteria with Host Mucosal Surfaces** 439
Rolf Freter

Index 459

Contributors

E.H. Beachey, Veteran's Administration Hospital, Memphis, Tennessee, U.S.A.

T.M. Buchanan, Departments of Medicine and Pathbiology, University of Washington, Seattle, Washington, U.S.A.

A.M. Collier, Department of Pediatrics, School of Medicine, University of North Carolina, North Carolina, U.S.A.

J.P. Duguid, Bacteriology Department, Ninewells Hospital, Dundee, U.K.

Madilyn Fletcher, Department of Environmental Sciences, University of Warwick, Coventry, U.K.

R. Freter, Department of Microbiology, University of Michigan, Ann Arbor, Michigan, U.S.A.

T.K. Gartner, Department of Biology, Memphis State University, Memphis, Tennessee, U.S.A.

R.J. Gibbons, Forsyth Dental Center, Harvard School of Dental Medicine, Boston, Massachusetts, U.S.A.

S. Hamada, Osaka University Dental School, Osaka, Japan.

G.W. Jones, University of Michigan Medical School, Ann Arbor. Michigan, U.S.A.

Barbara Lippincott, Department of Biological Sciences, Northwestern University, Evanston, Illinois, U.S.A.

J.A. Lippincott, Department of Biological Sciences, Northwestern University, Evanston, Illinois, U.S.A.

I. Ofek, Veteran's Administration Hospital, Memphis, Tennessee, U.S.A.

D.C. Old, Bacteriology Department, Ninewells Hospital, Dundee, U.K.

W.A. Pearce, Department of Medicine, University of Washington, Seattle, Washington, U.S.A.

D.R. Phillips, St Jude Children's Research Hospital, Memphis, Tennessee, U.S.A.

D.C. Savage, Department of Microbiology, School of Basic Medical Sciences, University of Illinois, Urbana, Illinois, U.S.A.

H.D. Slade, Department of Microbiology and Immunology, The Medical and Dental Schools, Northwestern University, Chicago, U.S.A.

J. van Houte, Forsyth Dental Center, Harvard School of Dental Medicine, Boston, Massachusetts, U.S.A.

M.E. Ward, Department of Microbiology, Faculty of Medicine, Southampton General Hospital, Southampton, U.K.

P.J. Watt, Department of Microbiology, Faculty of Medicine, Southampton General Hospital, Southampton, U.K.

A.J. Wicken, The University of New South Wales, Kensington, New South Wales, Australia.

Preface

Bacteria adhere to and colonize almost any surface. Within minutes after sub-merging a solid object in seawater or freshwater, the surface becomes colonized by adherent micro-organisms, and the earliest organisms to adhere are bacteria. Adherent colonies of bacteria have also been observed on particles of sand, soil, other bacteria, plant tissues, and a variety of animal tissues. Shortly after birth, the skin and the mucosal surfaces of the upper respiratory tract and the gastro-intestinal tract of animals and man become heavily colonized by a variety of adherent bacteria which persist in varying numbers as indigenous parasites. The apparent symbiotic balance between the host and his indigenous parasites occasionally is upset by the invasion of harmful bacteria which adhere to and colonize these surfaces. Pathogenic bacteria may also adhere to and colonize normally sterile surfaces such as the mucosa of the genito-urinary tract and the lower respiratory tract, and occasionally even endothelial surfaces of the cardiovascular system, resulting in the development of serious infectious diseases.

Although marine microbiologists have been aware for a long time that bacteria must stick to surfaces in order to avoid being swept away by moving streams of water, not until recently has it been widely recognized that adherence must be an important ecological determinant in the colonization of specific sites in plants and animals, and in particular an important early event in the pathogenesis of bacterial infections in animals and man. It is true that Dr G. Guyot as early as 1908 (*Abl. Bakt.*, I. *Abt. orig.* **47**, 640) reported studies on the adhesiveness of bacterial cells for blood erythrocytes, and that some 20 years ago Dr Duguid (see Chapter 7) had already demonstrated the mannose sensitivity of the adherence of several genera and species of gram-negative bacteria to erythrocytes and intestinal epithelial cells. Nevertheless, the study of the mechanisms of bacterial adherence did not really 'catch on' until about 10 years ago when Dr R.J. Gibbons and his colleagues began reporting a series of elegant studies showing the selective nature of the adherence of bacteria to the various niches of the oral cavity and dental surfaces (see Chapter 3). Largely because of these studies, bacterial adherence has grown into one of the most active, if not the most exciting, areas of study in the field of microbial ecology and infectious diseases.

This book compiles for the first time into one volume a series of monographs prepared by invited experts on the subject of bacterial adherence. A particular emphasis is placed on the possible role of bacterial adhesive properties in the colonization and ecology of a variety of surfaces, both animate and inanimate. The mechanisms by which bacteria interact with these surfaces, be they smooth surfaces or tissue cell membranes, are examined in detail, the implicit premise being that the interaction of bacteria with surfaces involves specific molecular ligands (or adhesins as some authors prefer to call them) on the surfaces of bacteria, which interact by a specific 'lock and key' mechanism with receptor molecules on the surfaces to be colonized. Moreover, it is presumed that the isolation and identification of the adhesive molecular structures on both surfaces will suggest new approaches to the control of certain bacterial infectious diseases. One obvious approach would be to try to block the adherence of harmful bacteria by the application of ligand or receptor materials once these substances or their analogues have been identified. Indeed, limited success has already been reported using such approaches and is discussed by several of the contributors to this volume.

The first ten chapters of this book are devoted to detailed studies of the chemical and molecular basis of the association of bacteria with animal tissues. The adhesive properties of marine micro-organisms are discussed by Fletcher in Chapter 11 and that of plant micro-organisms by Drs Lippincott and Lippincott in Chapter 12. This is followed by a more general view of recognition systems among eukaryotic cells elegantly discussed by Phillips and Gartner. In a more philosophical vein, Dr Freter, in the final chapter, discusses prospects for the future both with regard to the possible means one might use to interfere with the adherence process to prevent infectious diseases and with regard to the directions future work in the field should take.

It is hoped that this book will be of interest to a wide audience including microbiologists, physicians, dentists, cell biologists, biochemists, and many others. It is hoped in particular that the various contributions will serve to arouse the interest of students of the biological sciences to pursue specific problems which remain to be unraveled in this rapidly growing area of microbial ecology and infectious diseases.

As with most multi-author books, the present book on *Bacterial Adherence* has some unavoidable overlaps among chapters and in a few instances there may even be some duplication. However, it was considered more desirable to allow each chapter to stand on its own rather than to shorten the book by only a few pages.

My editorial duties on this book were made much easier by the untiring secretarial assistance of Mrs Johnnie Smith who among other things, gently prodded the contributors to submit their articles. My special thanks to Dr Uli Schwarz, who granted me beautiful laboratory and office spaces at the Max Planck Institute für Virusforschung in Tübingen, W. Germany, during a sabbatical leave from the Veterans Administration Hospital and University of Tennessee in Memphis. Many of my editorial duties on this volume were discharged while there. Many thanks to my longtime colleague and friend, Dr I. Ofek, for lively and stimulating discussions and for many helpful suggestions. The comments and suggestions offered by many of the contributing authors are also deeply appreciated.

May, 1979   Edwin H. Beachey, M.D.

1 General Concepts and Principles of Bacterial Adherence in Animals and Man

ITZHAK OFEK and EDWIN H. BEACHEY

1.1	Introduction	*page*	3
1.2	Repulsive and attractive forces		5
	1.2.1 Surface hydrophobicity and net surface charge		5
	1.2.2 Ligand—receptor interactions		6
	1.2.3 Ligand and receptor accessibility		10
1.3	Genetic and phenotypic variables which infleunce bacterial adherence		11
	1.3.1 Genetic variants of bacteria		11
	1.3.2 Phenotypic variants of bacteria		12
1.4	Host cell-associated variables		13
	1.4.1 Variables associated with epithelial cells		13
	1.4.2 Comparison of the adherence of bacteria to epithelial and other host cells		14
	1.4.3 Bacterial attachment to phagocytic cells		14
	1.4.4 Resistance to phagocytosis		15
1.5	Exogenous factors which influence bacterial adherence		16
1.6	Relationship of bacterial adherence to infections		17
	1.6.1 Infectivity versus adherence		18
	1.6.2 Changes in adhering abibility of bacteria *in vivo*		23
	1.6.3 Blocking of bacterial adherence *in vivo*		24
	References		26

Acknowledgements
Research support for previously unpublished data included in this review was
provided by program-directed research funds from the United States Veterans
Administration and by Research Grants AI-13550 and AI-10085 from the United
States Public Health Service. E.H.B. is the recipient of a Medical Investigatorship
award from the U.S. Veterans Administration.

We thank Mr D.C. Williams for drawing the figures, Dr Barry I. Eisenstein for
stimulating discussions and review of the manuscript, Mrs Carol Beachey for
editorial assistance and Mrs Johnnie Smith for secretarial assistance in preparing the
manuscript.

Bacterial Adherence
(*Receptors and Recognition,* Series B, Volume 6)
Edited by E.H. Beachey
Published in 1980 by Chapman and Hall, 11 New Fetter Lane, London EC4P 4EE
© 1980 Chapman and Hall

1.1 INTRODUCTION

Adherence of bacteria to tissue surfaces has recently gained increasing attention as an important initial event in the pathogenesis of bacterial infections (Ofek and Beachey, 1980). The infectious process in animals and man can be envisioned as a stepwise process in which bacteria must first adhere to a tissue surface. The failure to adhere would result simply in their being swept away in the fluids which constantly bathe the tissue surfaces. Adherence of pathogenic organisms is then followed by colonization and eventual invasion of the surface either by a toxin produced by the colonizing organisms or by the bacteria themselves. In the deeper tissues, the attachment of the bacteria to phagocytic cells results in their ingestion and destruction. Organisms whose surfaces are not recognized by antibody and complement or phagocytic cells can multiply unimpeded to produce systemic infections. These steps in the infectious process are depicted diagramatically in Fig. 1.1.

Although the fact that bacteria adhere to specific tissue surfaces has been well-documented, the precise role of adherence in tissue colonization and the development of infectious diseases has been difficult to define primarily because the interaction between bacteria and the various tissue surfaces represent cell—cell interactions which are physicochemically so complex that the specificity of the interactions are difficult to assess in a single model system. The study of a single modality such as adherence may give only partial clues as to the complex mechanisms of the interaction of bacteria with specific tissue surfaces in the intact host. For example, Freter (1978; see also Chapter 14, this volume) has pointed out that *V. cholera* infections of the gastro-intestinal tract involve a number of complex mechanisms including bacterial motility, chemotactic attraction, penetration of the mucus gel on the intestinal villi, adhesion to receptors in the mucus gel, chemotaxis into deeper intervillous spaces, adherence of the organisms to the epithelial surface and finally the elabotation of an injurious toxin.

Since the intitial cell—cell interaction is a surface phenomenon, investigators have been searching for molecules on the surfaces of the prokaryotic and eukaryotic cells that might attract and bind each other in a specific way. Thus far, most of these studies have been performed *in vitro*. Suspensions of eukaryotic cells (e.g. epithelial cells, erythrocytes, monolayers of tissue culture cells, excised tissues) are exposed to a standardized concentration of bacteria for a given period of incubation. The non-adherent bacteria are then removed by filtration or by repeated differential centrifugations. The advantage of using these *in vitro* systems is that one can readily assess the effects of various conditions of incubation or of various defined substances upon the adherence process. The disadvantage is that many of the physiological barriers of the intact tissue surfaces as outlined above are bypassed. Therefore, the

3

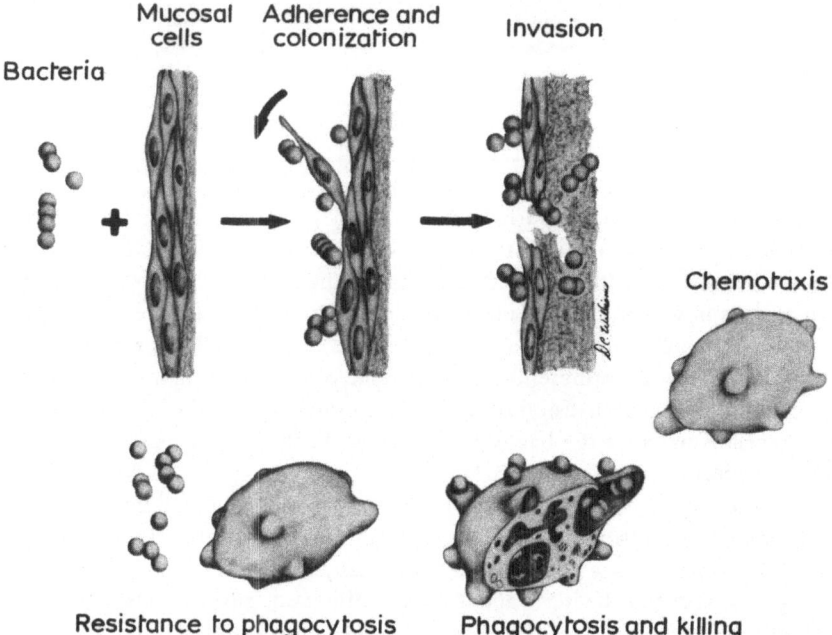

Fig. 1.1 Schematic of the proposed sequential steps in the infectious process in animals and man. In the first step, bacteria adhere to and multiply on the tissue surface, composed of epithelial cells in this diagram. In order to maintain colonization, the organisms must replenish the newly exposed epithelial cells as old cells with adherent bacteria are lost by desquamation (top center; see also Section 3.4 and Fig. 3.7). Injury to the epithelial cell barrier either by toxins produced by the bacteria themselves, or by other agents such as viruses, allows the bacteria to invade into deeper tissues (top right). In response to chemotactic stimuli produced by the invading organisms in conjunction with antibodies and complement, phagocytic cells (far right) migrate into the area. If the organisms lack surface components to mask their ligands, or if the host is immune and coats the bacteria with opsonic antibody, the organisms are recognized by receptors on the phagocytic cells for the bacterial ligand itself or for the Fc portion of the opsonic antibody, respectively. Once attached, the organisms are ingested and destroyed (bottom right). If, however, the organisms produce capsular material to which the host is not immune and which masks surface ligands, the organisms are not recognized by the phagocytic cells (bottom left) and are thereby able to multiply unhindered to produce a systemic infectious disease.

data obtained from *in vitro* experiments must be interpreted with caution as to their *in vivo* significance. Nevertheless, with this caution in mind, *in vitro* methods may be useful in identifying the molecules which cause the cells to stick to each other once they do come interpreted.

In this chapter we will deal with some of the general aspects of bacterial adherence with special emphasis on the epithelial cell. Epithelial cells cover all external surfaces of animals and humans and therefore represent the first cells that the colonizing organisms must encounter.

1.2 REPULSIVE AND ATTRACTIVE FORCES

The surface charges of eukaryotic and prokaryotic cells are, in sum, both negative. Yet there are both long-range and short-range attractive forces which act to overcome the repulsive force between the like-charged surfaces. The long-range attractive forces operating according to the DLVO theory are elegantly discussed by Watt and Ward in Section 9.5.2 and will, therefore, not be dealt with further here. Rather we focus on non-specific and specific short-range forces.

1.2.1 Surface hydrophobicity and net surface charge

Several methods have been developed to measure surface charge and surface hydro-phobicity of bacteria in relation to adherence. Brinton (1965) measured the electro-phoretic mobility of bacteria and found that the type 1 fimbriated *E. coli*, which adhere readily to cells, are less mobile than the non-fimbriated organisms which lack adhesive properties (see also Chapter 8, this volume). The importance of net surface charge was perhaps best demonstrated by Ward and his colleagues (see Section 10.5.1) who demonstrated that the adherence of non-fimbriated gonococci could be increased to the level of fimbriated organisms by chemically modifying the surface charges of the non-fimbriated bacteria. It is suggested that fimbriae increase adherence by simply counteracting repulsive electrostatic forces.

The hydrophobicity of the bacterial surface can be measured by partition of the bacterial suspension between biphasic aqueous solutions (Stendahl *et al.*, 1973), or by measuring the degree of adsorption of the test strain to hydrophobic gels consist-ing of Sepharose beads covalently linked to various hydrophobic residues (Smythe *et al.*, 1978). Perers *et al.* (1977) and Kihlstrom and Edebo (1976) found that the degree of hydrophobicity of *S. typhimurium* and *E. coli*, as measured by the partition method, correlated with the ability of the organisms to adhere to tissue culture cells. The greater the hydrophobicity the greater was the adherence. Smyth *et al.* (1978) found that K-88 positive strains of *E. coli* which agglutinate erythrocytes, adsorb strongly to hydrophobic ligands on Sepharose beads as opposed to the isogenic mutant which lacked K-88 antigen and did not agglutinate erythrocytes (see also Section 9.5.5).

It is possible that hydrophobic molecules on the bacterial surface allow the bacteria to approach the negatively charged epithelial cell and thereby enable special binding molecules on each of the cell surfaces to interact with each other to form specific bonds of high affinity (Fig. 1.2). By analogy, concanavalin A, which

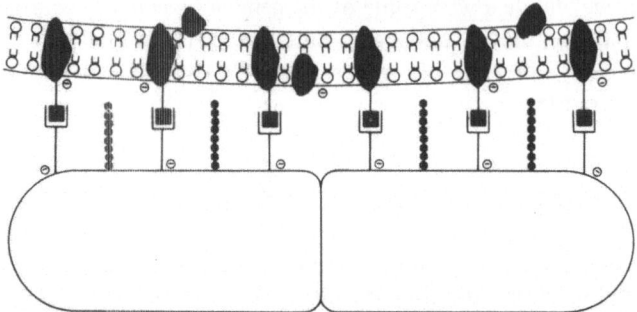

Fig. 1.2 Attachment of a bacterium (bottom) via specific ligands (Ψ) to complementary receptor (■) on the membrane (top) of an animal cell. In order to overcome the net negative charge (⊖) on both the bacterial and cell membrane surfaces, hydrophobic molecules (♦) on the surface of the bacteria are attracted toward the hydrophobic phospholipid molecules (ᑫ) in the lipid bilayer membrane. The irregular black structures represent protein and glycoprotein molecules incorporated into the animal cell membrane.

binds specifically to mannose residues, has been shown (Lis and Sharon, 1977) to possess hydrophobic sites which facilitate the interaction of the lectin with the membrane receptor. Once bacteria become attached to the cells, the avidity of attachment is dependent upon the formation of many bonds which, for practical purposes, are irreversible since the probability for all of the binding sites to become simultaneously unbound is very small (Bell, 1978).

1.2.2 Ligand–receptor interactions

There is much evidence to suggest that bacteria possess molecules on their surfaces capable of binding in a stereospecific fashion with complementary molecules on the surface of tissue cells of the host (Jones, 1977; Ofek *et al.*, 1978) (see Fig. 1.2). In this chapter, the binding molecules on bacteria are called ligands and those on host cells, receptors. The interaction of the bacterial ligands with host cell receptors can be compared to the well-known antigen–antibody or plant lectin–sugar interactions (Sharon and Lis, 1972). Thus, the specificity of the interaction can be demonstrated by

(1)   inhibiting the interaction with large excesses of 'haptens' either identical to or resembling the native ligand or the native receptor (Fig. 1.3),
(2)   chemical or enzymatic treatment of the bacteria or tissue cells to abolish or alter the specific surface structures involved in adherence, and
(3)   blocking the ligand or receptor with specific antibodies directed against antigens composing these structures.

The demonstration of the ability of bacterial or host cell surfaces to bind the

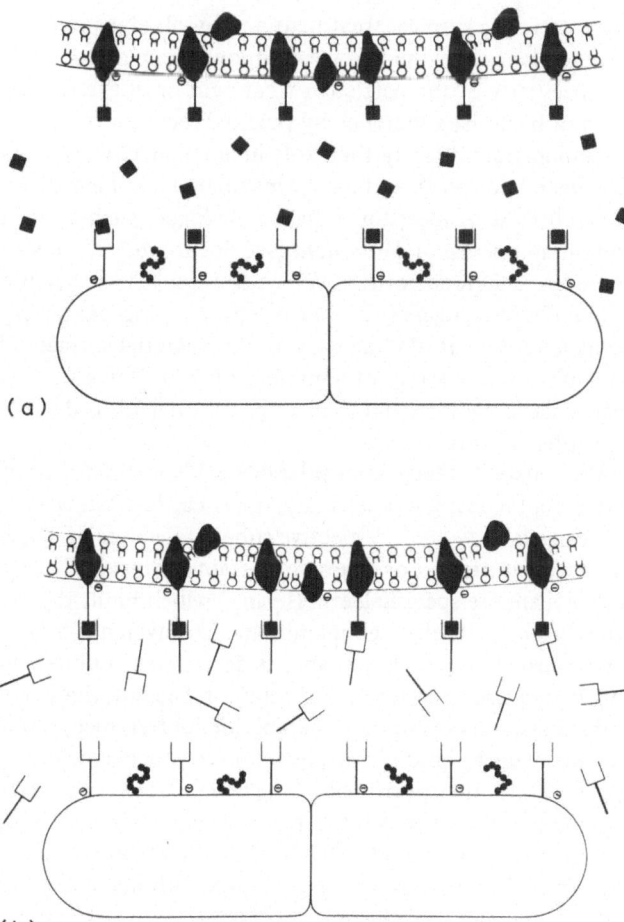

Fig. 1.3 Specific blockade of bacterial adherence by an excess of (a) isolated receptor or receptor analogue material (■), or by (b) isolated ligand or ligand analogue material (Ψ). (See also Fig. 1.2).

purified receptor or ligand, respectively, in a specific manner provides additional evidence that the surface substances are involved in receptor—ligand recognition. Binding assays with the purified materials can be used to determine the affinity and number of binding sites for the ligand or receptor on the respective cell surfaces. Bacteria may produce substances which bind to specific receptors on host cells but the role of such substances in adherence has yet to be ascertained. For example, although lipopolysaccharide of gram-negative organisms binds to specific receptors on host cells (Springer *et al.*, 1974), its role in mediating adherence of micro-organisms to animal cells has not been established. The interpretation of results obtained by enzymatic or chemical treatment and by blocking experiments with antibodies as to the identity of the surface components involved in adherence depends on the specificity of the particular treatment or of the blocking antibody.

The inhibition of bacterial adherence by purified receptor or ligand material would serve to demonstrate directly their role in adherence. Unfortunately, in most cases neither the ligand nor receptor have been isolated and identified and, therefore, one must resort to the use of receptor or ligand analogues such as, for example, L-fucose to prevent adherence of *V. cholera* (see Section 8.3.3., this volume) and D-mannose to prevent adherence of *E. coli* (see Section 7.2, this volume). Particles which contain immobilized sugars (or other receptor analogues) on their surfaces can serve to demonstrate sugar binding sites on the bacterial surface. These principles have now been applied to the study of a number of organisms and the evidence obtained for particular organisms and their respective ligands is described in detail in subsequent chapters in this volume.

As will become apparent, there is considerable evidence that a number of bacteria, including *E. coli*, streptococci gonococci, mycoplasma, *V. cholera* as well as marine and plant parasites, adhere to their respective substrata via specific ligand interactions with complementary receptors. For example, certain strains of *E. coli* have been shown to possess a mannose-specific lectin (ligand) which binds the organisms to mannose residues on the epithelial cell membrane. The evidence is as follows: First, of the many sugars tested only mannose and its derivatives inhibited adherence (Ofek *et al.*, 1977a; see also Chapter 9, this volume). Second, the chemical modification of cell membrane sugars by sodium metaperiodate treatment abolished the ability of the treated cells to bind *E. coli*, but not streptococci which adhere to different receptors. Third, the specific blocking of mannose residues, but not other sugar residues, on the epithelial cells by lectins prevented adherence of *E. coli* but again, not of streptococci (Ofek *et al.*, 1977a). Fourth, mannose-binding activity of whole *E. coli* cells was demonstrated by aggregation with mannan-containing yeast cells (Ofek *et al.*, 1977a; Ofek and Beachey, 1978). Fifth, the fraction of organisms which adhered to epithelial cells and was displaced by α-methyl mannoside possessed much higher mannose-binding activity as compared to the unfractionated and non-adherent organisms (Ofek and Beachey, 1978). Furthermore, the degree of mannose binding activity, as monitored by aggregometry with mannan-containing yeast cells, correlated with the ability of the organisms to adhere to epithelial cells (Ofek and Beachey, 1978). Finally, type I fimbriae-containing solutions were shown to possess cell membrane-binding activity (Salit and Gotschlich, 1977a and b) and to inhibit adherence of *E. coli* to animal cells (Isaacson *et al.*, 1978).

In general, the chemical nature of bacterial ligands was found to be (a) protein in certain strains of *E. coli* (Salit and Gotschlich, 1977a; Eshdat *et al.*, 1978; Evans and Pauley, 1978; Jones, 1977), gonococci (Buchanan and Pearce, 1976 and Chapter 10), and mycoplasma (Hu *et al.*, 1977; see also Chapter 6, this volume); (b) the lipid portion of glycolipid molecules such as lipoteichoic acid (LTA) in group A streptococci (Ofek *et al.*, 1975; Beachey and Ofek, 1976) — although not conclusive, LTA may be the ligand of other LTA-bearing gram-positive bacteria since it has been found to inhibit the epithelial adherence of various strains of α-hemolytic streptococci (unpublished observations) as well as of *Staphylococcus aureus* (Aly *et al.*, 1979); (c) sugars and other carbohydrates in certain plant parasites (Chapter 12) and cariogenic streptococci (Chapters 3 and 4). The ligands may be associated with surface structures called fimbriae in certain gram-negative bacteria (Chapter 10, this volume) and with fibrillae in gram-positive bacteria (Gibbons, 1977).

Thus far, sugar moieties are the only residues identified as bacterial receptors on epithelial cells (Ofek *et al.*, 1978; see also Section 10.3.4). These are residues of mannose for certain *E. coli* strains (Ofek *et al.*, 1977a; Ofek and Beachey, 1978), fucose for *V. cholera* (Chapter 8), sialic acid for *Mycoplasma sp.* (Chapter 6) and β-galactosyl residues for K-88 positive *E. coli* (Gibbons *et al.*, 1975). It is possible that *N*-acetyl-D-galactosamine residues serve as site for the attachment for *Leptotrichia buccalis* since this sugar alone of the many tested inhibited hemagglutination induced by the micro-organisms (Kondo *et al.*, 1976). The sugar residues are almost certainly present either as glycolipids or glycoproteins in the cell membrane. There is no conclusive evidence which indicates that proteins or peptides in the epithelial cell membrane are involved in the specific recognition of bacteria. Bacterial surfaces, however, were found to bind proteins or glycoproteins (Kashket and Guilmette, 1978; Gibbons and Quershi, 1978) and in some cases via specific binding sites on the bacterial surfaces (Christensen *et al.*, 1976). Thus, as with bacterial antibodies, such protein preparations may inhibit bacterial adherence by either combining with the ligand in a stereospecific fashion (competitive inhibition), by steric hindrance, or by simply modifying the surface hydrophobicity and charge of the bacteria.

Recent studies (Simpson *et al.*, 1979) in our laboratory suggest that certain proteins may harbor determinants which bind specifically to bacterial ligands and inhibit epithelial adherence. Albumin is known to possess binding sites for fatty acids (Peters, 1975). Ester-linked fatty acids of lipoteichoic acid (LTA) were shown to mediate the binding of LTA to cell membranes and to serve on the streptococcal surface as a ligand which binds the organisms to epithelial cells (Ofek *et al.*, 1975; Beachey and Ofek, 1976). We investigated the ability of albumin binding sites for fatty acids to bind ester-linked fatty acids of LTA and thus to block the cell membrane binding of LTA. Albumin covalently linked to Sepharose 4B bound intact but not deacylated [^{3}H]-LTA. The bound [^{3}H]-LTA was specifically eluted by unlabelled LTA, albumin and the methyl ester of palmitic acid. The bound [^{3}H]-LTA was eluted with 50% ethanol, but not with increasing ionic concentration. The binding of [^{3}H]-LTA to erythrocytes was inhibited by albumin, but not other serum protein fractions (Beachey *et al.*, 1979b; see also Section 5.2.1). The epithelial adherence of streptococci was inhibited by as little as 50 μg ml^{-1} of albumin. Fatty acid-free albumin was a more potent inhibitor than the albumin of Cohn fraction V, indicating that the binding sites of albumin for fatty acids are involved in blocking the adherence. Epithelial cell attachment of streptococci was restored by removing the albumin from the mixture, indicating that inhibition of adherence by albumin is reversible and does not involve any irreversible alterations on either type of cell.

These data suggest that the fatty acid binding sites of the albumin molecule are able to bind ester-linked fatty acids of LTA and thereby inhibit both binding of LTA itself and of streptococci whose adherence appears to be mediated by LTA. It is possible that certain membrane proteins or glycoproteins contain LTA binding sites which serve as sites for the adherence of streptococci to epithelial cells. The idea is strengthened by (a) the finding of Bartelt and Duncan (1978) that treatment of tissue culture cells with trypsin abolished the ability of the cells to bind streptococci, (b) the finding that epithelial cells (Simpson *et al.*, 1978) as well as other host cells (Beachey *et al.*, 1979a and b) possess a single population of binding sites for LTA, and (c) the finding that only the right side-out, but not the inside-out erythrocyte membrane binds LTA (Chiang *et al.*, 1979), indicating that components associated with the membrane rather than the membrane itself binds the ester-linked fatty acid of LTA. The ultimate proof, however, that a protein(s) of cell membranes is involved in streptococcal adherence awaits isolation from the cell membrane of a protein which binds LTA in a stereospecific fashion.

1.2.3 Ligand and receptor accessibility

Recent studies on ligand—receptor binding in cell—cell interactions revealed an additional factor of potential importance. Sharom *et al.* (1977) isolated concanavalin A (con A) receptors from human erythrocytes and incoporated them into artificial membranes (liposomes). By comparing the ability of erythrocytes and liposomes to bind con A, the authors concluded that agglutination by the lectin did not reflect a difference in the number of receptors or lipid fluidity, but rather a difference in receptor organization and presentation. Is it possible that the organization of the receptors on epithelial cells and the ligands on bacterial cells play a role in adherence?

Other investigators have presented evidence that blocking moieties may be important. In preliminary studies it was found by Selinger and Reed (1979) that the formation of capsule on the pneumococcal surface impairs the ability of the organisms to adhere to epithelial cells presumably by masking bacterial ligands as depicted diagramatically in Fig. 1.4. Salit and Gotschlich (1977b) found that only after trypsinization of human erythrocytes were the cells agglutinated by *E. coli* and their fimbriae, indicating that a protein structure blocks the binding of the *E. coli* to mannose residues on the erythrocyte membrane. Brecher *et al.* (1978) showed that a strain of *Actinomyces viscosus* adhered to saliva-coated hydroxylapatite presumably via fibrillae on the surface of the organisms. A mutant derivative which possessed visible fibrillae, however, lacked adhering ability apparently because these structures were masked with extracellular material, composed of heteropolysaccharides.

To summarize the foregoing, bacteria may be envisaged as small particles which are negatively charged and coated with ligand molecules with the capacity to bind in a stereospecific fashion to receptor molecules on the negatively charged surface of tissue membranes (see Fig. 1.2). Hydrophobic molecules counteract the repulsive

Fig. 1.4 Interference with the adherence of a bacterium to a tissue cell membrane. Capsular material (●) masks both the hydrophobic molecules (♦) and the specific ligands (Ψ) which ordinarily serve to attract and bind the cells to each other.

forces of the negatively charged surfaces and thereby permit specific binding between the ligand and receptor molecules. Bacterial adherence must be considered as a cell—cell (prokaryotic—eukaryotic) interaction rather than simply a molecule-cell interaction. Cell—cell interactions involve the simultaneous binding of a large number of ligand molecules with a large number of receptor molecules to form many independent bonds, and for this reason the attachment of bacteria becomes virtually irreversible except in the presence of large excesses of free receptor or ligand molecules (see Fig. 1.3). Although limited, there is evidence to suggest that such interactions are dependent upon spatial arrangement and accessibility of the ligands and receptors on the respective cells (see Fig. 1.4).

1.3 GENETIC AND PHENOTYPIC VARIABLES WHICH INFLUENCE BACTERIAL ADHERENCE

The ability of bacteria to adhere is dependent upon the ability of the organisms not only to synthesize the ligand but also to express it in an accessible configuration on the surface, and, as pointed out above, surface components other than the ligand may influence adherence, either by altering the hydrophobicity of the organism, or by masking the ligand itself. Synthesis and expression of the bacterial ligand appears to be under genetic and phenotypic control.

1.3.1 Genetic variants of bacteria

There is evidence to suggest that the formation of the ligand in certain bacteria is

under the control of genes located in plasmids. For example, the ability of *E. coli*
to produce K88 antigen in animal strains or colonizing factor in certain human strains
of the organisms is transferable to nonproducing strains with extrachromosomal DNA
(see Chapter 7, this volume). Plasmid-dependent adherence has been found in human
pathogens of *E. coli* (Williams *et al.*, 1978) as well as in certain plant parasites
(Matthysse *et al.*, 1978). The genetic alterations, of course, may affect surface
components other than the ligand as well and thereby indirectly affect the expression
of the ligand on the bacterial surface. The latter type of variation was shown to occur
in *Proteus mirabilis* which appears to adhere to epithelial cells via surface fimbriae
(Silverblatt, 1974; Silverblatt and Ofek, 1978a and b). It was found that rough
mutants which are defective at various stages of lipopolysaccharide biosynthesis are
unable to form or express fimbriae, or to adhere to epithelial cells under growth
conditions which promote both activities in the parent strain (Ofek and Rottem, 1978).
Genetic alterations were shown also to affect the hydrophobic properties of the
bacterial surfaces and hence the ability to adhere to animal cells (Perers *et al.*, 1977;
Kihlstrom and Edebo, 1976).

1.3.2 Phenotypic variants of bacteria

Phenotypic variations are more complex because they involve many factors such as
laboratory manipulations, and the variations may or may not exist *in vivo*. The source
of isolation, the phase and conditions of growth, and numbers of laboratory passages
are among the factors which influence adherence. Certain aspects of these variables
for particular species of bacteria will be discussed in detail in subsequent chapters in
this volume. Here we wish to emphasize that conditions which affect adherence may
differ among strains of the same or different species of bacteria. Certain streptococci
were found (Ellen and Gibbons, 1973) to adhere best during the logarithmic phase
of growth while certain strains of *E. coli* adhered best during the stationary phase
of growth (Duguid and Gillies, 1957; Ofek *et al.*, 1977a).

Despite the foregoing evidence, not much is known about the synthesis, assembly
and expression of bacterial ligands. Limited studies have been reported recently on
the phenotypic control of fimbriae-mediated hemagglutinating activity. The adhesion
of *Salmonella typhimurium* was found to be cyclic AMP-dependent and, therefore,
under the control of catabolite repression (Saier *et al.*, 1978). These findings are in
agreement with the observations that glucose suppresses hemagglutinating activity
and fimbriae formation, presumably through its catabolite (Ottow, 1975).

The comparison of a parent strain with its laboratory derivative in adherence and
antigenic composition, in certain instances, has shed some light on the mechanism
of bacterial adherence. Sparling (1979) recently cautioned the validity of such
comparisons since a genotypic (and phenotypic) variant may lack or acquire several
properties or molecules besides the one being tested by the investigator. With this
caution in mind, the use of genotypic and phenotypic variants have helped to focus
the studies on the possible role of certain surface molecules in bacterial adherence

and, perhaps more importantly, on the role of adherence in the infectious process (see Section 1.5).

1.4 HOST CELL-ASSOCIATED VARIABLES

Since it is known that bacteria vary in their ability to colonize different epithelial surfaces, the source of host cells is of utmost importance in the study of bacterial adherence. The variables involved are determined by the host and cannot be controlled by laboratory manipulations as easily as in the case of bacteria. Because of the relative inaccessibility of certain tissues, however, it is often useful to study adherence mechanisms using more readily accessible tissue cells bearing similar receptors for the bacteria under study. Measurable adherence to any host cell might be of value for assaying ligand production by the bacteria, provided the host cell chosen for study bears the receptors for the bacterial ligand. In the following, we will focus on epithelial cells and will discuss the variables in these cells as they relate to bacterial adherence.

1.4.1 Variables associated with epithelial cells

It is not clear why adherence of a particular bacterial pathogen varies from one epithelial tissue to another. The underlying molecular variables are not known. Decreased adherence could result in lack of receptor synthesis, masking of receptors, or ineffectual orientation of receptors for particular bacteria. Increased adherence may be associated with production of additional receptors for which the particular bacteria bears ligand on its surface.

Some evidence suggests that the mechanism of adherence of one epithelial cell may be different from that of another epithelial cell for the same organism. Svanborg-Eden *et al.* (1977) found that adherence to oral epithelial cells of uropathogenic *E. coli* is blocked by mannose, while adherence of the same strain to exfoliated urinary epithelial cells is not, suggesting that uroepithelial cells have receptors different from those of oral epithelial cells. The receptors of the uroepithelial cells apparently recognize a surface ligand other than the one which binds to mannose residues.

Since host cells are pleotropic in their receptor formation or expression (see Sections 1.4.2 and 1.6.1) and bacteria often display phenotypic variability in their ability to adhere, it is imperative to introduce standard controls in the study of bacterial adherence. If the variability of the host cell is under study, the bacteria must be standardized by assuring that the test strain was grown and its adherence studied at reproducible if not optimal conditions required for adherence. Conversely, if the variability of bacteria is under study, the choice of the target host cell must be such that its adhering characteristics are stable and well-defined.

1.4.2 Comparison of the adherence of bacteria to epithelial and other host cells

Studies of the attachment of bacteria to host cells other than those from epithelial
surfaces may be illuminating. Investigations of the mechanisms by which bacteria
adhere to endothelial cells has shed light on the pathogenesis of bacterial endocarditis
(Ramirez-Ronda, 1978; Scheld *et al.,* 1978; Gould *et al.,* 1975). It has been shown, for
example, that the endothelial surfaces of human heart valves bind large numbers of
streptococci, staphylococci and pneumococci, but very few gram-negative bacteria
(Gould *et al.,* 1975). This finding is consistent with the propensity of the gram-
positive, but not of the gram-negative, organisms to cause infections of the heart
valves (bacterial endocarditis). Streptococci appear to bind to the endothelial surfaces
by at least two mechanisms. In the first, the organisms adhere to undamaged
endothelial surfaces. In the second, associated with dextran production on the surface
of the streptococci, the organisms adhere to fibrin—platelet deposits on damaged
endothelial tissues and the ability to bind to the damaged tissues appears to correlate
with the infectivity of the organisms (Ramirez-Ronda, 1978; Scheld *et al.,* 1978).

Although the adherence of gonococci to a variety of host cells including oral
epithelial cells, tissue culture cells and fallopian tube organ cultures (see Chapter 9)
has been studied extensively, virtually nothing is known concerning the 'homing' and
adherence of these organisms to synovial tissues of articular joints. Yet, these tissues
are known to attract and bind the organisms during disseminated gonococcal infections.
Studies of the adherence of gonococci to synovial cells, therefore, should provide
useful insights into the pathogenesis of gonococcal arthritis (see also Section 10.3.4).

1.4.3 Bacterial attachment to phagocytic cells

Perhaps even more intriguing are studies of the attachment of bacteria to phagocytic
cells of the host, since attachment to these cells leads to ingestion and intracellular
killing of the organisms. Most studies in the past have focused on the resistance of
bacteria to attachment rather than on the attachment process itself. The resistance
of certain bacteria to the attachment to phagocytic cells enables the organisms to
escape phagocytosis and killing, and thus to proliferate and produce systematic
infections in the host (see Fig. 1.1).

Two mechanisms of phagocytic attachment of bacteria have been proposed, the
first being a model of hydrophobic interaction. VanOss (1978) measured the
contact angle formed by a sessile drop of saline on a monolayer of various micro-
organisms or phagocytic cells on a solid surface. The angle between the tangent to the
drop and the solid surface at the point where the solid, liquid, and air meet was
measured. It was determined that the higher the angle of contact the higher the
surface hydrophobicity. Furthermore, it was found that bacteria which were more
hydrophobic than the phagocytes, were readily phagocytosed while bacteria more
hydrophilic than the phagocytes were resistant to phagocytosis. The importance of
surface hydrophobicity in bacteria—phagocyte interaction was similarly emphasized

by studies which employed the partition method to measure surface hydrophobicity (Stjernstrom *et al.*, 1977; Stendahl *et al.*, 1973).

Thus, as with bacteria–epithelial cell interactions, the relative surface hydrophobicity of the bacteria and phagocytic cells is an important determinant of attachment. It is also likely that surface hydrophobic properties facilitate the interaction to permit more specific molecules on each of the cells to interact in a stereospecific fashion (see Fig. 1.2). Evidence for the latter hypothesis was obtained in studies of the interaction of *E. coli* and streptococci with mouse peritoneal macrophages and human polymorphonuclear leukocytes (PMN). Bar-Shavit *et al.* (1977) and Silverblatt *et al.* (1979) demonstrated that mannose residues on the phagocytic cells serve as sites of attachment for *E. coli*. Ofek and Beachey (1979) reported studies which suggest that LTA receptors on phagocytes serve to attach streptococci. The fimbriae of *Proteus mirabilis* seem to mediate attachment of bacteria to both epithelial and phagocytic cells (Silverblatt, 1974; Silverblatt and Ofek, 1978a and b). These studies support the hypothesis that a single mechanism of recognition of certain pathogens by host cells may either promote infection via adherence to and subsequent colonization of mucosal surfaces or terminate infection by attachment to and subsequent ingestion by phagocytic cells.

In contrast, Swanson and King (1978) found that the mechanism by which gonococci adhere to phagocytic cells is completely different from the mechanism by which they adhere to epithelial cells. Their attachment to epithelial cells appears to be mediated by fimbriae while attachment to phagocytes appears to be mediated by a trypsin-sensitive protein called leucocyte association factor located on the surface of the bacteria.

1.4.4 Resistance to phagocytosis

In view of the importance of the resistance to phagocytosis in the ability of bacterial pathogens to induce infectious diseases, possible mechanisms of resistance are considered. Resistance to phagocytic attachment in the absence of exogeneous substances (e.g. antibodies or complement components) may be due to (1) inability of the bacteria to form and express on its surface the ligand which is responsible for binding the organisms to the phagocytic membrane, or to (2) the formation and expression on their surface of other substances which either mask the ligand or alter the surface hydrophobicity and net surface charge.

There is circumstantial evidence that the second possibility exists among several bacterial pathogens. In *Neisseria gonorrhoeae* several investigators showed that resistance to phagocytic attachment is associated with fimbriated organisms (Jones and Buchanan, 1978; Buchanan *et al.*, 1978; Thongthai and Sawyer, 1973; Punsalang and Sawyer, 1973; Thomas *et al.*, 1973; Ofek *et al.*, 1974; Blake and Swanson, 1975; Swanson *et al.*, 1975; Witt *et al.*, 1976). As already mentioned, Swanson and King (1978, and King and Swanson, 1978) showed that the attachment of gonococci to phagocytic cells is dependent on formation and expression of a trypsin-sensitive leucocyte association factor. It is possible, therefore, that in some fimbriated gonococci the fimbriae mask the LA factor from binding the organisms to phagocytic cells.

The same may hold for streptococcus–phagocytic cell interactions. It has been recognized for a long time that M protein on the surface of streptococcal cells confers resistance to phagocytic attachment (reviewed by Stollerman, 1975). Recently, we have shown that lipoteichoic acid (LTA) receptors are involved in the phagocytic attachment of streptococci which lack M protein on their surface (Ofek and Beachey, 1979). Our results suggest the possibility that M protein masks surface LTA and thereby prevents the organism from binding to receptors on the phagocytic cell membrane.

The idea that bacteria resist phagocytosis or possess antiphagocytic properties by virtue of the ability of the organisms to secrete substances on their surfaces which mask a specific ligand is further supported by recent studies of Wilkinson *et al.* (1979). They showed that although both the encapsulated and unencapsulated derivative of *S. aureus* bound antibody and complement (C3) on their surface, only the un-encapsulated derivative was recognized and ingested by phagocytic cells via im-munoglobulin Fc and complement C3 receptors. The authors concluded that, in order for the coating antibodies or C3 to mediate attachment and opsonization, they must be in 'proper configuration' on the bacterial surface. The presence of capsule on the bacterial surface seems to interfere in the proper configuration of the coating antibody and thus to impede phagocytosis.

The masking effect by the surface substances, however, may not be preferential. For example, the polysaccharide capsule of pneumococci may prevent the putative ligand from binding the organisms to both epithelial and phagocytic cells (Selinger and Reed, 1979). In contrast, it seems that neither the M protein of streptococci nor the fimbriae of gonococci interfere in the epithelial adherence of these organisms (Alkan *et al.,* 1977; see also Chapter 9, this volume). The preferential effect may be due to differences in net surface charge, surface hydrophobicity or arrangement and presentation of receptors on epithelial cells, as opposed to phagocytic cells. Such differences may exist even among various phagocytic cells. For example, a particular test strain may resist attachment to macrophages but may readily attach to poly-morphonuclear leucocytes due to apparent differences in the surface charge and hydrophobicity of the two types of phagocytic cells (VanOss, 1978).

1.5 EXOGENOUS FACTORS WHICH INFLUENCE BACTERIAL ADHERENCE

The exogenous factors which influence bacterial adherence are summarized in Table 1.1. The modalities listed are useful in attempts to identify the chemical composition and structural features of the bacterial ligands and host cell receptors involved in adherence. Antibodies directed against specific surface components of

Table 1.1 Exogenous factors which influence bacterial adherence

Factors	Possible mechanism of influence
1. Blocking substances	
(a) bacterial antibodies	Inhibition of adherence by competitive inhibition or steric hindrance
(b) receptor and ligand analogues	Competitive inhibition
2. Enzymes and chemical reagents	Alterations of binding sites or hydrophobicity and net surface charge of bacterial or epithelial surfaces

3. Sublethal concentrations of antibiotics	Decreased adherence by:
	  (a) enhanced release of ligand from bacteria
	
	  (b) suppression of the formation and expression of ligand on bacterial surface

either type of cell may block adherence either by competitive inhibition or by steric hindrance, keeping in mind that inhibition by antibody-induced clumping of bacteria must be excluded as the cause of reduced adherence (Freter, 1978). Although the inhibition of adherence by specific antibodies does not itself indicate that the antigen is involved in the specific interaction with host cell receptors, the information gained in such experiments may serve to narrow the field of candidates for the specific ligand or receptor involved in a particular case. As previously mentioned (Section 1.2.2), inhibition of bacterial adherence by purified receptor or ligand material would serve to demonstrate directly their role in adherence. However, in most cases neither the receptor or ligand have been identified. Nevertheless, much information may be gained by studying the effects of various receptor and ligand analogues. Treatment with proteolytic enzymes, neuraminidase (see Chapter 6, this volume), lectins, and chemical reagents such as sodium metaperiodate (Ofek *et al.*, 1977a) have also been used for identifying surface components involved in bacterial adherence. It is conceivable that some of the exogenous agents present *in vivo* (e.g. sugar residues, proteolytic enzymes, bacterial toxins) modulate bacterial adherence during the course of certain infections. The affects of antibiotics are discussed in Section 1.6.3.

1.6 RELATIONSHIP OF BACTERIAL ADHERENCE TO INFECTIONS

The efficacy of the cleansing mechanisms operating on various mucosal surfaces to prevent colonization with bacteria has been discussed elsewhere (Ofek and Beachey,

1980, Gibbons, 1977). Briefly, the mucosal and endothelial surfaces are constantly bathed by secretions (urinary and blood flow, mucus gels) which together with mechanical—anatomical mechanisms (sneezing, coughing, ciliary action and peristalsis) sweep those surfaces free of pathogenic bacteria. Although it is logical that the bacteria must adhere to host cells to counteract the normal cleansing mechanisms, it remains to be established whether or not bacterial adherence is a prerequisite for infection. In the following paragraphs we shall examine some of the evidence that the development of infectious diseases is indeed dependent upon bacterial adherence.

1.6.1 Infectivity versus adherence

Two approaches have been used to study the relationship between adherence and infections. In the first approach, laboratory bacterial derivatives (phenotypic or genotypic) which vary in their ability to adhere to certain host cells *in vitro* have been assayed for their ability to produce experimental infections in laboratory animals or human volunteers. Although the studies outlined in Table 1.2 show a correlation, but not cause and effect (see Sparling, 1979), they suggest a central role of adherence in infections. We shall describe one study in human volunteers which serves to demonstrate the role of adherence in the infectious process. Diarrhea caused by enterotoxigenic *E. coli* is induced by both heat-labile and heat-stable enterotoxins. Satterwhite *et al.* (1978) demonstrated that an isogenic derivative of an enterotoxigenic strain of *E. coli* which lacked adhering ability and colonizing factor, but retained its ability to produce enterotoxin, was unable to induce diarrhea in human volunteers. In contrast, the parent strain, which possessed colonizing factor and adhering ability was able to produce the disease (Satterwhite *et al.*, 1978). In addition, volunteers infected with the non-adherent derivative strain shed the organisms in their stools for a much shorter period of time than did volunteers infected with the adherent variant. These studies suggest that, while both the parent and the mutant derivative produce the diarrhea-inducing enterotoxin, the bacteria must be able to adhere to the intestinal epithelial cells in order to multiply, colonize the mucosal surface and then produce sufficient amounts of enterotoxin to cause diarrhea. The adherence, therefore, appears to be important only at the initial step of the infectious process preventing the rapid loss of the organism from the gut.

The second approach has been aimed at correlating the known propensity of certain bacteria to cause particular infections *in vivo* with their ability to adhere to the affected tissue *in vitro* (Table 1.3) or at correlating the known predisposition of susceptible individuals to develop infections caused by specific pathogens with the ability of their affected tissues to bind the test pathogen (Table 1.4). In general such studies have demonstrated a high degree of correlation between adherence and infection. Only in a few cases has good *in vitro* adherence been associated with low infectivity. In no case was poor adherence associated with a high degree of infectivity.

It is not surprising that adherence should be a necessary, but not the only,

Table 1.2 Relationship between epithelial cell adherence *in vitro* and bacterial infectivity *in vivo*

Organisms	Bacterial variants	Relative adherence *in vitro*	Relative infectivity *in vivo*	Literature references
Gonococci	T1* (fimbriated)	Good	High	Chapter 9
	T4 (non-fimbriated)	Poor	Low	
E. coli (entero-toxigenic)	CF† positive	Good	High	Satterwhite, *et al.,* 1978
	CF-negative	Poor	Low	
Streptococci	Dextran producer	Good	High	Scheld, *et al.,* 1978
	Dextran non-producer	Poor	Low	
Salmonella	Fimbriated	Good	High	Chapter 7
	Non-fimbriated	Poor	Moderate	
E. coli	K88+	Good	High	Chapter 7
	K88−	Poor	Low	
Proteus mirabilis	Fimbriated	Good	High	Silverblatt, 1974; Silverblatt and Ofek, 1978a and b
	Non-fimbriated	Poor	Low	

* T1, type 1 and T4, type 4 colony morphology
† CF, colonization factor

requirement for the development of infectious diseases, because bacterial virulence is multifactorial. Other properties of bacteria such as growth, motility, chemotaxis, resistance to lethal host defenses (e.g. bactericidal antibodies), elaboration of toxins or tissue-penetrating ability all play a role in virulence and irfectivity. Thus, while receptors may be available on many different host cells for the attachment of a particular pathogen, the ability of that pathogen to cause infections may be limited to only certain site(s) *in vivo*. It is even possible that a bacterial pathogen may lose permanently its adhering ability once isolated and passaged in artificial media. Methods need to be developed to recognize laboratory variants which possess adhering ability and can be selectively monitored as is the case for type 1 and type 4 colonies of *Neisseria gonorrhoeae* (see Chapter 9, this volume).

Studies of phenotypic and genotypic variants of host cells which usually become colonized by bacteria have provided additional evidence that specific cell receptors are involved in the recognition of pathogenic organisms. It has been shown, for example, that the intestinal brush borders of pigs, which are innately resistant to

Table 1.3 Correlation between site of infection and *in vitro* adherence

Type of infection	Source of target cell	Organisms	Relative adherence to target cell (*in vitro*)	Relative incidence of infection	References
Endocarditis	Heart valve endothelium	*S. viridans* and Staphylococci	Good	High	Gould *et al.*, 1975
		E. coli	Poor	Low	
		Pseudomonas	Good	High (in heroin addicts)	
Mucosal colonization	Buccal mucosa	*S. mitis*	Poor	Low	Gibbons, Chapter 4
		S. salivarious	Good	High	Candy *et al.*, 1978; Ofek *et al.*, 1977a
		E. coli	Good	Low	
Pyoderma	Skin epithelium	*S. pyogenes* 'skin' strain	Good	High	Alkan *et al.*, 1977
		'pharyngeal' strain	Low	Moderate	
Pharyngitis	Buccal cells	*S. pyogenes* 'skin' strains	Moderate	Low	Alkan *et al.*, 1977
		'pharyngeal' strains	Good	High	

(*continued on next page*)

Table 1.3 Correlation between site of infection and *in vitro* adherence (*continued*)

Type of infection	Source of target cell	Organisms	Relative adherence to target cell (*in vitro*)	Relative incidence of infection	References
Cervicitis and vaginitis	Vaginal cells	Anerobic bacteria	Poor	Low	Mardh and Westrom, 1976
		N. gonorrhoeae	Good	High	
		S. agalactiae (group B)	Moderate	None	
		C. vaginale	Moderate	Intermediate	
Pyelonephritis	Uroepithelial cells	*E. coli* Pyelonephritic strains	Good	High	Svanborg-Eden *et al.*, 1977 and Svanborg-Eden *et al.*, 1978
		Non-pyelonephritis strains	Poor	Low	

Table 1.4 Correlation between the predisposition to develop particular infections and *in vitro* adherence to target host cells

Predisposed subjects	Bacteria	Target cells	Target cells from:	Relative *in vitro* adherence	References
Females with recurrent urinary tract infections	*E. coli*	Vaginal and periurethral epithelial cells	Subject Normal	High Low	Kallenius and Winberg, 1978; Fowler and Stamey, 1978
Individuals with damaged heart valves (i.e. rheumatic heart disease)	Streptococci	Endothelium of heart valves	Damaged Normal	High Low	Ramirez-Ronda, 1978
Individuals with viral infections	*S. sanguis* and *S. agalactiae*	Tissue culture cells	Virus-infected culture Uninfected culture	High Low	Sanford *et al.*, 1978
Staphylococcal carriers	*S. aureus*	Nasal epithelium	Carrier Non-carrier	High Low	Aly *et al.*, 1979
Genetic variants of pigs	*E. coli* K88	Intestinal epithelium	K88-resistant phenotype K88-susceptible phenotype	Low High	Sellwood *et al.*, 1975
Primates	*N. gonorrhoeae*	Organ cultures of oviducts	Human Nonprimate	High Low	Johnson *et al.*, 1977

colonization and infection by K88 antigen producing *E. coli*, do not bind the organisms *in vitro* (Table 1.4). These genetically resistant pigs appear to lack a receptor in their intestinal cells for K88-producing *E. coli*. Similarly, it has been found that the erythrocytes of certain black Africans, inherently resistant to malaria infestations, lack the Duffy Fy^a and Fy^b antigenic determinants. It has now been shown that these determinants serve as specific receptors for malaria parasites (Miller *et al.*, 1976). Thus, individuals whose erythrocytes lack these receptors can be considered to be genetically immune to malaria.

The transient inability of the buccal epithelial cells from newborn infants during the first 2 days of life to bind streptococci has been shown to parallel the low level of oral colonization by bacteria during neonatal life (Ofek *et al.*, 1977b). Whether the reduced binding and colonization is due to the absence or to the masking of epithelial cell receptors for bacteria remains to be determined.

1.6.2 Changes in adhering ability of bacteria *in vivo*

During certain infections, the infecting bacteria may change in their ability to adhere to tissue cells. This is well illustrated by the studies of Silverblatt (1974) and Silverblatt and Ofek (1978a and b) of ascending urinary tract infections of rats due to *Proteus mirabilis*. These bacteria become heavily fimbriated during the stationary phase of growth *in vitro* and heavily fimbriated organisms adhere well while lightly fimbriated organisms adhere poorly (Table 1.5). The heavily fimbriated organisms when injected into the bladder of rats were found to be capable of producing renal infections in significantly greater numbers than lightly fimbriated organisms (Silverblatt, 1974). Electron microscopy, however, revealed that the majority of the organisms reaching the renal cortex were non-fimbriated while those remaining in the renal pelvis retained heavy fimbriation. In another experiment, either heavily or lightly fimbriated organisms were introduced intravenously rather than intravesicularly into rats (Silverblatt and Ofek, 1978a and b). After one hour, quantitative bacterial counts revealed that equal numbers of the two types of organisms reached the renal cortex. Five hours later, however, the heavily fimbriated organisms had been cleared while the non-fimbriated organisms persisted (Table 1.5). These studies demonstrated first of all that fimbriated organisms with good adhering ability are able to undergo a change to a low degree of fimbriation and presumably a concomittantly low degree of adherence *in vivo*. Secondly, fully fimbriated organisms are cleared more rapidly from the blood stream and deeper tissues than are non-fimbriated organisms.

The question as to why fimbriated organisms are cleared from the blood stream and deeper tissues more rapidly than non-fimbriated organisms was partially answered by *in vitro* phagocytosis experiments (Silverblatt and Ofek, 1978a and b). Fimbriated organisms were found to be highly susceptible to phagocytosis while non-fimbriated bacteria were found to be resistant (Table 1.5). Taken together these findings suggested that the adaptability of the bacteria to the pressures of a particular micro-environment is a mechanism necessary for survival. In the renal pelvis, the organisms

Table 1.5 Infectivity and adhering characteristics* of *Proteus mirabilis*

	Organisms	
Property	Heavily fimbriated	Lightly fimbriated
Adherence to epithelial cells	+	−
Infectivity in rats (ascending infection)	+	−
Resistance to phagocytosis	−	+
Fimbriation *in vivo* †		
Pelvic mucosa	+	−
Renal cortex	−	+
Clearance from renal cortex ‡	+	−

* Adapted from Silverblatt (1974), and Silverblatt and Ofek (1978a)
† Determined following inoculation of heavily fimbriated organisms into the bladder of rats (adapted from Silverblatt and Ofek, 1978a)
‡ Determined following i.v. inoculation of either heavily or lightly fimbriated organisms in rats (adapted from Silverblatt and Ofek, 1978b)

need fimbriae and high adhering ability to defend themselves against being swept away by the flow of urine. In the deeper tissues and blood stream, however, the possession of fimbriae would be suicidal because binding of the organisms to phagocytic cells would result in their ingestion and eventual destruction.

There is some evidence to suggest that similar bacterial adaptation occurs during the infectious process in man. Craven and Frasch (1978) found, for example, that meningococci isolated from the pharyngeal cavity of humans who were chronic carriers of the organism adhered very well to isolated pharyngeal cells. In contrast, specimens of the same organisms isolated from the blood of patients with meningococcal meningitis adhered very poorly to pharyngeal cells, again suggesting that a low degree of adhering ability somehow serves to protect the organisms against destruction once they reach the blood stream.

1.6.3 Blocking of bacterial adherence *in vivo*

If adherence is indeed an essential step in bacterial infectious diseases, the obvious question is whether interruption of this step is feasible and whether such measures would prevent infections. Although only a few studies have been performed to answer these questions, the results have been tantalizing.

Along these lines, Brinton *et al.* (1978) and Nagy *et al.* (1978) have prepared vaccines from purified fimbriae of gram-negative bacteria. Although they claim some degree of protectivity, it is arguable whether the antibodies induced by the fimbrial vaccines prevent infection solely by preventing adherence of the organisms (Morgan *et al.*, 1978). To date, there is no evidence to indicate that antibody to a known bacterial ligand is more protective *in vivo* than is antibody directed against any other surface antigen (Freter, 1978). Moreover, the presence of anti-ligand antibodies in the host may not always provide protection against infection even though the antibody blocks adherence *in vitro*. For example, most human sera contain antibodies against streptococcal LTA which inhibit streptococcal adherence *in vitro* (Beachey and Ofek, 1976) yet the LTA antibodies apparently do not protect against streptococcal infections (Stollerman, 1975). The presence of secretory IgA antibodies which inhibit adherence may be more relevant (Svanborg-Eden and Svennerholm, 1978; Williams and Gibbons, 1972; Tramont, 1977), but more studies are needed to assess the role of such antibodies in preventing adherence and infections (Freter, 1978; see also Chapter 4).

Recently, attempts have been made to block adherence by a receptor analogue *in vivo*. Mice were infected via the urinary bladder with a strain of *E. coli* which possesses a mannose-specific lectin and whose epithelial adherence is inhibited by mannose (Ofek *et al.*, 1977a; Ofek and Beachey, 1978). Aronson *et al.* (1979) showed that the frequency with which *E. coli* adhered to bladder cells and colonized the bladder of mice *in vivo* was markedly reduced by suspending the organisms in a solution containing methyl α-mannopyranoside (αMM) before instilling them into the bladder. Animals whose bladders were instilled with *E. coli* suspended in methyl α-glucopyranoside rather than αMM or with *Proteus mirabilis* suspended in αMM developed no fewer infections than control animals instilled with organisms in buffer alone. Similar preliminary results were obtained in a rabbit model of gastrointestinal infection where colonization by *E. coli* was specifically blocked by D-mannose, but not other sugars (Hirschberger *et al.*, 1977). It remains to be seen whether receptor analogues can prevent bacterial colonization and infections in humans.

Of special interest is the recent demonstration that sublethal concentrations of antibiotics can alter the ability of certain bacteria to adhere to epithelial cells (see Table 1.1). Penicillin G causes an enhanced loss of lipoteichoic acid, the ligand which binds group A streptococci to host cells. Concomitantly, the treated organisms lose their adhesive properties (Alkan and Beachey, 1978). The same drug in certain strains of *E. coli* inhibits the expression of the mannose-specific ligands by distorting cell wall biosynthesis (Ofek *et al.*, 1979). Streptomycin and other aminoglycoside antibiotics suppress the formation and expression of the mannose-specific ligand in *E. coli* probably by acting on the bacterial ribosomes to induce abnormal protein synthesis (Eisenstein *et al.*, 1979a and b). Whether or not such sublethal concentrations intermittently reach the mucosal surfaces during the course of therapy for bacterial infections remains to be investigated. In any case, the use of certain antibiotics whose mode of action is well-known to inhibit specifically the normal production of the ligand involved in adherence may serve as a useful model system to elucidate the genetic and biochemical modulation of bacterial factors responsible for colonization of host tissue surfaces.

Alkan, M.L. and Beachey, E.H. (1978), *J. clin. Invest.*, **61**, 671–677.

Alkan, M., Ofek, I. and Beachey, E.H. (1977), *Infect. Immun.*, **18**, 555–557.

Aly, R., Litz, C., Shinefield, H. and Lancaster, M. (1979), *Clin. Res.*, **27**, 40A.

Aronson, M., Medalia, O., Schori, L., Mirelman, D., Sharon, N. and Ofek, I. (1979), *J. Infect. Dis.*, **139**, 329–332.

Bar-Shavit, Z., Ofek, I., Goldman, R., Mirelman, D. and Sharon, N. (1977), *Biochem. biophys. Res. Commun.*, **78**, 455–460.

Bartelt, M.A. and Duncan, J.L. (1978), *Infect. Immun.*, **20**, 200–208.

Beachey, E.H., Dale, J., Grebe, S., Ahmed, A., Simpson, W.A. and Ofek, I. (1979a), *J. Immunol.*, **122**, 189–195.

Beachey, E.H., Dale, J.B., Simpson, W.A., Evans, J.D., Knox, K.S., Ofek, I. and Wicken, A.J. (1979b), *Infect. Immun.*, **23**, 618–625.

Beachey, E.H. and Ofek, I. (1976), *J. exp. Med.*, **143**, 759–771.

Bell, G.I. (1978), *Science*, **200**, 618–627.

Blake, M. and Swanson, J. (1975), *Infect. Immun.*, **11**, 1402–1404.

Brecher, S.M., van Houte, J. and Hammond, B.F. (1978), *Infect. Immun.*, **22**, 603–614.

Brinton, C.C., Jr. (1965), *Trans. N.Y. Acad. Sci.*, **27**, 1003–1054.

Brinton, C.C., Bryan, J., Dillon, J., Guerina, N., Jacobson, L.J., Labik, A., Lee, S., Levine, A., Lim, S., McMichael, J., Polen, S., To, A.C. and To, S.C. (1978), In: *Immunobiology of Neisseria gonorrhoeae.* (Brooks, G.F. *et al.*, eds), pp. 155–178, American Society for Microbiology, Washington.

Buchanan, T.M., Chen, K.C.S., Jones, R.B., Hildebrandt, J.F., Pearce, W.A., Hermodson, M.A., Newland, J.C. and Luchtel, D.L. (1978), In: *Immunobiology of Neisseria gonorrhoeae.* (Brooks, G.F. *et al.*, eds), pp. 145–154, Microbiology, Washington.

Buchanan, T.M. and Pearce, W.A. (1976), *Infect. Immun.*, **13**, 1483–1489.

Candy, D.C.A., Chadwick, J., Leung, T., Phillips, A., Harris, J.T. and Marshall, W.C. (1978), *Cancet*, **2**, 1157–1158.

Chiang, T.M., Alkan, M.L. and Beachey, E.H. (1979), *Infect. Immun.*, **26**, (in press).

Christensen, P., Johansson, B.G. and Kronvall, G. (1976), *Acta path. microbiol. scand*, Sect. C, **84**, 73–76.

Craven, D.C. and Frasch, C.E. (1978), In: *Immunobiology of Neisseria gonorrhoeae.* (Brooks, G.F. *et al.*, eds), pp. 250–252, American Society for Microbiology, Washington.

Duguid, J.P. and Gillies, R.R. (1957), *J. path. Bact.*, **74**, 397–411.

Eisenstein, B.I., Ofek, I. and Beachey, E.H. (1979a), *Clin. Res.*, **26**, 675A.

Eisenstein, B.I., Ofek, I. and Beachey, E.H. (1979b), *J. clin. Invest.*, **63**, 1219–1228.

Ellen, R.P. and Gibbons, R.J. (1973), *Infect. Immun.*, **9**, 85–91.

Eshdat, Y., Ofek, I., Yashouv-Gan, Y., Sharon, N. and Mirelman, D. (1978), *Biochem. biophys. Res. Commun.*, **85**, 1551–1559.

Evans, D.G. and Pauley, J.A. (1978), Annual meeting of the American Society of Microbiology, Abstr. B45.

Fowler, J.E., Jr. and Stamey, T.A. (1977), *J. Urol.*, **117**, 472.

Freter, R. (1978), Cholera and Related Diarrheas–Molecular Aspects of a Global Health Problem, *Proc. Nobel Symposium*, **43**, Basel. (In press).

Gibbons, R.J. (1977), In: *Microbiology 1977.* (Schlessinger, D., ed.), pp. 395—406, American Society of Microbiology, Washington.

Gibbons, R.A., Jones, G.W. and Sellwood, R. (1975), *J. gen. Microbiol.,* **86**, 228—240.

Gibbons, R.J. and Qureshi, J.V. (1978), *Infect. Immun.,* **22**, 665—671.

Gould, K., Ramirez-Ronda, C.H., Holmes, R.K. and Sanford, J.P. (1975), *J. clin. Invest.,* **56**, 1364—1370.

Hirschberger, M., Mirelman, D. and Thaler, M.M. (1977), *Gastroenterology,* **72**, 1069 (Abstr.).

Hu, P.C., Collier, A.M. and Baseman, J.B. (1977), *J. exp. Med.,* **143**, 1328—1343.

Isaacson, R.E., Fusco, P.C., Brinton, C.C. and Moon, H.W. (1978), *Infect. Immun.,* **21**, 392—397.

Johnson, A.P., Taylor-Robinson, D. and McGee, Z.A. (1977), *Infect. Immun.,* **18**, 833—839.

Jones, G.W. (1977), In: *Microbial Interactions.* Receptors and Recognition Series B, Vol. 3, pp. 139—176, (Reissing, J.L., ed.), Chapman and Hall, London.

Jones, R.B. and Buchanan, T.M. (1978), *Infect. Immun.,* **20**, 732—738.

Kallenius, G. and Winberg, J. (1978), *Lancet,* **2**, 540.

Kashket, S. and Guilmette, K.M. (1978), *Caries Res.,* **12**, 170—172.

Kihlstrom, E. and Edebo, L. (1976), *Infect. Immun.,* **14**, 851—857.

King, G.J. and Swanson, J. (1978), *Infect. Immun.,* **21**, 575—584.

Kondo, W., Sato, M. and Ozawa, H. (1976), *Archs. Oral Biol.,* **21**, 363—369.

Lis, H. and Sharon, N. (1977), *The Antigens,* Vol. IV. pp. 429—529, Academic Press, New York.

Mardh, P.A. and Westrom, L. (1976), *Infect. Immun.,* **13**, 661—666.

Matthysse, A.G., Wyman, P.M. and Holmes, K.V. (1978), *Infect. Immun.,* **22**, 516—522.

Miller, L.H., Mason, S.J., Clyde, D.F. and McGuiness, H.H. (1976), *N. Eng. J. Med.,* **295**, 302—304.

Morgan, R.L., Isaacson, R.E., Moon, H.W., Brinton, C.C. and To, C.C. (1978), *Infect. Immun.,* **22**, 771—777.

Nagy, B., Moon, H.W. Isaacson, R.E., To, C.C. and Brinton, C.C. (1978), *Infect. Immun.,* **21**, 269—274.

Ofek, I. and Beachey, E.H. (1978), *Infect. Immun.,* **22**, 247—254.

Ofek, I. and Beachey, E.H. (1980), *Advances in Internal Medicine.* (Stollerman, G.H., ed.), Year Book Medical Publishers, Chicago.

Ofek, I. and Beachey, E.H. (1979), In: *Pathogenic Streptococci.* (Parker, M.T., ed.), pp. 44—46, Redbooks Ltd., Chertsey, Surrey.

Ofek, I., Beachey, E.H. and Bisno, A.L. (1974), *J. Infect. Dis.,* **129**, 310—315.

Ofek, I., Beachey, E.H., Eisenstein, B.I., Alkan, M.L. and Sharon, N. (1979), *Infect. Dis.,* (In press).

Ofek, I., Beachey, E.H., Eyal, F. and Morrison, J.C. (1977b), *J. Infect. Dis.,* **135**, 267—274.

Ofek, I., Beachey, E.H., Jefferson, W. and Campbell, G.L. (1975), *J. exp. Med.,* **141**, 990—1003.

Ofek, I., Beachey, E.H. and Sharon, N. (1978), *Trends Biochem. Sci.,* **3**, 159—160.

Ofek, I., Mirelman, D. and Sharon, N. (1977a), *Nature,* **265**, 623—625.

Ofek, I. and Rottem, S. (1978), *FEMS Microbiology Letters,* **4** (4), 229—232.

Ottow, J.C.G. (1975), *A. Rev. Microbiol.,* **29**, 79—107.

28   *Bacterial Adherence*

Perers, L., Andaker, L., Edebo, L., Stendahl, O. and Tagesson, C. (1977), *Acta path. microbiol. scand.* Sect. **B, 85**, 308–316.

Peters, T., Jr. (1975), *The Plasma Proteins.* Vol. II (Putman, F. W., ed.), pp. 133–181, Academic Press, New York.

Punsalang, A.P., Jr. and Sawyer, W.D. (1973), *Infect. Immun.*, **8**, 255–263.

Ramirez-Ronda, C.H. (1978), *J. clin. Invest.*, **62**, 805–811.

Saier, M.H., Jr., Schmidt, M.R. and Leibowitz, M. (1978), *J. Bact.*, **134**, 356–358.

Salit, I.E. and Gotschlich, E.C. (1977a), *J. exp. Med.*, **146**, 1169–1181.

Salit, I.E. and Gotschlich, E.C. (1977b), *J. exp. Med.*, **146**, 1182–1194.

Sanford, B.A., Shelokov, A. and Ramsay, M.H. (1978), *J. Infect. Dis.*, **137**, 176–181.

Satterwhite, T.K., DuPont, H.L., Evans, D.G. and Evans, D.J., Jr. (1978), *Lancet,* **2**, 181–184.

Scheld, W.M., Valone, J.A. and Sande, M.A. (1978), *J. Clin. Invest.*, **61**, 1394–1404.

Selinger, D. and Reed, W.P. (1979), *Infect. Immun.*, **23**, 545–548.

Sellwood, R., Gibbons, R.A., Jones, G.W. and Rutter, J.M. (1975), *J. med. Microbiol.*, **8**, 405–411.

Sharom, F.J., Barrat, D.G. and Grant, C.W.M. (1977), *Proc. natn. Acad. Sci. U.S.A.*, **74**, 2751–2755.

Sharon, N. and Lis, H. (1972), *Science,* **177**, 949–959.

Silverblatt, F.J. (1974), *J. exp. Med.*, **140**, 1696–1711.

Silverblatt, F.J., Dreyer, J. and Schauer, S. (1979), *Infect. Immun.*, **24**, 218–223.

Silverblatt, F.J. and Ofek, I. (1978a), *J. Infect. Dis.*, **138**, 664–667.

Silverblatt, F.J. and Ofek, I. (1978b), *Infections of the Urinary Tract*, (Kass, E.H., ed.), pp. 49–59, The University of Chicago Press, Chicago and London.

Simpson, W.A., Dale, J.B., Ofek, I. and Beachey, E.H. (1979), *Clin. Res.*, **27**, 801A.

Simpson, W.A., Ofek, I., Dale, J.B., Sarasohn, C., Morrison, J. and Beachey, E.H. (1978), *Abstract of the Proc. 18th Intersci. Conf. Antimicrobial Agents Chemother.* No. 255.

Smythe, C.J., Jonsson, P., Olsson, E., Söderlund, O., Rosengren, J., Hjerten, S. and Wadström, T. (1978), *Infect. Immun.* **22**, 462–472.

Sparling, P.F. (1979), *Microbiology.* (Schlessinger, D., ed.), pp. 249–254, American Society of Microbiology, Washington.

Springer, G.F., Adye, J.E., Bezkorovainy, A. and Jingensons, B. (1974), *Biochemistry*, **13**, 1379–1389.

Stendahl, O., Tagesson, C. and Edebo, M. (1973), *Infect. Immun.*, **8**, 36–41.

Stjernstrom, I., Magnusson, K.E., Stendahl, O. and Tagesson, C. (1977), *Infect. Immun.*, **18**, 261–265.

Stollerman, G.H. (1975), *Rheumatic Fever and Streptococcal Infection.* Grune and Stratton, New York.

Svanborg-Eden, C., Ericksson, B. and Hanson, L.A. (1977), *Infect. Immun.*, **18**, 767–774.

Svanborg-Eden, C., Eriksson, B., Hanson, L.A., Jodal, U., Kaijser, B., Janson, C.L. Lindberg, U. and Olling, S. (1978), *J. Pediatr.*, **93**, 398–403.

Svanborg-Eden, C. and Svennerholm, A.M. (1978), *Infect. Immun.*, **22**, 790–797.

Swanson, J., King, G. and Zeligs, B. (1975), *Infect. Immun.*, **11**, 453–459.

Swanson, J. and King, G. (1978), In: *Immunology of Neisseria gonorrhoeae.* (Brooks, G.F. *et al.*, ed.), pp. 221–226, American Society for Microbiology, Washington.

Thomas, D.W., Hill, T.C. and Tyeryar, F. Jr. (1973), *Infect. Immun.*, **8**, 98.

Thongthai, C. and Sawyer, W.D. (1973), *Infect. Immun.*, **7**, 373–379.

Tramont, E.C. (1977), *J. clin. Invest.*, **59**, 117–124.

VanOss, C.J. (1978), *A. Rev. Microbiol.*, **32**, 19–39.

Wilkinson, B.J., Peterson, P.K. and Quie, P.G. (1979), *Infect. Immun.*, **23**, 502–508.

Williams, R.C. and Gibbons, R.J. (1972), *Science*, **177**, 697–699.

Williams, P.H., Sedwick, M.I., Evans, P.J., Turner, R.H. and McNeish, A.D. (1978), *Infect. Immun.*, **22**, 393–402.

Witt, K., Veale, D.R. and Smith, H. (1976), *J. med. Microbiol.*, **9**, 1–12.

2   Adherence of Normal Flora to Mucosal Surfaces

DWAYNE C. SAVAGE

2.1	Introduction	*page*	33
2.2	Indigenous micro-organisms adhering to gastro-intestinal mucosal surfaces		34
	2.2.1 The mucosal surfaces		34
	2.2.2 Methods for study of microbial adherence to surfaces in the gastro-intestinal tract		36
	2.2.3 Indigenous micro-organisms reported to adhere to gastro-intestinal surfaces		40
2.3	Mechanisms by which indigenous micro-organisms adhere to mucosal epithelial surfaces		49
	2.3.1 Squamous epithelium		50
	2.3.2 Columnar epithelium		52
2.4	Summary and conclusions		56
	References		56

Acknowledgements
The author's research discussed herein has been supported by the National Dairy Council and the National Institute of Allergy and Infectious Diseases.

Bacterial Adherence
(*Receptors and Recognition,* Series B, Volume 6)
Edited by E.H. Beachey
Published in 1980 by Chapman and Hall, 11 New Fetter Lane, London EC4P 4EE
© 1980 Chapman and Hall

2.1 INTRODUCTION

Micro-organisms of many genera and species have been found associating closely with epithelial surfaces of mucous membranes in various areas of the bodies of animals of numerous types including man (Gibbons and Van Houte, 1975; Jones, 1977; Savage, 1977a). Some of these microbial types are known to adhere to substances on or in the mucosal epithelia with which they associate. In this chapter, I discuss some such micro-organisms confining the discussion to micro-organisms of the 'normal flora' (see definition below) found in the gastro-intestinal tracts of man, some other mammals and certain birds. Micro-organisms of the normal flora are known to adhere to mucosal surfaces in the buccal cavity of humans (Gibbons and Van Houte, 1975) and undoubtedly adhere to such surfaces in the upper respiratory (Jones, 1977; Aly *et al.*, 1977) and urogenital (Davis *et al.*, 1977a; Larson *et al.*, 1978) tracts of humans and other higher animals, and the gastro-intestinal tracts of lower animals of several types (Breznak and Pankratz, 1977; Tannock and Savage, 1974a). In most cases, however, little is known in detail about the phenomena, except for those involving organisms in the buccal cavity (Gibbons and Van Houte, 1975). Some of these are discussed elsewhere in this volume (Gibbons, Chapter 4; Slade, Chapter 5).

The term 'normal flora' is usually used collectively to describe various microbial types found by culture or microscopy on the skin and mucous membranes and in certain body cavities of animals healthy or not (Slack and Snyder, 1978). It is used as well as a synonym for 'indigenous microbiota', meaning collectively those autochthonous microbial residents of habitats on particular body surfaces or in particular body cavities in normal animals (Savage, 1977a). These definitions do not necessarily describe the same micro-organisms. The first suggests that all microbial types found on or in or cultured from certain surfaces or cavities are truly normal residents of habitats in the areas. Much recent evidence supports a concept, however, that many micro-organisms isolatable at any given time from an open ecosystem, such as the gastro-intestinal tract, do not satisfy criteria for autochthony and must be regarded as transients in the system (Savage, 1977a). Transients could enter a habitat in the gastro-intestinal ecosystem in ingested material (including feces in a coprophagic animal) or even by passing down from habitats above the one being sampled. Certain such transients may temporarily colonize habitats and even adhere to epithelial surfaces in perturbed ecosystems (Savage, 1977c).

In the discussion to follow, therefore, I intend 'normal flora' to be taken as a synonym for 'indigenous microbiota', acknowledging, however, that often I must assume that the organisms in question will satisfy criteria for autochthony; in most cases, rigorous tests of the hypothesis have not been conducted (Savage, 1977a). The discussion has two goals. In reaching them, I shall have (1) summarized evidence that indigenous micro-organisms adhere to substances on or a part of the membranes of cells in the epithelium of mucosae in various areas of the mammalian gasto-intestinal tract, and (2) summarized and evaluated evidence concerning mechanisms by which the organisms adhere to substances on or a part of the membranes of mucosal epithelial cells.

2.2 INDIGENOUS MICRO-ORGANISMS ADHERING TO GASTRO-INTESTINAL MUCOSAL SURFACES

2.2.1 The mucosal surfaces

(a) *Architecture; secretions*

Indigenous micro-organisms adhere by complex mechanisms to substances in mucosal surfaces. Such complex mechanisms evolved, undoubtedly, because the gastro-intestinal canal is complex both architecturally and physiologically. Indeed, the mucosal surfaces themselves are exceedingly complex. A detailed discussion of gastro-intestinal anatomy, histology and physiology is beyond the scope of this chapter. Nevertheless, some information must be presented as background for the material to follow.

The gastro-intestinal tract of mammals and birds has five major areas: esophagus, stomach, small intestine, cecum and large intestine. Depending upon the animal species, any of these areas may be further compartmentalized or divided into subareas. In mammals, three basic variations on an overall theme can be recognized, ruminant, cecal and 'straight tube'. In the ruminant the stomach is ramified into compartments (Bauchop, 1977). In animals with a cecum, a blind pouch can be found extending to the side of the intestine from the distal end of the small intestine to the proximal end of the large bowel (McBee, 1977). In chickens the 'stomach' consists of a storage compartment (crop), proventriculus and gizzard (Fuller and Turvey, 1971); two ceca are present (Bauchop, 1977; McBee, 1977).

The esophagus is lined with stratified squamous epithelium that may or may not be keratinized (Savage, 1979b). Likewise, some 'gastric' compartments, such as the crop in chickens (Suegara *et al.,* 1975), part of the stomach in rodents (Savage *et al.,* 1968; Savage and Blumershine, 1974), and the rumens of cattle and sheep (McCowan *et al.,* 1978) are lined with stratified squamous epithelium that is usually keratinized. Gastric compartments not lined with squamous epithelium and the small and large intestines (including the cecum) are lined with a single layer of columnar cells. In the small intestine, the mucosa is organized so that the epithelium covers finger- or leaf-shaped villi that protrude into the lumen. Villi are not found in the stomach or large intestine, although in both areas the mucosa may fold when the lumen is empty. In all areas, the epithelium lines depressions or pits in the mucosa. These, the Crypts of Lieberkuhn, are located at the bases of the villi in the small bowel, and are spaced periodically in the mucosa of the stomach and large bowel (Savage, 1979b).

The columnar epithelium consists of cells of a variety of types, including secreting-absorbing cells (enterocytes) and goblet (mucous) cells. The predominating cell type in most areas is the enterocyte. These cells and goblet cells are produced through mitosis in the Crypts of Lieberkuhn. They move on a basement membrane up the crypt wall and out onto the lumenal surface. Mitosis normally cannot be detected in the cells after they pass out of the crypt onto the lumenal surface proper. The cells continue to glide slowly on the basment membrane until they reach areas where they shed into the lumen. These areas, called extrusion zones, are located on the tips of the villi in the small bowel and between crypts on the mucosal surface in the stomach and large intestine. The process by which the epithelial cells are created by mitosis in the crypts and migrate to extrusion zones, where they are discarded, is a mechanism for normal renewal of the epithelium. Squamous epithelium is also renewed by a process in which outer cells (often keratinized) are shed as new cells are formed under them (Gibbons and Van Houte, 1975).

On the columnar enterocytes the cytoplasmic membrane exposed to the lumen is ramified into rod-shaped structures called microvilli or 'brush-border'. The membranes of the microvilli are covered by a thin coat called the glycocalyx. This coat, composed of glycoproteins and glycolipids (Ito, 1969; Roseman, 1974), separates the membranes of the microvilli from the lumen of the tract. The glycocalyx is covered in most areas, but especially in the crypts, by mucus. The mucus, composed of glycoproteins of various types, some of which are sulfated and acidic, is synthesized by the goblet cells (Roseman, 1974). Such secretions may or may not be found on the surfaces of squamous epithelia.

In the small bowel, villi move in such a way that the mucus and lumenal content move downstream. The action is enhanced by peristalsis in which the entire mucosa moves vigorously due to muscular contraction. In the stomach and large bowel, the lumenal content and mucus are also propelled, generally in a distal direction, by peristalsis (Misiewicz, 1976).

Thus, except possibly for the stratified squamous epithelium in the esophagi, crops of chickens, parts of the stomach of some monogastric animals, and rumens of ungulates, most gastro-intestinal 'surfaces' are dynamic three-dimensional compartments. These compartments are composed of mucinous glycoproteins, flowing on a glycocalyx composed of glycoproteins and glycolipids overlying the epithelial cell membrane that is convoluted into microvilli. The membranes themselves (and the glycocalyx) are presumably in a dynamic fluid state with continuous movement and transposition of their macromolecular constituents (Cherry, 1975). This 'compartment' is made even more dynamic by the secretive and absorptive functions of the epithelial cells (McColl and Sladen, 1975) and movement engendered by epithelial replacement, villous motility and peristalsis.

As shall be seen, presumably because of this dynamism and complexity, some microbial types that associate with the surfaces have evolved specialized structures for adhering to the glycocalyx or membrane proper of the epithelial cells. Such organisms have been detected particularly in areas of the tract where mobility moves lumenal contents and mucus at rates more rapid than those at which the microbes reproduce. In other areas, however, especially where lumenal contents may not move along at rates more rapid than those at which the micro-organisms multiply, the organisms may or may not adhere to the epithelial glycocalyx or cell membranes. In such areas, they may colonize the glycoprotein constituents in the mucinous 'compartment' overlying the membranes, (Savage, 1977a). As also shall be seen, this latter possibility complicates considerably experimental efforts to learn the mechanisms by which particular microbial types associate with particular surfaces.

2.2.2 Methods for study of microbial adherence to surfaces in the gasto-intestinal tract

Experimental efforts to study the mechanisms by which micro-organisms adhere to surfaces in the gastro-intestinal ecosystem have been attempted with the epithelial surfaces intact in living animals and also with mucosal cells or tissues removed from animals and maintained *in vitro*. In studies with intact animals, microscopic techniques, including visible and ultra-violet light (immunofluorescence) and scanning and transmission electron microscopy (Savage, 1977b), have been used with success to demonstrate microbes associating with stomach and gut surfaces. Such methods are highly limited, however, as tools for study of the mechanisms by which the organisms adhere to a given surface. As is discussed later, work with transmission electron microscopy can and has (Savage, 1977b; Savage *et al.*, 1971; Brooker and Fuller, 1975; Fuller *et al.*, 1978; Takeuchi and Savage, 1973) yielded information suggestive that certain microbial types adhere to particular surfaces by macromolecules of certain classes. Likewise, findings made with the scanning electron microscope are suggestive that some microbial types adhere to intestinal epithelial surfaces by filaments extending from the microbial cells to the surface (Savage, 1977b). Information gained in such a way can be at best only indicative of the mechanism by which the organisms adhere, however, and may even be misleading because of shrinkage due to drying and other artifacts induced during preparation of specimens (Clark and Glagov, 1976).

Indeed, for similar reasons, microscopic techniques can be misleading about whether or not micro-organisms are even adhering to surfaces. Drying artifacts in methods used for preparing specimens for any microscopic examination, but especially for electron microscopy, can distort the relationships of microbial cells with epithelial surfaces, making the cells appear closer to or even further away from the surface than they are in the living animal. Possibly even worse, washing techniques used in such preparatory methods may flush micro-organisms from a surface. Undoubtedly, organisms colonizing mucus are most susceptible to being washed away in such procedures (Blumershine and Savage, 1974). However, even microbial types adhering weakly to the glycocalyx or microvillous membranes could disappear from their habitat during preparation. Thus, microscopy must be used with great care in efforts to assess what microbial types are associating with which substances on a mucosal surface.

Likewise, methods for culturing micro-organisms are considerably limited as tools for assessing which types of organisms adhere to mucosal surfaces. Microscopic studies of gastro-intestinal mucosal surfaces in several animal species have revealed many microbial types that have yet to be cultured *in vitro* (Phillips *et al.*, 1978; Savage, 1977a; Savage and Blumershine, 1974; Snellen and Savage, 1978). Moreover, even when the microbial type can be cultured *in vitro*, estimates of its population levels in the lumen vs. on the epithelial surface are limited in accuracy by clumping of the organisms, washing and tissue grinding procedures, quality and type of media used, and other less well-defined considerations. Thus, microbial culture methods must also be used with care as tools for assessing the types (and the numbers) of micro-organisms associating with mucosal surfaces.

Because of the limitations on what can be learned in experiments with intact animals, numerous investigators have attempted to study the mechanisms by which microbes adhere to epithelial surfaces using models involving mucosal cells or tissues maintained and exposed to the microbes in some type of *in vitro* culture system. Such models have been particularly successful as tools for study of the mechanisms by which certain oral (Gibbons and Van Houte, 1975) and respiratory (Aly *et al.*, 1977; Ofek *et al.*, 1977) bacterial pathogens adhere to mucosal surfaces. Success has also been claimed in studies where such models have been used to study mechanisms by which certain bacterial pathogens adhere to intestinal surfaces (Jones, 1977). Likewise, and as shall be discussed in more detail subsequently (see Section 2.3), such models have enjoyed limited success as tools for study of the mechanisms by which indigenous micro-organisms adhere to gastric surfaces involving keratinized stratified squamous epithelium (Fuller, 1975; Suegara *et al.*, 1975; Kotarski and Savage, 1977, and manuscript in preparation). In general, however, such models have been of little value as tools for study of the mechanisms by which indigenous microbes adhere to gastro-intestinal surfaces involving columnar epithelial cells.

Three main types of systems have been used for assaying the numbers of micro-organisms adhering to tissues or cells maintained and exposed to the microbial cells *in vitro*. In one method (Gibbons and Van Houte, 1975), microbial cells adhering to individual mucosal cells are detected by light microscopy. Usually, both the number of tissue cells with adherent bacteria and the number of adherent bacteria per tissue cell are estimated. Often the tissue cells are categorized by number of bacteria on them. The second method involves estimating by cultural methods the number of viable microbial cells adhering to mucosal tissue or cells (Suegara *et al.*, 1979). In the third, the tissues or cells are exposed to micro-organisms labeled with some radioactive isotope (usually ^{3}H or ^{14}C). The organisms adhering to the tissue cells are then detected by liquid scintillation counting (Perers *et al.*, 1977; Suegara *et al.*, 1979; Kotarski and Savage, manuscript in preparation).

Each of the models has some unique assets and disadvantages. The first model has the advantage of permitting direct observation and counting of microbial cells associating with dispersed tissue cells. (The method is not satisfactory for cells organized into tissues.) However, direct counting of the cells by microscopic means has some problems. One is that both dead and live microbial and tissue cells are counted. Another is that the method is prone to all the subjective vageries of any direct technique of counting cells, including distribution of cells on the microscope slide; spatial configuration of cells being counted (clumps, folded tissue cells, etc.); visual acuity, perception and choice of the observer; etc. Thus, the limits of accuracy in such systems must be inherently broad and should be tested for each experiment.

In the second model, where the level of the population of micro-organisms adhering to mucosal tissues or cells is estimated by culture methods, the system has the advantage of being somewhat more objective than model one. Moreover, viable cells only are estimated, permitting experiments on viability as a factor in adherence. The system has some singular disadvantages, however, the most important of which is the significant error surrounding each individual estimation (due to inefficiency in tissue washing and grinding, pipetting errors, inefficient growth of the micro-organisms on the media used, etc.). Such error requires rather elaborate statistics and often a large number of replicate samples for the investigator to have confidence in differences seen. Most importantly, if the microbial population levels estimated for a test group differ from those estimated for a control group by not much more than one $\log_{10}$, then observers may have difficulty developing confidence in their assay system (Kotarski and Savage, 1977, and manuscript in preparation).

The disadvantages with models one and two have led some investigators to employ some variant of model three, where the number of microbial cells labeled with a radioactive isotope and adhering to tissue cells is estimated by liquid scintillation counting. These systems enjoy several advantages. Any particular assay is relatively fast. Thus, experiments can be done efficiently with several replicate groups without hardship (eye strain) for the investigator. Most importantly, however, if the systems are accomplished with care and proper control (Kotarski and Savage, manuscript in preparation), then they can be highly objective ways of assessing the level of microbial populations adhering to tissues *in vitro*. However, the systems suffer from several disadvantages. As with the first model, living and dead microbial or tissue cells cannot be discriminated. Most importantly, however, the models can be misleading. Radioactive isotope may leak from intact microbial cells or be released from lysing cells and be incorporated into tissues. Under such circumstances, microbial cells may appear to be adhering to tissues, when they are not. Such phenomena require careful control.

In addition to their unique assets and problems, all three models share some common advantages and disadvantages. Common to all is the obvious advantage that, in each system, the microbial or tissue cells can be manipulated chemically in studies of the factors mediating their adherence to each other. A common disadvantage is that washing procedures influence the accuracy of estimates of the levels of adhering microbial populations. Most critically, however, microbial cells grown *in vitro* may differ from cells growing in an animal in properties mediating their adherence to surfaces. Moreover, tissues or cells maintained *in vitro* may differ from such cells *in vivo* in properties involved in adherence. Put another way, microbial cells may well adhere to tissue cells *in vitro* but do so by mechanisms different from those mediating adherence *in vivo* (see also Sections 1.1 and 8.1).

Microbial (particularly bacterial) envelopes are well known to be influenced in molecular structure by the composition of the media in which the organisms are grown (Smith, 1977). Such an influence can take two forms. In the first, the organisms may simply incorporate into their envelopes constituents of the medium in which they are grown. In this event, the cells do not change in genotype, although their apparent surface phenotype may differ depending upon the composition of the medium. In the second case, the organisms may alter not only in surface phenotype, but also in genotype as it concerns direction or regulation of the synthesis of envelope and other surface components. Such genetic changes are common in micro-organisms grown transfer after transfer in artificial culture media. The net effect of these phenomena is that micro-organisms in a population grown *in vitro* may well present to animal cells maintained *in vitro* surfaces that differ markedly from those that organisms of the same type growing *in vivo* present to animal cells *in vivo*. This is not to say that the microbial cells would not adhere to the tissues *in vitro*. As stated above, all or a fraction of the population may well adhere. It is to emphasize, rather, that organisms may adhere by mechanisms different from those by which they adhere *in vivo*.

These problems of interpretation concerning the microbial surfaces may be further complicated by changes in the epithelial 'surfaces' themselves. As noted above, epithelial 'surfaces', especially those involving columnar cells can be visualized as three-dimensional compartments. At least theoretically, micro-organisms can adhere to any of the components in the compartment from mucinous glycoproteins to the microvillous membranes themselves. (As shall be seen later, this possibility may be more than theoretical.) When mucosal tissues or cells are maintained *in vitro*, however, the three-dimensional nature of the compartment may not be preserved. For example, mucin may not remain in a layer on the surface of enterocytes maintained *in vitro*. As a consequence, micro-organisms that normally colonize mucin *in vivo* would be presented with an abnormal circumstance.

As noted above, model one is used for assessing the number of microbes adhering to dispersed tissue cells. Squamous cells can be washed from gastric mucosa and dispersed *in vitro*. Thus, model one has been used in studies of the mechanisms by which bacteria adhere to such cells from animals that have squamous epithelium in their stomachs (Fuller, 1975; Suegara *et al.*, 1975). Models two and three are more appropriate, however, for studies with columnar epithelial cells from areas of the gastro-intestinal tract not lined with squamous epithelium. These latter models do not require dispersed cells as does model one and thus permit use of intact mucosal epithelium. When whole mucosal tissue is used, the three-dimensional compartment on the surface of the epithelium may be preserved at least to some extent. Nevertheless, when such models are used in studies of the mechanisms of adherence, investigators must be cautious indeed about extrapolating from their findings to the situation in the living animal.

2.2.3 Indigenous micro-organisms reported to adhere to gastro-intestinal surfaces

Micro-organisms of numerous types have been seen in close association with epithelial surfaces in all parts of the gastro-intestinal tracts of putatively normal laboratory rodents and chickens. Likewise, they have been observed associating with surfaces in various areas of the tracts of mammals other than rodents and including man (Tables 2.1 and 2.2). Most such information has been derived in studies in which scanning and transmission electron microscopy were used to detect the organisms on the surfaces. Thus, in most cases, little is known about the mechanisms by which the microbes overcome movement and epithelial turnover in the tract and maintain themselves in the epithelial habitats. In fact, in most cases, evidence is not available concerning the structure(s) to which the organisms adhere on the surface, or in many instances, even whether they 'adhere' to any structure at all.

As discussed above, a reasonable theoretical case can be made that microbes can colonize mucus on a particular surface and only appear in microscopic preparations to adhere to epithelial cells underlying the mucus. To remain in the mucous layer, the microbes would have to multiply at a rate exceeding that at which the mucus flows away downstream (Savage, 1977a). Under such circumstances the organisms could be said to be adhering to mucinous glycoproteins. Nevertheless, their ability to remain as a population on a given surface implies that they are not trapped in the mucin as it flows along. Indeed, they may be able to move about in the mucous layer (Allweiss *et al.*, 1977; see also Section 8.3.2) and even digest its constituents and use them as carbon, nitrogen and energy sources (Savage, 1978a). Thus, an hypothesis that micro-organisms remain in epithelial habitats by adhering to mucin has some complexities that are difficult to rationalize. Still, when a particular microbial type is found with regularity on a surface in a region of the gastro-intestinal tract (i.e. the stomach or small bowel and particularly the latter), where the lumenal content and presumably the mucin flows at a fairly rapid rate, then an hypothesis that the organism adheres to some structure on the surface must be regarded as a reasonable one.

Such an hypothesis has been made concerning some of the microbial types that associate with gastro-intestinal surfaces in laboratory rodents, chickens and certain other animal types (Savage, 1977a). In particular, micro-organisms that alter the architecture of the mucosal surfaces with which they associate must adhere in some way to the glycocalyx or to the underlying membranes of the epithelial cells (Table 2.2). For example, the segmented, filamentous bacteria observed on the epithelium of the small bowels of laboratory rodents and chickens and, under certain circumstances, dogs, undoubtedly adhere to the membranes of the columnar epithelial cells themselves (Table 2.2; Fig. 2.1). Similarly, rod-shaped bacteria

Table 2.1 Bacteria or yeasts believed to be indigenous and reported to associate with gastro-intestinal epithelia without obviously altering their architecture[a]

Microbial type	Area of tract	Animal type	Reference
Streptococci	Esophagus[b]	Mice, rats	Savage et al., 1968, Scott and Gardner, 1976
		Guinea pigs	Yoon and Tannock, 1978
	Stomach[b]	Mice	Dubos et al., 1965; Savage et al., 1968
		Swine	Fuller et al., 1978
Lactobacilli	Stomach[b]	Mice	Dubos et al., 1965; Savage et al., 1968; Roach et al., 1977
		Rats	Brownlee and Moss, 1961; Dubos et al., 1965; Savage, 1969b; Morotomi et al., 1975; Scott and Gardner, 1976; Watanabe et al., 1977
		Hamsters	Kunstyr, 1974
		Swine	Dubos et al., 1965; Tannock and Smith, 1970; Fuller et al., 1978
	Crop[b]	Chickens	Fuller and Turvey, 1971
Spirochetes	Stomach[c]	Cats, dogs	Kasai and Kobayashi, 1919; Vial and Orrego, 1960; Sedar and Friedman, 1961
Torulopsis sp. (Yeasts)	Stomach[c]	Mice	Savage and Dubos, 1967
		Rats	MacKinnon, 1959; Savage and Dubos, 1967; Scott and Gardner, 1976

Table 2.1 Bacteria or yeasts believed to be indigenous and reported to associate with gastro-intestinal epithelia without obviously altering their architecture[a] (*continued*)

Microbial type	Area of tract	Animal type	Reference
Tcrulopsis sp. (Yeasts)	Stomach[c]	Guinea pigs	Yoon and Tannock, 1978
Gram-positive and negative cocci and rods	Small intestine[c]	Humans	Plaut *et al.*, 1967; Peach *et al.*, 1978
Rods and cocci	Small intestine[c]	Chickens	Salanitro *et al.*, 1978; Untawale *et al.*, 1978
Fusiform-shaped bacteria	Cecum[c]	Mice	Savage *et al.*, 1968; Syed *et al.*, 1970
	Colon[c]		Savage *et al.*, 1971; Savage and Blumershine, 1974; Geissinger and Abandowitz, 1977; Leach *et al.*, 1973; Tannock and Savage, 1974b; Ducluzeau *et al.*, 1977
		Rats	Davis *et al.*, 1977a
Spirochetes	Cecum[c] (crypts)	Rats	Davis *et al.*, 1972; Gustafsson and Maunsbach, 1971; Henrikson, 1973
	Colon[c]	Dogs	Turek and Meyer, 1978
Spiral-shaped bacteria; spirochetes	Cecum[c]	Mice	Savage *et al.*, 1971
	Colon[c]		Savage and Blumershine, 1974; Geissinger and Abandowitz, 1977; Lee and Phillips, 1978
		Rats	Leach *et al.*, 1973; Phillips *et al.*, 1978

Table 2.1 Bacteria or yeasts believed to be indigenous and reported to associate with gastro-intestinal epithelia without obviously altering their architecture[a] (*continued*)

Microbial type	Area of tract	Animal type	Reference
Spiral-shaped bacteria; spirochetes	Colon[c]	Dogs	Leach *et al.*, 1973; Davis *et al.*, 1977b
Gram-positive and gram-negative bacteria; rods and cocci	Colon[c]	Chickens	Fuller and Turvey, 1971; Salanitro *et al.*, 1978
		Humans	Nelson and Mata, 1970; Moore *et al.*, 1978; Peach *et al.*, 1978; Hartley *et al.*, 1979
		Dogs	Davis *et al.*, 1977b
Rod, coccal and spiral-shaped micro-organisms	Rumen[b]	Sheep	Bauchop *et al.*, 1975
		Cattle	McCowan *et al.*, 1978

a   As assessed in most cases by electron microscopy; adapted from Table 1 in Savage, 1979a

b   Keratinized stratified squamous epithelium

c   Secreting columnar epithelium

Table 2.2 Bacteria believed to be indigenous and reported to associate with and alter the architecture of intestinal epithelia[a]

Microbial type	Area of tract	Animal type	Reference
Segmented, filamentous bacteria	Small intestine[b]	Mice	Hampton and Rosario, 1965; Blumershine and Savage, 1974; Davis and Savage, 1974; Abrams, 1977; Owen and Nemantic, 1978; Blumershine and Savage, 1978
		Rats	Reimann, 1965; Savage, 1969a; Davis and Savage, 1974; Chase and Erlandson, 1976; Davis and Savage, 1976; Davis, 1976; Snellen and Savage, 1978; Garland et al., 1978; Phillips et al., 1978
		Chickens	Fuller and Turvey, 1971
		Dogs	Davis et al., 1977b
Rod-shaped bacteria	Colon[b]	Rats	Wagner and Bartnett, 1974; Phillips et al., 1978
Spirochetes and spiral-shaped bacteria	Colon[b]	Rhesus monkeys	Takeuchi and Zeller, 1972
Spirochetes	Colon[b]	Humans	Lee et al., 1971; Minio et al., 1973

[a] As assessed in most cases by electron microscopy; adapted from Table 2 in Savage, 1979a

[b] Secreting columnar epithelium

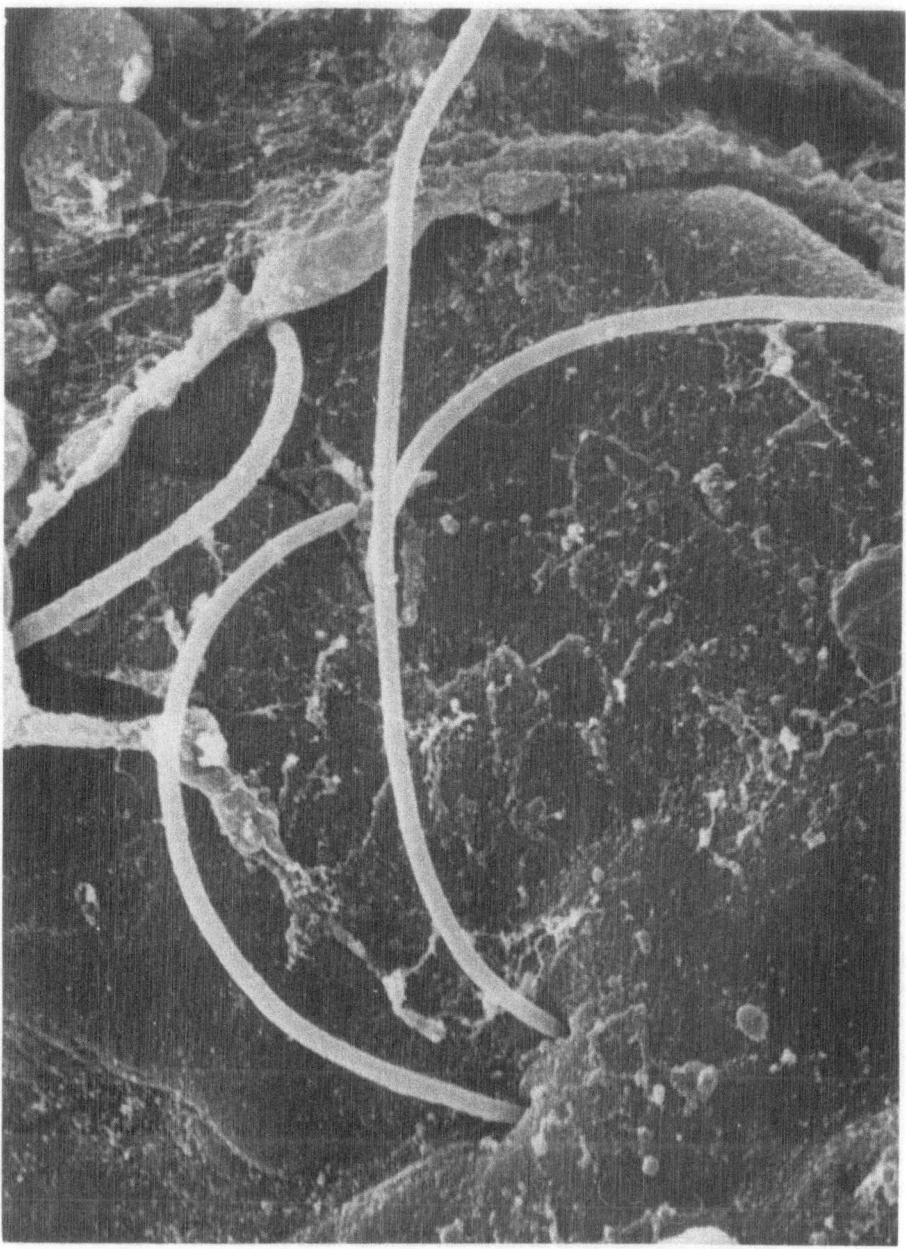

Fig. 2.1 Indigenous segmented, filamentous bacteria viewed with the scanning electron microscope on the surface of the columnar epithelium covering the villi in the small intestine of an adult rat. Note that the bacteria adhere to the surface by one end inserted into a 'hole' in the epithelium. × 2700.

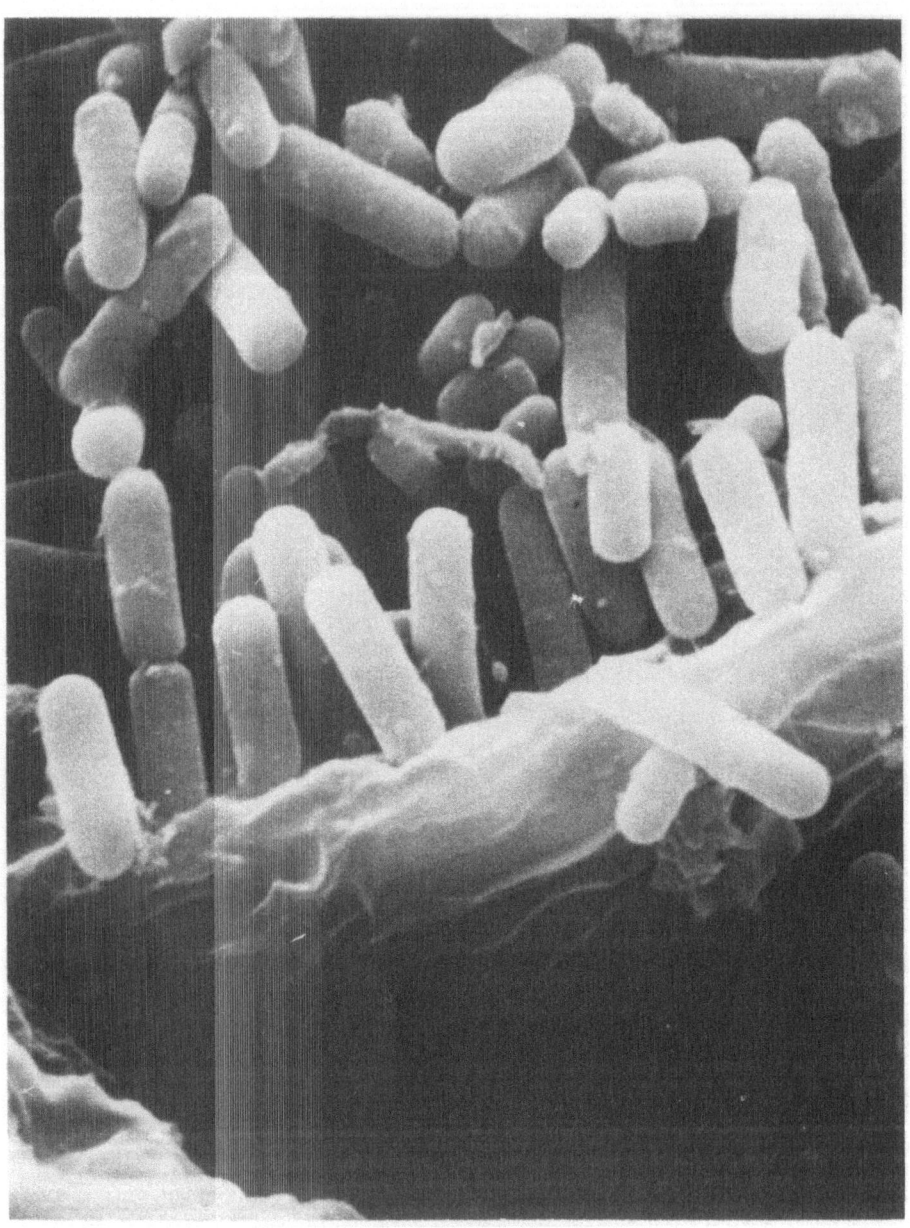

Fig. 2.2 Indigenous microbial community (composed principally of
Lactobacillus species) as viewed with the scanning electron microscope
of the keratinized stratified squamous epithelium in the stomach of an
adult mouse. x 10 000 (Savage and Blumershine, 1974).

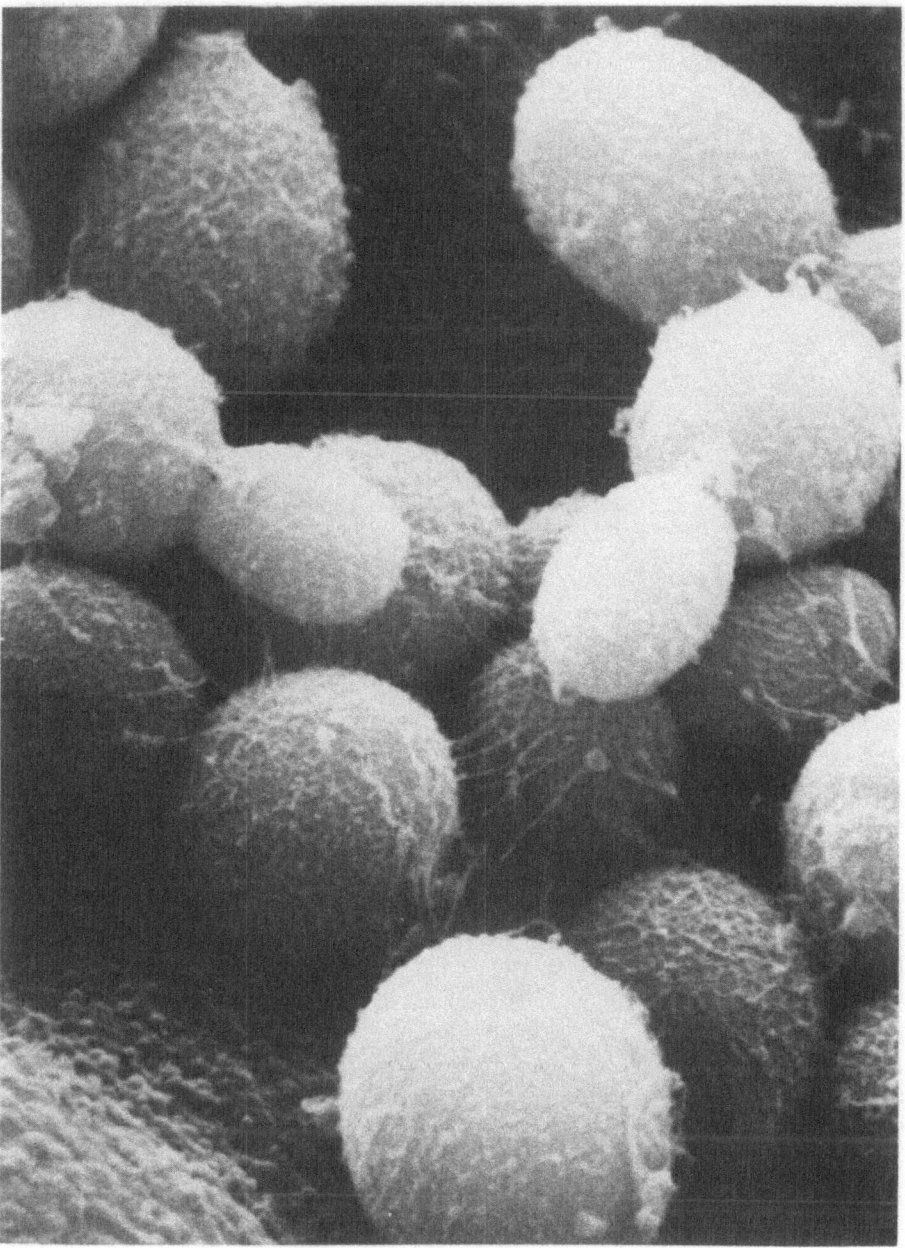

Fig. 2.3 Indigenous yeast, *Torulopsis pintolopesii,* as viewed with the scanning electron microscope on the columnar epithelium in the stomach of an adult mouse. x 5380.

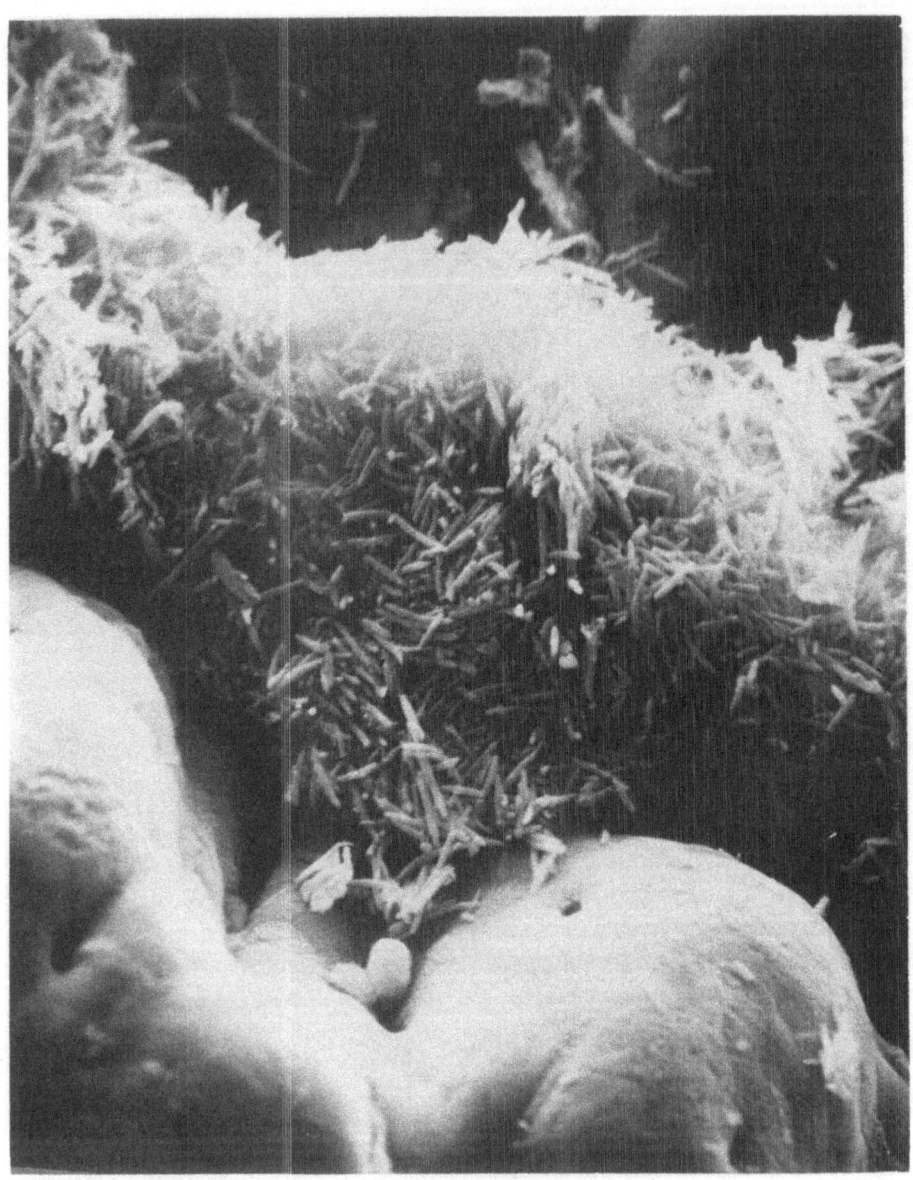

Fig. 2.4 Layer of indigenous bacteria as viewed with the scanning electron microscope on the columnar epithelium in the colon of an adult mouse. × 1100 (Savage and Blumershine, 1974).

reported to be present in the colons of rats and spirochetes and spiral-shaped bacteria that colonize cecal crypts in rats and the colonic mucosa in rhesus monkeys and possibly also man surely adhere by one end to the microvillous membrane (or its glycocalyx) of the epithelial cells (Table 2.2).

The case is much less clear, however, for the microbial types reported to associate with and not obviously alter the architecture of the gastro-intestinal mucosa (Table 3.1). Without a doubt, the lactic acid bacteria that associate with the keratinizing squamous epithelium in the esophagi and stomachs of rodents, the crops of chickens and the pars oesophagia of swine adhere to the keratinized cells on the surface of the epithelium (Fig. 2.2). By contrast, none of the micro-organisms reported to associate with and not alter the architecture of mucosa covered with columnar secreting epithelium (Table 2.1) can be said, with certainty, to adhere to particular structures on those surfaces. As shall be discussed subsequently, some of the microbial types, for instance the *Torulopsis* species that colonizes the gastric secreting epithelium in rodents (Fig. 2.2) and some bacterial species that colonize the colonic epithelium in laboratory mice (Fig. 2.4), may adhere by some means to structures on those surfaces. In no other case, however, is it known from experimentation that organisms reported to associate with and not alter the architecture of columnar epithelia also adhere to some structure on the surfaces.

In fact, at this time, a dilemma exists for investigators wishing to make such an assessment. As discussed above, microscopy is not always too useful as a tool for assessing whether or not microbes seen in association with a surface in an intact animal, actually adhere to some structure on it. Moreover, also as discussed, microbial cells may adhere to the glycocalyx or membranes of tissues or cells maintained *in vitro*, when they do not do so *in vivo*. Thus, particularly when columnar secreting epithelium is involved, and especially when the epithelial cells are not altered architecturally, significant difficulties arise for investigators attempting to determine even whether or not a particular microbial type actually adheres to a structure on the surface. Needless to say, such a problem considerably complicates any studies of the mechanisms by which an organism adheres to a surface.

2.2 MECHANISMS BY WHICH INDIGENOUS MICRO-ORGANISMS ADHERE TO MUCOSAL EPITHELIAL SURFACES

As must be obvious by now, at this time little is known about the mechanisms by which indigenous micro-organisms adhere to gastro-intestinal mucosal surfaces. Experimental efforts to study such mechanisms were started only within the past decade and have proceeded slowly (Savage, 1975).

Nevertheless, some information available can be discussed as evidence of the state of the art. The information can be organized most easily in reference to the two types of epithelium, squamous and columnar, found in the gastro-intestinal canals of higher animals.

2.3.1 Squamous epithelium

As noted above (Table 2.1), lactic acid bacteria of various types are known to associate with the stratified squamous epithelium in the esophagi of guinea pigs, the esophagi and stomachs of laboratory rodents, the crops of chickens and the pars oesophagia of swine. Also, as noted earlier, convincing evidence supports an hypothesis that these lactic acid bacteria adhere in some way to the keratinized cells. In transmission electron micrographs taken of preparations of mucosal tissues from the crops of chickens (Brooker and Fuller, 1975) and the stomachs of rodents (Takeuchi and Savage, 1973), the microbial cells can be seen to be separated from the keratinized cells by substances with a generally filamentous ultra-structure (Fig. 3.5). Substances with such ultra-structure are believed to mediate adherence of bacteria to animate and inanimate surfaces (McCowen *et al.*, 1978). No mucus in quantity or even glycocalyx can be seen on the epithelial surface at least in mice and rats (Savage, unreported observation; Takeuchi, A., personal communication). Moreover, the organisms are difficult to wash from the tissues (Dubos *et al.*, 1965; Savage and Blumershine, 1974). Thus, they undoubtedly adhere in some way to the outer surface of the keratinized cells, and multiply in the habitat at a rate sufficient to maintain the population in spite of sluffing of the epithelial cells during normal turnover processes.

The mechanisms by which the lactic acid bacteria adhere to the tissue cells involve a degree of specificity. *Lactobacillus* strains isolated from rats and mice, but not strains isolated from some other mammalian types and from birds of various types, adhere *in vitro* to keratinized cells removed from the forestomachs of rats (Suegara *et al.*, 1975). By contrast, *Lactobacillus* strains isolated from chicken and ducks, but not from mammals, also adhere *in vitro* to crop epithelial cells from chickens (Fuller and Turvey, 1971). Although derived from experiments involving tissue cells exposed to bacterial cells *in vitro*, these findings are consistent with the observation that *Lactobacillus* strains isolated from rodents (rats, mice), but not from humans, swine, chickens or inanimate sources, colonize the gastric non-secreting (keratinizing squamous) epithelium in mono-associated ex-germfree mice (Savage, 1979a). Thus, the findings support an hypothesis that particular *Lactobacillus* strains recognize the surfaces of cells of particular animal types.

Within a given animal type, however, *Lactobacillus* strains able to adhere to the keratinized gastric epithelium may also be able to adhere to other mucosal surfaces. A *Lactobacillus* strain isolated from a mouse that adheres to keratinized gastric epithelium from mice adheres equally well *in vitro* to columnar gastric epithelium from the animals (Kotarski and Savage, 1977; and manuscript in preparation). Possibly, therefore, substances on the keratinized cells that recognize the surfaces of the bacterial cells may be present in epithelial cell surfaces not normally colonized by the bacteria. It follows then, that factors other than the ability of a particular *Lactobacillus* strain to adhere to a particular surface may dictate which gastro-intestinal epithelial habitat the strain can colonize (Savage, 1977a, 1979a).

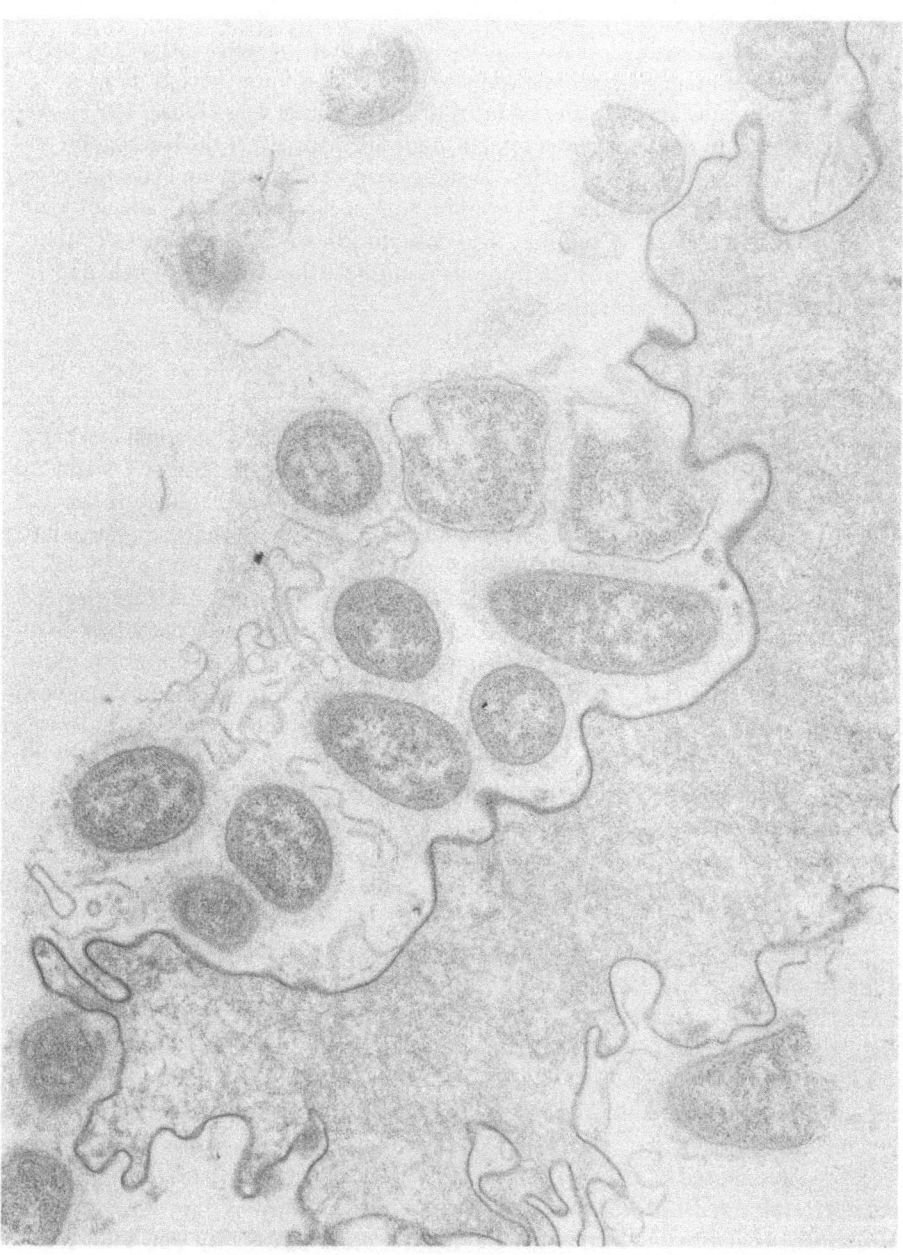

Fig. 2.5 Indigenous lactic acid bacteria localized on keratinized squamous cells in the stomachs of adult mice; note the materials with filamentous ultra-structure between the bacterial and epithelial cell; transmission electron microscopy. x 23 000 (Microphotograph courtesy of A. Takeuchi).

This hypothesis receives more attention later.

Specificity of the type discussed above suggests that the outer surfaces of the squamous cells contain substances serving as recognition units ('receptors') for complementary substances on the surfaces of the bacterial cells (Jones, 1977). At this time, such putative substances remain undefined for either the lactobacilli or the gastric cells (Savage, 1979a). The substances on the bacteria are believed to be acidic polysaccharides, or at least, to contain such macromolecules (Takeuchi and Savage, 1973; Fuller, 1975), but may also contain some macromolecules of other types (Suegara *et al.*, 1975). These findings require further study (Kotarski and Savage, manuscript in preparation).

2.3.3 Columnar epithelium

As noted, micro-organisms that associate with columnar gastro-intestinal epithelia may or may not obviously alter the architecture of the mucosa (Tables 2.1 and 2.2). Also as noted, in no case where the mucosal architecture remains intact, is the evidence available sufficient to support strongly an hypothesis that the microbial cells adhere to substances on or in the epithelial cell membranes.

A case in point is the yeast, *Torulopsis pintolopesii*, known to colonize the columnar epithelium of the gastric mucosa in mice and rats (and reported to colonize the same epithelium in guinea pigs) (Table 2.1). In preparations examined by light (Savage and Dubos, 1967), transmission electron (Takeuchi, personal communication) and scanning electron (Savage, 1978c) microscopy, these organisms can be seen to associate intimately with the surface of the epithelial cells (Fig. 2.3). In such a location, however, they are undoubtedly colonizing (i.e., multiplying in) a layer of gastric mucin. Indeed, the organisms can be seen in such a layer in frozen histological sections of mouse or rat stomach examined by light microscopy (Savage and Dubos, 1967). Thus, microscopy does not make clear that the organisms adhere to the epithelium.

Nevertheless, the yeast colonizes the keratinized stratified squamous epithelium in the stomachs of monoassociated gnotobiotic (i.e. ex-germfree) mice (Suegara *et al.*, 1979) and in conventional mice treated with anti-bacterial drugs to remove from the surface the resident lactic acid bacteria (Savage, 1969a). Thus, in analogy with the lactobacilli, apparently cells of the organism can stick to the surfaces of squamous cells *in vivo*. Moreover, the cells adhere to the epithelial surface of mucosal tissues removed from every major area of the mouse gastro-intestinal tract including the stomach exposed to the organisms *in vitro* (Suegara *et al.*, 1979). Such evidence must be interpreted with care, however, because it was derived from experiments involving cells and tissues maintained *in vitro*. *T. pintolopesii* may well adhere to the microvillous membranes of gastric columnar cells *in vivo*. Unfortunately, that hypothesis is yet to be supported in some rigorous way.

Because cells of *T. pintolopesii* can stick *in vitro* to surfaces from all regions of the murine gastro-intestinal tract, but can colonize *in vivo* only gastric surfaces, a question can be raised concerning forces that limit it to the gastric surfaces *in vivo*. As with the lactobacilli in the stomach, factors other than the yeast's capacity to adhere to the epithelium may dictate what habitat it colonizes in the animal (Artwohl and Savage, 1979).

Another example of an organism that colonizes columnar epithelium without changing its architecture is a *Clostridium* sp. isolated from the colon of a mouse (Tannock and Savage, 1976). This organism associates with the colonic and cecal epithelium in gnotobiotic mice (Savage, 1977b; Tannock and Savage, 1976) and is believed to be an indigenous organism that forms part of the microbial community on the epithelium of the murine large bowel (Table 2.2; Fig. 2.3; Blumershine and Savage, 1974; Savage *et al.*, 1971).

Early evidence, largely from studies with scanning electron microscopy of preparations of intestinal mucosa from mice, suggested that bacteria in that natural community adhered weakly to membranes of the colonic epithelial cells via long thin filaments (Fig. 2.6; Blumershine and Savage, 1974). That hypothesis was reinforced

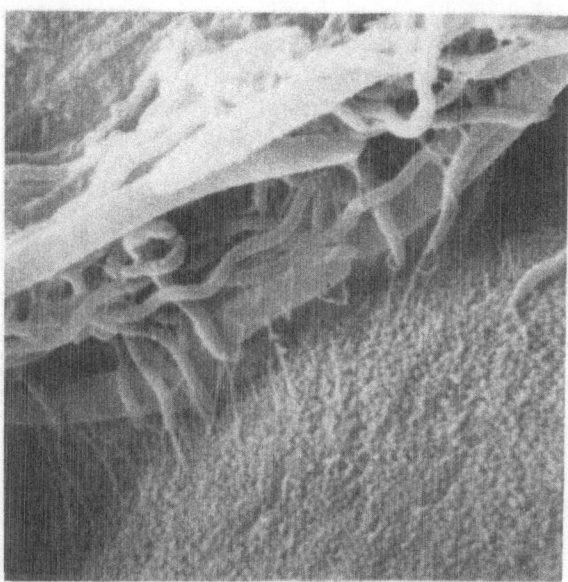

Fig. 2.6 Bottom of layer of indigenous bacteria as viewed by scanning electron microscopy on the colonic epithelium of a mouse (see Fig. 2.4). Note the filaments extending between the bacteria and the surface. x 9400 (Savage and Blumershine, 1974).

when filaments were seen, also in preparations examined by scanning electron microscopy, extending between the cells of the *Clostridium* species and the microvillous surface in gnotobiotic mice associated with that organism and one other microbial species (Savage, 1977b). These bacteria can be readily washed from their surface, in contrast, for example, to the lactic acid bacteria adhering to gastric squamous epithelium (Dubos *et al.*, 1965; Savage and Blumershine, 1974). Therefore, they may well adhere weakly via filaments to the glycocalyx of the epithelium. Again, however, no rigorous evidence exists supporting an hypothesis that the organisms adhere to some component of the colonic surface.

In contrast, micro-organisms that alter the mucosa with which they associate (Table 2.2), clearly adhere to some components of the epithelial cells. Examples are the segmented, filamentous prokaryotes known to colonize the epithelium of the small bowels of laboratory mice and rats (Fig. 2.1). The organisms are Gram-variable (Savage, 1969a) but have ultra-structure characteristic of gram-positive bacteria (Davis and Savage, 1974; Chase and Erlandsen, 1976). Unfortunately, they have never been cultured *in vitro* or identified taxonomically. Therefore, what is known about the mechanisms by which they maintain themselves on the epithelium is derived from evidence from microscopic studies of preparations of rodent small intestines.

At least two morphological forms of the filaments have been described; one type is segmented; the other shows no obvious segmentation in preparations examined by scanning electron microscopy (Blumershine and Savage, 1978). Both types adhere by one end to the epithelium, the adhering end inserting into a socket that forms in the microvillous membrane of an epithelial cell (Fig. 2.7; Snellen and Savage, 1978).

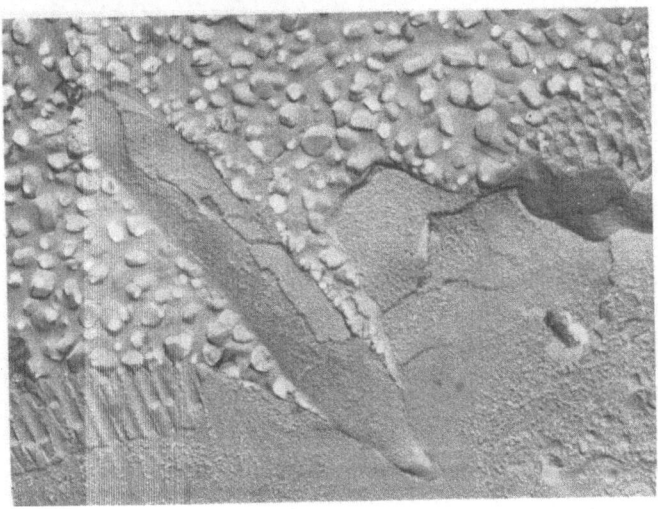

Fig. 2.7 Indigenous segmented, filamentous bacterium attached to an epithelial cell in the ileum of an adult rat (see Fig. 2.1) as viewed by transmission electron microscopy in a frozen and fractured preparation. The filament and attachment site have been fractured longitudinally. Note the fusion of the microvilli to the immediate left and their apparent destruction to the right of the filament. × 12 000 (Snellen and Savage, 1978).

In the segmented form of the organism, the segment in the socket differs considerably in structure from other segments in the filament (Blumershine and Savage, 1978; Chase and Erlandsen, 1976; Davis and Savage, 1974; Davis and Savage, 1976).

The membrane of the epithelial cell is intact around the end of the bacterium. Thus, the organism does not penetrate into the cytoplasm of the epithelial cell (Snellen and Savage, 1978). However, the membrane around the site differs in ultra-structure from adjacent microvillous membranes. Microvilli around the site are attenuated and fused (Fig. 2.7; Blumershine and Savage, 1978; Snellen and Savage, 1978). Even the cytoplasm adjacent to the site in the epithelial cells differs in ultra-structure from that remote from the site. Therefore, the interaction between the attaching end of the microbe and the epithelial cell leads to marked changes in the ultra-structure of the latter.

These changes can be interpreted to be due to a shift in the fluidity of the epithelial cell membranes in the attachment socket, and even perhaps the adjacent cytoplasm, from a fluid to a less fluid, even crystalline, state (Snellen and Savage, 1978). A socket surrounded by crystalline membranes and gelled cytoplasm could be more stable as a site of attachment of the organism than would normally fluid membranes and cytoplasm. Rod-shaped bacteria adhering to the epithelium in the colons of rats (Wagner and Barrnett, 1974; Phillips *et al.*, 1978), spirochetes in the cecal crypts of rats (Davis *et al.*, 1972) and spirochetes and spiral-shaped bacteria adhering to the epithelium in the colons of monkeys (Takeuchi, 1972) and even man (Lee *et al.*, 1971; Minio *et al.*, 1973), may induce changes in the epithelial cells similar to those described above for the filamentous bacteria. Nothing is known about the nature of the substance(s) that could mediate such changes. Little doubt exists, however, that microbes that may produce such substances adhere strongly to the brush-border membranes of the mucosal cells. Much further research is needed, however, before the mechanisms by which the organisms adhere are understood.

Whatever the mechanisms that mediate their adherence, the segmented filament-ous prokaryotes undoubtedly multiply in their habitat on the small bowel epithelium (Blumershine and Savage, 1978). To survive on the surface they must continuously repopulate new epithelial cells being made during normal turnover of the epithelium (Savage, 1979b). The mechanisms by which the non-segmented forms of the organisms reproduce on the surface are not yet clear. At least two mechanisms may have been detected, however, for the segmented form. In one case, certain segments in the filament remote from the attachment site differentiate into structures resembling the attaching segment (Blumershine and Savage, 1978). These structures may fragment away from the attached filament. If they contact the brush-border before being swept away down the lumen, then presumably they initiate a new attachment socket.

In the second case, many segments of the filaments produce intracellular prokaryotic bodies that strikingly resemble the segments involved in attachment. These intra-cellular structures arise probably from mechanisms similar to those involved in endospore formation (Blumershine and Savage, 1978; Chase and Erlandsen, 1976; Davis and Savage, 1974, 1976; Snellen and Savage, 1978). When the bodies are 'mature', the 'mother' cells in which they form may lyse, releasing them onto the epithelium. Similarly to the differentiated filaments described earlier, the bodies presumably can initiate new attachment sockets in the microvilli. Again, however, these hypotheses can be tested adequately only when the organisms are isolated and can be studied in pure culture.

2.4 SUMMARY AND CONCLUSIONS

Indigenous micro-organisms of numerous types associate with epithelial surfaces in the gastro-intestinal tracts of mammals and birds. Some such microbial types, e.g. *Lactobacillus* sp. in the stomachs of laboratory rodents and crops of chickens, adhere to the surface with which they associate without obviously altering it architecturally. In such cases, the bacteria adhere to the surfaces of the epithelial cells by macro-molecules on their surfaces, possibly polysaccharides, interacting in specific ways with receptor macromolecules on the epithelial surface. Other such microbial types, e.g. segmented, filamentous bacteria in the rodent small bowel, adhere to the membranes of the epithelial cells with which they associate and alter them architecturally. In such cases, the microbes have specialized segments or ends for adhering to the membranes, and probably elaborate substances that stabilize the membranes and cytoplasm at the site in the epithelial cell to which they attach. In no case, however, whether or not the epithelium is altered architecturally, is the mechanism known by which the micro-organisms adhere to the mucosal epithelia.

REFERENCES

Abrams, G.D. (1977), *Am. J. clin. Nutr.,* **30**, 1880–1886.
Allweiss, B., Dostal, J., Carey, K.E. and Freter, R. (1977), *Nature,* **266**, 448–450.
Aly, R., Shinefield, H.I., Strauss, W.G. and Maibach, H.I. (1977), *Infect. Immun.,* 17, 546–549.
Artwohl, J.E. and Savage, D.C. (1979), *Appl. Environ. Microbiol.,* **37**, 697–703.
Bauchop, T. (1977), In: *Microbial Ecology of the Gut,* (Clarke, R.T.J. and Bauchop, T. eds.), Academic Press, London, New York, San Francisco, pp. 223–250.
Bauchop, T., Clarke, R.T.J. and J.C. Newhook, (1975), *Appl. Microbiol.,* **30**, 668–675.
Blumershine, R.V. and Savage, D.C. (1978), *Microb. Ecol.,* **4**, 95–103.
Breznak, J.A. and Pankratz, H.S. (1977), *Appl. Environ. Microbiol.,* **33**, 406–426.
Brooker, B.E. and Fuller, R. (1975), *J. Ultrastruct. Res.,* **52**, 21–31.
Brownlee, A. and Moss, W. (1961), *J. Path. Bact.,* **82**, 513–516.
Chase, D.G. and Erlandsen, S.L. (1976), *J. Bact.,* **127**, 572–583.
Cherry, R.J. (1975), *FEBS Letters,* **55**, 1–7.
Clark, J.M. and Glasgov, S. (1976), *Science,* **192**, 1360–1361.

Davis, C.P. (1976), *Appl. Environ. Microbiol.*, **31**, 304–312.
Davis, C.P. and Savage, D.C. (1974), *Infect. Immun.*, **10**, 948–956.
Davis, C.P. and Savage, D.C. (1976), *Infect. Immun.*, **13**, 180–188.
Davis, C.P., Balish, E. and Uehling, D. (1977a), *Scanning Electron Microscopy*, **II**, 269–274.
Davis, C.P., Cleven, D., Balish, E. and Yale, C.E. (1977b), *Appl. Environ. Microbiol.*, **34**, 194–206.
Davis, C.P., Mulcahy, D., Takeuchi, A. and Savage, D.C. (1972), *Infect. Immun.*, **6**, 184–192.
Dubos, R., Schaedler, R.W., Costello, R. and Hoet, P. (1965), *J. exp. Med.*, **122**, 67–76.
Ducluzeau, R., Ladire, M., Cullut, C., Raibaud, P. and Abrams, G.D. (1977), *Infect. Immun.*, **17**, 415–424.
Fuller, R. (1975), *J. gen. Microbiol.*, **87**, 245–250.
Fuller, R. and Turvey, A. (1971), *J. Appl. Bacteriol.*, **34**, 617–622.
Fuller, R., Barrow, P.A. and Brooker, B.E. (1978), *Appl. Environ. Microbiol.*, **35**, 582–591.
Garland, C.D., Stark, A.E., Lee, A. and Dickson, M.R. (1978), In: *Microbial Ecology*, (Loutit, M.W. and Miles, J.A.R. eds), pp. 240–243, Springer-Verlag, Berlin, Heidelberg, New York.
Geissinger, H.D. and Abandowitz, H.M. (1977), *Trans. Am. Micros. Soc.*, **96**, 254–257.
Gibbons, R.J. and Van Houte, J. (1975), *Ann. Rev. Microbiol.*, **29**, 19–44.
Gustafsson, B.E. and Maunsbach, A.B. (1971), *Z. Zellforsch.*, **120**, 555–578.
Hampton, J.C. and Rosario, B. (1965), *Lab. Invest.*, **14**, 1464–1481.
Hartley, C.L., Neumann, C.S. and Richmond, M.H. (1979), *Infect. Immun.*, **23**, 128–132.
Henrikson, R.C. (1973), *Z. Zellforsch.*, **140**, 445–449.
Ito, S. (1969), *Fedn. Proc. Fedn. Am. Socs. exp. Biol.*, **28**, 12–15.
Jones, G.W. (1977), In: *Microbial Interactions (Receptors and Recognition)*, (Reissig, J.L. ed.), pp. 141–176, Chapman and Hall, London.
Kasai, K. and Kobayashi, R. (1919), *J. Parasitol.*, **6**, 1–11.
Kotarski, S.F. and Savage, D.C. (1978), *Abstr. A. Meet. Am. Soc. Microbiol.*, 1978, 15.
Kunstyr, I. (1974), *Zbl. Vet. Med.*, **21**, 553–561.
Larsen, B., Markovetz, A.J. and Galask, R.P. (1978), *Appl. Environ. Microbiol.*, **35**, 444–450.
Leach, W.D., Lee, A. and Stubbs, R.P. (1973), *Infect. Immun.*, **7**, 961–972.
Lee, A. and Phillips, M. (1978), *Appl. Environ. Microbiol.*, **35**, 610–613.
Lee, F.D., Kraszewski, A., Gordon, J., Howie, J.G.R., McSeveney, D. and Harland, W.A. (1971), *J. Br. Soc. Gastroenterol.*, **12**, 126–133.
MacKinnon, J.E. (1959), *Mycopathologia*, **10**, 207–208.
McBee, R.H. (1977), In: *Microbial Ecology of the Gut*, (Clark, R.T.J. and Bauchop, T. eds), pp. 185–222, Academic Press, London, New York, San Francisco.
McColl, I. and Sladen, G.E.G., (ed.) (1975), *Intestinal Absorption in Man*, Academic Press, London, New York, San Francisco.

McCowan, R.P., Cheng, K.-J., Bailey, C.B.M. and Costerton, J.W. (1978), *Appl. Environ. Microbiol.*, **35**, 149−155.

Minio, F., Tonietti, G. and Torsoli, A. (1973), *Rendic. Gastroenterol.*, **5**, 183−195.

Misiewicz, J.J. (1976), In: *Disorders of the Gastro-intestinal Tract, The Science and Practice of Clinical Medicine*, Vol. 1, (Dietschy, J.M. ed.), pp. 27−29, Grune and Stratton, New York.

Moore, W.E.C., Cato, E.P. and Holdeman, L.V. (1978), *Am. Clin. Nutr.*, **31**, S33−S42.

Morotomi, M., Watanabe, T., Suegara, N., Kawai, Y. and Mutai, M. (1975), *Infect. Immun.*, **11**, 962−968.

Nelson, D.P. and Mata, L.J. (1970), *Gastroenterology*, **58**, 56−61.

Ofek, I., Beachey, E.H., Eyal, F. and Morrison, J.C. (1977), *J. Infect. Dis.*, **135**, 267−274.

Owen, R.L. and Nemanic, P. (1978), *Scanning Electron Microscopy*, **II**, 367−378.

Peach, S., Lock, M.R., Katz, D., Todd, I.P. and Tabaqchali, S. (1978), *Gut*, **19**, 1034−1042.

Perers, L., Andaker, L., Edebo, L., Stendahl, O. and Tagesson, C. (1977), *Acta path. microbiol., scand. Sect. B.*, **85**, 308−316.

Phillips, M., Lee, A. and Leach, W.D. (1978), *Aust. J. exp. Biol. Med. Sci.*, **56**, 649−662.

Plaut, A.G., Gorbach, S.L., Nahas, L., Weinstein, L., Spanknebel, G. and Levitan, R. (1967), *Gastroenterology*, **53**, 868−873.

Reimann, H.A. (1965), *J. Am. Med. Ass.*, **192**, 100−102.

Roach, S., Savage, D.C. and Tannock, G.W. (1977), *Appl. Environ. Microbiol.*, **33**, 1197−1203.

Roseman, S. (1974), In: *Biology and Chemistry of Eucaryotic Cell Surfaces*, pp. 317−335, Academic Press, New York, London.

Salanitro, J.P., Blake, I.G., Muirhead, P.A., Maglio, M. and Goodman, J.R. (1978), *Appl. Environ. Microbiol.*, **35**, 782−790.

Savage, D.C. (1969a), *J. Bact.*, **97**, 1505−1506.

Savage, D.C. (1969b), *J. Bact.*, **98**, 1278−1283.

Savage, D.C. (1975), *Microbiology*, 1975, (Schlessinger, D., ed.), pp. 120−123, Am. Soc. Microbiol., Washington, D.C.

Savage, D.C. (1977a), *A. Rev. Microbiol.*, **31**, 107−133.

Savage, D.C. (1977b), *Microbiology*, 1977, (Schlessinger, D., ed.), pp. 422−426, Am. Soc. Microbiol., Washington, D.C.

Savage, D.C. (1977c), In: *Microbial Ecology of the Gut*, (Clark, R.T.J. and Bauchop, T. eds), pp. 277−310, Academic Press, London, New York, San Francisco.

Savage, D.C. (1978a), *Am. J. clin. Nutr.*, **31**, S131−S135.

Savage, D.C. (1978b), *Micro-organisms and the Gastro-intestinal Tract, Gastro-intestinal Microecology: One Opinion*, In: *Microbial Ecology*, (Loutit, M.W. and Miles, J.A.R., eds), pp. 234−239, Springer-Verlag, Berlin, Heidelberg, New York.

Savage, D.C. (1979a), *Am. J. clin. Nutr.*, **32**, 113−118.

Savage, D.C. (1979b), In: *Absorption of Micro-organisms to Surfaces*, (Bitton, G. and Marshall, K.C. eds), Wiley-Interscience, New York, In Press.

Savage, D.C. and Blumershine, R.V. (1974), *Infect. Immun.*, **10**, 240−250.

Savage, D.C. and Dubos, R.J. (1967), *J. Bact.*, **94**, 1811−1816.

Savage, D.C., Dubos, R. and Schaedler, R.W. (1968), *J. exp. Med.,* **127**, 67–76.

Savage, D.C., McAllister, J.S. and Davis, C.P. (1971), *Infect. Immun.,* **4**, 492–502.

Scott, A. and Gardner, I.C. (1976), *Microbios Letters,* **2**, 157–162.

Sedar, A.W. and Friedman, M.H.F. (1961), *J. Biophys. Biochem. Cytol.,* **11**, 349–363.

Slack, J.M. and Snyder, I.S. (1978), *Bacteria and Human Disease,* Year Book Medical Publishers, Inc., Chicago and London, 39–46.

Smith, H. (1977), *Bact. Rev.,* **41**, 475–500.

Snellen, J.E. and Savage, D.C. (1978), *J. Bact.,* **134**, 1099–1107.

Suegara, N., Morotomi, M., Watanabe, T., Kawai, Y. and Mutai, M. (1975), *Infect. Immun.,* **12**, 173–179.

Suegara, N., Siegel, J.E. and Savage, D.C. (1979), *Infect. Immun.,* **24**, In Press.

Syed, S.A., Abrams, D.G. and Freter, R. (1970), *Infect. Immun.,* **2**, 376–386.

Takeuchi, A. and Savage, D.C. (1973), *Abstr. A. Meet. Am. Soc. Microbiol.,* 1973, 115.

Takeuchi, A. and Zeller, J.A. (1972), *Infect. Immun.,* **6**, 1008–1018.

Tannock, G.W. and Savage, D.C. (1974a), *Infect. Immun.,* **9**, 475–476.

Tannock, G.W. and Savage, D.C. (1974b), *Infect. Immun.,* **9**, 591–598.

Tannock, G.W. and Savage, D.C. (1976), *Infect. Immun.,* **13**, 172–179.

Tannock, G.W. and Smith, J.M.B. (1970), *J. comp. Pathol.,* **80**, 359–367.

Turek, J.J. and Meyer, R.C. (1978), *Infect. Immun.,* **20**, 853–855.

Untawale, G.G., Pietraszek, A. and McGinnis, J. (1978), *Proc. Soc. exp. Biol. Med.,* **159**, 276–280.

Vial, J.D. and Orrego, H. (1960), *J. Biophysic. Biochem. Cytol.,* **7**, 367–381.

Wagner, R.C. and Barrnett, R.J. (1974), *J. Ultrastruct. Res.,* **48**, 404–413.

Watanabe, T., Morotomi, M., Suegara, N., Kawai, Y. and Mutai, M. (1977), *Microbiol. Immun.,* **21**, 183–191.

Yoon, C.S. and Tannock, G.W. (1978), *Can. J. Microbiol.,* **24**, 1099–1101.

3 Bacterial Adherence and the Formation of Dental Plaques

RONALD J. GIBBONS
and
JOHANNES van HOUTE

3.1	Introduction	*page*	63
3.2	Plaque development		65
	3.2.1 Gross observations		65
	3.2.2 Heterogeneity of plaques		69
3.3	Supragingival plaque formation		71
	3.3.1 Significance of bacterial adhesive interactions		71
	3.3.2 The acquired pellicle		73
	3.3.3 Models for studying bacterial plaque formation		74
3.4	Parameters involved in plaque initiation		77
3.5	Bacterial attachment to teeth		81
	3.5.1 Mechanisms of bacterial attachment		81
	3.5.2 Influence of the acquired pellicle		83
	3.5.3 Inhibition of bacterial adherence by oral fluids		85
	3.5.4 Influence of preformed plaque		88
3.6	Accumulation of bacteria on teeth		89
	3.6.1 Role of bacterial polymers in the accumulation of *Strep. mutans* on teeth		89
	3.6.2 Role of bacterial polymers in the accumulation of bacteria other than *Strep. mutans*		94
	3.6.3 Direct interspecies interactions		96
	3.6.4 Role of host-derived polymers		99
	References		99

Bacterial Adherence
(*Receptors and Recognition,* Series B, Volume 6)
Edited by E.H. Beachey
Published in 1980 by Chapman and Hall, 11 New Fetter Lane, London EC4P 4EE
© 1980 Chapman and Hall

3.1 INTRODUCTION

The oral cavity contains a variety of types of surfaces which are colonized by bacteria; these include keratinized and non-keratinized mucosal surfaces and the teeth. In individuals with a healthy dentition, only the outer enamel layer covering the crowns of teeth is exposed to bacteria. However, in individuals with chronic periodontal disease, the alveolar bone which supports the teeth is progressively lost and apical migration of the gingiva occurs. This results in exposure of the cementum which covers the roots of the teeth.

In general, the highest numbers of bacteria are found on the teeth and on the tongue dorsum; the buccal and palatal mucosa are more sparsely populated. Desquamation limits colonization on mucosal surfaces but large bacterial accumulations (often 100–200 mg) can develop on the teeth, especially in areas which are protected from oral cleansing (Fig. 3.1, Table 3.1). Such areas include occlusal fissures, surfaces between the teeth, areas along the gingival margin, and subgingival pockets; these sites are also difficult to keep clean by tooth brushing. The flow of

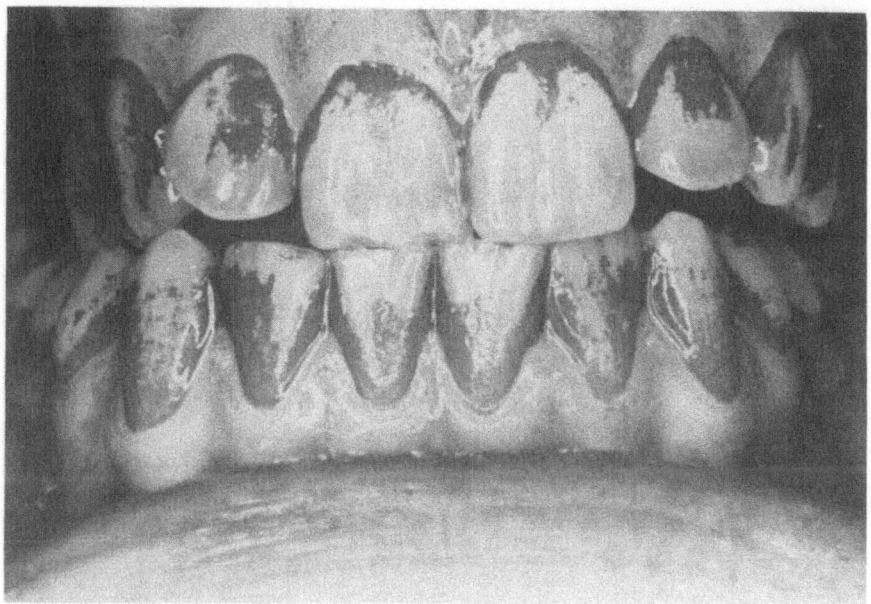

Fig. 3.1 Dental plaque present on the teeth of an individual who refrained from toothbrushing for 72 hrs. The plaque has been stained with erythrocin for better visualization.

Table 3.1 Bacterial concentrations on oral surfaces and in saliva

	Coronal plaque	Gingival plaque	Saliva	Tongue epithelial cells	Cheek epithelial cells
Direct microscopic count	2.5×10^{11}/g	1.7×10^{11}/g	–	100/cell	10–20/cell
Total cultivable count—anaerobic incubation	4.6×10^{10}/g	4.0×10^{10}/g	1.1×10^{8}/ml	–	–
Total cultivable count—aerobic incubation	2.5×10^{10}/g	1.6×10^{10}/g	4.0×10^{7}/ml	–	–

Data from Richardson and Jones, 1956; Socransky et al., 1963; Gibbons et al., 1964; Hoffman and Frank, 1966; Gibbons and van Houte, 1971.

fluids derived from the salivary glands (parotid, submaxillary, sublingual, and minor glands) and from the gingival crevice area i.e. crevicular fluid which is a serous inflammatory filtrate, contributes to the cleansing forces operative in the mouth. Mechanical forces arising from mastication, swallowing and other oro-facial movements are also thought to play a role.

The bacterial masses which form on the teeth are termed dental plaques (Fig. 3.1). Plaques consist of densely packed bacteria which are embedded in an amorphous material called the plaque matrix. The matrix is thought to impart structural integrity to the microcosm and is believed to consist predominantly of extracellular polymers synthesized by bacteria and of macromolecules derived from saliva and crevicular fluid.

Interest in dental plaques stems from the fact that these bacterial masses are responsible for the development of dental decay and various types of periodontal diseases. Dental diseases are the most prevalent bacterial infections that afflict humans and the cost associated with their treatment is of major economic concern.

3.2 PLAQUE DEVELOPMENT

3.2.1 Gross observations

The enamel surfaces of teeth are covered by a thin membranous film which is termed the acquired pellicle (Fig. 3.2). The pellicle is usually $0.1-0.2$ μm thick and is essentially free of bacteria. It is thought to consist primarily of glycoproteins and other macromolecular components of oral fluids which have become selectively adsorbed to the enamel mineral.

Shortly after a tooth is cleaned, clusters of bacteria can be found on the acquired pellicle (Saxton, 1973). Some of these clusters represent remnants of plaque since its removal by commonly used oral hygiene procedures is usually incomplete, especially in protected areas of the teeth (Björn and Carlsson,1964); other clusters represent newly deposited bacteria derived from oral fluids or surfaces. Initially, the clusters tend to be well separated (Fig. 3.3) and are found in highest concentrations near the gingival crevice and in fissures or tooth imperfections. They gradually increase in size as a result of bacterial proliferation and eventually develop into macroscopically evident colonies. Simultaneously, new bacteria adsorb to the surface and eventually a continuous bacterial layer is formed, which progressively increases in thickness. Histologic sections of plaque formed on epoxy crowns have demonstrated the presence of columns of similar bacteria extending from the inside towards the plaque periphery (Fig. 3.4; Listgarten *et al.*, 1975). This columnar development is probably due in part to a much more rapid rate of growth of bacterial cells at the plaque periphery and to the physical presence of adjacent organisms. Plaques can become several hundred microns thick on some sites on teeth but eventually their thickness becomes limited by abrasive forces.

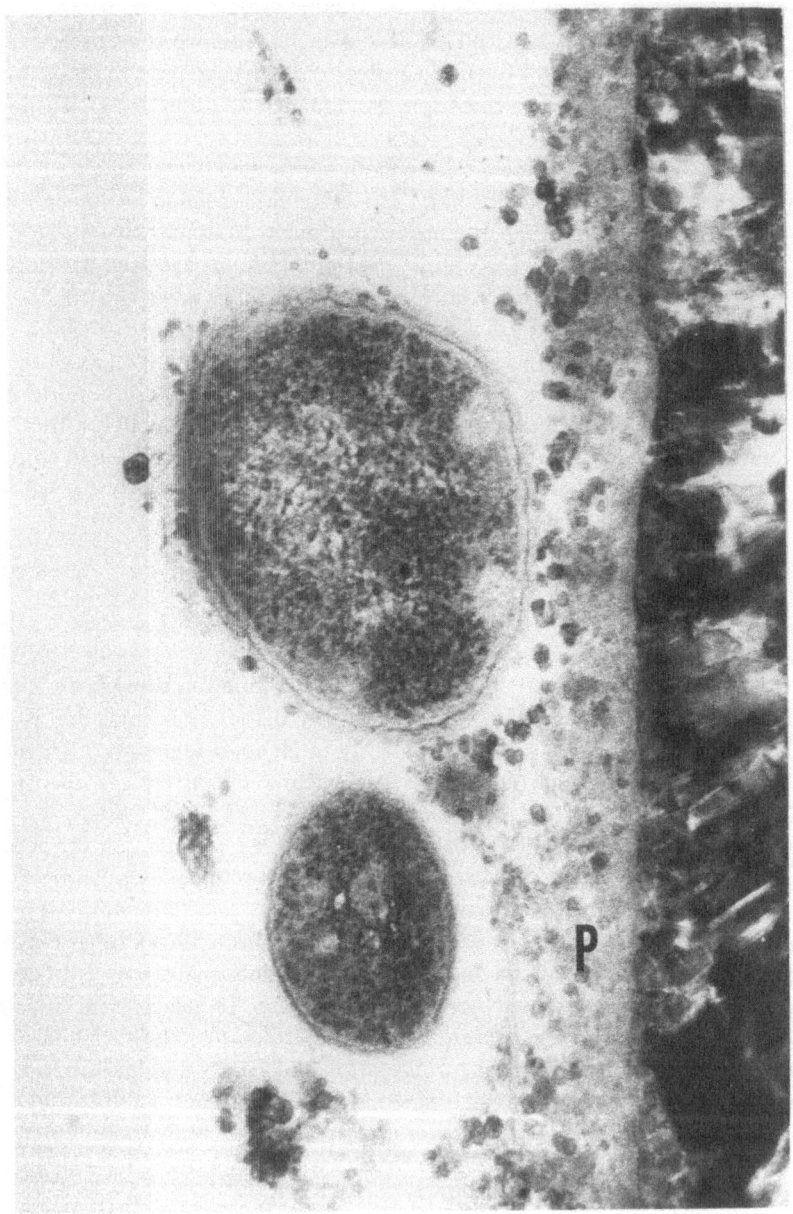

Fig. 3.2 Electron micrograph of a natural human tooth demonstrating the attachment of bacteria to the acquired pellicle (P). This photograph was generously provided by Dr R.M. Frank (Frank and Houver, 1970), Strasbourg, France.

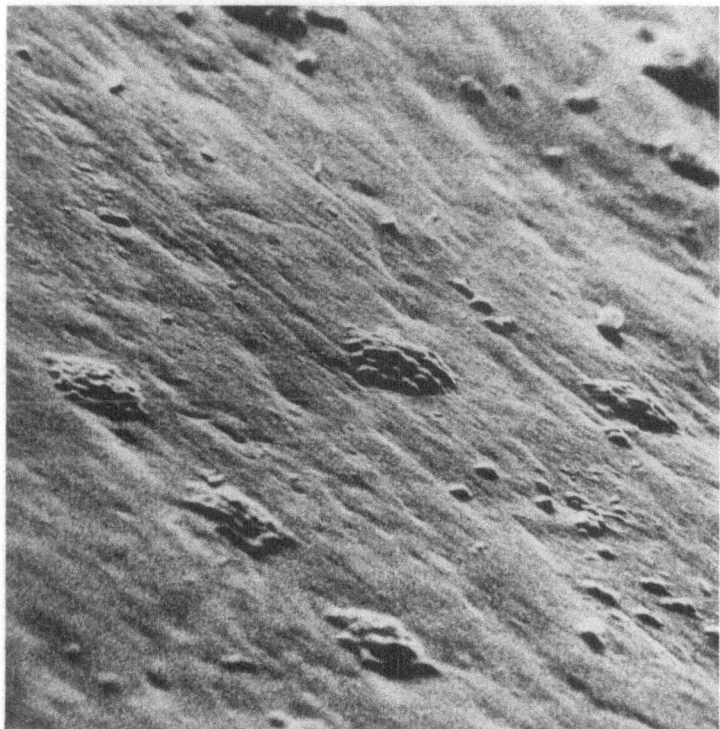

Fig. 3.3 Scanning electron micrograph of a natural human tooth 17 h after cleaning showing clusters of bacteria. This photograph was kindly provided by Dr C.A. Saxton, Isleworth, England.

Because of its protected nature, plaques form quickly in the gingival crevice area (Fig. 3.1). Failure to effectively remove gingival plaque may lead to inflammation of the gums and, in time, to a pathologically-deepened gingival sulcus or periodontal pocket. Bacterial accumulations developing in such pockets are called subgingival plaques. Microscopic studies have indicated that subgingival plaques associated with periodontitis, a chronic type of periodontal disease involving gingival inflammation and destruction of alveolar bone, frequently consist of a band of gram-positive organisms attached to the root surface with an overlying loosely adherent zone of gram-negative organisms (Fig. 3.5). Many motile forms may also be present, especially at the very depth of the pocket. Subgingival plaques associated with periodontosis, a rapidly progressing and highly localized form of periodontal disease affecting young adults, are morphologically different from those related to perio-dontitis. They consist of more sparsely distributed bacteria embedded in an amorphous matrix and they lack the adherent zone of gram-positive bacteria (Listgarten, 1976).

In supragingival plaques, gram-positive bacteria are predominant and the band of

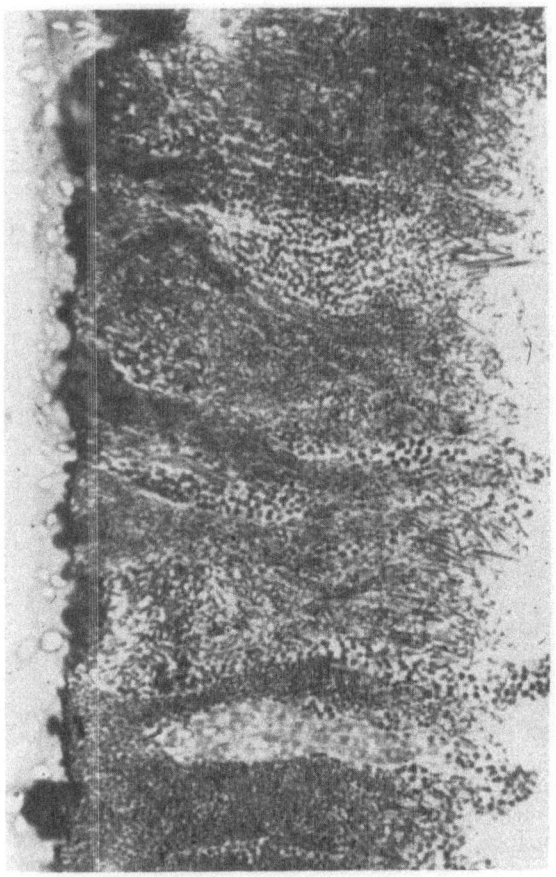

Fig. 3.4 Section of human dental plaque allowed to develop for one day on an epoxy crown showing the columnar arrangement of bacteria. This photograph was kindly provided by Dr M. Listgarten (Listgarten *et al.*, 1975), Philadelphia, Pa, U.S.A.

gram-positive organisms in subgingival plaques therefore probably represents an extension of growth of supragingival organisms. The high concentrations of gram-negative organisms found in subgingival sites may reflect the fact that this environment is more protected from oral cleansing mechanisms and, consequently, organisms with weaker adherent capabilities can colonize and persist. Their colonization may also be influenced by the unique prevailing growth conditions which include a high degree of anaerobiosis and a more nutrient-enriched milieu because of the presence of crevicular fluid.

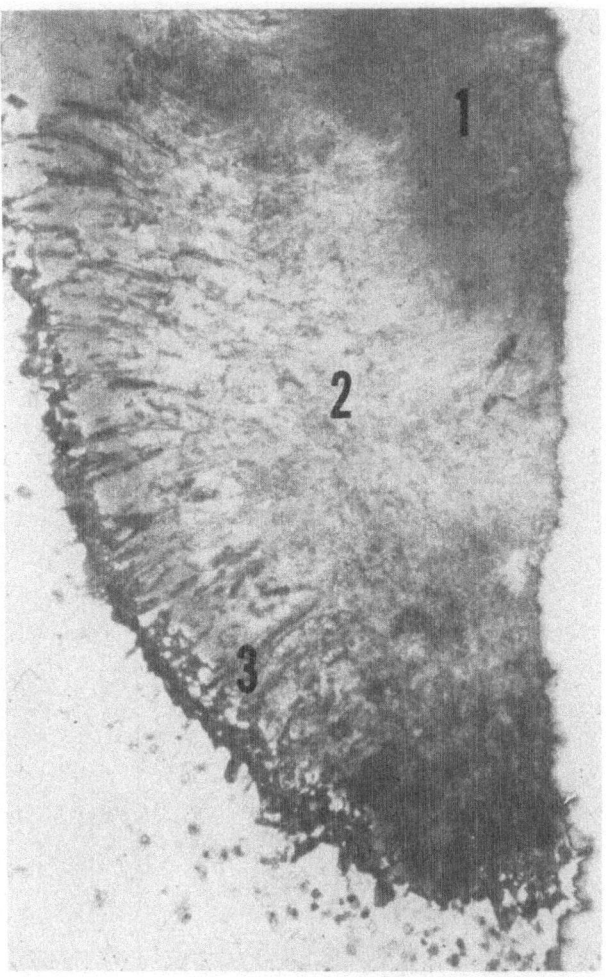

Fig. 3.5 Histologic section of a human gingival sulcus with two-month-old subgingival plaque. Three zones of bacterial growth can be distinguished. Zone 1 consists predominantly of gram-positive bacteria attached to the tooth surface. Zone 2 consists mainly of gram-negative organisms located deeper in the sulcus. Zone 3, which is adjacent to the gingival epithelium, consists of many spirochetes and other motile forms. Photo kindly provided by Dr M. Listgarten (Listgarten *et al.*, 1975), Philadelphia, Pa., U.S.A.

3.2.2 Heterogeneity of plaques

In addition to differences in the bacterial composition of supra- and subgingival plaques, differences exist between supragingival plaques developing on various areas

Table 3.2 Proportions of *Strep. mutans* of cultivable flora in multiple plaque samples obtained from a single incipient (white spot) caries lesion and from the surrounding sound enamel

Subject	Samples from white spot area						Samples from sound surface area					
	1	2	3	4	5	6	1	2	3	4	5	6
K.E.	20.8	2.9	2.3	8.3	10.7	5.8	0.08	0.3	0.6	0.6	0.2	
R.W.	6.6	1.1	10.2	0.01	1.0	8.3	0.01	1.0	0.01	0.01	0.9	1.2
S.K.	67.6	92.7	34.5	73.8			0.4	0.5	0.5			
L.G.	3.8	1.3	0.05	28.0	2.4	0.2	0.2	0.03	0.2	0.03	1.7	0.5

Data from Duchin and van Houte, (1978b).

of the teeth. For example, plaques in fissures often contain much higher proportions of *Streptococcus mutans* than plaques present on the buccal or lingual surfaces of teeth (Ikeda *et al.*, 1973; Gibbons *et al.*, 1974; Shklair *et al.*, 1974). Even plaque on a single tooth surface can be quite heterogeneous. For example, the proportions of *Strep. mutans* can vary more than 1000-fold between different samples of plaques taken a few mm apart from single incipient caries lesions or from sound enamel surfaces (Table 4.2; Duchin and van Houte, 1978b). The columnar arrangement of plaque discussed previously further illustrates its heterogeneous nature (Fig. 3.4).

Differences have also been demonstrated in the composition of bacterial plaques associated with different dental diseases. Thus, plaques associated with the initiation of carious lesions on enamel surfaces contain significantly higher proportions of *Strep. mutans* than plaques associated with sound surfaces (for review see Gibbons and van Houte, 1975b; Loesche, 1977). Similarly, *Actinomyces* and other filamentous organisms are abundant in plaques associated with cemental or root surface carious lesions (Jordan and Hammond, 1972; Sumney and Jordan, 1974) while increased proportions of gram-negative organisms, including *Veillonella, Fusobacterium, Selenomonas* and *Bacteroides* sp. are present in plaques associated with gingivitis (van Palenstein-Helderman, 1975). Subgingival plaques in patients with periodontosis contain mainly 5 types of gram-negative rods which include *Capnocytophaga* and *Actinobacillus* sp., (Slots, 1976; Newman and Socransky, 1977), whereas subgingival plaques associated with advanced periodontitis contain high proportions of *Bacteroides melaninogenicus, Fusobacterium, Eikenella, Selenomonas* and *Campylobacter* sp. (Slots, 1977; Socransky, 1977). The differences in plaque composition associated with various types of dental diseases suggest that each disease is a distinctive infectious process caused by different bacteria.

In addition to differences in the microbial composition of plaques, the rate of plaque formation can vary between individuals and between areas of the mouth. For example, increased rates of plaque formation have been associated with areas of gingivitis (Saxton, 1973), possibly because the increased flow of crevicular fluid which accompanies gingival inflammation may provide more nutrients or promote bacterial attachment. Plaque has also been reported to develop more slowly in 4 to 5-year old children than in 23 to 29-year old adults (Matsson, 1978) whereas older individuals (65 to 78 years of age) form plaque at the highest rate (Holm-Pedersen *et al.*, 1975).

3.3 SUPRAGINGIVAL PLAQUE FORMATION

3.3.1 Significance of bacterial adhesive interactions

It was thought in the past that plaque formed by the deposition of salivary mucins onto the tooth surface which entrapped bacteria (Winkler and Backer-Dirks, 1958; Jenkins, 1968, 1972). The mucin was hypothesized to be precipitated from saliva

Table 3.3 Proportions of prominent oral bacteria on various surfaces of the mouth and in saliva*

Organism	Plaque near gingival crevice	Coronal plaque	Tongue dorsum	Buccal mucosa	Saliva
Streptococcus salivarius	< 0.5	< 0.5	<20	11	20
Streptococcus mitis	8	15	8	60	20
Streptococcus sanguis	8	15	4	11	8
Streptococcus mutans	< 1	0–50	< 1	< 1	< 1
Enterococci	0–10	< 0.1	< 0.01	< 0.1	< 0.1
Gram-positive filaments†, including *Actinomyces, Nocardia, Rothia, Corynebacterium, Bacterionema, Leptotrichia* and others.	35	42	20	?	15
Lactobacillus species	< 1	< 0.005	< 0.1	< 0.1	< 1
Neisseria species†	< 0.5	< 0.5	< 0.5	< 0.5	< 1
Veillonella species†	10	2	12	1	10
Bacteroides melaninogenicus	6	< 1	< 1	< 1	< 1
Bacteroides oralis	5	5	4	?	?
Fusobacterium species†	3	4	1	?	< 1
Hemophilus species†	+	?	+	+	5
Spirochetes†	2	< 0.1	< 0.1	< 0.1	< 0.1
Campylobacter	5	1	< 0.5	< 0.5	?

Data from Socransky *et al*, 1963; Gibbons *et al*, 1964; Gordon and Jong, 1968; Socransky and Manganiello, 1971; Gibbons and van Houte, 1975.

* Data are expressed as a percentage of total flora cultivable on anaerobically incubated blood agar.

† The oral distribution of individual species has not been determined.

by acids produced by bacterial fermentation, by surface denaturation, or through the action of neuraminidase (Leach, 1963) which cleaves terminal sialic acid residues from the oligosaccharide side chains of mucin and thereby decreases its solubility. Further production of acids or neuraminidase by entrapped bacteria on the teeth would augment this process. Though it was known that the bacterial composition of plaques differed significantly from that of saliva (Table 3.3), this was assumed to be due to differences in the rates of growth of various bacterial species on the tooth surface. Thus, *Strep. salivarius*, which is found in high proportions on the tongue dorsum and in saliva but in only low proportions in plaques, was assumed to grow at a lower rate than predominant supragingival plaque organisms such as *Strep. sanguis* or *Strep. mitis.* On the other hand, detailed nutritional studies have indicated that the common oral *Streptococcus* species possess generally similar and relatively simple growth requirements (Carlsson, 1970; 1972; Rogers, 1973). This does not support the hypothesis that selective growth is responsible for population differences between plaques and saliva.

It is now thought that the formation of dental plaques requires two types of specific bacterial adherent interactions. First, bacteria attach selectively to the acquired pellicle and secondly, they accumulate via specific interactions with components comprising the matrix of plaques (Gibbons and van Houte, 1973, 1975a). These adhesive processes are thought to entail chemical and/or physical interactions between components of the bacterial cell surface and constituents comprising the enamel pellicle or plaque matrix. Because cell surface components can vary considerably among various bacterial species, it may be expected that different bacteria will vary in their attachment to the enamel pellicle and accumulate to different extents on teeth.

3.3.2 The acquired pellicle

Since the acquired pellicle represents the outer surface of teeth to which bacteria attach, its properties and composition are of interest. Enamel is composed primarily of a complex calcium phosphate salt termed hydroxyapatite (Ca_{10} $(PO_4)_6$ $(OH)_2$). A small amount (0.3–2%) of organic material is located between the hydroxyapatite crystals and comprises the enamel matrix. Human enamel may also contain varying quantities of sodium, magnesium, chlorine, fluorine and other inorganic elements, generally in quantities of less than 1% (Weatherell and Robinson, 1973). However, some of these, such as fluoride, can significantly influence the properties of enamel, including its solubility (Brudevold, 1975). The enamel surface has the properties of calcium ions, phosphate and hydroxyl groups, and often fluoride ions; bound water molecules are also thought to be present (Arends and Jongbloed, 1977; Pruitt, 1977; Rölla, 1977). Because both phosphate groups and calcium atoms are exposed, hydroxyapatite is amphoteric and can bind both acidic and basic proteins (Bernardi *et al.,* 1972). However, more phosphate groups than calcium atoms are exposed on the outer surface of hydroxyapatite crystals and therefore dental enamel

has a net negative charge. Acidic glycoproteins, such as salivary mucins, are thought
to bind to calcium atoms in the hydration shell (Stern layer) and the presence of
anions such as phosphate or fluoride tends to reduce adsorption by competing for
available calcium sites (Bernardi, 1973). Basic components bind to phosphate groups
and can be desorbed by calcium and other cations.

   In vitro studies have shown that a variety of salivary proteins and glycoproteins
adsorb to powdered hydroxyapatite to different extents (Ericson, 1967; Hay, 1967).
Because of their strong negative charge, mucins adsorb avidly to hydroxyapatite
and are present in pellicles formed within 2 hours on human teeth *in vivo* (Sönju
et al., 1974). Salivary proteins rich in tyrosine, proline or histidine have also been
found to adsorb strongly to hydroxyapatite (Oppenheim *et al.*, 1971; Hay, 1973,
1975). Hydrolysates of pellicles have a high content of acidic amino acids
(Armstrong, 1966, 1967; Mayhall, 1970); this further supports the idea that acidic
proteins and peptides are selectively adsorbed to the enamel surface. Indirect immunofluorescent studies have also indicated that lgA and lysozyme are regular
constituents of the pellicle; amylase and lgG are less consistently detected and
albumen and fibrinogen are encountered infrequently (Orstavik and Kraus, 1973).
Muramic and di-aminopimelic acids are not generally detected, indicating that
bacteria are not present in significant quantities (Mayhall, 1970).

   Pellicles naturally present on teeth cannot be removed easily (Meckel, 1967).
They can be abraded off by vigorous pumicing or floated off by decalcifying teeth
in strong acids; natural pellicles are therefore highly insoluble membranous films.
However, experimental pellicles formed by exposing powdered hydroxyapatite
or teeth to saliva for a few hours can be mostly desorbed by phosphate solutions
(McGaughey and Stowell, 1967). Consequently, it is thought that pellicle components become modified over time after their adsorption to teeth. These changes could
entail the formation of hydrogen or other types of bonds between adsorbed molecules
as well as their partial degradation by enzymes present in oral fluids or elaborated
by oral bacteria. Experimental pellicles formed by exposing hydroxyapatite to
saliva for times ranging from 1 hour to several days adsorb different numbers of
bacteria (Clark and Gibbons, 1977) and have different perm-selective properties
(Zahradnik *et al.*, 1976). The acquired pellicle is believed to help protect the teeth
from decalcification when acidic foods and beverages are ingested (Meckel, 1967);
it also influences the caries process by altering the diffusion of hydrogen ions from
the plaque and calcium and phosphate ions from the tooth (Zahradnik *et al.*, 1977).

3.3.3 Models for studying bacterial plaque formation

The observation that strains of *Strep. mutans* form adherent deposits on glass
surfaces (Fig. 3.6), extracted teeth, or steel wires immersed in sucrose broth
cultures attracted considerable attention because this could serve as a model of
plaque formation (for review see Chapter 5, Slade, 1976). Subsequently, several
investigators used glass surfaces to assess the relative ability of various oral bacteria

Fig. 3.6 Inverted culture vessels emptied after growth of a strain of
Streptococcus mutans in broth containing glucose (left) or sucrose
(right). Note the adherent deposit of *Streptococcus mutans* on the
glass surface which formed only in sucrose broth (right) and is associated
with the synthesis of extracellular glucans.

to form plaque-like deposits *in vitro*. Quantitative methods of measuring bacterial
accumulations on glass or wires have also been described (McCabe *et al.*, 1967;
Mukasa and Slade, 1973). However, such models lack the presence of an acquired
salivary pellicle which alters the selectivity of bacterial attachment; they also fail
to distinguish effectively between adherent interactions involved in bacterial
attachment to the surface and those required for cell accumulation (Clark and
Gibbons, 1977; Clark *et al.*, 1978a). Moreover, many bacteria present in high
proportions in human plaques do not form such accumulations *in vitro* (Slade, 1976).
Consequently, data derived from such models cannot be easily interpreted in terms
of plaque formation in the human mouth.

The adsorption of bacteria to hydroxyapatite surfaces which have been pretreated with human saliva to form an experimental pellicle represents an approach
which more closely mimics the *in vivo* situation. Powdered human enamel
(Hillman *et al.*, 1970), slabs of bovine teeth (Orstavik *et al.*, 1974), slices of whale
dentin (Olsson and Krasse, 1976), powdered synthetic hydroxyapatite (Liljemark

and Schauer, 1975), hydroxyapatite discs (Clark and Gibbons, 1977), and spheroidal hydroxyapatite beads (Clark *et al.,* 1978a) have been used as adsorbents in such studies. In general, the selectivity of bacterial adsorption to saliva-treated hydroxyapatite surfaces parallels that observed in experiments performed directly in the mouths of humans. In addition, bacterial adsorption to hydroxyapatite surfaces which contain salivary pellicles can be described in physical chemical terms (Gibbons *et al.,* 1976; Clark *et al.,* 1978a).

Variables which influence the adsorptive behavior of bacteria to solid surfaces include the culture medium used for propagation, the phase of growth, and the bacterial cell concentration. Environmental parameters which influence adsorption include the pH, the ionic strength, the composition of the suspending fluid, the incubation time and temperature, and the degree of agitation (for review see Marshall *et al.,* 1971; Daniels, 1972; Hattori and Hattori, 1976). These parameters should be controlled to simulate the conditions in the mouth as closely as possible in *in vitro* models for studying bacterial adherent interactions involved in plaque formation.

It has been possible to study the adsorption of bacteria to teeth and other oral surfaces directly in the mouths of humans and experimental animals. In one approach, teeth were thoroughly cleaned to reduce their bacterial populations to negligible levels; after a short period of time (30–120 min) in which normal activities were resumed, the proportions of different organisms which adsorbed to the teeth were compared with their proportions in saliva (van Houte *et al.,* 1970). With this method, the ability of naturally occurring bacteria to attach to teeth is studied directly. However, it can only be used to study organisms which are present in easily measurable concentrations in saliva. In a second *in vivo* approach, mixtures of antibiotic-resistant strains of indigenous bacteria have been introduced into the mouths of volunteers (Gibbons and van Houte, 1971; Liljemark and Gibbons, 1971; 1972; van Houte *et al.,* 1971, 1972). After a brief period of time, the proportions of each labeled organism recovered from the teeth and other oral surfaces were determined by use of antibiotic-containing media and compared to those present in the original mixture.

Use of antibiotic-resistant mutants permits the study of bacterial attachment to preformed dental plaques (Ruangsri and Orstavik, 1977; Slots and Gibbons, 1978) or to other surfaces which harbor high concentrations of indigenous organisms such as the tongue dorsum. Potential drawbacks of such experiments are that the adherence of strains cultivated in artificial media may differ from that of naturally occurring organisms (Gibbons, 1977) and the properties of the mutants may not be identical to those of the wild-type strains (Bammann *et al.,* 1978). It is also not clear whether organisms which are artificially introduced into the mouth have adequate opportunity to interact with components of oral fluids, which can influence their attachment. Moreover, strains of oral bacteria maintained in the laboratory frequently, though not invariably, lose the ability to adsorb to oral surfaces (Williams and Gibbons, 1975); similar observations have been made for other bacteria (Suegara *et al.*, 1975). Clearly, caution must be used in extrapolating observations made with laboratory-propagated strains to the *in vivo* situation and key observations should be confirmed with freshly isolated bacterial strains.

3.4 PARAMETERS INVOLVED IN PLAQUE INITIATION

The first step in plaque formation involves the selective adsorption of bacteria from oral fluids to the pellicle surface. The selectivity of this process has been documented well. For example, a variety of bacteria have been found to adsorb to different extents to saliva-treated hydroxyapatite surfaces *in vitro* (Hillman *et al.*, 1970; Orstavik *et al.*, 1974; Olsson and Krasse, 1976; Clark *et al.*, 1978a). Similarly, exposure of cleaned, pellicle-covered teeth to either naturally occurring bacteria or to artificially introduced, antibiotic-resistant strains *in vivo* has also indicated a high order of selectivity (Table 3.4).

The actual numbers of cells of a given bacterial species which attach to a cleaned tooth surface also depend upon their concentration available for adsorption. Bacterial attachment to teeth is not highly efficient and only a small percentage of the available bacteria becomes associated with the tooth surface. For example, experiments involving humans with defined salivary concentrations of naturally occurring or artificially introduced strains of oral streptococci have shown that approximately 10^4 cells of *Strep. mutans* per ml of saliva must be present before one cell can be recovered from a cleaned smooth tooth surface (van Houte and Green, 1974). In the case of *Strep. sanguis*, about 10^3 organisms per ml of saliva are required because of the organism's greater avidity.

Studies with artificial fissures which were placed in the mouths of human volunteers have indicated that concentrations of *Strep. mutans* of 10^2 per ml of saliva or higher will usually result in colonization (Svanberg and Loesche, 1977). Similarly, the fissures of newly erupted teeth readily become colonized by *Strep. mutans* when the organism is present in saliva in concentrations of 10^2-10^3/ml (Duchin and van Houte, 1978a).

Collectively, the available observations suggest that the salivary concentrations of this organism required to initiate colonization of fissures is probably lower than that required to initiate colonization of smooth tooth surfaces. This probably explains why *Strep. mutans* can be more readily isolated from the fissures of molar teeth of humans and experimental animals than from buccal or lingual smooth surfaces. Presumably, fissures as well as smooth surfaces are exposed to similar salivary bacterial concentrations in the mouth, but more organisms are removed by oral cleansing mechanisms from smooth surfaces. It is also possible that capillary and other hydrostatic forces may contribute to the greater susceptibility of fissures to colonization by *Strep. mutans* (O'Brien, 1976), but data to substantiate this are not available.

Studies of the attachment of several indigenous bacteria have shown that their

Table 3.4 Correlation between the experimentally observed adherence of oral bacteria and their proportions found indigenously

	Relative indigenous proportions			Experimentally observed adherence		
	Plaque on teeth	Tongue dorsum	Buccal mucosa	Tooth surface	Tongue dorsum	Buccal mucosa
Strep. salivarius	Low	High	Mod.	Low	High	Mod.
Strep. mitis	High	Mod.	High	High	Mod.	High
Strep. sanguis	High	Mod.	Mod.	High	Mod.	Mod.
Strep. mutans	Low to high*	Low	Low	Low	Low	Low
Lactobacilli	Low	Low	Low	Low	Low	Low
Veillorella	Low	High	Low	Low	High	Low
Neisseria	Low	Low	Low	Low	Low	Low
B. melaninogenicus ss *asaccharolyticus*	Mod.†	Low	Low	Mod.†	Low	Low

* High under the influence of dietary sucrose.
† Gingival crevice plaque.

Data derived from van Houte et al., (1970, 1971, 1972); Gibbons and van Houte (1971); Liljemark and Gibbons (1971, 1972); Gibbons (1974); Slots and Gibbons (1978).

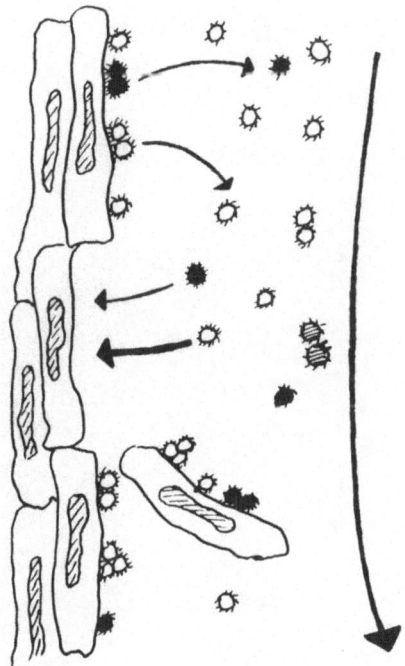

Fig. 3.7 Diagram illustrating bacterial colonization on a mucosal surface. A cyclical process involving bacterial attachment, proliferation, dislodgement and re-attachment is thought to be required for persistent colonization. Desquamation of epithelial cells requires continuous bacterial re-attachment and amplifies the influence of bacterial adherence. The organism with the lower affinity for epithelial cells will become eliminated over time.

relative adherence correlates with their natural distribution in the mouth (Table 3.3). This has lead to the realization that attachment *per se* is an important ecological determinant, which influences the bacterial colonization of host tissues; it has also led to an appreciation of the recognition capabilities of the bacterial cell surface.

The influence of bacterial adherence on the population levels of a given organism can vary from site to site within the mouth. Because of desquamation, adherence probably exerts a greater selective pressure in the case of mucosal surfaces than in the case of teeth. Desquamation continuously exposes new epithelial surfaces to which bacteria must attach if they are to persist at that site (Figs. 3.7 and 3.8). Although new epithelial cells could become colonized by the spread of growth of bacteria from adjacent cells, microscopic observations give the impression that this probably does not occur with high frequency on many oral epithelial surfaces. For example, buccal and palatal epithelial cells generally harbor 10–20 bacteria per cell and few cells have regions with confluent bacterial layers (Table 3.1). Most

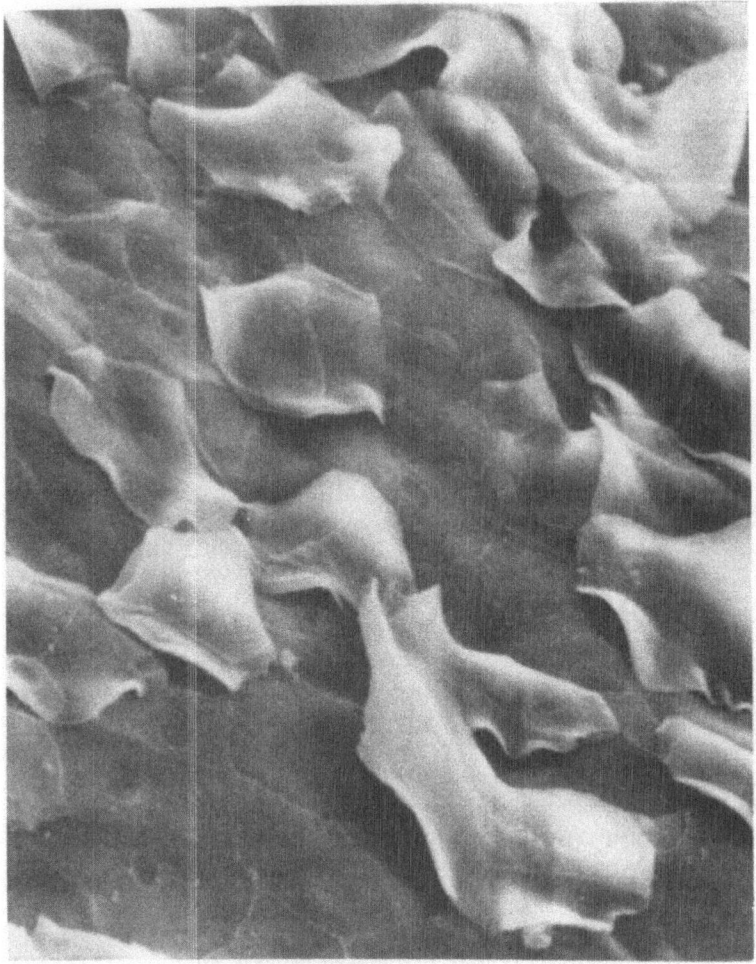

Fig. 3.8 Scanning electron micrograph of human gingiva. A large proportion
of surface epithelial cells appears to be in the process of desquamating.

epithelial cells therefore probably become colonized by the attachment of bacteria
transiently present in oral fluids. Persistent colonization of a given bacterial species
on a mucosal surface would require some of the progeny to become continuously
removed and be available for reattachment (Fig. 3.7). Thus, saliva generally contains
10^7 to 10^8 bacteria per ml which represents the wash-off from oral surfaces. The
constant desorption of cells of a given bacterial species from tissue surfaces also
increases their probability of being transmitted to new hosts and therefore fosters
the perpetuation of the species.

In contrast to continuously shedding mucosal surfaces, large populations of bacteria can accumulate on the teeth; this undoubtedly contributes to the high vulnerability of teeth to microbial attack. However, some bacterial cell turnover occurs in dental plaque. For example, antibiotic-resistant strains of *Strep. mutans* inoculated into the mouths of humans may colonize the teeth temporarily, but they usually disappear after a few weeks or months (Jordan *et al.*, 1972; Edman *et al.*, 1975). It seems likely that bacteria which have attached to the surface of thick, preformed, plaques would be quickly eliminated by the prevailing cleansing forces; such organisms would also be expected to be more affected by antibodies present in oral fluids. Furthermore, multiplication of underlying organisms would continuously move organisms on the plaque periphery away from the tooth surface.

Under natural conditions, *Strep. mutans* colonizes the teeth in a highly localized and persistent manner. It may be consistently detected in high proportions on some, but not on other, surfaces, even within the same mouth (Ikeda *et al.*, 1973; Gibbons *et al.*, 1974; Shklair *et al.*, 1974). The inability of *Strep. mutans* to spread to certain tooth surfaces is thought to be due to its relatively feeble ability to attach to teeth, and to its low numbers generally found in saliva; its persistence on a given tooth surface, therefore, suggests that cells of the organism have become attached directly to the pellicle surface.

3.5 BACTERIAL ATTACHMENT TO TEETH

3.5.1 Mechanisms of bacterial attachment

The adsorption of bacteria to solid surfaces has been studied by marine and soil microbiologists (Marshall, 1971, 1976; Daniels, 1972; Hattori and Hattori, 1976). Zobell (1943) and subsequent investigators (Corpe, 1970; Marshall *et al.*, 1971; Chapter 12) noted that the initial phase of attachment of marine bacteria to glass surfaces involved a loose association. This was followed by a phase in which cells became more firmly attached and this process was often associated with the synthesis of mucilagenous or holdfast material. Marshall and co-workers (1971) considered the initial phase to be reversible and theorized that the organisms were located approximately 100 Å from the surface and were in a state of equilibrium in which they were attracted by van der Waals forces and repelled by the negative electrostatic energies of the bacterial and the adsorbent surfaces. Furthermore, the attachment of the organisms would become more firm or 'irreversible' if extracellular polymeric material was synthesized which could bridge the space between the bacterial cell and the adsorbent surface and effectively link both together. This concept is consistent with the electron microscopic observations of Robertson (1959) that the distance between contacts of mammalian cells is generally 100–200 Å and that polymeric material, frequently of polysaccharide nature, is often present in this space. Thus, polymeric bridging may be important for promoting firm attachment of a variety of cells. A reduction of the negative charges on the opposing surfaces by the addition of cations or by lowering the pH decreases the gap between them and facilitates the development of stable adhesion.

Early studies of plaques suggested that the bacteria were held together primarily by electrostatic forces (Silverman and Kleinberg, 1967). This was based upon the observations that delating agents such as EDTA would cause a partial dispersal of plaque and that addition of calcium ions would promote reaggregation. More recently, Rölla and co-workers (Rölla, 1976; 1977; Rölla *et al.*, 1977, 1978) have suggested that negatively-charged components on the bacterial surface might bind to calcium atoms on the pellicle surface in a similar manner as acidic proteins can bind to calcium atoms of hydroxyapatite. Furthermore, they suggested that lipoteichoic acids, which contribute to the negative charge on the surface of many gram-positive bacteria, may be the components responsible for the binding of oral streptococci to teeth (Rölla *et al.*, 1977, 1978). It was also hypothesized that the synthesis of extracellular glucans by *Strep. mutans* might entrap additional lipoteichoic acids which would increase the negative charge of the organism and thereby promote its attachment.

A number of objections may be raised which question whether bacteria attach to teeth solely by electrostatic attractions. Electrostatic forces appear to be primarily involved in bacterial adsorption to hydroxyapatite crystals; this process is significantly affected by pH changes over the range of 5–7 (Hillman *et al.*, 1970). However, adsorption of many bacteria to pellicle-covered hydroxyapatite is much less affected by pH (Hillman *et al.*, 1970) and appears to involve different mechanisms (Clark *et al.*, 1978a). Olsson and co-workers (1976a, 1976b) determined the electrophoretic mobilities and ζ potential of 28 strains of oral streptococci. They observed that cells of different species which possessed the same ζ potential nevertheless adsorbed differently to slices of whale dentin, especially in the presence of saliva. They concluded that while long-range electrostatic forces may influence the initial association of bacteria with the surface, the type-specific properties of the bacterial cell surface become the important determinants of adherence after these forces are overcome (Olsson *et al.*, 1976b). In addition, there does not appear to be a clear relationship between the presence of lipoteichoic acids and bacterial attachment. Thus, lacto-bacilli, enterococci, *Strep. salivarius*, and *Strep. mutans* possess lipoteichoic acids (Knox and Wicken, 1973; Wicken and Knox, 1975), yet they adsorb relatively poorly to teeth or to saliva-treated hydroxyapatite. In contrast, cells of *Strep. mitis* (Rosan, 1978) and *Actinomyces viscosus* (Hamada *et al.*, 1976; Wicken *et al.*, 1978) do not contain detectable levels of lipoteichoic acid, but they adsorb in high numbers. Moreover, the binding properties of extracted lipoteichoic acids do not correlate with the binding properties of intact bacteria. For example, lipoteichoic acid derived from *Strep. mutans* adsorbs to erythrocytes (Knox and Wicken, 1973), yet intact cells of this organism do not (Gibbons and Quershi, 1976). These observations are analogous to those recently made by Alkan *et al.* (1977) that Group A streptococci isolated from skin have different adherent properties than those derived from the throat; however, lipoteichoic acids derived from these strains have similar adherent properties.

The remarkable specificity involved in the attachment of bacteria to different tissues, organs, and hosts is clearly difficult to explain on the basis of simple electrostatic attractions. Rather, a number of observations suggest that bacteria possess a recognition system which is superimposed upon such forces. This recognition system appears to be comparable to that possessed by mammalian cells and can identify and interact with surface components present on specific oral and other host tissues (Chapter 2). The adsorption of some oral bacteria to saliva-treated hydroxyapatite can be inhibited by certain sugars which strongly suggests the involvement of a lectin-receptor type interaction. Thus, adsorption of strains of *Leptotrichia buccalis* is inhibited by lactose and *N*-acetyl-D-galactosamine (Kondo *et al.*, 1976, 1978). Further, adsorption of *Actinomyces naeslundii* strain 12104 is inhibited by fructose and sucrose as well as by galactose and lactose; adsorption of *Strep. mutans* strain H12 is inhibited by hexosamines and other amines while adsorption of *Actinomyces viscocus* strain Ly-7 was not affected by any of 15 sugars tested (Qureshi and Gibbons, 1978). The bacterial cell surface is well known to possess components with lectin-like properties. These include fimbriae, sugar-binding proteins and various surface-bound enzymes and strong evidence suggests that such components mediate the attachment of bacteria to carbohydrate receptors on mammalian cell surfaces (Gibbons, 1977; Ofek *et al.*, 1978). It is therefore reasonable to believe that such components participate in the attachment of bacteria to the pellicle coating of teeth and probably serve as polymeric bridges. This view is also supported by the observation that several oral species can specifically bind to salivary glycoproteins which comprise part of the acquired pellicle.

3.5.2 Influence of the acquired pellicle

In vitro studies have shown that bacteria attach differently to hydroxyapatite crystals than to saliva-treated hydroxyapatite. The presence of adsorbed salivary components on synthetic hydroxyapatite or enamel powder prepared from human teeth often decreases the number of *Strep. mutans* and *Strep. salivarius* cells which attach, while it increases the number of cells of *Strep. mitis, A. viscosus*, or *A. naeslundii* which adsorb (Hillman *et al.*, 1970; Liljemark and Schauer, 1975; Clark *et al.*, 1978a). Similarly, higher numbers of *Strep. mutans* cells can be recovered from pellicle-free than from pellicle-covered teeth following *in vivo* exposure to this organism (Ruangsri and Orstavik, 1977).

Studies of the adsorption of oral bacteria to untreated and saliva-treated hydroxyapatite suggest that it follows the kinetics of a Langmuir adsorption isotherm (Gibbons *et al.*, 1976; Clark *et al.*, 1978a). Analyses of the isotherms have indicated that the number of binding sites on untreated hydroxyapatite fall within a narrow range for several *Streptococcus* and *Actinomyces* species (Clark *et al.*, 1978a). This suggests that their adsorption to these crystals occurs by a similar mechanism and supports the idea that they are binding by electrostatic forces to calcium atoms on the hydroxyapatite surface. On the other hand, the number of available binding sites for these organisms was found to differ greatly in the case of saliva-treated hydroxyapatite surfaces. This suggests that different organisms interact with different receptor molecules present in the salivary pellicle. This is also indicated by experiments which have shown that cells of different oral streptococci do not compete with each other for binding sites on saliva-treated hydroxyapatite (Liljemark and Schauer, 1977). It seems clear that the presence of a pellicle on hydroxyapatite greatly increases the selectivity of bacterial adsorption.

Analyses of the isotherms obtained with streptococci and actinomyces have also indicated that the strength of the adsorption bond is generally weaker in the case of saliva-treated hydroxyapatite (Clark *et al.*, 1978a). Further, at high bacterial cell concentrations such as those present in saliva, the number of organisms which attach to saliva-treated hydroxyapatite is more related to the number of available binding sites than to the strength of the adsorption bonds.

Earlier studies have shown that *Strep. sanguis, Strep. mitis, Strep. mutans* and other oral bacteria aggregate when they are suspended in saliva (Gibbons and Spinell, 1970; Hillman *et al.*, 1970). Subsequent investigations of these so-called salivary agglutinins (Kashket and Donaldson, 1972; Ericson *et al.*, 1975) have indicated that they include high molecular weight mucinous glycoproteins (Hay *et al.*, 1971; Levine *et al.*, 1978) some of which are blood-group reactive (Gibbons and Qureshi, 1976), lysozyme (Pollock *et al.*, 1976), and immunoglobulins. There is evidence that the aggregating factors differ for various oral species. For example, saliva treated with varying quantities of hydroxyapatite looses the ability to aggregate some, but not other, bacterial species (Ericson and Magnusson, 1976). The repeated adsorption of saliva with high concentrations of cells of a particular bacterial species may render it unreactive for that species, but not necessarily for other species (Kashket and Donaldson, 1972). Recent studies have further shown that neuraminidase treatment of of purified salivary glycoproteins rendered them unreactive with cells of two strains of *Strep. sanguis* though they still reacted with cells of *Strep. mutans* (Levine *et al.*, 1978). McBride and Gisslow (1977) have also demonstrated that only certain strains of *Strep. sanguis* are aggregated by preparations of whole saliva, and this does not appear to correlate with the presence of various cell wall antigens delineated by Rosan (1973). Treatment of the saliva with certain proteases or with neuraminidase also resulted in loss of aggregating activity. Mixed gangliosides containing sialic acid were also potent inhibitors of aggregation which further suggests a direct involvement of sialic acid residues in the aggregation of these strains.

Bacteria that aggregate in the presence of saliva often, but not invariably, adhere in higher numbers to saliva-treated hydroxyapatite or to teeth than to untreated hydroxyapatite (Hillman *et al.*, 1970). This suggests that some salivary aggregating factors can selectively adsorb to hydroxyapatite and become part of the acquired pellicle; their subsequent interaction with bacteria probably accounts for the adherence-promoting effects of the pellicle. Some organisms aggregate with salivary components, but adhere poorly to saliva-treated hydroxyapatite. Evidently, the salivary components involved do not adsorb to a sufficient extent to hydroxyapatite in these cases (Ericson and Magnuson, 1976). The relationship between salivary aggregating components and bacterial adherence to saliva-coated hydroxyapatite surfaces has been studied in some detail with *Leptotrichia buccalis* (Kondo *et al.*, 1976, 1978). Strains of this species adhere in higher numbers to saliva-treated than to untreated enamel powder; they also agglutinate in the presence of whole saliva. Saliva-induced bacterial aggregation as well as bacterial adherence to saliva-treated enamel powder are inhibited by lactose and *N*-acetylgalactosamine, suggesting that the same salivary components are involved in both processes. Similar conclusions have been arrived at in other studies. For example, salivary components responsible for the aggregation of a *Streptococcus* strain could be removed from saliva via adsorption with hydroxyapatite (Orstavik, 1978). Pre-incubation of saliva with cells of this organism removed pellicle-forming components responsible for its attachment whereas pellicles consisting of components eluted from saliva-aggregated bacteria greatly enhanced attachment of the organism. The eluted salivary fractions were rich in lysozyme, but they lacked detectable quantities of immunoglobulins.

Blood group-reactive salivary mucins adsorb well to hydroxyapatite and comprise part of the acquired pellicle present on human teeth *in vivo* (Sönju *et al.*, 1974). Different strains of oral streptococci have been found to specifically interact with different subfractions of blood group-reactive salivary mucin (Gibbons and Qureshi, 1978). Thus, mucin preparations adsorbed by one streptococcal strain no longer contained blood group-reactive molecules which could interact with that strain, but they retained components capable of interacting with other strains. Therefore, it has been suggested that these molecules represent some of the receptors involved in bacterial attachment to teeth. Other evidence supporting this view is that binding of blood group-reactive mucins to a *Strep. mutans* strain can be inhibited by various amines, but not by certain neutral sugars (Gibbons and Qureshi, 1978). Amines are also effective in inhibiting adsorption of the organism to saliva-treated hydroxyapatite. It is of interest to note that preparations of c antigen, glycerol teichoic acid, or crude glucosyltransferase do not significantly inhibit mucin binding to this strain, and hence these components are unlikely to be the streptococcal ligands involved.

3.5.3 Inhibition of bacterial adherence by oral fluids

Mechanical processes such as mastication, movement of the tongue and cheeks, and the flow of saliva have been traditionally considered as important oral cleansing mechanisms. Recent investigations indicate, however, that constituents of oral fluids, especially saliva, may more specifically inhibit the attachment of bacteria and thereby reduce their populations on oral surfaces. Thus, a variety of oral bacteria attach in much lower numbers to saliva-treated hydroxyapatite when they are suspended in saliva than when they are suspended in buffer (Orstavik *et al.*, 1974; Clark and Gibbons, 1977). Components of saliva may bind to specific ligands on the bacterial surface which are responsible for bacterial attachment to similar salivary

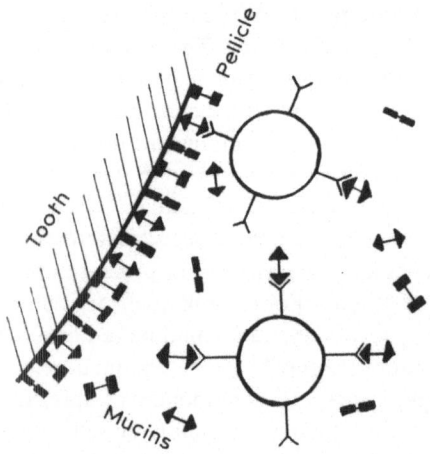

Fig. 3.9 Diagrammatic representation of bacteria interacting with the acquired pellicle covering the enamel. The pellicle is composed of mucins and other salivary macromolecules which have become adsorbed to the enamel mineral. Bacterial cells possess lectin-like ligands on their surface which bind to certain types of mucin molecules present in the pellicle. Mucins in saliva can also interact with these ligands and block attachment of the organisms to teeth.

components present in the pellicle (Fig. 3.9). Salivary components could also bind to bacterial surface components near the ligands and sterically interfere with attachment. Thus, salivary mucins with blood group-substance reactivity strongly inhibit attachment of cells of *B. melaninogenicus* to saliva-treated hydroxyapatite (Slots and Gibbons, 1978). Mucins also inhibit attachment of oral streptococci to epithelial surfaces (Williams and Gibbons, 1975), whereas antibodies of the IgA class have been shown to impair the attachment of oral streptococci to buccal epithelial cells *in vitro* (Williams and Gibbons, 1972). In addition, enzymes such as lysozyme can bind to bacterial cells and induce their aggregation; this probably also promotes the clearance of bacteria from the mouth.

Perhaps the strongest evidence which suggests that salivary components actually influence bacterial colonization has been obtained in studies of *Bacteroides melaninogenicus* (Slots and Gibbons, 1978). *B. melaninogenicus* cells are piliated and when suspended in buffer attach to a variety of surfaces including saliva and serum-treated hydroxyapatite, epithelial cells and various gram-positive bacteria. However, when suspended in saliva, significant attachment only occurs to *Actinomyces* and certain other gram-positive organisms. *In vivo* experiments have shown that the presence of preformed dental plaques, which contain gram-positive bacteria, greatly fosters the attachment and colonization of this organism (Table 3.5).

Overall, the cleansing effects of saliva must be considered to be very effective.

Table 3.5 Recovery of *B. melaninogenicus* SUBSP *asaccharolyticus* 381-R cells from oral surfaces following their introduction into the mouth

Sites sampled	Approximate area sampled (mm^2)	# *B. melaninogenicus* cells recovered mm^{-2}			
		Subject 1		Subject 2	
		10 min	150 min	10 min	150 min
Preformed plaque	25	200	468	37 200	1044
Clean teeth	25	0	0	9	0
Tongue dorsum	350	10	1	857	1
Buccal mucosa	350	0	0	4	0
Saliva*		5.0×10^5*		5.6×10^6*	

* Values for saliva are expressed per ml.
Data from Slots and Gibbons (1978).

This is indicated by the fact that the numbers of bacteria on oral surfaces such as the buccal mucosa are surprisingly low (Table 3.1) in spite of their continuous exposure to oral fluids which contain about 10^8 organisms per ml. It would appear that those bacterial cells that do attach to teeth and epithelial surfaces *in vivo* possess some ligands which are free of salivary components (Fig. 3.9) (Magnusson and Ericson, 1976). Since bacteria coated with salivary components do not attach well to saliva-treated hydroxyapatite surfaces, it seems that salivary components are not able to interact strongly with an already-formed pellicle. This is also suggested by the fact that pellicle formation in germ-free rats and in humans is self-limiting.

Humans are well known to differ greatly with respect to the rate at which plaque forms on their teeth and there is evidence which suggests that this may be due to differences in the efficiency of oral cleansing mechanisms in some instances. For example, the total bacterial recoveries from thoroughly cleaned teeth exposed for a few hours to the oral environment can vary up to 100-fold between individuals (Ruangsri and Orstavik, 1977). In these studies, the same individuals were also found to vary with respect to the persistence of *Strep. mutans* cells on the teeth 24 hours after their application and this was directly related to the total numbers of bacteria that became adsorbed (Ruangsri and Orstavik, 1977). Thus, a high oral cleansing activity was associated with a low persistence of *Strep. mutans* cells and a low rate of plaque formation. Similarly, Magnusson and co-workers (1976) have reported that the concentration of an aggregating factor in 'resting' whole saliva for *Strep. mutans* cells was inversely related to the rate of plaque formation; however, no correlation was noted for stimulated whole saliva.

3.5.4 Influence of preformed plaque

Various evidence suggests that the presence of plaque on the teeth can alter the attachment and colonization of bacteria. This is not surprising since the surface of plaque is different from that of the acquired pellicle. Besides chemical differences, the total surface area of plaque available for bacterial attachment is much higher due to the presence of filamentous and other bacteria extending outwards. Bacteria located on the outer surface of a thick plaque are also more prone to removal by cleansing forces than bacteria in the deeper layers. In this regard, it has been shown that *Strep. mutans* cells applied to the surface of preformed plaques are removed more readily over a 24-hour period than those applied to cleaned tooth surfaces (Ruangsri and Orstavik, 1977). Further, *Strep. mutans* has greater difficulty in colonizing artificial mylar fissures inserted in teeth *in vivo* which are precolonized by indigenous bacteria than those which are sterile (Svanberg and Loesche, 1977). It is also more difficult to implant strains of *Strep. mutans* on the teeth of rodents which are already infected with this organism than on the teeth of uninfected rodents (van Houte *et al.*, 1977).

Besides hindering the colonization of some bacteria, preformed plaque may also favor colonization of others as discussed in the case of *B. melaninogenicus* (Slots and Gibbons, 1978). In addition, Listgarten (1976) has observed that a band of gram-negative bacteria is commonly found on the surface of zones of gram-positive organisms in subgingival plaques (Fig. 3.5). This is consistent with the view that the colonization of other gram-negative organisms may also be enhanced by preformed plaque.

3.6 ACCUMULATION OF BACTERIA ON TEETH

3.6.1 Role of bacterial polymers in the accumulation of *Strep. mutans* on teeth

Once bacteria are attached to the acquired pellicle and initiate growth, additional adhesive interactions must occur to permit cell accumulation. Both bacterially derived polymers and salivary components may play an important role in this process. Considerable information exists about the mechanisms involved in the accumulation of *Strep. mutans* on teeth due to its suspected importance as an etiologic agent of dental caries. This area has been the topic of several recent reviews (Gibbons and van Houte, 1975a, 1975b; van Houte, 1976).

Initial studies demonstrated that *Strep. mutans* could form large plaque accumulations on the teeth of rats or hamsters fed diets rich in sucrose (Fig. 3.10), but not on the teeth of animals fed diets with other carbohydrates. It was subsequently found that *Strep. mutans* could synthesize extracellular polysaccharides from sucrose but not from other common carbohydrates. These polymers were associated with the formation of large plaque accumulations by this organism. The extracellular polysaccharides formed consist of different types of glucans and fructans (Guggenheim, 1970b). One type of glucan consists of glucose units which are predominantly α 1, 6-linked, though varying proportions of α 1, 3 and α 1, 4 bonds are usually present; these glucans are similar to classical dextrans (Fig. 3.11). *Strep. mutans* also forms glucans which are predominantly α 1, 3 linked and tend to be highly insoluble. These polymers have been named mutan (Guggenheim, 1970a). Both types of glucans have been detected in human dental plaque (Hotz *et al.*, 1972). Fructans are also synthesized by most *Strep. mutans* strains, though they have not been studied extensively. The fructan formed by one strain examined contained predominantly β-2, 1 linkages, and is therefore similar to inulin (Rosell and Birkhed, 1974). The enzymes responsible for the synthesis of glucans and fructans are glucosyl and fructosyl transferases, respectively (Fig. 3.11). These enzymes are elaborated constitutively and they transfer glucosyl or fructosyl moieties from sucrose to a primer molecule. The energy required for this process is derived from the energy-rich disaccharide bond of sucrose. The glucosyl- and fructosyltransferases usually exist as aggregates and this has greatly hindered their isolation and purification. Consequently, it is uncertain whether separate enzymes are responsible for the

Fig. 3.10 Molar teeth of a hamster fed a diet rich in sucrose and infected with *Streptococcus mutans*. Note the large streptococcal masses and the decay developing in the second molar. Photograph kindly provided by Dr P.H. Keyes, Bethesda, Md., U.S.A.

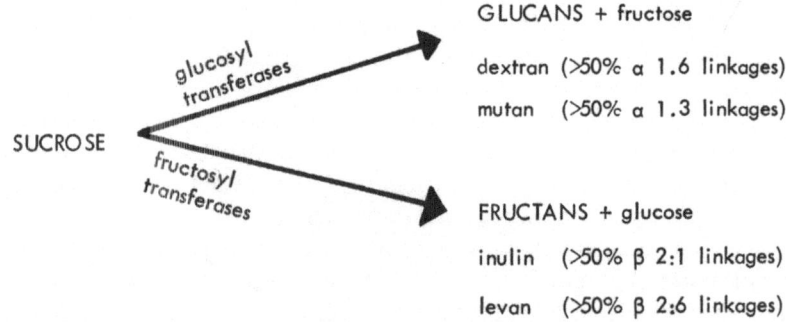

Fig. 3.11 Glucosyl and fructosyltransferases elaborated constitutively by *Streptococcus mutans* synthesize extracellular glucans and fructans, respectively, from sucrose with the liberation of fructose and glucose. The glucans consist of predominantly α 1, 6-linked dextrans, which contain varying proportions of α 1, 3 and α 1, 4 bonds, and of predominantly α 1, 3-linked mutan. Mutans may also contain varying proportions of α1, 6 and α 1, 4 bonds. Fructans similar to inulin or levan may also be produced.

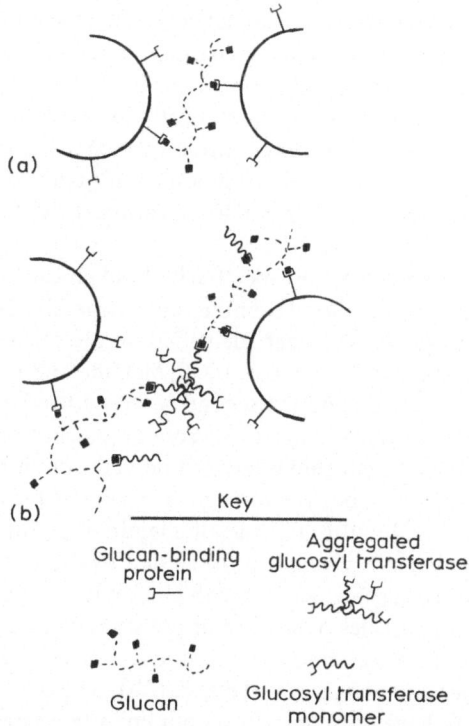

Fig. 3.12 (a) Cells of *Streptococcus mutans* 6715 grown in the absence
of sucrose or dextran contain glucan-binding proteins on their surface.
Though glucosyltransferase is synthesized by such organisms, it does
not bind to their surface. These cells will aggregate if high-molecular
weight dextran is added.

(b) When grown in the presence of sucrose, glucosyltransferase
synthesizes glucan molecules which bind to glucan-binding proteins on
the surface of *Streptococcus mutans* 6715 cells. Aggregates and monomers
of glucosyltransferase become associated with the surface of the organisms
by binding to bound glucan molecules. Large streptococcal masses develop
under such conditions.

synthesis of each type of linkage or whether glucan fragments formed by the action
of glucan hydrolases can be incorporated into growing molecules (Germaine and
Schachtele, 1976; Germaine *et al.,* 1977).

Considerable evidence suggests that glucans play an important role in the
accumulation of *Strep. mutans* cells on surfaces (for review see Gibbons and van
Houte, 1975a; van Houte, 1976). Thus, many *Strep. mutans* strains will form large
plaque-like accumulations on metal wires, glass or extracted teeth immersed in broth
cultures which contain sucrose (Fig. 3.6) (Slade, 1976). However, such accumulations

are not formed when sucrose is replaced by glucose, fructose, or other common dietary carbohydrates and they can be prevented or partially removed by glucan hydrolases (dextranases and mutanases). Also, the proportions of *Strep. mutans* in plaques on the teeth of humans or hamsters can be dramatically increased by the presence of dietary sucrose. Further, mutants of *Strep. mutans* which are defective in the synthesis of insoluble glucans do not form significant amounts of plaque *in vitro* or on the teeth of rodents and such strains have diminished cariogenicity (de Stoppelaar *et al.*, 1971; Tanzer and Freedman, 1978).

Cells of *Strep. mutans* are known to specifically bind glucan molecules and this frequently results in their aggregation (Gibbons and Fitzgerald, 1969). At least two types of glucan-binding ligands have been identified on the surface of *Strep. mutans* cells (Fig. 3.12) (Germaine and Schachtele, 1976; McCabe and Hamelik, 1978). One appears to be glucosyltransferase and the second is thought to be a glucan-binding protein. Organisms grown in the absence of sucrose or glucan lack cell-bound glucosyltransferase, but they aggregate weakly upon addition of preformed glucan due to its interaction with the glucan-binding protein (Fig. 3.12a). McCabe and Hamelik (1978) have isolated multiple forms of a lectin-like protein from the culture liquor of *Strep. mutans* which is thought to be similar to this cell surface ligand; this material was complexed to varying degrees with glucosyltansferase. Organisms grown in the presence of small quantities of sucrose or preformed glucan possess high levels of cell-bound glucosyltransferase and such streptococci possess an increased capacity to bind glucan molecules and to aggregate (Fig. 3.12b). Though glucosyltransferase molecules are thought to have only one glucan binding site, aggregates of the enzyme, possibly complexed with glucan-binding lectin, apparently bind to glucan molecules, which in turn are bound to glucan-binding ligands on the organism's surface. This probably accounts for the increased capacity of such organisms to bind glucan.

Mutants of *Strep. mutans* with impaired glucan synthesis have been isolated which no longer form tenaciously adhering plaque-like deposits *in vitro*, even though they aggregate in the presence of glucan (Tanzer *et al.*, 1974; Freedman and Tanzer, 1974; Tanzer and Freedman, 1978). A second class of mutants has been obtained which can no longer agglutinate in the presence of nanogram quantities of glucan, but is still able to form plaques *in vitro* and in experimental animals in the presence of sucrose; these mutants also induce dental caries. It has been suggested, therefore, that glucan-induced agglutination and plaque formation are dissociable traits (Tanzer and Freedman, 1978). However, these mutants do aggregate in the presence of higher concentrations of glucan and the *in vitro* and *in vivo* conditions used for assessing plaque formation may not have permitted the alteration to become expressed.

Cells of several bacterial species have been shown to form adherent masses on glass surfaces when incubated with sucrose and glucosyltransferase preparations derived from *Strep. mutans* or *Strep. salivarius* (Mukasa and Slade, 1973; Slade 1976; McCabe and Donkersloot, 1977; Hamada *et al.*, 1978). This is probably due to non-specific entrapment of cells in the glucan synthesized because glucosyltransferase does not bind firmly to their surface. It has been proposed that such a process may promote the accumulation of bacteria other than *Strep. mutans* in plaques (Slade, 1976; Hamada *et al.*, 1978), but evidence to support this is lacking. Strains of *Strep. sanguis* synthesize glucosyltransferase but most do not form plaque *in vitro* in the presence of sucrose. This could be due to a lack of glucan-binding protein and/or to a difference in the types of polysaccharides produced.

It should be noted that strains considered as *Strep. mutans* have a number of phenotypic similarities (Carlsson, 1968; Edwardsson, 1968; Guggenheim, 1968). All strains synthesize extracellular glucans from sucrose and have many other morphologic and biochemical characteristics in common. In addition, all wild-type strains are thought to preferentially colonize the teeth and to induce dental caries in experimental animals. However, DNA homology studies have indicated that *Strep. mutans* strains are genetically heterogenous and can be divided into 5 genetic groups (Coykendall, 1974; Coykendall *et al.*, 1976). In addition, 7 serotypes have been delineated which relate to these genetic groups (Bratthall, 1970; Perch *et al.*, 1974). It is not clear at the present time whether *Strep. mutans* cells of different genetic groups accumulate on the teeth by similar mechanisms. For example, recent *in vitro* studies have suggested that the accumulation of strains of serotypes c and e (genetic group 1) occurs in the presence of either sucrose or glucose, but that the accumulation of strains belonging to the other genetic groups is sucrose-dependent (Tinanoff *et al.*, 1978).

(a) *The initial attachment of Strep. mutans cells to teeth*
Besides enhancing the accumulation of *Strep. mutans* cells, glucan synthesis has also been considered to be essential for the initial attachment of this organism to teeth. This was suggested by the observation that dietary sucrose favored the implantation of *Strep. mutans* on the teeth of rodents (Krasse, 1965; Guggenheim *et al.*, 1966). In addition, with few exceptions, *Strep. mutans* would only form plaque deposits on solid surfaces *in vitro* when sucrose was present (McCabe *et al.*, 1967; Slade, 1976). Other studies showed that antibiotic-resistant *Strep. mutans* cells could become established on previously cleaned teeth of humans who frequently ingested sucrose (Krasse *et al.*, 1967).

More recent studies have shown that *Strep. mutans* strains belonging to a number of different serotypes can establish on the teeth of rats fed diets in which sucrose is undetectable (van Houte *et al.*, 1976). Colonization occurred almost exclusively in the fissures. It was also found that the minimum infective dose of all but a few strains was similar for rats fed diets with sucrose or glucose. Apparently, glucan synthesis is not essential for the initial attachment and persistent colonization of *Strep. mutans* on the teeth of rats. This conclusion is supported by the observation that the establishment of a mutant strain of *Strep. mutans*, which lacks cell-associated glucosyltransferase, on the teeth of rats does not differ from that of the parental strain (Clark *et al.*, 1978b).

In vitro studies have also shown that *Strep. mutans* cells can attach to hydroxyapatite treated with human saliva and that the synthesis of extracellular glucan prior to, during, or after cell attachment does not favorably influence the cell numbers that

become attached; in fact, sucrose exposure tended to decrease the number of
streptococci which adsorbed (Clark and Gibbons, 1977). In the same study, exposure
of *Strep. mutans* cells adsorbed to smooth human tooth surfaces *in vivo* to sucrose
also failed to promote their persistence. The presence of *Strep. mutans* in plaque
of humans with hereditary fructose intolerance or sucrose deficiency who consume
very low amounts of sucrose (van Houte and Duchin, 1975; Hoover and Newbrun,
1978) is also consistent with the idea that its attachment to teeth does not require
sucrose.

Although the evidence strongly suggests that glucan synthesis is not essential for
colonization of the teeth by *Strep. mutans*, it is uncertain whether glucans that may
be present in the acquired pellicle or in performed plaque can enhance its initial
attachment. *In vitro* experiments carried out in the absence of saliva have shown that
glucan present on extracted human teeth, glass or hydroxyapatite surfaces significantly
increases the number of *Strep. mutans* cells that attach (Gibbons and Fitzgerald, 1969;
Kuramitsu, 1974; Liljemark and Schauer, 1975; Chapter 4). On the other hand,
similar numbers of *Strep. mutans* cells adsorb to saliva-treated hydroxyapatite which
has or has not been exposed to glucans (Clark, 1978). In addition, topical application
of dextran to the teeth of rats does not alter the minimum infective dose of *Strep.
mutans* (van Houte and Upeslacis, 1976). However, soluble glucan synthesized by
glucosyltransferase preparations derived from *Strep. mutans* has a much lower
affinity for saliva-treated hydroxyapatite than for untreated hydroxyapatite
(Clark, 1978). It is questionable, therefore, whether sufficient glucan was actually
associated with the surfaces in the studies; it is also not known whether the glucans
used were of the proper type.

3.6.2 Role of bacterial polymers in the accumulation of bacteria other than *Strep. mutans*

Strains of *Actinomyces viscosus* and *Actinomyces naeslundii* will form accumulations
on wires and other solid surfaces when grown in the presence of a variety of carbohydrates (McCabe *et al.*, 1967; Slade, 1976). This indicates that they are capable of
synthesizing components which permit them to accumulate *in vitro*. Similarly,
strains of *A. viscosus* readily form plaques in animals fed diets containing glucose,
sucrose, or starch (Fig. 3.13) (Jordan *et al.*,1969; van der Hoeven *et al.*, 1974). In
addition, the minimum infective dose and the total populations of *A. viscosus*
recoverable from the teeth are similar for rats fed diets devoid of carbohydrate or
supplemented with glucose or sucrose (Brecher and van Houte, 1978). These data
indicate that extracellular glucans are not required for the accumulation of these
organisms on the teeth.

Strains of *A. viscosus* have been found to synthesize an array of other surface-
associated components which might foster their accumulation. These include an
extracellular heteropolysaccharide slime composed largely of *N*-acetylglusosamine

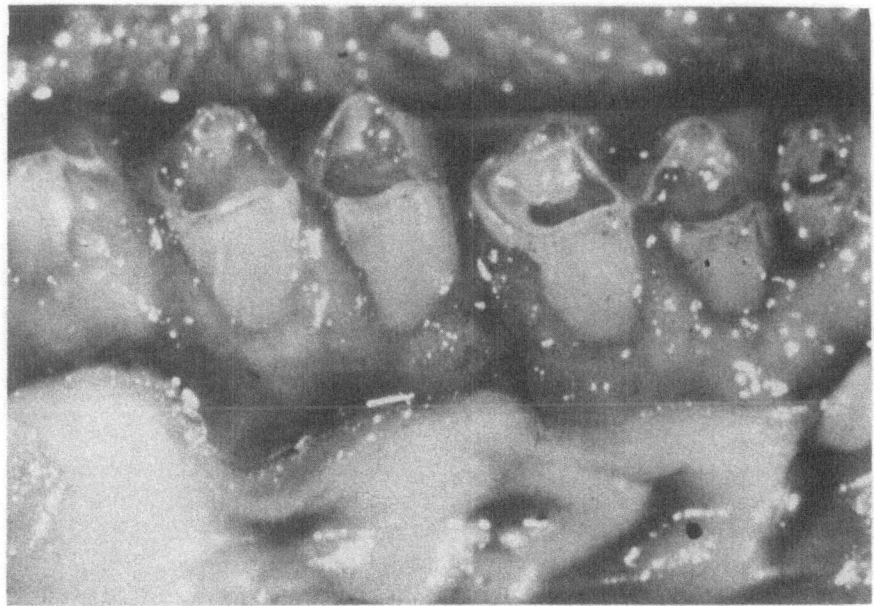

Fig. 3.13 Molar teeth of a hamster infected with *Actinomyces viscosus* showing massive accumulations of gingival plaque. Photograph kindly provided by Dr P.H. Keyes, Bethesda, Md., U.S.A.

and lesser amounts of glucose and galactose (Rosan and Hammond, 1974; van der Hoeven, 1974), a fatty acid-substituted amphipathic heteropolysaccharide which contains mannose, glucose and galactose (Wicken *et al.*, 1978) and surface appendages which resemble fimbriae (Brecher *et al.*, 1978; Cisar *et al.*, 1978a; Girard and Jacius, 1974). Fimbriae have also been observed on cells of *A. naeslundii* (Ellen *et al.*, 1978). Extracellular fructans may be produced (Howell and Jordan 1967) but they are not thought to be important for accumulation of the organism since they are synthesized specifically from sucrose. It has also been suggested that spontaneous agglutination of *Actinomyces* cells at low pH could promote their accumulation on teeth (Miller *et al.*, 1978).

The role of heteropolysaccharides and fimbriae in promoting accumulation of *Actinomyces* cells has not yet been clarified. Some data suggest that these surface components may influence attachment to various surfaces. For example, mechanical removal of fimbriae from *A. naeslundii* cells by blending significantly impairs their attachment to erythrocytes or human mucosal epithelial cells (Ellen *et al.*, 1978). In other studies, a strain of *A. viscosus* was found to form gingival plaque and induce periodontal disease in gnotobiotic rats. This strain possessed numerous fimbriae and attached in high numbers to saliva-treated hydroxyapatite. However, an avirulent mutant produced excessive amounts of heteropolysaccharide slime which completely embedded the fimbriae; this slime was apparently responsible for the organism's inability to attach to saliva-treated hydroxyapatite *in vitro* or to colonize the gingival crevice area of rats and induce periodontal pathology (Brecher *et al.*, 1978).

Most strains of *Strep. sanguis* and *Strep. salivarius* do not form plaque accumulations *in vitro* even though extracellular glucans may be synthesized (Carlsson, 1968; Niven *et al.*, 1941). These organism generally lack the ability to bind extracellular glucans and they do not aggregate in the presence of glucans as do cells of *Strep. mutans* (Gibbons and Fitzgerald, 1969). Presumably, most *Strep. sanguis* and *Strep. salivarius* cells lack a glucan-binding protein on their surface. Studies with humans have indicated that large fluctuations in the dietary sucrose content do not appreciably influence the populations of *Strep. sanguis* or *Strep. salivarius* recoverable from the teeth (Carlsson, 1965; 1967; de Stoppelaar *et al.*, 1970). This also suggests that glucan synthesis does not significantly alter the colonization of these organisms.

Strains considered as *Strep. sanguis* are quite heterogeneous (Carlsson, 1968; Rosan, 1973) and the possibility exists that the accumulation of some types *in vivo* may be favored by dietary sucrose. This is also suggested by the observations that a few strains of *Strep. sanguis* as well as *Strep. salivarius* form plaque on wires or glass surfaces immersed in sucrose broth cultures. In these instances, plaque formation may be mediated by extracellular glucans. This view is supported by the observation that glucan hydrolase preparations inhibit *in vitro* accumulations of *Strep. salivarius* (Kelstrup and Gibbons, 1970).

3.6.3 Direct interspecies interactions

Several observations indicate that there are some bacteria in plaque which accumulate by directly adhering to cells of dissimilar species. Sections of dental plaque examined by electron microscopy frequently give the impression that morphologically dissimilar cells are directly attached to each other (Fig. 3.14) (Gibbons and van Houte 1973). An extreme example of this is the so-called 'corncob' structure which consists of a central filamentous organism covered by coccal forms (Figs. 3.15a and b) (Jones 1972). Electron microscopic studies of sections of 'corncobs' have shown that the coccal forms possess thin surface fibrils which seem to be directly attached to the cell surface of filaments (Listgarten *et al.*, 1973). 'Corncobs' have been isolated by micromanipulation and the filaments have been identified as *Bacterionema matruchotii* and the coccal forms as streptococci (Mouton *et al.*, 1977); these arrangements have also been reproduced *in vitro* (Takazoe *et al.*, 1978).

The adhesive interactions involved in 'corncob' formation are highly specific because only a few of many streptococcal strains that have been tested form 'corncobs' with *B. matruchotii in vitro* and other types of filamentous organisms do not produced similar epiphytic arrangements (Mouton *et al.*, 1977; Takazoe *et al.*, 1978). 'Corncobs' seem to be especially prevalent at the periphery of dental plaque along the gingival margin (Mouton *et al.*, 1977).

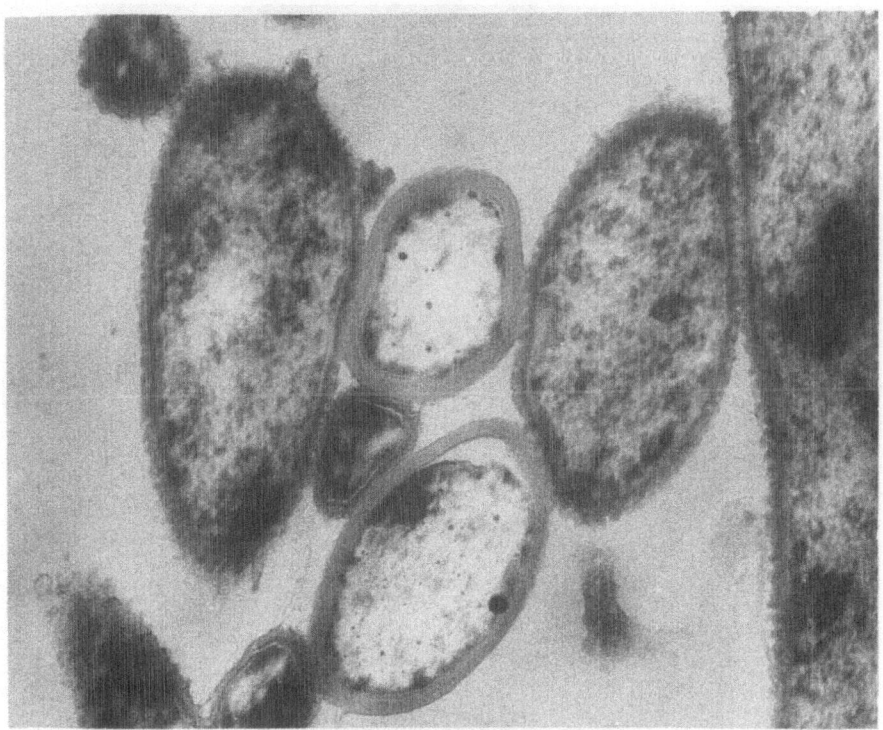

Fig. 3.14 Electron micrograph of a section of human dental plaque showing different types of bacterial cells which appear to be directly attached to each other. Photograph generously provided by Dr Z. Skobe, Boston, Mass., U.S.A.

In vitro experiments have shown that a variety of oral bacterial species can co-aggregate when they are mixed together (Gibbons and Nygaard, 1970; Ellen and Balcerzak-Roczkowski, 1973; Kelstrup and Funder-Nielsen, 1974). As in the case of 'corncob' structures, the formation of these co-aggregates is highly specific. For example, only certain strains of *Strep. sanguis* will aggregate when mixed with certain strains of *A. naeslundii*. The mechanism of co-aggregation occurring between *A. viscosus* strain T14 and *Strep. sanguis* strain 34 has recently been investigated (McIntire *et al.*, 1978). This process was independent of glucan synthesis, required calcium, and involved the interaction of a protein or a glycoprotein present on the cell surface of *A. viscosus* and a carbohydrate on the cell surface of *Strep. sanguis*. Electron microscopic observations further suggested an involvement of the fimbriae-like surface appendages of *A. viscosus* (Cisar *et al.*, 1978b). The formation of co-aggregates was also found to be strongly inhibited by galactosides and hence the fimbriae of *A. viscosus* appear to be reacting with receptors containing these sugars on *Strep. sanguis* cells.

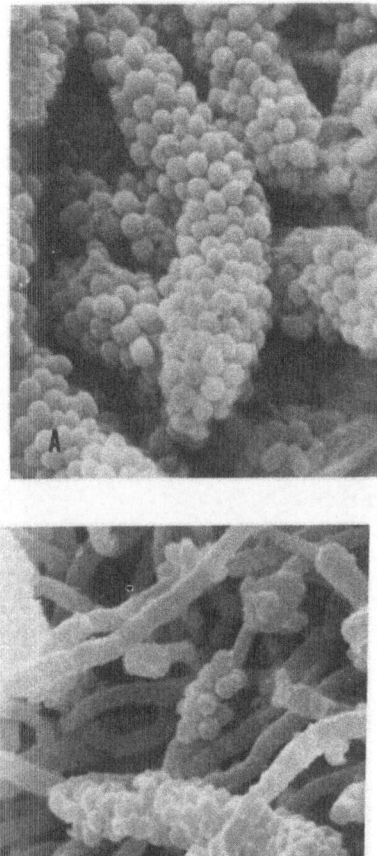

Fig. 3.15(a) Scanning electron micrograph of a sample of human dental plaque showing 'corncob' structures.

(b) Scanning electron micrograph showing cocci attached to the surface of filamentous bacteria forming 'corncob' structures. Photographs were kindly provided by Dr Z. Skobe, Boston, Mass., U.S.A.

Under some conditions, *Strep. sanguis* and *Strep. mutans* cells can also bind to the surface of *A. viscosus* cells via glucans and form large co-aggregates (Bourgeau and McBride, 1976). Thus, strains of glucose-grown streptococci will not co-aggregate with *A. viscosus* cells unless they are first mixed with high molecular weight dextran. *A. viscosus* cells are able to bind dextran and then, in turn, bind the glucose-grown streptococci. Dextrans isolated from *Strep. sanguis, Strep. mutans,* and *Leuconostoc* species are all effective in promoting streptococcal binding to *A. viscosus* cells (Bourgeau and McBride, 1976). The dextran-binding receptor on *A. viscosus* cells appears to be a protein which is functionally similar to the glucan-binding protein possessed by *Strep. mutans* cells.

Another type of interaction occurs between cells of *A. viscosus* and *Veillonella alcalescens.* Bladen *et al.* (1970) observed that *V. alcalescens* did not form plaque-like accumulations on wires *in vitro*, but could attach to preformed accumulations of *Actinomyces* cells. Also, as mentioned earlier, cells of *Bacteroides melaninogenicus* subspecies *asacharolyticus* and other gram-negative organisms found in subgingival plaques, including *B. ochraceus, F. nucleatum* and *Eikenella corrodens* will attach to the surface of strains of *A. israelii, A. naeslundii,* and *A. viscosus* (Slots and Gibbons, 1978). Evidence that such interactions may be of ecological significance has been discussed previously.

3.6.4 Role of host-derived polymers

Host-derived polymers may also be involved in the accumulation of some bacteria on teeth (Gibbons and van Houte, 1973). Salivary components, including high molecular weight mucins, immunoglobulins and lysozyme have been demonstrated to bind to the surfaces of bacteria and induce their aggregation. Such components would be expected to bind to the surfaces of bacteria proliferating on the teeth and foster their accumulation. Additionally, they could become part of the plaque matrix by being present on the surface of bacteria which become attached to the teeth.

An involvement of components of oral fluids in the accumulation of plaque bacteria is supported by evidence that the plaque matrix contains macromolecules derived from saliva and crevicular fluid (for review see Jenkins, 1968). Further, lgA and lgG have been shown to each comprise about 1% of the dry weight of pooled plaque and serum albumin is also readily demonstrable (Taubman, 1974). Other studies have shown that samples of pooled human plaque can be partially dispersed by EDTA and that the high molecular weight components which are released can aggregate *Strep. mitis* cells upon the addition of calcium (Hay *et al.*, 1971). In these studies it was also found that an aggregating factor for *Strep. mitis* present in saliva resembled the extracted matrix components in chromatographic behavior and in its requirement for calcium.

REFERENCES

Alkan, M., Ofek, I.and Beachey, E.H. (1977), *Infect. Immun.*, **18**, 555–557.
Arends, J. and Jongbloed, W.L. (1977), *Swed. Dent. J.*, **1**, 215–224.
Armstrong, W.G. (1966), *Nature*, **210**, 197–198.
Armstrong, W.G. (1967), *Caries Res.*, **1**, 89–103.

Bammann, L.L., Clark, W.B. and Gibbons, R.J. (1978), *Infect. Immun.*, **22**, 721–726.

Bernardi, G. (1973), Colleques internationaux du centre national de la recherche scientifique No. 230. Physicochimie et cristallographie des apatites d'intérêt biologique.

Bernardi, G., Giro, M.G. and Gaillard, C. (1972), *Biochim. biophys. Acta*, **278**, 409–420.

Björn, H. and J. Carlsson (1964), *Odont. Revy.*, **15**, 23–28.

Bladen, H., Hageage, G., Pollock, F. and Harr, R. (1970), *Archs. oral Biol.*, **15**, 127–133.

Bourgeau, G. and McBride, B.C. (1976), *Infect. Immun.*, **13**, 1228–1234.

Bratthall, D. (1970), *Odont. Revy.*, **21**, 143–152.

Brecher, S.M. and van Houte, J. (1978), Unpublished data.

Brecher, S.M., van Houte, J. and Hammond, B.F. (1978), *Infect. Immun.*, **22**, 603–614.

Brudevold, F. (1975), *Improving Dental Practice Through Preventive Measures.*, The C.V. Mosby Company, U.S.A.

Carlsson, J. (1965), *Odont. Revy.*, **16**, 336–347.

Carlsson, J. (1967), *Odont. Revy.*, **18**, 173–178.

Carlsson, J. (1968), *Odont. Revy.*, **19**, 137–160.

Carlsson, J. (1970), *Caries Res.*, **4**, 305–320.

Carlsson, J. (1972), *Archs. oral Biol.*, **17**, 1327–1332.

Cisar, J.O., McIntire, F.C. and Vatter, A.E. (1978b), *Adv. exp. Med. Biol.*, **107**, 695–702.

Cisar, J.O., Vatter, A.E. and McIntire, F.C. (1978a), *Infect. Immun.*, **19**, 312–319.

Clark, W.B. (1978), *DMS Thesis*, Harvard School of Dental Medicine, Boston, Mass.

Clark, W.B., Bammann, L.L. and Gibbons, R.J. (1978a), *Infect. Immun.*, **19**, 846–853.

Clark, W.B., Bammann, L.L. and Gibbons, R.J. (1978b), *Infect. Immun.*, **21**, 681–684.

Clark, W.B. and Gibbons, R.J. (1977), *Infect. Immun.*, **18**, 514–523.

Corpe, W.A. (1970), *Adhesion in Biological Systems*, Academic Press, New York.

Coykendall, A.L. (1974), *J. gen. Microbiol.*, **83**, 327–328.

Coykendall, A.L., Bratthall, D., O'Connor, K. and Dvarskas, R.A. (1976), *Infect. Immun.*, **14**, 667–670.

Daniels, S.L. (1972), *Dev. Ind. Microbiol.*, **13**, 211–253.

de Stoppelaar, J.D., König, K.G., Plasschaert, A.J.M. and van der Hoeven, J.S. (1971), *Archs. oral. Biol.*, **16**, 971–975.

de Stoppelaar, J.D., van Houte, J. and Backer-Dirks, D. (1970), *Caries Res.*, **4**, 114–123.

Duchin, S., and van Houte, J. (1978a), *Infect. Immun.*, **20**, 120–125.

Duchin, S., and van Houte, J. (1978b), *Archs. oral. Biol.*, **23**, 779–786.

Edman, D.C., Keene, H.J., Shklair, I.L. and Hoerman, K.C. (1975), *Archs. oral Biol.*, **20**, 145–148.

Edwardsson, S. (1968), *Archs. oral Biol.*, **13**, 637–646.

Ellen, R.P. and Balcerzak-Raczkowski, I.B. (1973), *J. Periodont. Res.*, **12**, 11–20.

Ellen, R.P., Walker, D.L. and Chan, K.H. (1978), *J. Bact.*, **134**, 1171–1175.

Ericson, T. (1967), *Caries Res.,* **1**, 52–58.

Ericson, T. and Magnusson, I. (1976), *Caries Res.,* **10**, 8–18.

Ericson, T.II., Pruitt, K. and Wedel, H. (1975), *J. oral Path.,* **4**, 307–323.

Frank, R.M. and Houver, G. (1970), *Dental Plaque,* E. and S. Livingston, Edinburgh.

Freedman, M.L. and Tanzer, J.M. (1974), *Infect. Immun.,* **10**, 189–196.

Germaine, G.R., Harlander, S.K., Leung, W.S. and Schachtele, C.F. (1977), *Infect. Immun.,* **16**, 637–648.

Germaine, G.R. and Schachtele, C.F. (1976), *Infect. Immun.,* **13**, 365–372.

Gibbons, R.J. (1974), *Anaerobic Bacteria: Role in Desease,* Thomas, Springfield, Illinois.

Gibbons, R.J. (1977), *Microbiology,* 395–406.

Gibbons, R.J., dePaola, P.F., Spinell, D.M. and Skobe, Z. (1974), *Infect. Immun.,* **9**, 481–488.

Gibbons, R.J. and Fitzgerald, R.J. (1969), *J. Bact.,* **98**, 341–346.

Gibbons, R.J., Moreno, E.C. and Spinell, D.M. (1976), *Infect. Immun.,* **14**, 1109–1112.

Gibbons, R.J. and Nygaard, M. (1970), *Archs, oral Biol.,* **15**, 1397–1401.

Gibbons, R.J. and Qureshi, J.V. (1976), *Microbial Aspects of Dental Caries,* Information Retrieval, Inc., Washington, D.C.

Gibbons, R.J. and Qureshi, J.V. (1978), *Infect. Immun.,* **22**, 665–671.

Gibbons, R.J. and Spinell, D.M. (1970), *Dental Plaque,* E and S Livingston, Edinburgh.

Gibbons, R.J., Socransky, S.S., deAraujo, W.C. and van Houte, J. (1964), *Archs. oral Biol.,* **9**, 365–370.

Gibbons, R.J. and van Houte, J. (1971), *Infect. Immun.,* **3**, 567–573.

Gibbons, R.J. and van Houte, J. (1973), *J. Periodont.,* **44**, 347–360.

Gibbons, R.J. and van Houte, J. (1975a), *A. Rev. Microbiol.,* **29**, 19–44.

Gibbons, R.J. and van Houte, J. (1975b), *A. Rev. Med.,* **26**, 121–136

Girard, A.F. and Jacius, B.H. (1974), *Arch. oral Biol.,* **19**, 71–79.

Gordon, D.F. and Jong, B.B. (1968), *Appl. Microbiol.,* **16**, 428–429.

Guggenheim, B. (1968), *Caries Res.,* **2**, 147–163.

Guggenheim, B. (1970a), *Helv. Odon. Acta,* **14**, 89–108.

Guggenheim, B. (1970b), *Int. Dent.J.,* **20**, 675–678.

Guggenheim, B., König, K.G., Herzog, E. and Mühlemann, H.R. (1966), *Helv. Odont. Acta,* **10**, 101–112.

Hamada, S., Tai, S. and Slade, H. (1976), *Infect. Immun.,* **14**, 903–910.

Hamada, S., Kobayashi, Y. and Slade, H. (1978), *Microbiol. Immun.,* **22**, 279–282.

Hattori, T. and Hattori, R. (1976), *Crit. Rev. Microbiol.,* **4**, 423–461.

Hay, D.I. (1967), *Arch. oral Biol.,* **12**, 937–946.

Hay, D.I. (1973), *Arch. oral Biol.,* **18**, 1531–1541.

Hay, D.I. (1975), *Arch. oral Biol.,* **20**, 553–558.

Hay, D.I., Gibbons, R.J. and Spinell, D.M. (1971), *Caries Res.,* **5**, 111–123.

Hillman, J.D., van Houte, J. and Gibbons, R.J. (1970), *Arch. oral Biol.,* **15**, 899–903.

Hoffman, H. and Frank, M.E. (1966), *Acta Cytol.,* **10**, 272–285.

Holm-Pedersen, P., Agerback, N. and Theilade, E. (1975), *J. Clin. Periodont.,* **2**, 14–24.

Hoover, C.I. and Newbrun, E. (1978), *J. Dent. Res.,* **57**, (Special Issue A) Abst. 555.

Hotz, P., Guggenheim, B. and Schmid, R. (1972), *Caries Res.,* **6**, 103–121.

Howell, A. and Jordan, H.V. (1967), *Arch. oral Biol.,* **12**, 571–573.

Ikeda, T., Sandham, H.J. and Bradley, Jr., E.L. (1973), *Arch. oral Biol.,* **18**, 555–566.

Jenkins, G.N. (1968), *Caries Res.,* **2**, 130–138.

Jenkins, G.N. (1972), *Int. Dent. J.,* **22**, 350–362.

Jones, S.J. (1972), *Arch. oral Biol.,* **17**, 613–616.

Jordan, H.V., Englander, H.R., Engler, W.O. and Kulczyk, S. (1972), *J. Dent. Res.,* **51**, 515–518.

Jordan, H.V. and Hammond, B.F. (1972), *Arch. oral Biol.,* **17**, 1333–1342.

Jordan, H.V., Keyes, P.H. and Lim, S. (1969), *J. Dent. Res.,* **48**, 824–831.

Kashket, S. and Donaldson, C.S. (1972), *J. Bact.,* **112**, 1127–1133.

Kelstrup, J. and Funder-Nielsen, T.D. (1974), *J. biol. Buccale,* **2**, 347–362.

Kelstrup, J. and Gibbons, R.J. (1970), *Caries Res.,* **4**, 360–377.

Knox, K.W. and Wicken, A.J. (1973), *Bact. Rev.,* **37**, 215–257.

Kondo, W., Sato, M. and Ozawa, H. (1976), *Arch. and Biol.,* **21**, 363–369.

Kondo, W., Sato, M. and Sato, N. (1978), *Archs. oral Biol.,* **23**, 453–458.

Krasse, B. (1965), *Arch. oral Biol.,* **10**, 215–221.

Krasse, B., Edwardsson, S., Svensson, I. and Trell, L. (1967), *Arch oral Biol.,* **12**, 231–236.

Kuramitsu, H.K. (1974), *Infect. Immun.,* **9**, 764–765.

Leach, S.A. (1963), *Nature,* **199**, 486–489.

Levine, M.J., Herzberg, M.C., Levine, M.S., Ellison, S.A., Stinson, M.W., Li, H.C. and Van Dyke, T., (1978), *Infect. Immun.,* **19**, 107–115.

Liljemark, W.F., and Gibbons, R.J. (1971), *Infect. Immun.,* **4**, 264–268.

Liljemark, W.F. and Gibbons, R.J. (1972), *Infect. Immun.,* **6**, 852–859.

Liljemark, W.F. and Schauer, S.V. (1975), *Arch. oral Biol.,* **20**, 609–615.

Liljemark, W.F. and Schauer, S.V. (1977), *J. Dent. Res.,* **56**, 157–165.

Listgarten, M.A. (1976), *J. Periodontal.,* **47**, 1–18.

Listgarten, M.A., Mayo, H. and Amsterdam, M. (1973), *Arch. oral Biol.,* **18**, 651–656.

Listgarten, M., Mayo, H. and Tremblay, R. (1975), *J. Periodontol.,* **46**, 10–26.

Loesche, W.J. (1977), *Oral Sci. Rev.,* **9**, 65–102.

Magnusson, I. and Ericson, T. (1976), *Caries Res.,* **10**, 273–286.

Magnusson, I., Ericson, T. and Pruitt, K. (1976), *Caries Res.,* **10**, 113–122.

Marshall, K.C. (1971), *Soil Biochem.,* **2**, 409–445.

Marshall, K.C. (1976), *Interfaces in Microbial Ecology,* Harvard University Press, Cambridge.

Marshall, K.C., Stout, R. and Mitchell, R. (1971), *J. gen. Microbiol.,* **68**, 337–348.

Matsson, L. (1978), *J. Clin. Periodont.,* **5**, 24–34.

Mayhall, C. (1970), *Arch. oral Biol.,* **15**, 1327–1341.

McBride, B.C. and Gisslow, M.T. (1977), *Infect. Immun.,* **18**, 35–40.

McCabe, R.M., Keyes, P.H. and Howell, A. (1967), *Arch. oral Biol.,* **12**, 1653–1656.

McCabe, R.M. and Donkersloot, J.A. (1977), *Infect. Immun.,* **18**, 726–734.

McCabe, R.M. and Hamelik, R.M. (1978), *Adv. exp. Med. Biol.,* **107**, 749–760.

McGoughey, C. and Stowell, E. (1967), *Arch. oral Biol.,* **12**, 815–828.

McIntire, F.C., Vatter, A.E., Baros, J. and Arnold, J. (1978), *Infect. Immun.,* **21**, 978–987.

Meckel, A.H. (1967), *Caries Res.,* **2**, 104–114.

Miller, C.H., Palenik, C.J. and Stamper, K.E. (1978), *Infect. Immun.*, **21**, 1003–1009.
Mouton, C., Reynolds, H. and Genco, R.J. (1977), *J. biol. Buccale*, **5**, 321–332.
Mukasa, H. and Slade, H.D. (1973), *Infect. Immun.*, **8**, 555–562.
Newman, M.G. and Socransky, S.S. (1977), *J. Periodont. Res.*, **12**, 120–128.
Niven, C.F., Smiley, K.L. and Sherman, J.M. (1941), *J. biol. Chem.*, **140**, 105–109.
O'Brien, W.J. (1976), *Microbial Aspects of Dental Caries*, Information Retrieval, Inc., Washington, D.C.
Ofek, I., Beachey, E.H. and Sharon, N. (1978), *Trends in Biochemical Science*, **43**, 159–160.
Olsson, J., Glantz, P.O. and Krasse, B. (1976a), *Arch. oral Biol.*, **21**, 605–609.
Olsson, J., Glanz, P. and Krasse, B. (1976b), *Scand. J. Dent. Res.*, **84**, 240–242.
Olsson, J. and Krasse, B. (1976), *Scand. J. Dent. Res.*, **84**, 20–28.
Oppenheim, F.G., Hay, D.I. and Franzblau, C. (1971), *Biochemistry*, **10**, 4233–4238.
Orstavik, D. (1978), *Arch. oral Biol.*, **23**, 167–173.
Orstavik, D. and Kraus, F.W. (1973), *J. oral Path.*, **2**, 68–76.
Orstavik, D., Kraus, F.W. and Henshaw, L.C. (1974), *Infect. Immun.*, **9**, 794–800.
Perch, B., Kjems, E. and Ravn, T. (1974), *Acta path. microbiol. scand.*, **83**, 357–370.
Pollock, J.J., Iacono, V.J., Bicker, H.G., MacKay, B.J., Katona, L. and Taichman, L.B. (1976), *Microbial Aspects of Dental Caries*, Information Retrieval, Inc., Washington, D.C.
Pruitt, K.M. (1977), *Swed. dent. J.*, **1**, 225–240.
Qureshi, J.V. and Gibbons, R.J. (1978), Unpublished data.
Richardson, R.L. and Jones, M. (1956), *J. dent. Res.*, **37**, 697–709.
Robertson, J.D. (1959), *Biochem. Soc. Symposia*, **16**, 3–43.
Rogers, A.H. (1973), *Arch. oral Biol.*, **18**, 227–232.
Rölla, G. (1976), *Microbial Aspects of Dental Caries*, Information Retrieval, Inc., Washington, D.C.
Rölla, G. (1977), *Swed. dent. J.*, **1**, 241–251.
Rölla, G., Iverson, O.J. and Bonesvoll, P. (1978), *Adv. exp. Med. Biol.*, **107**, 607–617.
Rölla, S., Robrish, S.A. and Bowen, W.H. (1977), *Acta path. microbiol. scand.* Sect. B., **85**, 341–346.
Rosan, B. (1973), *Infect. Immun.*, **7**, 205–211.
Rosan, B. (1978), *Science*, **201**, 918–920.
Rosan, B. and Hammond, B.F. (1974), *Infect. Immun.*, **10**, 304–308.
Rosell, K. and Birkhed, D. (1974), *Acta chem. scand. Ser. B.*, **28**, 589.
Ruangsri, P. and Orstavik, D. (1977), *Caries Res.*, **11**, 204–210.
Saxton, C.A. (1973), *Caries Res.*, **7**, 102–119.
Shklair, I.L., Keene, H.J. and Cullen, P. (1974), *Arch. oral Biol.*, **19**, 199–202.
Silverman, G. and Kleinberg, I. (1967), *Arch. oral Biol.*, **12**, 1387–1405.
Slade, H.D. (1976), *Immunologic Aspects of Dental Caries*, Information Retrieval, Inc., Washington, D.C.
Slots, J. (1976), *Scand. J. dent. Res.*, **84**, 1–10.
Slots, J. (1977), *Scand. J. dent. Res.*, **85**, 114–121.
Slots, J. and Gibbons, R.J. (1978), *Infect. Immun.*, **19**, 254–264.
Socransky, S.S. (1977), *J. Periodont.*, **48**, 497–504.
Socransky, S.S., Gibbons, R.J., Dale, A.C., Bortnick, L., Rosenthal, E. and Macdonald, J.B. (1963), *Arch. oral Biol.*, **8**, 275–280.

104 *Bacterial Adherence*

Socransky, S.S. and Manganiello, A.D. (1971), *J. Periodont.*, **42**, 485–496.
Sönju, T., Christensen, T.B., Kornstad, L. and Rölla, G. (1974), *Caries Res.*, **8**, 113–122.
Suegara, N., Morotomi, M., Watanabe, T., Kawai, Y. and Mutai, M. (1975), *Infect. Immun.*, **12**, 173–179.
Sumney, D.L. and Jordan, H.V. (1974), *J. dent. Res.*, **53**, 343–351.
Svanberg, M.L. and Loesche, W.J. (1977), *Arch. oral Biol.*, 441–447.
Takazoe, I., Matsukubo, T. and Katow, T. (1978), *J. dent. Res.*, **57**, 384–387.
Tanzer, J.M. and Freedman, M.L. (1978), *Adv. exp. Med. Biol.*, **107**, 661–672.
Tanzer, J.M., Freedman, M.L., Fitzgerald, R.J. and Larson, R.H. (1974), *Infect. Immun.*, **10**, 197–203.
Taubman, M.A. (1974), *Arch. oral Biol.*, **19**, 439–446.
Tinanoff, N., Tanzer, J.M. and Freedman, M.L. (1978), *Infect. Immun.*, **21**, 1010–1019.
van der Hoeven, J.S. (1974), *Caries Res.*, **8**, 193–210.
van der Hoeven, J.S., Mikx, F.H.M., König, K.G. and Plasschaert, A.J.M. (1974), *Caries Res.*, **8**, 211–223.
van Houte, J. (1976), *Microbial Aspects of Dental Caries*, Information Retrieval, Inc., Washington, D.C.
van Houte, J., Burgess, R.C. and Onose, H. (1976), *Arch. oral Biol.*, **21**, 561–564.
van Houte, J. and Duchin, S. (1975), *Arch. oral Biol.*, **20**, 771–773.
van Houte, J., Gibbons, R.J. and Banghart, S.B. (1970), *Arch. oral Biol.*, **15**, 1025–1034.
van Houte, J., Gibbons, R.J. and Pulkkinen, A.J. (1971), *Arch. oral Biol.*, **16**, 1131–1142.
van Houte, J., Gibbons, R.J. and Pulkkinen, A.J. (1972), *Infect. Immun.*, **6**, 723–729.
van Houte, J. and Green, D.B. (1974), *Infect. Immun.*, **9**, 624–630.
van Houte, J. and Upeslacis, V.N. (1976), *J. dent. Res.*, **55**, 216–222.
van Houte, J., Upeslacis, V.N. and Edelstein, S. (1977), *Infect. Immun.*, **16**, 203–212.
van Palenstein-Helderman, W.H. (1975), *J. Periodont., Res.*, **10**, 294–305.
Weatherell, J. and Robinson, C. (1973), *Biological Mineralization*, John Wiley.
Wicken, A.J., Broady, K.W., Evans, J.D. and Knox, K.W. (1978), *Infect. Immun.*, **22**, 615–616.
Wicken, A.J. and Knox, K.W. (1975), *Science*, **187**, 1161–1167.
Williams, R.C. and Gibbons, R.J. (1972), *Science*, **177**, 697–699.
Williams, R.C. and Gibbons, R.J. (1975), *Infect. Immun.*, **11**, 711–718.
Winkler, K.C. and Backer-Dirks, O. (1958), *Int. dent. J.*, **8**, 561–585.
Zahradnik, R.T., Moreno, E.C. and Burke, E.J. (1976), *J. dent. Res.*, **55**, 664–670.
Zahradnik, R.T., Propas, D. and Moreno, E.C. (1977), *J. dent. Res.*, **56**, 1107–1110.
Zobell, C.E. (1943), *J. Bact.*, **46**, 39–59.

4 Mechanisms of Adherence of *Streptococcus mutans* to Smooth Surfaces *in vitro*

SHIGEYUKI HAMADA
and
HUTTON D. SLADE

4.1	Introduction	*page*	107
4.2	*In vitro* adherence of *S. mutans* and the effect of sucrose		107
	4.2.1 General considerations		107
	4.2.2 Adherence in the presence of cell-bound glucosyltransferase and sucrose		109
	4.2.3 Adherence and the quantity of glucan synthesized		110
	4.2.4 The GTase binding site on *S. mutans*		112
	4.2.5 The necessity for new glucan synthesis in *in vitro* adherence of *S. mutans* cells to smooth surface		113
	4.2.6 The development of *in vivo* and *in vitro* plaque		114
4.3	Dextran/glucan-induced agglutination		116
	4.3.1 General considerations		116
	4.3.2 The dextran/glucan binding site		116
4.4	A model of dextran/glucan-induced agglutination, GTase binding and subsequent adherence		117
4.5	Properties of glucosyltransferases from *S. mutans* and *S. sanguis*		120
	4.5.1 Comparison of the properties of GTase		120
	4.5.2 Glycoprotein nature and aggregation of GTase from *S. mutans*		120
	4.5.3 Intracellular and extracellular GTase of *S. mutans*		121
4.6	Structural aspects of polysaccharides produced by oral streptococci		124
	4.6.1 Structure of the glucans		124
	4.6.2 Structure of the fructans		126
4.7	Reduction of bacterial adherence to tooth surfaces		127
	4.7.1 Inhibition of adherence by glucanases		127
	4.7.2 Inhibition of adherence by immunization		128
4.8	Concluding remarks and summary		130
	References		131

Bacterial Adherence
(*Receptors and Recognition,* Series B, Volume 6)
Edited by E.H. Beachey
Published in 1980 by Chapman and Hall, 11 New Fetter Lane, London EC4P 4EE
© 1980 Chapman and Hall

4.1 INTRODUCTION

It has been demonstrated that *S. mutans* of human and animal origin produces dental caries in various experimental animals, if the animals were fed a caries-inducing high-sucrose diet. Other oral streptococcal species such as *S. sanguis, S. salivarius, S. milleri* also cause dental caries in experimental animals, but only sporadically (Gibbons and Van Houte, 1975a; Scherp, 1971; Hamada *et al.*, 1978a). The two virulence factors of *S. mutans* responsible for caries induction are ascribed to the ability to colonize tooth surfaces and subsequent acid production when fermentable sugars are available in excess (Scherp, 1971).

The aim of this chapter is to analyze current concepts of the adherence mechanism of oral streptococci which regulate the significant populations on tooth surfaces. Special attention will be focused on *S. mutans* because of its etiological importance. Various aspects of this literature have been discussed in several recent reviews (Gibbons and Van Houte, 1973, 1975a, b; Scherp, 1971; Slade, 1976, 1977; see also Chapter 3) and in books (McHugh, 1970; Lasslo and Quintana, 1973; Stiles *et al.*, 1976).

4.2 *IN VITRO* ADHERENCE OF *S. MUTANS* AND THE EFFECT OF SUCROSE

4.2.1 General considerations

Adherence of *S. mutans* to smooth surfaces mainly correlates with its ability to utilize sucrose for the synthesis of adherent water-insoluble glucans. The newly synthesized glucans strengthen the initial attachment of *S. mutans* cells with solid surfaces, thus resulting in the firm colonization and accumulation of *S. mutans* on solid surfaces including tooth surfaces.

In *in vitro* systems, *S. mutans* can form heavy adherent deposits of a glucose polymer on surfaces of various solid supports, i.e. wire, glass, HA and extracted or artificial teeth when serially incubated in broth media containing sucrose, but not most other sugars (Jordan and Keyes, 1966; McCabe *et al.*, 1967; Tanzer and McCabe, 1968; Gibbons and Nygaard, 1968; Sudo *et al.*, 1975; Hamada *et al.*,1975).

Figure 4.1a shows the scanning electron microscopic figures of the artificial dental plaque formed on the cleaned extracted tooth surface by serial transfers of a pure culture of *S. mutans* in 5% sucrose-containing broth. Outlines of the bacterial growth are obscured by glucan. A higher magnification (Fig. 4.1b) shows details of the cell-glucan structures (Hamada *et al.*, 1975).

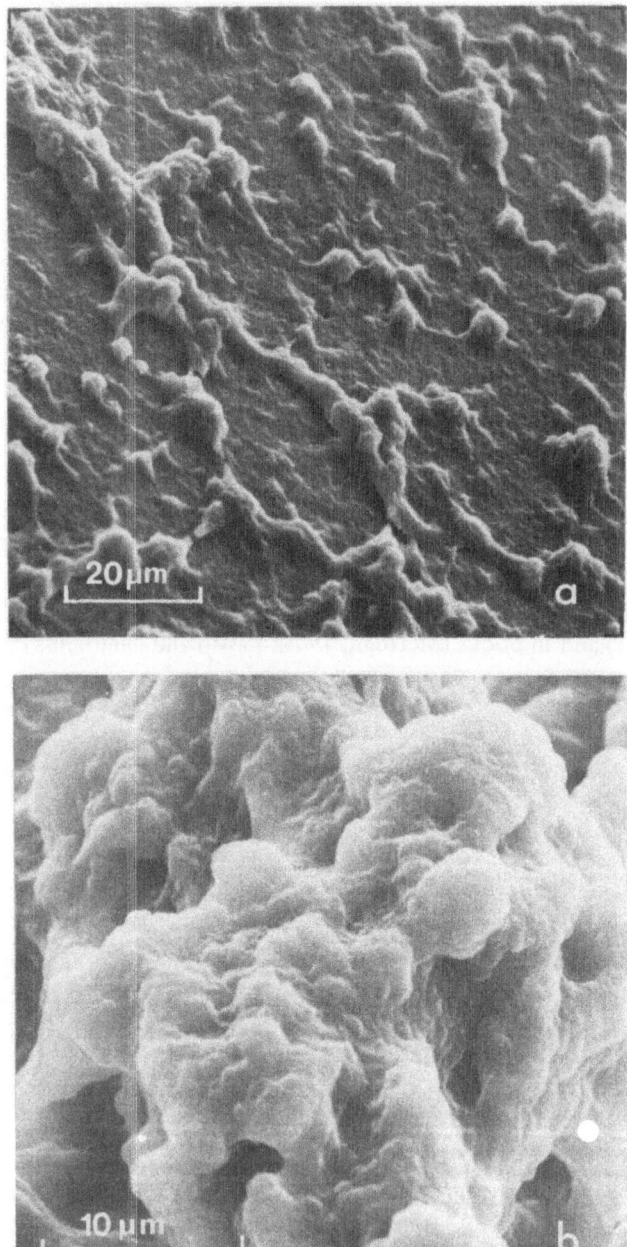

Fig. 4.1 Artificial plaque formation by *S. mutans* strain OMZ176 on the enamel surface of an extracted tooth. (a) and (b) show lower and higher magnifications, respectively. (Reproduced from Hamada *et al.* (1975) with permission of the American Society for Microbiology, Washington, D.C.).

4.2.2 Adherence in the presence of cell-bound glucosyltransferase (GTase) and sucrose

A series of studies using glass surfaces has contributed many details of the mechanism of the *in vitro* adherence of *S. mutans* and other species of oral bacteria. The initial experiments showed that viable non-growing suspensions of *S. mutans* adhered to a glass surface in the presence of sucrose (Mukasa and Slade, 1974a). The adherence was mainly due to the synthesis of glucan by cell-associated GTase. To inactivate the enzyme, cell suspensions were held at 65°C for two 1-hour periods, or at 100°C for 20 minutes. Upon the addition of GTase and sucrose the heat-treated cells adhered

Table 4.1 Binding of glucosyltransferase (GTase) and adherence of various gram-positive and gram-negative bacteria

Bacterial species	Cells + GTase, 1 h at 37°C, wash cells, add sucrose, incubate		Cells + GTase + sucrose, mix, incubate	
	% adherence	relative adherence	% adherence	relative adherence
S. mutans B13 (*d*)	73	100	72	100
S. sanguis OMZ9	7	10	41	57
S. salivarius	44	6	9	13
S. pyogenes group A (Richards)	0	0	52	75
Strept. group E (K129)	0	0	42	58
Lact. plantarum ATCC8014	0	0	65	80
Actinomyces viscosus OMZ105	1	2	65	90
Fusobacterium nucleatum ATCC25586	0	0	73	101
Leptotrichia buccalis ATCC19616	0	0	76	106
E. coli OMZ116	6	8	46	64
Proteus mirabilis OMZ109	11	14	73	101
Serratia marcescens OMZ144	8	11	50	69
Neisseria perflava OMZ143	9	12	66	92

GTase from *S. mutans*, strain B13, type *d* (Modified from Slade, 1976).

in significant numbers to glass. The addition of glucan synthesized by *S. mutans* or soluble dextran (T2000 Pharmacia) to these heat-treated cells did not result in adherence. The binding of GTase by these cells is indicated by the addition of enzyme to the cells, removal of free enzyme by washing and addition of sucrose to the cells. Adherence resulted upon incubation.

Further experiments have utilized these procedures for quantitating adherence and analyzing the mechanism. In method I the heat-treated cells, enzyme and sucrose were mixed simultaneously and incubated at 37°C for 17 hours. Table 4.1 shows that by this procedure 5 serotype strains of *S. mutans* and a number of gram-positive and gram-negative species became adherent ranging from 48–106% of the *S. mutans* control. In method II, the heat-treated cells plus enzyme were mixed and incubated, the cells washed, enzyme added and the mixture incubated. In contrast to method I, when the same bacterial strains were tested by this procedure the only strains showing adherence were those of *S. mutans* (Table 4.1). These strains showed a relative adherence of 75–114%, whereas the other non-*mutans* species were 12% or less.

Several other methods have been used to measure the *in vitro* adherence of oral bacteria to glass surfaces. Olson *et al.* (1972) described the development of an assay method for the adherence of *S. mutans* in which cells were grown in 1 ml broth containing 1% sucrose at a 30° angle for 12 h. After removal of the non-adherent cells and gentle washing of the cells which remained adhered to the glass surface, 2 ml 0.5N sodium hydroxide was added to disperse the cells into an even suspension. Turbidity was read at a wave length of 540 nm. This assay method was used to demonstrate that immune sera against *S. mutans* whole cells inhibited the adherence of the cells to the glass surfaces.

Schachtele *et al.* (1975, 1976a) presented a method for adherence which involved monitoring the attachment of ^{3}H-thymidine-labeled bacteria to glass scintillation vials. Cells equivalent to 100 μg protein/35 000 cpm/ml/vial were mixed with buffered sucrose in scintillation vials and incubated for 18 h at 37°C. Adherence, however, was relatively low (*ca.* 15%) under the conditions described. The adherence of *S. mutans* was dependent on sucrose, and was inhibited by dextranase. Exogenous dextrans produced no adherence in the absence of sucrose.

More recently, Newbrun *et al.* (1977) described a similar assay method using ^{3}H-thymidine-labeled cells of *S. mutans* OMZ176 (serotype *d*) and obtained more than 50% adherence of cells after incubation with sucrose for more than 3 h. They also reported that incorporation of maltose and low-molecular weight dextran inhibited the adherence of *S. mutans* to the surface of glass scintillation vials.

4.2.3 Adherence and the quantity of glucan synthesized

Measurements have been made of the quantities of glucan synthesized by these species under the same conditions as used in method II above (Table 4.2). It is seen that the *S. mutans* strains (except serotype *b*, FA1) synthesized significant quantities of glucan. The type *a* and *g* strains were most active. Non-heat-treated cells did not

Table 4.2 Synthesis of cell-bound glucan by serotype strains of *S. mutans* and various gram-positive and- negative bacteria.

Bacterial species	Strain	Glucan formed (cpm)					
		Heat-treated cells		Lyophilized cells			
		No GTase	+ GTase	No GTase	+ GTase		
S. mutans	HS6 (*a*)	260	3803	46 837	49 741		
	HS1 (*a*)	128	2711	20 675	22 748		
	FA1 (*b*)	63	227	233	625		
	Ingbritt (*c*)	116	3537	5 083	10 130		
	B13 (*d*)	105	2466	6 321	8 009		
	MT703 (*e*)	90	5418	4 531	8 760		
	MT557 (*f*)	61	1981	1 606	5 818		
	OMZ65 (*g*)	48	3380	20 519	20 959		
S. sanguis	OMZ9	318	541	2 350	3 100		
S. mitis	ATCC903	437	704	650	560		
S. bovis	8177	94	197	67	316		
Streptococcus group E	K129	97	290	260	480		
L. plantarum	ATCC8014	511	838	330	580		
A. viscosus	OMZ105	243	572	420	2 640		
A. naeslundii	ATCC12164	111	200	1 320	1 890		
E. coli	E16	239	2286	280	530		
	OMZ116	95	250	970	780		
S. marcescens	OMZ144	269	1130	380	560		

(Reproduced from Hamada *et al.* (1978a) with permission of the American Society for Microbiology).

show additional glucan synthesis after the cells were exposed to enzyme (except type *f*, MT557). *S. sanguis* showed significant glucan synthesis although considerably lower than the *S. mutans* strains. The two *Actinomyces* strains showed a limited capacity to bind enzyme and synthesize glucan. Table 5.2 also shows that heat-treated *S. mutans* cells (except FA1) synthesized considerably more glucan than the non-*mutans* strains after binding the GTase. In the case of *E. coli* (E16) and *S. marcesens* (OMZ144), the heat treatment appeared to have modified the cell surface so that limited binding of GTase occurred.

4.2.4 The GTase binding site on *S. mutans*

The chemical nature of the GTase binding site on the cell surface of *S. mutans* is of considerable interest. The binding site was first studied using heat-treated cells of *S. mutans* strain HS6 (serotype *a*) which had been grown in Todd-Hewitt broth. Antiserum specific for the *a-d* site on the type *a* antigenic polysaccharide was found to inhibit the binding of the enzyme and cell adherence. Anti-*a* specific serum, except in 5-fold larger quantities, did not inhibit the binding of the enzyme and subsequent adherence. These antisera did not contain polyglycerophosphate or dextran/glucan antibodies (Mukasa and Slade, 1973, 1974a). Seven serotype specific carbohydrate antigens of *S. mutans* have been identified (Slade, 1977).

The inhibition of anti-*a-d* serum is considered to be due to a restriction of access of GTase to the binding site of the cell surface which may be located in close proximity to the type antigen. Also, anti-glucan serum inhibited the adsorption of the enzyme by masking the binding site (Mukasa and Slade, 1974a). In addition, it was found that treatment of the cells with dextranase and proteolytic enzymes resulted in a significant inhibition of the synthesis of insoluble glucan and of adherence. Similar effects of enzyme treatment have been obtained with serotype *e S. mutans* cells (Hamada and Slade, 1976c). These results indicate that the binding site is blocked by glucan antibody and destroyed by dextranase, and the *a-d* site is adjacent to, or a part of, the sites (Slade, 1977).

More recently, it has been found that if *S. mutans* B13 (serotype *d*) was grown in sucrose-free organic medium, brain heart infusion broth or chemically defined FMC medium developed by Terleckyj *et al.* (1975), the cells retained very weak cell-associated GTase activity, and failed to adhere even in the presence of sucrose (Hamada and Torii, 1978; Hamada and Slade, 1979). If sucrose was incorporated in the above described media or if commercially available Todd-Hewitt broth or trypticase soy broth was employed for cultivation of *S. mutans,* GTase was detected mostly on the cell surface where cell-bound, water-insoluble glucan was synthesized. These findings indicate that commercial products of Todd-Hewitt broth and Trypticase soy broth contained trace amounts of sucrose from which *S. mutans* synthesized small quantities of cell-bound water-insoluble glucan by the action of GTase at the cell surface. Once *S. mutans* cells formed cell-bound glucan, the cells could bind extracellularly excreted GTase to the surface glucan. The hypothesis has been supported by several experiments (Hamada and Torii, 1978).

(1)   Cells grown in broth containing sucrose plus dextranase possessed no cell-associated GTase. Dextranase is known to prevent cell-bound glucan formation (Hamada *et al.*, 1975; Hamada and Slade, 1976; Mukasa and Slade, 1974a).

(2)   Heat-treated cells grown in sucrose broth bound cell-free GTase. The degree of GTase-binding by heat-treated cells or intrinsic cell-associated GTase activity increased with the concentration of sucrose in the broth medium in which cells had been grown (Hamada and Slade, 1979).

(3)   Pretreatment of the heat-treated cells with dextranase, NaOH or antiglucan serum resulted in loss of GTase-binding ability of the cells. Antiserum against type-specific antigen markedly agglutinated the cells and inhibited subsequent cell adherence to smooth surfaces, but did not inhibit the binding of GTase to the cell surface, which indicated no participation of type *d* antigenic cell wall polysaccharide ·in the binding of GTase (Hamada and Torii, 1978).

(4)   Furthermore, the fact that cell-free insoluble glucan synthesized from sucrose bound cell-free GTase renders strong support to the hypothesis. The GTase did not bind to other insoluble polysaccharides, i.e. cellulose, starch, amylopectin, dextrin, inulin, etc. (Hamada and Torii, 1978). The pre-incubation of high-molecular weight dextrans such as T250, T500 and T2000 to the binding mixtures markedly decreased binding of GTase to the cells, whereas low-molecular weight dextrans T10 and T18 did not alter the binding ability (Hamada and Torii, 1979). It should be noted that the binding or adsorption of GTase to insoluble glucans of *S. mutans* has recently been reported, but without special reference to the adherence ability of *S. mutans* cells and/or glucans (Robrish *et al.*, 1972; McCabe and Smith, 1973, 1977; Koga and Inoue, 1978; Kuramitsu and Ingersoll, 1978).

4.2.5 The necessity for new glucan synthesis in *in vitro* adherence of *S. mutans* cells to smooth surfaces

As described briefly in the preceding section (4.2), active *de novo* glucan synthesis by GTase from sucrose is required in order for adherence of *S. mutans* cells to occur (Mukasa and Slade, 1973). Hamada (1977) has reported that *S. mutans* B13 cells which had been incubated with 1% sucrose solution for various time intervals (0.5−18 h) did not produce significant adherence when incubated in buffer without sucrose, indicating that preformed glucan itself had no adherent ability. When sucrose was again added to the washed cell suspension, which had been pre-incubated with 1% sucrose, the cells became adherent as new glucan was synthesized.

A time course of the adherence of the cell to glass surfaces and cell-associated water-insoluble (^{14}C) glucan synthesis is shown in Fig. 4.2. The adherence of the cells to glass in the presence of sucrose increased linearly up to 7 h, after which little increase was observed. Slowing in the linear increase in glucan synthesis occurred between 2−6 h although adherence was optimal during this period.

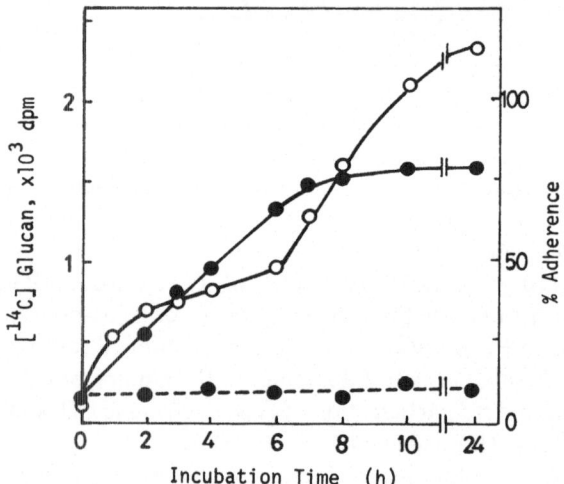

Fig. 4.2 Kinetics of adherence of *S. mutans* B13 cells and synthesis of cell-bound, water-insoluble (^{14}C)-glucan from (^{14}C)-glucose labeled sucrose (○—○) as a function of time. Adherence of non-growing cells to glass surface was carried out in the presence (●—●) or absence (●— — —●) of sucrose. (Reproduced from Hamada (1977) with permission of the Faculty Press, Cambridge).

On the other hand, Kuramitsu (1974) has reported that cells of 2 of 4 serotype strains of *S. mutans* adhered to a glucan layer that had been preformed on a glass surface by cell-free GTase and sucrose. By the use of antibody to GTase, approximately one-half of this adherence was shown to involve the binding of the cells to the GTase on the surface of the glucan layer (Hamada and Slade, 1976). Further investigations (Hamada, 1977; Koga and Inoue, 1978) have shown that the adherence of *S. mutans* cells to precoated glucan is also mediated by additional glucan synthesis from residual GTase and sucrose in the preformed glucan layer. These results indicate that the association between glucan and a putative receptor on the cell-surface of *S. mutans* (Gibbons and Fitzgerald, 1969) does not completely account for the irreversible adherence of *S. mutans*.

4.2.6 The development of *in vitro* and *in vivo* plaque

It appears that *S. mutans* differs from other gram-positive and gram-negative species in its mechanism of adherence. The non-*mutans* species do not synthesize glucan nor adhere in significant numbers either in the presence or absence of GTase. Their ability to bind GTase is poor. The adherence of these species (Table 4.1) is of a non-specific type, most likely due to a 'trapping' of the cells in the glucan as it is synthesized by extracellular GTase. This procedure may be of quantitative significance in the *in vivo* development of the plaque to maturity on the surface of the pellicle-covered tooth. This may result from a random contact of the bacteria with the glucan on the surface or a covering of the bacteria on the surface with the spreading film of glucan synthesized by *S. mutans* cells (Hamada *et al.*, 1978b; Slade, 1977).

This process could follow the initial attachment of *S. mutans* to the tooth surface. Initial attachment may be due to electrostatic charges (Rölla, 1977) and/or a complex formation between a glycoprotein in the pellicle and glycoprotein on the streptococcal cell surface. GTase, which possesses some of the properties of a glycoprotein (Slade, 1977), may provide this mechanism when on the cell surface. The colonization of rats by *S. mutans* followed the synthesis of glucan by cells initially attached to the tooth surface. New glucan synthesis was concluded to be the principal mechanism of adherence as the plaque developed to maturity (Van Houte and Upeslacis, 1976). The multiplication of the streptococci, the synthesis and binding of GTase, the synthesis of cell-bound glucan and the entrapment of various bacterial species would all contribute to the development of plaque on the tooth surface. Species other than *S. mutans* following initial attachment could increase in number in the glucan mass and thereby maintain their attachment (Slade, 1977; Hamada *et al.*, 1978b).

A trapping effect, however, is probably not completely responsible for the varied types of bacteria in *in vivo* plaque. Results with untreated and saliva-coated human tooth powder have shown that the initial attachment of four oral bacterial species occurs in larger numbers than *S. mutans* (Hillman *et al.*, 1970). Other results indicate that the low numbers of *S. mutans* in human saliva may be irsufficient for their initial and firm attachment to smooth tooth surfaces (Van H ute and Green, 1974, see also Chapter 4 this volume). The limited numbers of *S. mutans* on these surfaces relate to their location in discrete microcolonies (Gibbons *et al.*, 1974). An increase in the numbers of these cells and subsequent continued attachment, however, are very likely related to a continuation of the synthesis of glucan (Van Houte and Upeslacis, 1976). The ability of *S. mutans* to synthesize large quantities of glucan may not require large numbers of these cells to form plaque. This species is usually present in much smaller numbers compared with the total oral anaerobic flora on blood agar (Gibbons and Van Houte, 1975b). The entrapment of various species in the glucan matrix from the saliva would appear to be a continuous process as the plaque develops to maturity.

The agglutination of *S. mutans* cells may also contribute to the development of *in vitro* and *in vivo* plaque. Agglutination is known to occur upon the addition of water-soluble, high-molecular-weight dextran or glucan (Gibbons and Fitzgerald, 1969; Wu-Yuan *et al.*, 1978) or lectins (Hamada *et al.*, 1977). The extent of *in vivo* plaque development by this process may be limited due to the few cells of *S. mutans* found free in the oral cavity (Gibbons and Van Houte, 1973). The agglutination of *Actinomyces* by glucan (Kelstrup and Funder-Nielsen, 1974; McBride and Bourgeau, 1975; Bourgeau and McBride, 1976), the aggregation of this organism with *S. mutans*, *S. sanguis*, and *S. mitis* (Ellen and Balcerzak-Raczkowski, 1977) and cell-to-cell attachment of these streptococci (Schachtele *et al.*, 1976a), indicate that these processes may also aid in the establishment of these species in the developing plaque. These cell-to-cell attachments however are increased in the presence of glucan.

4.3 DEXTRAN/GLUCAN-INDUCED AGGLUTINATION

4.3.1 General considerations

Non-growing viable cells of *S. mutans* are known to agglutinate when incubated with sucrose (Guggenheim and Schroeder, 1967) or upon the addition of high molecular weight dextran (Gibbons and Fitzgerald, 1969). Dextrans possessing molecular weights of 2×10^6 and 2×10^5 were effective, whereas one of 2×10^4 was not. No agglutination was obtained after the cells were heated at $100°C$ for 10 min. Strains of serotypes *a, f* and a streptomycin-resistant type *c* strain were positive, whereas the parent type *c* strain and strains of types *b* and *e* were negative. Recent results (Wu-Yuan *et al.*, 1978) show that *a, d* and *g* strains agglutinate with dextran T2000. Glucans from 5 serotype strains did not in all cases cause agglutination in these *a, d* and *g* strains, whereas strains of the remaining serotypes were weak or negative. Lectins also cause agglutination of *S. mutans* (Hamada *et al.*, 1977). In addition to *a, d* and *g* strains, type *f* is also agglutinated by Con A. RCA I and RCA II agglutinate *a, d* and *g* only. These results illustrate that the cell surface binding sites involved in agglutination are not present in all serotypes and indeed not in all strains of the same serotype. The dextran-synthesizing species *S. bovis, S. sanguis* and *L. mesenteroides* do not agglutinate (Gibbons and Fitzgerald, 1969).

Although *Actinomyces* does not produce glucan it is agglutinated by this polymer (McBride and Bourgeau, 1975). Glucan may also be involved in the co-aggregation of *Actinomyces* with *S. mutans, S. sanguis* and *S. mitis* (Ellen and Balcerzak-Raczkowski, 1977). McIntire *et al.*, (1978) however have shown that the co-aggregation of *A. viscosus* T14V and *S. sanguis* 34 is independent of dextran. A carbohydrate related to lactose in *S. sanguis* and a protein or glycoprotein in *A. viscosus* appear to be involved in the co-aggregation. Immunochemical data show that the fibrils on the surface of *A. viscosus* (Cisar *et al.*, 1978) mediate the co-aggregation.

4.3.2 The dextran/glucan binding site

The binding site of dextran/glucan on the surface of the *S. mutans* cell is of considerable interest. The inactivation by papain and periodate indicates that a glycoprotein is involved (Kelstrup and Funder-Nielsen, 1974). Inactivation by trypsin and dextranase also indicates the participation of a glycoprotein (Wu-Yuan *et al.*, 1978). Sodium dodecyl sulfate (SDS) and urea caused an irreversible destruction of the receptor sites, and divalent cations such as Ca^{2+} or Mg^{2+} appear to be required for activity (Kelstrup and Funder-Nielsen, 1974; McCabe *et al.*, 1976).

The blocking of the agglutination site by mannose or galactose (Kelstrup and Funder-Nielsen, 1974) and by Con A (Wu-Yuan *et al.*, 1978), which reacts with α-D-mannopyranosyl units, is further support for a glycoprotein structure. Inactivation of the site by antiserum to the *d*-specific polysaccharide and antiserum to glucan (from *S. mutans*) also supports this concept.

The use of ^{14}C-T20 and ^{3}H-T70 dextrans has shown that these polymers will bind to *S. mutans* (type *d*) B13 strain although no agglutination is obtained. Treatment of these cells with EDTA, SDS, trypsin or dextranase did not reduce the binding of T20 or T70 whereas the binding of ^{14}C-glucan was reduced about 80% (Wu-Yuan *et al.*, 1978).

The binding site for dextran/glucan on the cell surface of *S. mutans* does not appear to be GTase (Kelstrup and Funder-Nielsen, 1974; McCabe and Smith, 1975; Nalbandian *et al.*, 1974). However, if GTase becomes associated with glucan bound to the cell surface, the addition of glucan will cause agglutination (Germaine and Schachtele, 1976).

4.4 A MODEL OF DEXTRAN/GLUCAN-INDUCED AGGLUTINATION, GTase BINDING AND SUBSEQUENT ADHERENCE

Models of the streptococcal cell wall (Figs. 4.3 and 4.4) have been presented (Slade 1976, 1977, 1978) which help to picture the activities at the surface of *S. mutans* which may be responsible for agglutination, the binding of GTase, glucan synthesis and subsequent adherence in the presence of sucrose.

Numerous investigations support the view that adherence and agglutination are separate processes (Germaine and Schachtele, 1976; Kelstrup and Funder-Nielsen, 1974; Mukasa and Slade, 1973; Nalbandian *et al.*, 1974; Freedman and Tanzer 1974). Heat lability of the agglutination site and heat stability of the GTase binding site is a clear expression of a difference between the two. Also, the effect of Con A emphasizes this difference. The Con A-treated cells did not agglutinate with T2000, nor did they bind significant quantities of ^{14}C-glucan, however they were able to bind GTase and become adherent *in vitro* in the presence of sucrose (Hamada *et al.*, 1977). In contrast, several plant lectins have been shown to inhibit the *in vitro* adherence of *S. mutans* (Staat *et al.*, 1978). The model (Fig. 4.4) represents agglutination as requiring a glucan molecule associated with a protein (a glycoprotein). The latter is heat-sensitive. Fig. 4.4 expresses the difference between agglutination and adherence.

Germaine and Schachtele (1976) have presented a model which suggests that in agglutination the cross-linking of the *S. mutans* cells is mediated by (I) the interaction of cell-free GTase with multiple sites on individual dextran/glucan molecules attached to the bacterial surface, and (II) dextran-glucan molecules of sufficient size which are

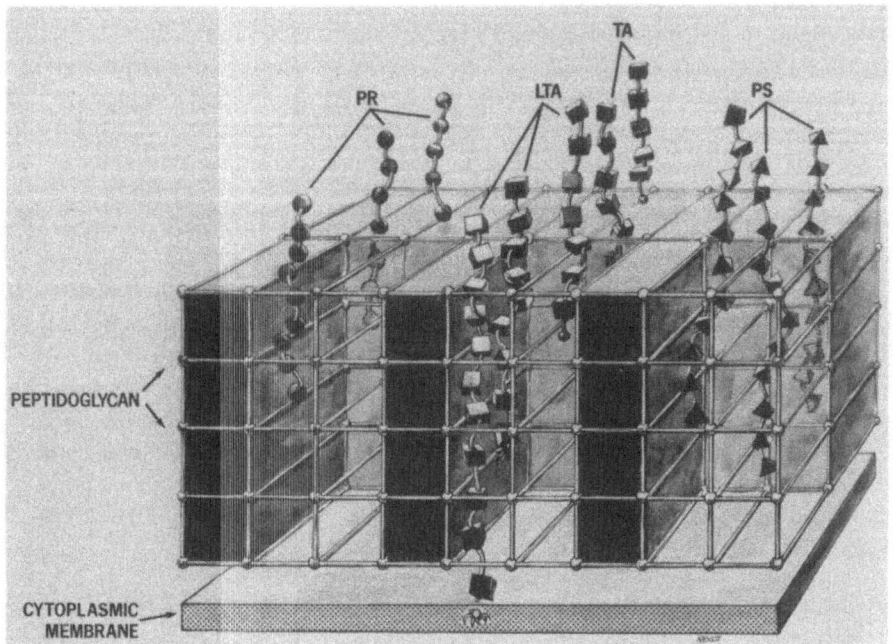

Fig. 4.3 Model of the cross-section of a streptococcal cell wall. The peptidoglycan is open in three areas in order to visualize more clearly the location
of the antigenic polymers within the peptidoglycan.

PR — protein, PS — polysaccharide, TA — teichoic acid, LTA — lipoteichoic
acid.

(Reproduced from *Secretory Immunity and Infection*, pp. 761—762, 1978,
with permission of the Plenum Publishing Co., New York).

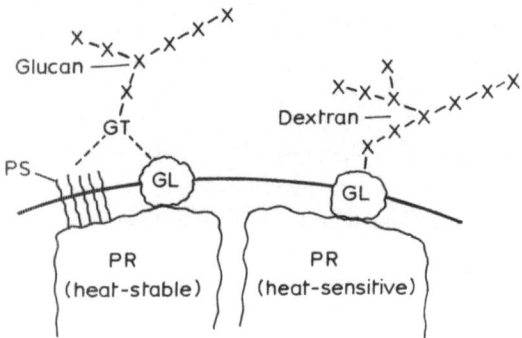

Fig. 4.4 Model representing binding sites for glucosyltransferase and dextran
on the cell wall surface of *S. mutans*.

PR — protein, PS — type-specific polysaccharide, GL — glucan, GT — glucosyltransferase.

(Reproduced from Slade (1977) with permission of the American Society for
Microbiology).

bound to a non-enzyme dextran acceptor on the cell surface. The evidence presented above supports the latter possibility. The inability of T20 or T70 to produce agglutination (Wu-Yuan *et al.*, 1979) may be due to a polymer size which is insufficient to bridge 2 or more cells.

Evidence is available however which indicates that a second binding site exists for dextran/glucan-induced agglutination. This site does not appear to contain glucan. B13 cells grown in an invertase-treated organic medium agglutinated to the same degree with T2000 as those cells grown in the non-treated culture medium. Cells grown in an invertase-treated or non-treated synthetic medium however, did not agglutinate (Wu-Yuan *et al.*, 1979). The existence of multiple dextran-binding proteins in *S. mutans* has been reported (McCabe and Hamelik, 1978).

Figure 4.4 illustrates diagramatically the evidence presented in Section 4.2. It is seen that GTase may bind to either the type-specific polysaccharide and/or glucan. The enzyme probably possesses a greater affinity for the glucan site than the polysaccharide due to the glycoprotein character of the enzyme (Slade, 1977). This reasoning is supported by the inability of *S. mutans* B13 cells grown on an invertase-treated or non-treated synthetic medium to bind GTase and produce *in vitro* adherence (Hamada and Slade, 1979). It is not likely that these cells possess any polymers related to glucan.

Figure 4.3 depicts the cell surface of *S. mutans* as possessing a mosaic character (Slade, 1978). The possible 'umbrella' effect of anti-type serum on the binding of GTase (Mukasa and Slade, 1974a) is in agreement with this type of organization of the cell surface. The evidence requires that both glucan and protein be a part of the binding site of GTase.

The possibility that *S. mutans* possesses multiple binding sites for glucans with differing specificities has been suggested (Mukasa and Slade, 1974a; Freedman and Tanzer, 1974; Janda and Kuramitsu, 1977). One specificity is that of molecular size (Gibbons and Fitzgerald, 1969; Wu-Yuan *et al.*, 1978). The presence of a binding site which possesses a specificity not related to glucan is now indicated (Wu-Yuan *et al.*, 1979). Other sites contain a heat-stable protein (McCabe *et al.*, 1977, 1978) or a heat-labile glycoprotein (Wu-Yuan *et al.*, 1978). The effectiveness of a surface site for the binding of glucan and agglutination is probably mediated by both the chemical composition and the steric arrangement of the polymers at that site.

Agglutinated *S. mutans* cells in the oral cavity may contribute to the numbers of *S. mutans* cells which initially attach to the pellicle covered smooth tooth surface. The quantitative nature of this process is not known, however the limited numbers of *S. mutans* in human saliva (Gibbons and Van Houte, 1975b) indicate that agglutination may not play a significant quantitative role in the initial *in vivo* attachment of these streptococci.

4.5 PROPERTIES OF GLUCOSYLTRANSFERASES FROM *S. MUTANS* AND *S. SANGUIS*

4.5.1 Comparison of the properties of GTase

Glucosyltransferase (EC 2.4.1.5) (GTase) is a constitutive enzyme synthesized by *S. mutans* and *S. sanguis*. The enzyme polymerizes the glucose moiety of sucrose to produce glucan(s). Wood (1967) first isolated a crude GTase from culture supernatant of *S. mutans* FA1. Subsequently, Carlsson *et al.*, (1969) partially purified GTase from culture supernatant of *S. sanguis* 804. Since then, the GTases from *S. mutans* and *S. sanguis* strains have been purified from the culture liquor by various investigators.

Some of the results so far reported are summarized in Table 5.3 for comparison. It is apparent that the properties of these GTase preparations are significantly different depending on the strain employed and its serotype, the purification methods, and on the culture media used for cultivating the organisms. General aspects of GTases of *S. mutans* have been reviewed recently (Ciardi, 1976; Newbrun, 1976).

4.5.2 Glycoprotein nature and aggregation of GTase from *S. mutans*

As shown in Table 5.3, purified GTase preparations have been found to contain significant quantities (30–40%) of carbohydrate, which may be a unique characteristics of the cell-free GTase. It appears that glucose is a main component of the carbohydrate as revealed by gas-liquid chromatography (Mukasa and Slade, 1974b) or as enzymatically determined in an acid hydrolysate of the enzyme (Scales *et al.*, 1975).

Addition of ConA to a purified FA1 GTase preparation did not produce visible precipitates (Scales *et al.*, 1975), whereas a similar addition to an enzyme fraction of GS5 GTase synthesizing insoluble glucan caused marked turbidity (Kuramitsu, 1974b). In this context, HS6 GTase activity was not inhibited by anti-dextran globulin, indicating that the glucose polymer in the enzyme is not dextran and may not directly function in the polymer synthesis (Mukasa and Slade, 1974b). This is supported by the fact that HA-chromatographed 6715 GTase preparations were entirely dependent upon added dextran for glucan synthesis. However, they possessed significant quantities of carbohydrate (Germaine *et al.*, 1974). There is the possibility of a glucan (or a levan) associated with the GTase preparations, because the culture media (Todd-Hewitt broth and Trypticase-Soy broth) employed in these studies have been found to contain trace amounts of sucrose and/or dextran (Spinell and Gibbons, 1974; Hamada and Torii, 1978).

The GTase of *S. mutans* exists in a number of aggregated forms which appear to be related to the synthesis of water-soluble or insoluble forms of glucan (Mukasa and Slade, 1974b; Kuramitsu, 1975). The molecular weight of the GTase preparations purified from complex media containing trace amounts of sucrose appears to be higher than those obtained from dialyzed or chemically defined media. The GTase from chemically defined medium could be disaggregated to a low-molecular-weight form by 1 M salt, whereas the Trypticase-Soy broth (TSB) enzyme was not significantly affected by the salt treatment and the activity was still contained in aggregated forms (Schachtele *et al.*, 1976b). Furthermore, it was found that the enzyme activity from FMC or fructose-grown cultures was primer-dependent in contrast to the enzyme from TSB grown cultures (Schachtele *et al.*, 1976b; Germaine and Schachtele, 1976). It was also found that the former enzyme formed salt-stable enzyme-dextran complexes in the presence of dextran T2000. It is calculated that *ca.*150 enzyme molecules are bound per dextran molecule, indicating that individual dextran molecule possesses multiple binding sites for GTase.

Recently, Harlander and Schachtele (1978) have reported that lysophosphatidyl-choline (LPC) and other phosphoglycerides can stimulate glucan production by *S. mutans* GTase, and the enhanced glucan synthesis by LPC and primer dextran are additive. GTase can be activated by binding of intact phosphoglyceride molecules to a site on the enzyme which is separate from either the glucosyl donor or glucosyl acceptor (primer) binding sites. These associations of GTase with phosphoglycerides may occur during and/or after excretion from *S. mutans* cultures, which could partially explain the aggregated form of the enzyme in culture broths, the multiple forms of 'purified' GTase and the variability of the enzyme elaborated under different cultural conditions.

4.5.3 Intracellular and extracellular GTase of *S. mutans*

It has been shown that GTase activity exists in *S. mutans* cultures both extracellularly and firmly associated with the cell surface and in relative proportions that may be changed in response to environmental factors. Robrish *et al.*, (1972) reported that almost all of the activities in Brain Heart Infusion broth cultures of three *S. mutans* strains were recovered in the centrifuged culture supernatant. Later studies (McCabe and Smith, 1973; Hamada and Torii, 1978) have shown that the recovery of enzyme from sucrose cultures is markedly decreased by high concentrations of sucrose in the culture media, and cell-free and cell-associated forms of GTase appear to be alternate states of the same enzyme protein. In this connection, Janda and Kuramitsu (1976) have demonstrated that most of the GTase activity associated with the cells is not a precursor for the extracellular GTase activity. Spinell and Gibbons (1974) also reported that growth of *S. mutans* in sucrose-containing medium promoted cell-associated GTase activity as compared to cell-free GTase activity in the supernatant.

Other groups (Kuramitsu, 1974b; Chassy *et al.*, 1976; Montville *et al.*, 1977), however, have demonstrated that some *S. mutans* had significant cell-associated GTase activity in media devoid of sucrose, and no major shift to a cell-associated form from a cell-free state was observed in sucrose cultures. It appears that differences exist among the various strains. It is of interest that Tween 80 was found to enhance the production of extracellular GTase in *S. mutans* OMZ176 (Umesaki *et al.*, 1977).

Table 4.3 Various properties of GTase prepared from culture supernatant of *S. mutans* and *S. sanguis*

Strain	Culture broth	Molecular weight (x 10^5 dalton)	Michaelis constant (mM)	Isoelectric point
S. mutans				
HS6 (*a*)*	Dialyzed complex	1.7	2.0	N.D.**
HS6 (*a*)	Todd-Hewitt	4–20	N.D.	N.D.
FA1 (*b*)	Trypticase-Soy	8	55	3.7
GS5 (*c*)	Todd-Hewitt + Dextran T10	0.45 1.9	1.7 2.1	4.3 6.2
OMZ176 (*d*)	Dialyzed complex	N.D.	0.98 –7.2	4.24 –6.03
6715 (*g*)	Trypticase-Soy	0.94	3	4.0
6715 (*g*)	Tryptone + Yeast extract + Tween 80	1.5	5	4.3
S. sanguis				
804	Dialyzed complex	N.D.	4.8	7.9
OMZ9	Chemically synthetic	1.0	2	N.D.

* Serotype
** Not described

Table 4.3 Various properties of GTase prepared from culture supernatant of *S. mutans* and *S. sanguis* (*continued*)

Strain	Culture broth	Optimum pH	Carbo-hydrate content	Water solubility of glucan	References
S. mutans					
HS6 (*a*)*	Dialyzed complex	5.75	N.D.	Insoluble + Soluble	Fukui *et al.* (1974)
HS6 (*a*)	Todd-Hewitt	5–7	+	Insoluble + Soluble	Mukasa and Slade (1974a, b)
FA1 (*b*)	Trypticase-Soy	6.0	+	N.D.	Scales *et al.* (1975)
GS5 (*c*)	Todd-Hewitt + Dextran T10	6.0	+	Insoluble + Soluble	Kuramitsu (1975)
OMZ176 (*d*)	Dialyzed complex	5–7	N.D.	Insoluble	Guggenheim and Newbrun (1969)
6715 (*g*)	Trypticase-Soy	5.5	+	N.D.	Chludzinski *et al.* (1974) Germaine *et al.* (1974)
6715 (*g*)	Tryptone + Yeast extract + Tween 80	N.D.	+	Insoluble	Ciardi *et al.* (1976, 1977)
S. sanguis					
804	Dialyzed complex	5.2–7.0	+	Jelly + Soluble	Carlsson *et al.* (1969) Newbrun (1971, 1976)
OMZ9	Chemically synthetic	5.8–6	N.D.	Soluble + Insoluble	Klein *et al.* (1976)

* Serotype
** Not described

In any case, it is clear that the cell-associated form of the enzyme plays a critical role in sucrose-dependent cell adherence to smooth surfaces. The problem will be discussed in more detail in Section 4.7.

4.6 STRUCTURAL ASPECTS OF POLYSACCHARIDES PRODUCED BY ORAL STREPTOCOCCI

Structural analyses of the polysaccharides, glucan and fructan, from selected strains of *S. mutans, S. sanguis* and *S. salivarius* have been performed by chemical and physical methods. The polysaccharides are synthesized from sucrose by the enzymatic action of GTase and fructosyltransferase (FTase) produced by these streptococcal species. The dextran-like glucans of *S. mutans* and *S. sanguis* play a role in the buildup of dental plaque on the surfaces of the teeth and the development of dental caries (Fitzgerald and Jordan, 1968; Gibbons and Banghart, 1967; Gibbons *et al.*, 1966). On the other hand, levan-like fructans serve as an extracellular storage substance which is capable of being split to fructose and metabolized by various bacterial species (Da Costa and Gibbons, 1968; Van Houte and Jansen, 1968).

4.6.1 Structure of the glucans

In the early phase of the investigations, it was erroneously presumed that glucans, based mainly on water-soluble fractions, from *S. mutans* cultures (Guggenheim and Schroeder, 1967; Gibbons and Banghart, 1968) and dental plaque (Critchley *et al.*, 1967) were 'dextrans'. The predominant linkage in dextran is α (1 → 6). Individual strains of micro-organisms such as *Leuconostoc mesenteroides* have been found to produce dextrans of different structure in which other types of glucosidic linkages are present; they are α (1 → 3), α (1 → 4) and α (1 → 2), and their proportions differ significantly among the various dextrans (Jeanes *et al.*, 1954; Wilham *et al.*, 1955). They showed a varying degree of branching at the 3 and 4 positions in the glucose residues. The dextran of simplest structure, which is produced by *L. mesenteroides* (strain NRRL B-512), possesses 95% of its glucose units linked α (1 → 6), with the remaining 5% linked α (1 → 3) (Lindverg and Svensson, 1968; Wilham *et al.*, 1955). Larm *et al.*, (1971) have reported that about 40% of the side chains contain only one D-glucose residue, at least 45% are two D-glucose units long and the rest are longer than two such glucose units.

Periodate oxidation studies (Lewicki *et al.*, 1971) of the glucans elaborated in sucrose-containing culture supernatant of *S. mutans* E49 (serotype *a*) indicated that ca. 70% of the D-glucose was α (1 → 6) linked, and 20% was α (1 → 3) linked. Similar results were obtained with the glucans prepared from *S. mutans* strains OMZ51 (serotype *b*) and Ingbritt (type *c*) and *S. sanguis* ATCC10558, indicating that about ∼ 80% of the glucose was in α (1 → 6) linkage (Sidebotham *et al.*, 1971). No significant structural differences between the glucans produced by *S. mutans* and *S. sanguis* were also found.

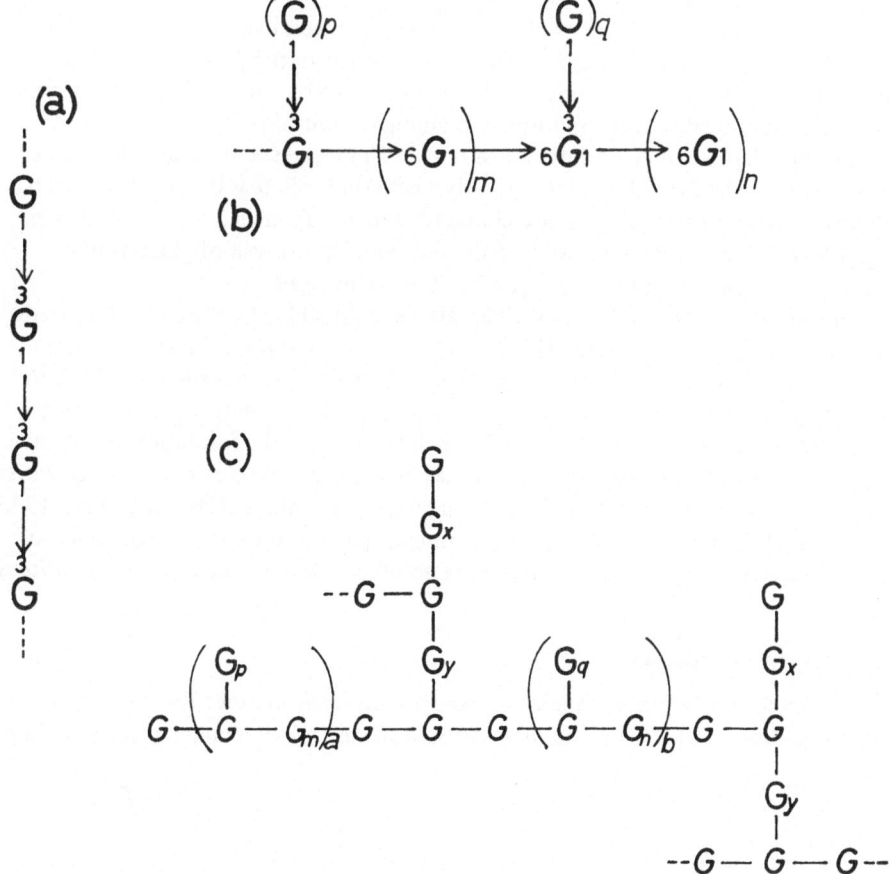

Fig. 4.5 Possible structure of water-insoluble (a), water-soluble (b), and native (c) glucans produced from sucrose by glucosyltransferases or oral streptococci. In (b), the values of p and q are usually 1 to 2. In (c), the relative values of a, b, m, n. p. q, x and y are dependent on the species or strain. (Modified from Hare *et al.* (1978)).

Significant portions of the glucan produced by *S. mutans* in sucrose culture were water-insoluble, which were removed by centrifugation with bacterial cells. To overcome the problem, glucosyltransferases forming water-soluble and water-insoluble glucans were obtained from culture supernatants and incubated with sucrose to permit the synthesis of cell-free water-insoluble glucans. Guggenheim (1970) has found that the insoluble glucans of *S. mutans* thus obtained are exclusively in α ($1 \rightarrow 3$) linear linkages, for which the term 'mutan' has been proposed (Fig. 5.5a). This result has been confirmed by other investigators (Ceska *et al.*, 1972; Baird *et al.*,

1973; Ebisu *et al.*, 1974; Hare *et al.*, 1978).

Recently, a study by Meyer *et al.*, (1978) using proton magnetic resonance has revealed that the average content of $\alpha (1 \to 6)$ linkages in the polymer fractions precipitating from solution during synthesis of glucans by various *S. mutans* strains was much lower than that of fractions remaining in solution.

It appears that a higher proportion of $\alpha (1 \to 3)$ linkages in the water-insoluble glucan explains the insolubility of this polysaccharide, whereas the soluble glucan as well as *L. mesenteroides* dextran are characterized as a branched $\alpha (1 \to 6)$ glucan (Fig. 4.5b). The water-insolubility and the extent of hydrolysis of glucans with dextranase decrease with the content of $\alpha (1 \to 3)$ linkages.

From the experimental data described above, a possible structure for the native glucan is illustrated in Fig. 4.5c. This generalization can explain the structure of glucans from *S. mutans* and *S. sanguis*. Typical *S. mutans* strains such as OMZ176 and K1R produce glucans in which $(x + y)$ are far greater than $(a + b)$, rendering glucans which are resistant to dextranase. Related to this, the formation of terminal $\alpha (1 \to 6)$ linkages in the insoluble glucan has been found to be essential for *S. mutans* to cause adherence to smooth surfaces by new glucan synthesis (Hamada *et al.*, 1975; Ebisu *et al.*, 1975). On the other hand, *S. sanguis* glucans possessing high values of $(m + n)$ and low values for $(x + y)$ are more soluble in water than those of *S. mutans*.

4.6.2 Structure of fructans

S. salivarius primarily produces a levan-like fructan from sucrose. The molecular weight is estimated at 2×10^7 by ultracentrifugation and viscometry (Newbrun and

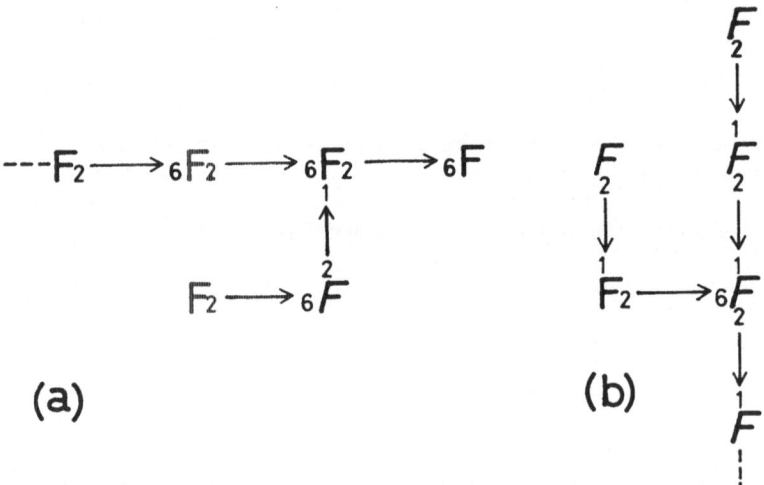

(a)   (b)

Fig. 4.6 Possible structure of fructans produced by *S. salivarius* (a) and *S. mutans* (b). (Modified from Ebisu *et al.* (1975)).

Baker, 1968) and 3×10^7 by light scattering (Stivala *et al.*, 1976). The fructan contains β-D-fructofuranose residues linked through positions 2 and 6, as well as 1, 2, and 6 (Fig. 4.6a) (Ebisu *et al.*, 1975; Hancock *et al.*, 1976). The β (2 → 6) fructofuranoside structure has also been found in fructans from other bacterial species (Feingold and Gehatia, 1957; Rapoport and Dedonder, 1966).

Some strains of *S. mutans* also synthesize fructans in addition to glucans from sucrose (Schachtele *et al.*, 1972; Robrish *et al.*, 1972; Hamada *et al.*, 1975; Mukasa and Slade, 1973). Baird *et al.* (1973) have first indicated that a 'water-soluble' fructan from an *S. mutans* strain has inulin-type β (2 → 1) fructofuranoside but not a β (2 → 6) fructofuranoside structure. Ebisu *et al.* (1975) also found a similar structure in the water-insoluble fructans from two strains of *S. mutans* (Fig. 4.6b). It appears that there is no definite correlation between water-solubility and differences in the chemical structure of these fructans.

4.7 REDUCTION OF BACTERIAL ADHERENCE TO TOOTH SURFACES

Procedures which suppress or eliminate *S. mutans* from tooth surfaces are expected to cause a reduction in the development of dental caries (Keyes and Shern), 1971; Parsons, 1974). Measures can be classified into two categories. (1) Application of antibiotics and antiseptics has been attempted with varying success to suppress specifically or non-sepecifically oral microflora (Fitzgerald, 1973; Loesche, 1976). (2) Another approach is to block the pathogenic process rather than to attack the organisms themselves. Thus, attempts are being made to inhibit adherence to tooth surfaces by administration of glucanase and by immunizing the susceptible host with various vaccines. In this section, the mechanisms of the latter two entities will be discussed in more detail.

4.7.1 Inhibition of adherence by glucanases

Fitzgerald *et al.* (1968) first demonstrated that application of a dextranase, an α (1 → 6) glucanase from *Penicillium funiculosum*, prevented plaque formation and dental caries in hamsters. However, Guggenheim *et al.* (1969, 1972) reported that dextranase had no anti-caries effect in their rat system while mutanase, an α (1 → 3) glucanase (Guggenheim and Haller, 1972) from *Trichoderma harzianum*, exerted a marked inhibitory effect. The difference in the result might be due to a difference in the animal system used, their indigenous flora, the activity of enzymes employed, and the manner of administration of diet and water.

Subsequent studies using monkeys (Bowen, 1971) and rats and hamsters (Hamada *et al.*, 1976a, 1976b), however, have confirmed earlier findings of Fitzgerald *et al.* (1968, 1973) that dextranase markedly inhibited caries development. Dextranase from *Spicaria violacea* prevented the formation of *in vitro* plaque as well as the formation of cell-bound water-insoluble glucan by *S. mutans* (Hamada *et al.*, 1975). These studies suggest that α (1 → 6) glucosidic linkages of water-insoluble glucan which exists at the interface between the bulk of the bacterial deposit and the outer tooth surface may be essential to anchor the bacterial deposit, i.e. dental plaque (Fitzgerald, 1973). The development of smooth surface caries was more effectively reduced by dextranase than that of pit-and-fissure caries. This may reflect inhibition of the sucrose-dependent adherence of *S. mutans* to smooth surfaces.

A significant proportion of oral bacteria in dental plaque have been found to elaborate dextranase *in vitro*. Bacteria so far shown to produce this enzyme are *S. mutans* (Mäkinen and Paunio, 1971; Staat and Schachtele, 1974; Giggenheim and Burckhardt, 1974; Dewar and Walker, 1975; Ellis and Miller, 1977), *Actinomyces israelii* (Staat and Schachtele, 1975), *Bacteroides ochraceus* (Schachtele *et al.*, 1975) and *Fusobacterium* sp. (Da Costa *et al.*, 1974). Schachtele *et al.* (1976b) reported that low levels of dextranase from either *A. israelii* or *B. ochraceus* reduced sucrose-dependent adherence of *S. mutans* to be a glass surface by 80%. It has also been shown that the presence of dextranase during the synthesis of glucans by *S. mutans* and *S. sanguis* from sucrose affects the amount and type of glucans synthesized (Walker, 1972; Ebisu *et al.*, 1974; Schachtele *et al.*, 1975; Hamada *et al.*, 1975; Linzer and Slade, 1976). However, the *in vivo* role of dextranase produced by oral bacteria has not been established.

4.7.2 Inhibition of adherence by immunization

Rabbit antisera elaborated against *S. mutans* whole cells have been found to inhibit the adherence of *S. mutans* to smooth surfaces *in vitro* by aggregating the bacteria (Olson *et al.*, 1972; Mukasa and Slade, 1973, 1974a; Genco *et al.*, 1974; Hamada and Torii, 1978; Hamada and Slade, 1976). Some of the whole cell antisera also possess activity which is inhibitory to extracellular GTase (Carlsson and Krasse, 1968; Evans and Genco, 1973; Evans *et al.*, 1976; Hamada and Slade, 1976; Smith and Taubman, 1977). Inhibition of adherence by anti-glucan antibody is also effective (Mukasa and Slade, 1974a).

It has been reported that antisera against crude or partially purified extracellular GTase preparations almost completely inhibited the GTase activity (Fukui *et al.*, 1974b; Linzer and Slade, 1976; Kuramitsu and Ingersoll, 1976; Smith and Taubman, 1977). Hamada *et al.* (1979) have recently found that the adherence of non-growing cells of *S. mutans* to a glass surface via new glucan synthesis by cell-associated GTase from sucrose was also significantly inhibited by antiserum to extracellular GTase. The anti-GTase serum did not cause cell aggregation, indicating the absence of antibody to the cell wall polysaccharide antigens.

Human salivary immunoglobulin IgA has been found to inhibit extracellular GTase activity prepared from a cariogenic strain of *S. sanguis* (Klein *et al.*, 1977, 1978). GTase activity from *S. mutans* HS6 (serotype *a*) was reported to be suppressed by secretory IgA or secretory component (Cole *et al.*, 1976). This result does not support the initial report of Fukui *et al.*, (1974a), which has described a seven-fold stimulation of GTase by either secretory IgA or secretory components. Related to this, Williams and Gibbons (1972) reported that human parotid secretory IgA interfered with bacterial adherence to epithelial surfaces *in vitro*, suggesting that the effect also operates *in vivo*. These results in *in vitro* studies may provide a rationale for *in vivo* immunological attempts of caries prevention.

Although an unsuccessful trial at active immunization of hamsters with *S. mutans* HS6 was reported by Fitzgerald and Keyes (1962), several studies have shown varying degrees of success in the reduction of dental caries. Since earlier studies have been extensively reviewed by Taubman and Smith (1974a, b) and Bowen (1976), attention will be given to a description of immunization studies in animals, which include *in vivo* adherence by the inoculated *S. mutans*.

In a study on the effects of immunization with *S. mutans* in rhesus monkeys maintained on a human type of cariogenic diet, Lehner *et al.* (1975a, b) have found that the incidence of smooth surface caries was reduced by subcutaneous or submucous immunization with a heat-treated *S. mutans* in Freund incomplete adjuvant. They also revealed a significant reduction in the numbers of *S. mutans* in plaque, crevicular fluid and in saliva of immunized animals, and there was an associated reduction in dental caries. Immunization also resulted in a delay in initial colonization and a slowing of the rate of colonization with *S. mutans* (Caldwell *et al.*, 1977). A significant increase in salivary antibodies to *S. mutans* could not be demonstrated. Since serum antibody titers were elevated in the immunized animals, protection seemed to correlate with serum IgG antibodies rather than salivary IgA antibodies.

In contrast to these studies, parotid ductal immunization of irus monkeys with killed *S. mutans* cells elicited high titers of both salivary IgA antibodies and serum IgG antibodies (Emmings *et al.*, 1975), and the immunized monkeys had fewer organisms on the infected surfaces than un-immunized control animals (Evans *et al.*, 1975). These authors concluded that the protective effect was due to the salivary IgA rather than to antibodies which might diffuse into crevicular fluid from the serum. Furthermore, rats injected with killed cells of *S. mutans* (Taubman and Smith, 1974c; McGhee *et al.*, 1975) or cell-free GTase (Taubman and Smith, 1977) in Freund complete adjuvant developed salivary antibodies and they always had lower caries scores than sham-immunized or non-immunized control groups. But it was difficult to demonstrate consistent reduction in recovery of *S. mutans* (Taubman and Smith, 1974a, b, 1977). More recently, however, Michalek *et al.* (1978) have found that the presence of specific salivary IgA antibodies in rats correlated with a reduction in the development of dental caries, accompanying a reduction in the levels of plaque scores and numbers of *S. mutans* in plaque. Krasse and Jordan (1977) reported that the hamsters, mice and rats in which the vaccines had been topically applied showed fewer numbers of bacteria than control animals one day after infection challenge.

As briefly summarized above, many investigators have reached a similar

conclusion that the presence of antibodies in the oral fluids correlated with a marked
reduction in the number of infected loci and the numbers of *S. mutans* on infected
tooth surfaces. However, the exact mechanism whereby antibody can interfere with
the pathogenic processes of dental caries caused by *S. mutans* is not well understood.
There are still many points which must be clarified with regard to caries immunity.

4.8 CONCLUDING REMARKS AND SUMMARY

Dental caries have recently been a severe social problem because of its ubiquitousness
in civilized populations. Among many oral bacterial species, *S. mutans* was identified
as a plaque-forming organism capable of producing dental caries in various experimental animals and most probably in humans. Extensive studies reveal that strains
of *S. mutans* are heterogenous on the basis of their immunological and genetical
specificities. However, they are relatively homogenous in terms of their physiological
properties including pathogenic potentials.

Special attention is paid in this chapter to the mechanism of, and factors which
relate to, the adherence of *S. mutans* to a smooth surface in the presence of sucrose.
Although the process of adherence is seemingly simple, the molecular mechanisms
of the events, even *in vitro*, are complex.

The adherence of *S. mutans* is mediated by the enzymatic action of GTases
capable of synthesizing water-insoluble adherent glucans. The GTases are found in
the cell-free form and in a cell-associated form. Sucrose seems to be critical in determining the ratio of the two forms of the enzyme, and the development of adherence
to smooth surfaces. Evidence indicates that salivary glycoprotein functions in the
initial attachment of *S. mutans* to the tooth surface.

S. mutans can be differentiated from most other bacteria by its ability to bind
GTase. The GTase binding site most probably contains glucan closely associated
with surface proteins of the cell. The fact that cell-free water-insoluble glucans bind
GTase and foster adherence in the presence of sucrose supports the above concept.
Related to this 'purified' GTase preparations have been found to contain significant
quantities of hexose.

Many strains of *S. mutans* are agglutinated upon addition of high molecular
weight dextran, but agglutination and sucrose-dependent adherence to smooth
surfaces appear to be separate traits. The latter property appears to be more closely
associated with cariogenicity. Certain clinical strains which are not agglutinated by
high molecular-weight dextran adhere in the presence of sucrose and produce caries
in animals; conversely there are mutant strains which were agglutinated by dextran
but do not adhere or produce caries in animals.

Blocking of the pathogenic process, i.e. adherence by *S. mutans* to tooth surfaces,
should be an effective measure for the prevention of caries. Application of various
glucanases and immunization procedures in laboratory animals have been shown to
be effective in reducing the incidence of caries and the degree of colonization by

S. mutans. Results of these studies should provide useful clues for future research and development, but it is apparent that many problems remain to be resolved. As yet, no highly effective measure for the prevention of dental caries is available.

REFERENCES

Baird, J.K., Longyear, V.M.C. and Ellwood, D.C. (1973), *Microbios.*, **8**, 143–150.
Beachey, E.H. and Ofek, I. (1976), *J. exp. Med.*, **143**, 759–771.
Bowen, W.H. (1971), *Br. dent. J.*, **131**, 445–449.
Bowen, W.H. (1976), In: *Immunologic Aspects of Dental Caries* (Bowen, W.H., Genco, R.J. and O'Brien, T.C., eds.), pp. 11–20, Information Retrieval, Washington, D.C.
Bourgeau, G. and McBride, B.C. (1976), *Infect. Immun.*, **13**, 1228–1234.
Caldwell, J., Challacombe, S.J. and Lehner, T. (1977), *J. med. Microbiol.*, **10**, 213–224.
Carlsson, J. and Krasse, B. (1968), *Arch. oral Biol.*, **13**, 849–852.
Carlsson, J., Newbrun, E. and Krasse, B. (1969), *Arch oral. Biol.*, **14**, 469–478.
Ceska, M., Granath, K., Norrman, B. and Guggenheim, B. (1972), *Acta chem. scand.*, **26**, 2223–2230.
Chassy, B.M., Beall, J.R., Bielawski, R.M., Porter, E.V. and Donkersloot, J.A. (1976), *Infect. Immun.*, **14**, 408–415.
Ciardi, J.E. (1976), In: *Immunologic Aspects of Dental Caries* (Bowen, W.H., Genco, R.J. and O'Brien, T.C., eds.), pp. 101–110, Information Retrieval, Washington, D.C.
Cisar, J.O., Vatter, A.E. and McIntire, F.C. (1978), *Infect. Immun.*, **19**, 312–319.
Cole, J.S., Clark, G.E. and Wistar, R. (1976), *J. Bact.*, **127**, 1595–1596.
Critchley, P., Wood, J.M., Saxton, C.A. and Leach, S.A. (1967), *Caries Res.*, **1**, 112–129.
Da Costa, T., Bier, L.C. and Gaida, F. (1974), *Arch. oral. Biol.*, **19**, 341–342.
Da Costa, T. and Gibbons, R.J. (1968), *Arch. oral. Biol.*, **13**, 609–617.
Dewar, M.D. and Walker, G.J. (1975), *Caries Res.*, **9**, 21–35.
Ebisu, S., Kato, K., Kotani, S. and Misaki, A. (1975), *J. Biochem.*, **78**, 879–887.
Ebisu, S., Misaki, A., Kato, K. and Kotani, S. (1974), *Carbohydr. Res.*, **38**, 374–381.
Ellen, R.P. and Balcerzak-Raczkowski, I.B. (1977), *J. periodontal Res.*, **12**, 11–20.
Ellis, D.W. and Miller, C.H. (1977), *J. dent. Res.*, **56**, 57–69.
Emmings, F.G., Evans, R.T. and Genco, R.J. (1975), *Infect. Immun.*, **12**, 281–292.
Evans, R.T., Emmings, F.G. and Genco, R.J. (1975), *Infect. Immun.*, **12**, 293–302.
Evans, R.T. and Genco, R.J. (1973), *Infect. Immun.*, **7**, 237–241.
Evans, R.T., Genco, R.J. and Emmings, F.G. (1976), *J. dent. Res.*, **55**, C127–C133.
Feingold, D.S. and Gehatia, M. (1957), *J. Polymer Sci.*, **23**, 783–790.
Fitzgerald, R.J. (1973), *J. Am. Dent. Ass.*, **87**, 1006–1009.
Fitzgerald, R.J., Fitzgerald, D.B. and Stoudt, T.H. (1973), In: *Germfree Research* (Heneghan, J.B., ed.), pp. 197–203, Academic Press, New York.
Fitzgerald, R.J. and Jordan, H.V. (1968), In: *Art and Science of Dental Caries Research* (Harris, R.S., ed.), pp. 79–86, Academic Press, New York.

Fitzgerald, R.J. and Keyes, P.H. (1962), Abstr. 40th Gen. Meet. Int. Ass. dent. Res., Abstr. #146.

Fitzgerald, R.J., Keyes, P.H., Stoudt, T.H. and Spinell, D.M. (1968), *J. Am. dent. Ass.*, **76**, 301–304.

Freedman, M.L. and Tanzer, J.M. (1974), *Infect. Immun.*, **10**, 189–196.

Fukui, K., Fukui, Y. and Moriyama, T. (1974a), *Infect. Immun.*, **10**, 985–990.

Fukui, K., Fukui, Y. and Moriyama, T. (1974b), *J. Bact.*, **118**, 805–809.

Genco, R.J., Evans, R.T. and Taubman, M.A. (1974), *Adv. exp. Med. Biol.*, **45**, 327–336.

Germaine, G.R., Chludzinski, A.M. and Schachtele, C.F. (1974), *J. Bact.*, **120**, 287–294.

Germaine, G.R. and Schachtele, C.F. (1976), *Infect. Immun.*, **13**, 365–372.

Gibbons, R.J. and Banghart, S.B. (1967), *Arch. oral. Biol.*, **12**, 11–24.

Gibbons, R.J. and Banghart, S.B. (1968), *Arch. oral. Biol.*, **13**, 697–701.

Gibbons, R.J., Berman, K.S., Knoettner, P. and Kapsimalis, B. (1966), *Arch. oral. Biol.*, **11**, 549–560.

Gibbons, R.J., De Paola, P.F., Spinell, D.M. and Skobe, Z. (1974), *Infect. Immun.*, **9**, 481–488.

Gibbons, R.J. and Fitzgerald, R.J. (1969), *J. Bact.*, **98**, 341–346.

Gibbons, R.J. and Nygaard, M. (1968), *Arch. oral. Biol.*, **13**, 1249–1262.

Gibbons, R.J. and Van Houte, J. (1973), *J. Periodontol.*, **44**, 347–360.

Gibbons, R.J. and Van Houte, J. (1975a), *A. Rev. Med.*, **26**, 121–136.

Gibbons, R.J. and Van Houte, J. (1975b), *A. Rev. Microbiol.*, **29**, 19–44.

Guggenheim, B. (1970), *Helv. odont. Acta*, **14**, 89–108.

Guggenheim, B. and Burckhardt, J.J. (1974), *Helv. odont. Acta*, **18**, 101–113.

Guggenheim, B. and Haller, R. (1972), *J. dent. Res.*, **51**, 394–402.

Guggenheim, B. and Newbrun, E. (1969), *Helv. odont. Acta*, **13**, 84–97.

Guggenheim, B., Regolati, B. and Mühlemann, H.R. (1972), *Caries Res.*, **6**, 289–297.

Guggenheim, B. and Schroeder, H.E. (1967), *Helv. odont. Acta*, **11**, 131–152.

Hamada, S. (1977), *Microbios Letters*, **5**, 141–146.

Hamada, S., Gill, K. and Slade, H.D. (1977), *Infect. Immun.*, **18**, 708–716.

Hamada, S., Mizuno, J., Murayama, Y., Ooshima, T., Masuda, N. and Sobue, S. (1975), *Infect. Immun.*, **12**, 1415–1425.

Hamada, S., Ooshima, T., Masuda, N., Mizuno, J. and Sobue, S. (1976a), *Jap. J. Microbiol.*, **20**, 321–330.

Hamada, S., Ooshima, T., Masuda, N. and Sobue, S. (1976b), *J. dent. Res.*, **55**, 552.

Hamada, S., Ooshima, T., Torii, M., Imanishi, H., Masuda, N., Sobue, S. and Kotani, S. (1978a), *Microbiol. Immunol.*, **22**, 301–314.

Hamad, S. and Slade, H.D. (1976), *J. dent. Res.*, **55**, C65–C74.

Hamada, S. and Slade, H.D. (1979), *Arch. oral Biol.*, In press.

Hamada, S., Tai, S. and Slade, H.D. (1978b), *Infect. Immun.*, **21**, 213–220.

Hamada, S., Tai, S. and Slade, H.D. (1979), *Microbiol. Immunol.*, **23**, 61–70.

Hamada, S. and Torii, M. (1978), *Infect. Immun.*, **20**, 592–599.

Hamada, S. and Torii, M. (1979), *J. gen. Microbiol.*, In press.

Hancock, R.A., Marshall, K. and Weigel, H. (1976), *Carbohydr. Res.*, **49**, 351–360.

Hare, M.D., Svensson, S. and Walker, G.J. (1978), *Carbohydr. Res.*, **66**, 245–264.

Harlander, S.K. and Schachtele, C.F. (1978), *Infect. Immun.*, **19**, 450–456.

Hillman, J.D., Van Houte, J. and Gibbons, R.J. (1970), *Arch. oral Biol.*, **15**, 899–903.

Janda, W.M. and Kuramitsu, H.K. (1976), *Infect. Immun.*, **14**, 191–202.

Janda, W.M. and Kuramitsu, H.K. (1977), *Infect. Immun.*, **16**, 575–586.
Jeanes, A., Haynes, W.C., Wilham, C.A., Rankin, J.C., Melvin, E.H., Austin, M.J., Cluskey, J.E., Fisher, B.E., Tsuchiya, H.M. and Rist, C.E. (1954), *J. Am. Chem. Soc.*, **76**, 5041–5052.
Jordan, H.V. and Keyes, P.H. (1966), *Arch. oral. Biol.*, **11**, 793–801.
Kelstrup, J. and Funder-Nielsen, T.D. (1974), *J. gen. Microbiol.*, **81**, 485–489.
Keyes, P.H. and Shern, R. (1971), *J. Am. Soc. prev. Dent.*, 1, 18–22.
Klein, J.P., Schöller, M. and Frank, R.M. (1977), *Infect. Immun.*, **15**, 329–331.
Klein, J.P., Schöller, M. and Frank, R.M. (1978), *Infect. Immun.*, **20**, 619–626.
Koga, T. and Inoue, M. (1978), *Infect. Immun.*, **19**, 402–410.
Krasse, B. and Jordan, H.V. (1977), *Arch. oral. Biol.*, **22**, 479–484.
Kuramitsu, H.K. (1974a), *Infect. Immun.*, **9**, 764–765.
Kuramitsu, H.K. (1974b), *Infect. Immun.*, **10**, 227–235.
Kuramitsu, H.K. (1975), *Infect. Immun.*, **12**, 738–749.
Kuramitsu, H.K. and Ingersoll, L. (1976), *Infect. Immun.*, **14**, 636–644.
Kuramitsu, H.K. and Ingersoll, L. (1978), *Infect. Immun.*, **20**, 652–659.
Larm, O., Lindberg, B. and Svensson, S. (1971), *Carbohydr. Res.*, **20**, 39–48.
Lasslo, A. and Quintana, R.P. (1973), *Surface Chemistry and Dental Integuments*, Charles C. Thomas, Springfield, IL.
Lehner, T., Challacombe, S.J. and Caldwell, J. (1975a), *Nature*, **254**, 517–520.
Lehner, T., Challacombe, S.J. and Caldwell, J. (1975b), *Arch. oral. Biol.*, **20**, 305–310.
Lewicki, W.J., Long, L.W. and Edwards, J.R. (1971), *Carbohydr. Res.*, **17**, 175–182.
Lindberg, B. and Svensson, S. (1968), *Acta chem. scand.*, **22**, 1907–1912.
Linzer, R. and Slade, H.D. (1976), *Infect. Immun.*, **13**, 494–500.
Loesche, W.J. (1976), *Oral Sci. Rev.*, **9**, 65–107.
Mäkinen, K.K. and Paunio, I.K. (1971), *Analyt. Biochem.*, **39**, 202–207.
McBride, B.C. and Bourgeau, G. (1975), *Arch. oral. Biol.*, **20**, 837–841.
McCabe, M. and Hamelik, R.M. (1978), *Secretory Immunity and Infection*, (McGhee, J.R., Mestecky, J. and Babb, J.L., eds.), pp. 749–759.
McCabe, M.M., Haynes, A.U. and Hamelik, R.M., (1976), In: *Microbial Aspects of Dental Caries* (Stiles, H.M., Loesche, W.J. and O'Brien, T.C., eds.), pp. 413–424.
McCabe, M.M. and Smith, E.E. (1973), *Infect. Immun.*, **7**, 829–838.
McCabe, M.M. and Smith, E.E. (1975), *Infect. Immun.*, **12**, 512–520.
McCabe, M.M. and Smith, E.E. (1977), *Infect. Immun.*, **16**, 760–765.
McCabe, R.M., Keyes, P.H. and Howell, A. Jr. (1967), *Arch. oral. Biol.*, **12**, 1653–1656.
McGhee, J.R., Michalek, S.M., Webb, J., Navia, J.M., Rahmen, A.F.R. and Legler, D. (1975), *J. Immunol.*, **114**, 300–305.
McHugh, W.D. (1970), *Dental Plaque*, D.C. Thomson, Ltd., Dundee.
McIntire, F.C., Vatter, A.E., Baros, J. and Arnold, J. (1978), *Infect. Immun.*, **21**, 978–988.
Meyer, T., Lamberts, B. and Egan, R.S. (1978), *Carbohydr. Res.*, **66**, 33–42.
Michalek, S.M., McGhee, J.R. and Babb, J.L. (1978), *Infect. Immun.*, **19**, 217–224.
Montville, T.J., Cooney, C.L. and Sinskey, A.J. (1977), *Infect. Immun.*, **18**, 629–635.

Mukasa, H. and Slade, H.D. (1973), *Infect. Immun.*, **8**, 555–562.

Mukasa, H. and Slade, H.D. (1974a), *Infect. Immun.*, **9**, 419–429.

Mukasa, H. and Slade, H.D. (1974b), *Infect. Immun.*, **10**, 1135–1145.

Nalbandian, J., Freedman, M.L., Tanzer, J.M. and Lovelace, S.M. (1974), *Infect. Immun.*, **10**, 1170–1179.

Newbrun, E. (1976), In: *Microbial Aspects of Dental Caries* (Stiles, H.M., Loesche, W.J. and O'Brien, T.C., eds.), pp. 649–664, Information Retrieval, Washington, D.C.

Newbrun, E. and Baker, S. (1968), *Carbohydr. Res.*, **6**, 165–170.

Newbrun, E., Finzen, F. and Sharma, M. (1977), *Caries Res.*, **11**, 153–159.

Olson, G.A., Bleiweis, A.S. and Small, P.A. Jr. (1972), *Infect. Immun.*, **5**, 419–427.

Parsons, J.C. (1974), *J. Peridontol.*, **45**, 177–186.

Rapoport, G. and Dedonder, R. (1966), *Bull. Soc. Chim. Biol.*, **48**, 1349–1357.

Robrish, S.A., Reid, W. and Krichevsky, M.I. (1972), *Appl. Microbiol.*, **24**, 184–190.

Rölla, G. (1977), *Caries Res.*, **11** (Suppl. 1), 243–261.

Scales, W.R., Long, L.W. and Edwards, J.R. (1975), *Carbohydr. Res.*, **42**, 325–338.

Schachtele, C.F., Harlander, S.K., Fuller, D.W., Zollinger, P.K. and Leung, W. L.S. (1976a) In: *Microbial Aspects of Dental Caries* (Stiles, H.M., Loesche, W.J. and O'Brien, T.C., eds.), pp. 401–412, Information Retrieval, Washington, D.C.

Schachtele, C.F., Harlander, S.K. and Germaine, G.R. (1976b), *Infect. Immun*, **13**, 1522–1524.

Schachtele, C.F., Loken, A.E. and Schmitt, M.K. (1972), *Infect. Immun.*, **5**, 263–266.

Schachtele, C.F., Staat, R.H. and Harlander, S.K. (1975), *Infect. Immun.*, **12**, 309–317.

Scherp, H.W. (1971), *Science*, **173**, 1199–1205.

Sidebotham, R.L., Weigel, H. and Bowen, W.H. (1971), *Carbohydr. Res.*, **19**, 151–159.

Slade, H.D. (1976), In: *Immunologic Aspects of Dental Caries* (Bowen, W.H., Genco, R.J. and O'Brien, T.C., eds.), pp. 21–38, Information Retrieval, Washington, D.C.

Slade, H.D. (1977), In: *Microbiology 1977* (Schlessinger, D., ed.), pp. 411–416, American Society for Microbiology, Washington, D.C.

Slade, H.D. (1978), *Secretory Immunity and Infection*, (McGhee, J.R., Mestecky, J. and Babb, J.L., eds.), pp. 761–763.

Smith, D. and Taubman, M.A. (1977), *Infect. Immun.*, **15**, 91–103.

Spinell, D.M. and Gibbons, R.J. (1974), *Infect. Immun.*, **10**, 1448–1451.

Staat, R.H., Doyle, R.J., Langley, S.D. and Suddick, R.P. (1978), Secretory Immunity and Infection (McGhee, J.R., Mestecky, J. and Babb, J.L., eds.), pp. 639–647.

Staat, R.H. and Schachtele, C.F. (1974), *I..fect. Immun.*, **9**, 467–469.

Staat, R.H. and Schachtele, C.F. (1975), *Infect. Immun.*, **12**, 556–563.

Stiles, H.M., Loesche, W.J. and O'Brien, T.C. (1976), *Microbial Aspects of Dental Caries*, Information Retrieval, Washington, D.C.

Stivala, S.S., Bahary, W.S., Long, L.W., Ehrlich, J. and Newbrun, E. (1976), *Biopolymers*, **14**, 1283–1292.

Sudo, S.Z., Gutfleisch, J.R., Schotzko, N.K. and Folke, L.E.A. (1975), *Infect. Immun.*, **12**, 576–585.

Tanzer, J.M. and McCabe, R.M. (1968), *Arch. oral. Biol.*, 13, 139—143.

Taubman, M.A. and Smith, D.J. (1974a) In: *Symposium on the Mechanisms of Exocrine Secretion* (Han, S., Sreebny, L. and Suddick, R., eds.), pp. 152—172, Univ. Michigan Press, Ann Arbor.

Taubman, M.A. and Smith, D.J. (1974b), *Infect. Immun.*, 9, 1079—1091.

Taubman, M.A. and Smith, D.J. (1977), *J. Immunol.*, 118, 710—720.

Terleckyj, B., Willett, N.P. and Shockman, G.D. (1975), *Infect. Immun.*, 11, 649—655.

Umesaki, Y., Kawai, Y. and Mutai, M. (1977), *Appl. environ. Microbiol.*, 34, 115—119.

Van Houte, J. and Green, D.B. (1974), *Infect. Immun.*, 9, 624—630.

Van Houte, J. and Jansen, H.M. (1968), *Arch. oral. Biol.*, 13, 827—830.

Van Houte, J. and Upeslacis, V.N. (1976), *J. dent. Res.*, 55, 216—222.

Walker, G.J. (1972), *J. dent. Res.*, 51, 409—414.

Wilham, C.A., Alexander, B.H. and Jeanes, A. (1955), *Arch. Biochem. Biophys.*, 59, 61—75.

Williams, R.C. and Gibbons, R.J. (1972), *Science*, 177, 697—699.

Wood, J.M. (1967), *Arch. oral. Biol.*, 12, 1659—1660.

Wu-Yuan, C., Tai, S. and Slade, H.D. (1978), *Secretory Immunity and Infection*, (McGhee, J.R., Mestecky, J. and Babb, J.L. eds.), pp. 737—748, Plenum Publishing Co., New York.

Wu-Yuan, C., Tai, S. and Slade, H.D. (1979), *Infect. Immun.*, In press.

5 Structure and Cell Membrane-Binding Properties of Bacterial Lipoteichoic Acids and their Possible Role in Adhesion of Streptococci to Eukaryotic Cells

ANTHONY J. WICKEN

5.1	Lipoteichoic acids	*page*	139
	5.1.1 Occurrence and structure		139
	5.1.2 Extraction and purification of lipoteichoic acids		141
	5.1.3 Location of lipoteichoic acids in the bacterial cell		143
5.2	Binding of lipoteichoic acids to eucaryotic cells		147
	5.2.1 Erythrocytes		147
	5.2.2 Other mammalian cells		151
5.3	Lipoteichoic acid and its possible involvement in the attachment of bacteria to the surfaces of animal cells		153
	References		156

Bacterial Adherence
(*Receptors and Recognition*, Series B, Volume 6)
Edited by E.H. Beachey
Published in 1980 by Chapman and Hall, 11 New Fetter Lane, London EC4P 4EE
© 1980 Chapman and Hall

5.1 LIPOTEICHOIC ACIDS

5.1.1 Occurrence and structure

The name teichoic acid first denoted a group of polyolphosphate-containing polymers isolated either from the cell walls of gram-positive bacteria or from the intracellular contents of disrupted organisms (Armstrong *et al.*, 1958). Lipoteichoic acids are distinguished from the earlier known cell wall teichoic acids on the basis of their struture. Teichoic acids are either glycosylated ribitol or glycerol phosphate polymers usually bearing D-alanyl ester substituents on hydroxyl groups and linked covalently to peptidoglycan (Archibald, 1974). Extension of the term to cover all bacterial cell wall and capsular polymers that contain glycerol or ribitol phosphate residues has been suggested (Baddiley, 1972) and results in a considerable structural diversity for this class of microbial component. Teichoic acids do not appear as constituents of all gram-positive bacteria and their presence is often dependent on culturing conditions— there are now several examples known of organisms grown under conditions of phosphate limitation replacing wall teichoic acid with acidic polysaccharides (Ellwood and Tempest, 1972). Lipoteichoic acids occur in a much wider range of gram-positive bacteria (Archibald, 1974; Knox and Wicken, 1973; Wicken and Knox, 1975a) and are much less subject to replacement by changed culturing conditions. Lipoteichoic acids are always of the glycerol phosphate polymer type covalently linked to a lipid moiety. The latter may take the form of a glycolipid, as in the lactobacilli and some streptococci (Wicken and Knox, 1970; Knox and Wicken, 1973; Wicken and Knox, 1975a; Wicken and Knox, 1975b), a phosphatidylglycolipid as in *Streptococcus faecalis* (Toon *et al.*, 1972; Ganfield and Pieringer, 1975), or simply fatty acid substitution of a glycerol moiety at one end of a polyglycerophosphate chain as has been reported in a mutant of *Bacillus licheniformis* (Button and Hemmings, 1976). Fig. 5.1 shows schematically the structures of a number of lipoteichoic acids ranging from species with non-glycosylated polyglycerophosphate chains (25–30 glycerophosphate units long) such as *L. casei* to heavily glycosylated chains (*Streptococcus faecalis*). D-Alanyl ester substituents are omitted for clarity but these are generally present on either glycerol or sugar hydroxyl groups. The lability of the ester bond makes for a variable degree of substitution with D-alanine depending on the severity of the extraction procedure used (see below). The lipid moiety of lipoteichoic acid is generally also found as a free lipid constituent of the plasma membrane. Fatty acid substitution generally reflects the overall fatty acid composition of the membrane lipids of a particular species and, in this sense, lipoteichoic acid preparations are inevitably heterogeneous.

Not all gram-positive species of bacteria contain lipoteichoic acids, *Actinomyces*

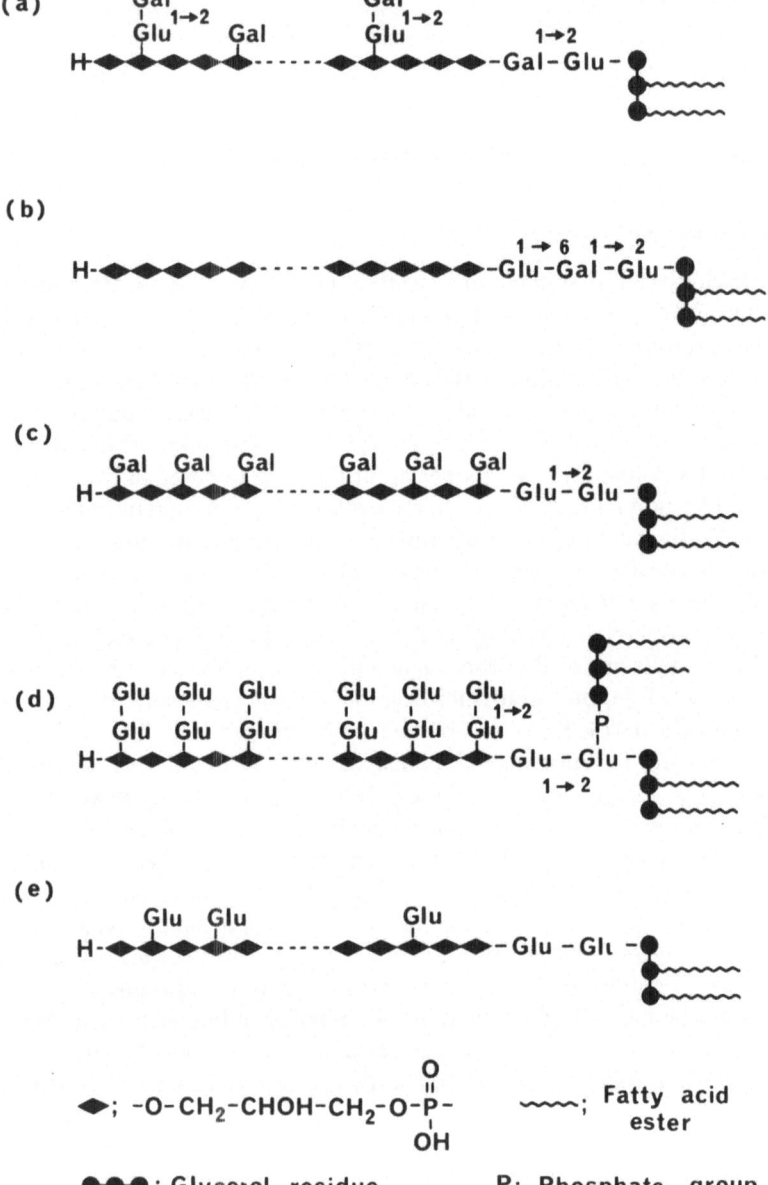

Fig. 5.1 Diagrammatic representation of the proposed structures of lipoteichoic acids from several gram-positive bacteria. (Bibliography in text).
(a) *Lactobacillus fermentum*;   (b) *L. casei*;   (c) *Streptococcus lactis*;
(d) *S. faecalis*;   (e) *L. helveticus*.

and *Micrococcus* being notable genera that lack this class of molecule. *Actinomyces* species, however, have been shown to contain fatty acid substituted heteropolysaccharides (Wicken *et al.*, 1978a) and *Micrococcus* species produce lipomannans which, in terms of an amphipathic molecular constituent, probably take the place of lipoteichoic acids in these genera (Owens and Salton, 1975; Powell *et al.*, 1975). Strains of *Streptococcus mitis* have also been reported to lack lipoteichoic acids recently (Rosan, 1978), but it is not known whether these organisms possess other amphipathic species. It seems likely that bacteria generally contain some form of amphipathic molecule at or near their surface (Wicken and Knox, 1978b). In gramnegative bacteria, this role is seen most readily in the lipopolysaccharides and lipoproteins of the outer membrane of the cell wall.

5.1.2 Extraction and purification of lipoteichoic acids

Early studies on teichoic acids involved extraction with cold 10% trichloracetic acid (Armstrong *et al.*, 1958; Archibald, 1974), which hydrolysed cell wall teichoic acid chains from their covalent association with peptidoglycan. A more extended survey (Baddiley and Davison, 1961) showed the presence of glycerol teichoic acids in similar trichloracetic acid extracts of the cytoplasmic contents of disrupted cells, after removal of cell wall fragments, in a wide variety of gram-positive organisms. Originally termed 'intracellular teichoic acids' since they clearly were not found in cell walls, these polymers were considered to be associated with the cell membrane, either still being associated with the cell membrane or released from it during protoplast formation from a number of species of gram-positive bacteria (Hay *et al.*, 1963; Smith and Shattock, 1964; Shockman and Slade, 1964). The nature of this projected membrane association remained unknown at the time and it should be noted that purification of intracellular or membrane teichoic acids gave products of the same order of molecular size as their trichloracetic acid-extracted cell wall counterparts.

Wicken and Knox (1971) first observed that extraction of the Group F antigen from *Lactobacillus fermentum* with less acidic phenol–water mixtures gave a glycerol teichoic acid of much higher apparent molecular weight than that extracted by the more conventional cold 10% trichloracetic acid. Treatment of the phenol– water-extracted product with trichloracetic acid converted it to the lower molecular weight form. Investigation showed the presence of a covalently-linked glycolipid in the phenol–water-extracted form which was hydrolysed by treatment with the more acidic conditions of trichloracetic acid extraction (Wicken and Knox, 1971). The term lipoteichoic acid was coined, in analogy to lipopolysaccharide, to describe this new class of bacterial amphipathic component. Association of these molecules in aqueous solution in the form of micelles was suggested to account for the high apparent molecular weight of lipoteichoic acids, removal of the lipid moiety resulting in lower molecular weight monomers. An earlier-studied 'intracellular teichoic acid', the group D antigen of *Streptococcus faecalis* (Wicken *et al.*, 1963, Wicken and Baddiley, 1963) was also shown to be a lipoteichoic acid (see Fig. 5.1) when isolated by the milder phenol—water extraction procedure (Toon *et al.,* 1972; Ganfield and Pieringer, 1975).

Since then, variations of the aqueous phenol extraction procedure have been used to obtain lipoteichoic acids from a variety of gram-positive bacteria (Coley *et al.,* 1975; Wicken and Knox, 1975a) and indeed the method appears applicable to the extraction of other amphophilic species from organisms where lipoteichoic acids are absent (Wicken *et al.,* 1978a). In a comparison of extraction conditions Wicken *et al.* (1973) showed that extraction of suspensions of disrupted or whole freeze-dried organisms with equal volumes of 90% aqueous phenol at 65–70°C, a procedure first devised for the extraction of lipopolysaccharides from gram-negative bacteria (Westphal *et al.,* 1952), afforded products that were low in contaminating protein. Extraction at lower temperatures gave higher contaminating protein but better preservation of D-alanyl substituents (Coley *et al.,* 1975). In both procedures, the long-chain fatty acid ester residues of the lipid moiety of lipoteichoic acids are retained. For complete extraction of lipoteichoic acid from small quantities of cells (milligram quantities) the addition of 0.01 M Mg^{2+} to the aqueous phase of the extraction milieu has been shown to be very effective (Kessler and Shockman, 1979).

It is probably misleading to suggest that there is any one variation of the generalized extraction procedure suitable for all organisms in all phases of growth, and a variety of conditions in terms of final yield and comparative purity of product should be examined with any new organism studied. Whatever variation is used, lipoteichoic acid is found in the aqueous phase of the extract along with varying amounts of contaminating polynucleotides, proteins and/or polysaccharides. Polynucleotide contamination can largely be eliminated by digestion with DNase and RNase prior to column chromatography. Digestion with broad-specificity peptidases such as pronase will greatly reduce, but not entirely eliminate, protein contamination of lipoteichoic acid preparations (Wicken and Knox, 1975a; Wicken *et al.,* 1973). Removal of contaminating polysaccharides when these are present provides an even greater problem since lipoteichoic acids readily form closely associated complexes with both proteins and polysaccharides that are not completely resolved by the simple gel-permeation chromatography generally used for partial purification of phenol—water extracts (Wicken and Knox, 1971, 1975; Coley *et al.,* 1975). Lectin- and hydrophobic-affinity· chromatography have been used with limited success in further purification of lipoteichoic acids (Wicken and Knox, 1975b, and unpublished observations).

More recently, the novel use of incorporation of lipoteichoic acid into liposomes and subsequent retrieval by organic solvent extraction has been reported to remove contaminating polysaccharide (Silvestri *et al.,* 1978), albeit the method is only applicable to the purification of lipoteichoic acid in milligram quantities. Ion-exchange chromatography also suffers from extreme limitations when used for lipoteichoic acid purification. A combination of hydrophobic and ionic interaction between lipoteichoic acids and ion-exchange resin generally makes for low and variable recoveries of lipoteichoic acid from such matrices (Wicken and Knox, 1975a), even in the presence of high salt concentrations and what can be eluted from ion-exchange resins may reflect fractions of relatively lower hydrophobicity that are not representative of the whole preparation. The incorporation of detergents such as sodium dodecylsulphate or Triton X100 into ion-exchange systems does allow for quantitative or near quantitative recovery of lipoteichoic acids from ion-exchange systems, but the researcher is then faced with the problem of complete removal of the detergent. The latter becomes a significant problem if the purpose of purification is to study biological properties of lipoteichoic acids that involve membrane inter-action. Clearly what is urgently needed is a method(s) for complete purification of lipoteichoic acids that (a) does not risk contamination of the preparation with membrane active chemicals and (b) affords yields of the polymer that can be considered representative of the starting material. Until this is achieved some doubt must always be exercised as to the real significance of the plethora of biological activities of lipoteichoic acids that is beginning to be reported in the literature (see Table 5.1). Some of these may be due to, or at least modified by, contaminating protein or polysaccharide in the lipoteichoic acid preparation studied. Equally, standardization of *in vivo* and *in vitro* biological test systems is required before the true roles of amphophiles such as lipoteichoic acids as biologically active products of bacteria can be properly assessed.

5.1.3 Location of lipoteichoic acids in the bacterial cell

The discovery of a membrane lipid component in lipoteichoic acids readily suggested the nature of the previously unknown method of attachment of these polymers to the plasma membrane — by simple intercalation of the fatty acid residues of the lipid moiety into the upper half of the bilayer of the membrane. Serological detection of lipoteichoic acid at the surface of some bacteria led to the proposal of a model (van Driel *et al.*, 1973), in which the long polar glycerophosphate chain of the lipoteichoic acid could penetrate the peptidoglycan network of the cell wall and, in some cases, be detectable as a surface antigen. In *Lactobacillus fermentum*, for instance, the lipoteichoic acid, which is also the group antigen, is the major surface immunogen when whole organisms are injected into rabbits (Knox *et al.*, 1970). Support for this model was obtained by the use of ferritin conjugated to goat antirabbit γ-globulin to detect the sites of interaction between lipoteichoic acid and rabbit antibody specific to the polyglycerolphosphate chain (van Driel *et al.*, 1973; Dickson and Wicken, 1974). Electron micrographs showed heavy labelling of the surface of whole organisms, protoplasts and membrane fragments while extension of the technique to immunochemical labelling after fixation and thin-sectioning (Dickson and Wicken, 1974) showed ferritin label extending from the upper surface of the membrane through the wall to the surface of the cell as demanded by the model. Space-filling models of the *Streptococcus faecalis* lipoteichoic acid (Ganfield and Pieringer, 1975) showed clearly that the spatial orientation of the four fatty acid hydrocarbon chains was such that the proposed intercalation into the upper half of the membrane bilayer was stereochemically possible. The lipid

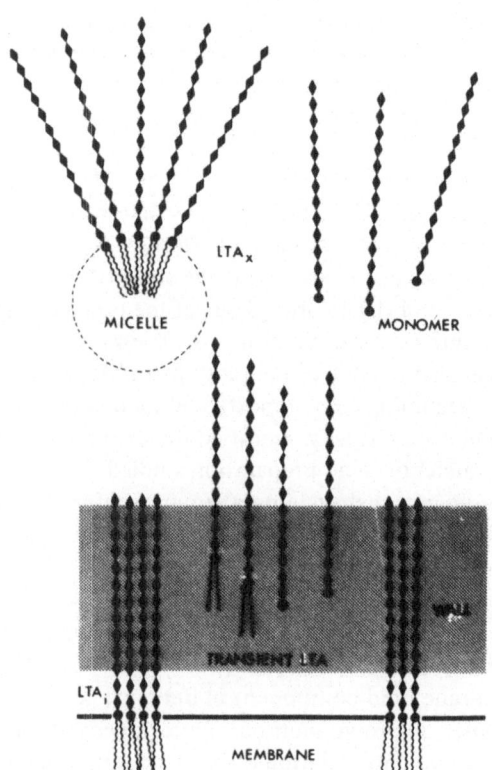

Fig. 5.2 Diagrammatic representation of the relationship of extracellular lipoteichoic acid (LTA_x) and intracellular lipoteichoic acid (LTA_i) to the wall and membrane of a generalized bacterial cell. LTA_x is represented in a fully acylated micellar as well as deacylated monomer form. Transient LTA_i, both acylated and deacylated, is depicted as being in the process of excretion from the cell. The LTA structures shown are abridged, glycosyl and D-alanyl substituents being omitted, to emphasize the amphipathic nature of the fully acylated molecule. Chains of diamonds represent polyglycerol phosphate; solid circles attached to wavy lines represent glycolipid; solid circles alone represent deacylated glycolipid. From Wicken and Knox (1977a) with permission of the American Society for Microbiology.

end of the molecule thus acts as a membrane 'anchor' for lipoteichoic acid.

While the model discussed above suggested a generalized location of lipoteichoic acid over the surface of the plasma membrane, a more specific site of location entirely in mesosomes was suggested (Huff *et al.*, 1974) from the analysis of mesosomal and membrane fractions from *Staphylococcus aureus.* Studies with *Streptococcus faecalis* (Joseph and Shockman, 1975) on the other hand did not show a preferential mesosomal location. The dangers of assignment of specific locations for lipoteichoic acids on the basis of sub-cellular fractions alone is compounded by the ready loss and perhaps relocation of lipoteichoic acid from membranes under conditions of low Mg^{2+} concentrations (Hay *et al.*, 1963). It is of course possible that localized concentrations of lipoteichoic acid may occur in regions of the cell and/or at different times during the cell cycle.

This somewhat static model of the location of lipoteichoic acid changed dramatically (Wicken and Knox, 1977a) when it was shown simultaneously in two laboratories that many organisms excreted lipoteichoic acids into the external environment (Markham *et al.*, 1975; Joseph and Shockman, 1975) and the presence of extracellular lipoteichoic acid readily explained the earlier known presence of erythrocyte-sensitizing common antigens in culture supernatants of many gram-positive organisms. Thus with these findings lipoteichoic acids were established as being located in two regions with respect to the bacterial cell (Fig. 5.2). Intracellular lipoteichoic acids (LTA_i) are regarded as normal plasma membrane components of most gram-positive bacteria, the glycolipid moiety being embedded in the upper half of the phospholipid bilayer of the cell membrane. As in the previous model, there is extension of the polar glycerolphosphate portion of the molecule as a cell surface component. Extracellular lipoteichoic acids (LTA_x) are found in the external environment and in many cases results from active excretion from the cell since it is found under conditions where there is no wall turnover or lysis of cells. A transient existence of LTA_i as a solely wall and surface component can also be postulated as part of the process $LTA_i \rightarrow LTA_x$. To further complicate the picture, from both the extracellular and cellular locations, lipoteichoic acids can be recovered in a high molecular weight micellar aggregate and also in a lower molecular weight deacylated monomer form which, lacking the fatty acid hydrocarbon chains, cannot undergo hydrophobic aggregation. The relative proportions of LTA_i to LTA_x in both locations as well as the proportion of acylated micellar form to deacylated monomer varies widely with the species of organism and its phase of growth (Joseph and Shockman, 1975; Markham *et al.*, 1975). In *Streptococcus faecalis*, extracellular LTA is entirely in the deacylated monomer form during logarithmic growth. Extracellular LTA from this organism has been shown to be derived directly from the cellular LTA and evidence for a deacylase in the membrane of this organism has been presented (Kessler and Shockman, 1978). In many of the oral streptococci on the other hand, lipoteichoic acid is excreted in both molecular forms even under balanced growth conditions in the chemostat (Wicken and Knox, 1977a; Knox and Wicken, 1978). With *Streptococcus mutans* BHT the amount of extracellular lipoteichoic acid is some 8—9-fold greater than the intracellular lipoteichoic acid at generation times of 14 hours (Wicken and Knox, 1977a). *Lactobacillus fermentum* shows a similar increase in excretion of lipoteichoic acid at generation times approaching that generally estimated to be likely growth rates of micro-organisms in the nutritionally limited environment of the oral cavity.

Precisely why micro-organisms should seemingly 'throw away' large amounts of energy in the form of excreted phosphate polymers remains a mystery. Of significance,

Table 5.1 Some biological properties of amphophiles

Property	LPS	LTA	LP	AcA	LM
Pyrogenicity	+	−	−		
Lethal toxicity	+	−			
Immunogenicity	+	+	+	+	+
Mitogenicity	+	+	+	+	
Eucaryotic membrane binding	+	+	+	+	+
Bone resorption	+	+		+	
Anticomplementary activity	+	+			
Stimulation Alt. Pathway C′	+	+			
Schwartzman reaction	+	+			
Hypersensitivity	+	+			
RES-stimulation	+	+			
Stimulation of non-specific immunity	+	+			
Macrophage lysosomal enzyme release	+	+			
Limulus assay	+	+			
Adjuvant activity	+	±			

LPS, lipopolysaccharide;   LTA, lipoteichoic acid;   LP, lipoprotein; AcA Actinomyces amphophile;   LM, lipomannan.

however, is the existance of extracellular lipoteichoic acid in its fully acylated form in the external environment of many micro-organisms. It should also be pointed out that in those organisms lacking lipoteichoic acids, such as the micrococci and the actinomycetes, the equivalent amphophiles, discussed above, are detectable in culture fluids, suggesting that in these cases excretion of amphophiles is also part of the cell cycle.

Current interest in amphophiles stems from their many and varied biological properties. The presently known biological activities of several amphophiles are compared with those of the oldest known amphophile, lipopolysaccharide, in Table 5.1 (for reviews see Wicken and Knox, 1977a, 1977b; 1978b). It will be readily seen that, with the exception of pyrogenicity and endotoxicity, lipoteichoic acids, the most studied of the other amphophiles, share a wide variety of biological properties with lipopolysaccharides and may mediate in disease processes through:

(i) Immunogenicity and cross-reaction antibodies;
(ii) Mitogenic lymphocyte stimulation;
(iii) Complement activation by the classical and/or alternative pathways and release of inflammatory factors;
(iv) Monocyte and macrophage activation with release of lysosomal enzymes;
(v) Stimulation of bone resorption, as occurs in periodontal disease.

While precise mechanisms of action of any one amphophile in producing such potentially wide-ranging effects in eucaryotic systems remains unknown, in all cases, the hydrophobic lipid portion of the molecule appears to be essential for biological activity. Destruction of the hydrophobic function by chemical or enzymatic deacylation removes the ability of amphophiles to interact with biological membranes and is accompanied by a loss of the biological properties listed in Table 5.1.

The existence of extracellular, acylated lipoteichoic acid, and the ability of micellar lipoteichoic acid to interact with a variety of different types of bacterial and mammalian macromolecules, cells and surfaces, has stimulated considerable interest in the possible role of this class of compound in such diseases as dental caries and periodontitis. Although currently little is known about the limits and preferences of such interactions, probably a broad range of substances can find their way into lipoteichoic acid-containing micelles. For example, it is known that lipoteichoic acid can interact with polysaccharides containing the *Streptococcus mutans* serotype a determinants (Silvestri *et al.*, 1978) and with extracellular glucans produced by sucrose-grown cultures of *S. mutans* (Melvaer *et al.*, 1974). Similar interactions have made it extremely difficult, as has been referred to earlier in this chapter, to remove polysaccharides, proteins and lipids from micellar lipoteichoic acid in order to obtain highly purified lipoteichoic acid. The binding of lipoteichoic acid, presumably through its ionized phosphate groups, to hydroxyapatite has suggested a role in adherence of dextran—lipoteichoic acid complexes to the tooth surface (Markham *et al.*, 1975; Rölla, 1976). Such interactions are discussed elsewhere in this book and beyond the scope of the present chapter which is concerned with the more hydro- phobic interactions of lipoteichoic acids.

5.2 BINDING OF LIPOTEICHOIC ACIDS TO EUCARYOTIC CELLS

5.2.1 Erythrocytes

The ready sensitization of erythrocytes with lipoteichoic acid and their agglutination with antisera specific to lipoteichoic acid was early used (Hewett *et al.*, 1970) as a semi-quantitative method of detection of lipoteichoic acids and assay of antisera. Lipoteichoic acids are readily bound to erythrocytes, without prior modification of the polymer or pre-treatment of the cells. The method has also been used to detect lipoteichoic acids in culture supernatants (Markham *et al.*, 1975; Knox and Wicken, 1978). Serial dilutions of dialysed culture fluid are used to sensitize erythrocytes and the maximum dilution at which maximum hemagglutination titer with standard antisera no longer occurs being a measure of the amount of lipoteichoic acid origin- ally present in the undiluted culture fluid. As is discussed later in this chapter, deacylated lipoteichoic acids do not bind to cell membranes and thus this method distinguishes between acylated and deacylated lipoteichoic acids. Sensitization of erythrocyte membranes seems to be a general property of amphophiles (Table 5.1) and the above method can be used to detect the presence of new amphophiles in culture supernatants, utilizing antisera prepared against the organisms grown in the same culture fluid as a detecting system. In this manner the amphophile from *Actinomyces* spp. was first detected (Wicken *et al.*, 1978a) and the method has been used to detect the presence of lipoteichoic acids in other filamentous organisms such as *Rothia dentiocariosa* (Wicken and Knox, unpublished observations).

Earlier studies on the sensitization of erythrocytes (Jackson and Moskowitz, 1966; Moskowitz, 1966) by a phenol—water extracted teichoic acid preparation, not known to be a lipoteichoic acid at that time, led to the conclusion that alanine ester residues were essential for binding to the erythrocyte membrane since loss of alanine under mild alkaline conditions resulted in loss of sensitization activity. However, the conditions used would also have hydrolysed fatty acid esters and the later studies with *Lactobacillus fermentum* lipoteichoic acid indicated that fatty acids rather than alanine residues are required for erythrocyte sensitization (Hewett *et al.*, 1970). Lipoteichoic acid preparations from hot phenol-water extracts are low in alanine substitution but will readily sensitize or absorb to erythrocyte membranes. Deacylated monomer lipoteichoic acid prepared either by treatment of micellar lipoteichoic acid under mild alkaline conditions (Knox and Wicken, 1973; Ofek *et al.*, 1975) or digestion with *Candida cylindracea* lipase (Wicken and Knox, unpublished observations) abolishes the binding capacity. Similarly, naturally occurring deacylated monomer lipoteichoic acid will not bind to erythrocyte surfaces. To further demonstrate the role of the fatty acid substituents in this process Beachey *et al.*, (1979) used ^{3}H-lipoteichoic acid from *Streptococcus pyogenes* as a more sensitive test of binding than hemagglutination with anti-lipoteichoic acid antisera. As expected, mild alkaline deacylation of the lipoteichoic acid removed the binding capacity, but this could be restored to nearly 50% of the original by reesterifying with stearoylchloride in *N, N*-dimethyl formamide and pyridine. Treatment of the deacylated lipoteichoic acid with the various reagents in the absence of stearoylchloride did not restore binding capacity. Thus, the main body of evidence would support strongly the contention that the fatty acid substituents of lipoteichoic acid are essential for their ability to absorb to the erythrocyte surface and is completely contrary to the report of Cooper *et al.* (1978) that lipid is non-essential to the process.

Adsorption of lipoteichoic acid to erythrocytes is presumably analogous to the adsorption of lipopolysaccharides to erythrocytes and other cell membranes, where the lipid component (lipid A) has been shown to play an essential role, probably by hydrophobic interaction with structures on the erythrocyte surface (Ciznar and Shands, 1971; Shands, 1973). The fatty acid content of lipoteichoic acids is of the order of 4—8%, considerably less than that for lipopolysaccharides but comparable to that found to be effective in rendering polysaccharides capable of absorbing to erythrocytes (Hammerling and Westphal, 1967) where the addition of 5% *O*-stearoyl groups to a number of polysaccharides gave optimal erythrocyte-sensitizing properties. Esterification of wall polysaccharides from streptococci of groups A and E (Pavlovskis and Slade, 1969; Slade and Hammerling, 1968) has also been used to absorb these antigens to erythrocytes. This is in further support of the idea that the

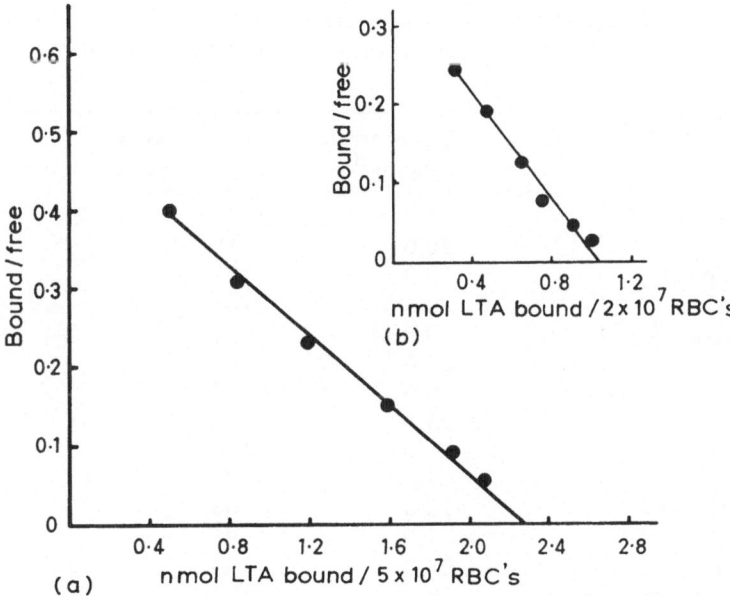

Fig. 5.3 Scatchard plot of the binding of *S. pyogenes* ^{3}H-lipoteichoic acid to human adult (a) and cord blood (b) erythrocytes. Cells were incubated with various concentrations of added (free) ^{3}H-lipoteichoic acid and the amount of bound lipoteichoic acid was determined after 2 h at 37°C. (From Beachey *et al.*, (1979b)).

interaction of amphophiles, whether natural or artificial, with erythrocyte membranes is essentially hydrophobic in nature.

Beachey *et al.* (1979b) have investigated the biological and biochemical character-istics of the binding of radioactive lipoteichoic acids from *Streptococcus pyogenes* and *S. faecalis* to human and sheep erythrocytes. Binding of ^{3}H-lipoteichoic acid from either source was found to be cell concentration-, time- and temperature-dependent. At 37°C, binding approached a maximum only after 2 hours of incubation. Estimation of the average number of LTA receptor sites per adult human erythrocyte as well as their binding affinity (Fig. 5.3) gave values of approximately 29 x 10^6 lipoteichoic acid binding sites per cell with a dissociation constant of 4.5 μM. Human cord erythrocytes gave a similar number of binding sites (30 x 10^6) but a higher dissociation constant of 31 μM. Similar analysis of the binding of lipoteichoic acid to sheep erythrocytes gave 7.2 x 10^6 binding sites and a dissociation constant of 1.6 μM which is of the same order of magnitude as found with adult human erythrocytes.

Cold-chase experiments with excesses of heterologous lipoteichoic acids displaced radiolabelled lipoteichoic acid from erythrocytes and, similarly, the binding of radiolabelled lipoteichoic acid to erythrocytes could be blocked by an excess of cold

Table 5.2 Effect of gangliosides, various sugars and enzymatic treatment on the binding of ^{3}H-lipoteichoic acid to human erythrocytes. Data from Beachey *et al.* (1979b)

Substance tested	Concentration (mg ml^{-1})	Percentage of control value
Albumin	5.0	47
	50.0	18
Ig (Cohn fraction II)	50.0	88
Ganglioside G$_{M2}$	0.5	41
Ganglioside G$_{M3}$	0.5	47
D-galactose	25.0	102
β-D-glucose	25.0	108
D-arabinose	25.0	91
L-fucose	25.0	107
D-mannose	25.0	104
N-acetyl-D-glucosamine	25.0	100
D-xylose	25.0	95
D-galactosamine	25.0	90
Methyl α, D-mannopyranoside	25.0	97
Neuraminidase	— *	123
Trypsin	10.0	104

* 10 U ml^{-1}.

heterologous as well as homologous lipoteichoic acid (Beachey *et al.*, 1979b). These results indicate that similar receptor sites on the erythrocyte surface are involved in the binding of lipoteichoic acids irrespective of their structure. This is not unexpected, given a hydrophobic interaction, when the overall heterogeneity of the fatty acid substitution of lipoteichoic acids is considered, i.e. probably lipoteichoic acids from different species are very similar as far as their hydrophobic 'ends' are concerned. In contrast, lipopolysaccharides from *Salmonella enteriditis* or *Serratia marcesens* failed to inhibit the binding of radiolabelled lipoteichoic acid or affect the kinetics of binding, indicating that these two amphophiles have different binding sites. That both amphipathic species were bound to erythrocytes was shown by hemagglutination with antisera specific to either amphophile and the titer was the same as with erythrocytes sensitized with either amphophile alone (Beachey *et al.*, 1979b). The hydrophobic region of lipopolysaccharide, lipid A, is of course structurally very different to the simpler glycolipids of lipoteichoic acids. Binding of lipoteichoic acid to human erythrocyte membranes only affected accessibility of A and B blood group antigens to their respective antibodies as occupation of lipoteichoic acid binding sites approached saturation.

The nature of the binding site for lipoteichoic acids was investigated (Beachey *et al.*, 1979b) by testing potential receptor analogues for their ability to inhibit lipoteichoic acid binding (Table 5.2). None of the nine different saccharides present on mammalian cell membranes (Roseman, 1975), or digestion with trypsin or neuraminidase inhibited lipoteichoic acid binding, suggesting that sugar moieties are probably not involved directly as structural elements in membrane receptors for lipoteichoic acids. Inhibition by both the gangliosides GM_2 and GM_3 may reflect hydrophobic interaction between lipoteichoic acid and the common lipid moiety of the gangliosides, unlike, for instance, cholera toxin which recognises specific sugar sequences on individual gangliosides (Wiegandt, 1976). This, in turn, may also reflect the apparent steric interference, as lipoteichoic acid becomes saturating, of blood group antigens, the latter now being recognised as glycolipids (Dejter-Suznski, 1978). The inhibitory effect of serum albumin (Table 5.2) would appear to be due to the high affinity of albumin for a variety of substances including fatty acids (Spector, 1975) and lipoteichoic acids (Wicken and Knox, unpublished observations), thereby neutralizing the ability of lipoteichoic acid to react hydrophobically with its receptor on the erythrocyte membrane.

More recent studies of the binding of group A streptococcal lipoteichoic acids to isolated human erythrocyte membranes (Chiang *et al.*, 1979) showed similar characteristics of binding to the studies, discussed above, with intact erythrocytes. Consistent with the idea of a single population of binding sites for lipoteichoic acids, autoradiofluorography of SDS-gel electrophoreses of total erythrocyte membrane proteins showed association of bound lipoteichoic acid with a single protein band, designated Band 6 (Steck, 1974). An erythrocyte receptor for lipopolysaccharides has been isolated and partially characterized as a electrophoretically homogeneous lipoglycoprotein (Springer *et al.*, 1974), which would appear to be distinct from the protein fraction binding lipoteichoic acids.

5.2.2 Other mammalian cells

While most detailed studies of lipoteichoic acid binding to eucaryotic cells have been carried out using erythrocytes, the spontaneous binding of lipoteichoic acid to oral mucosal cells (Beachey, 1975; Ofek *et al.*, 1975), human platelets (Beachey *et al.*, 1977) and a variety of other mammalian cells (Stewart and Martin, 1962) has also been reported. Like erythrocyte binding, fully acylated micellar lipoteichoic acid will bind but not the deacylated monomer form.

The binding of lipoteichoic acid to platelets is of interest in that while, like erythrocyte binding, the reaction was time-, temperature- and concentration-dependent with a dissociation constant of 19 μM, the reaction was much more rapid reaching a maximum in 10 minutes. While the kinetics of binding to platelets was also consistent with a single population of binding sites, these were patchy in their distribution over the platelet surface (Fig. 5.4) as judged by immunochemical labelling (Beachey *et al.*, 1977). The well known platelet aggregation phenomenon and release of serotonin was not induced by binding of lipoteichoic acid to the platelet surface, but lipoteichoic acid would inhibit aggregation induced by collagen

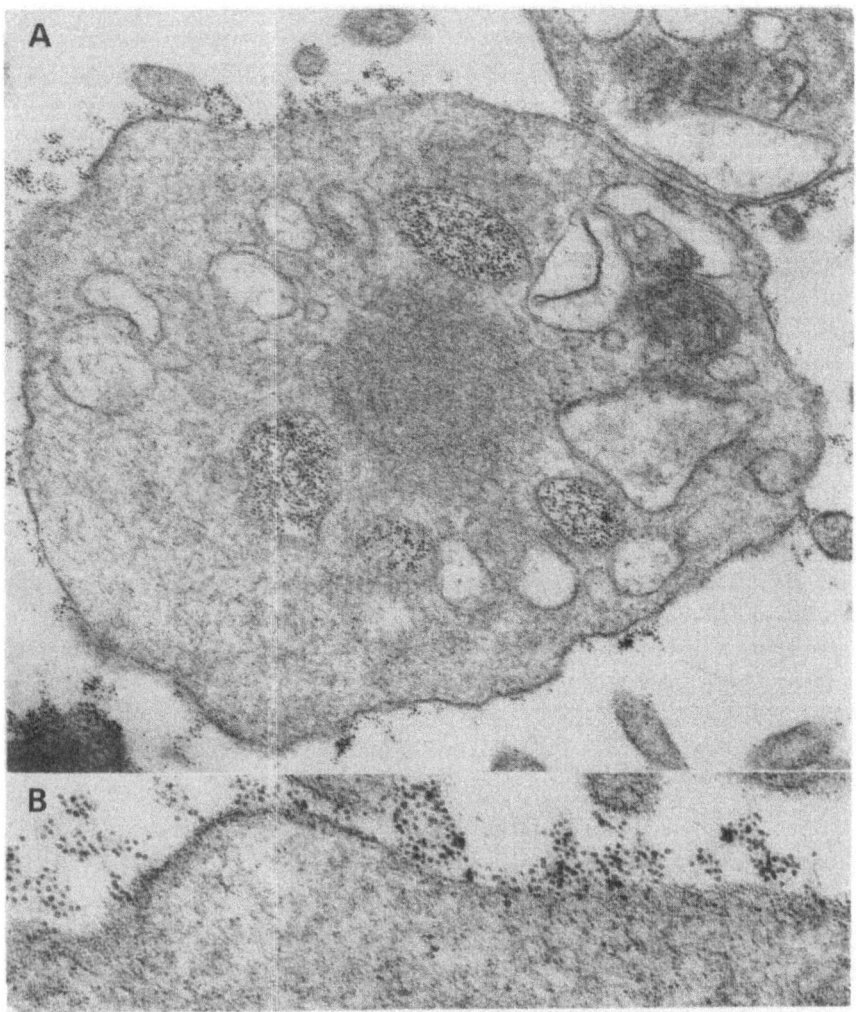

Fig. 5.4 Electron micrograph of ultrathin section of human platelets treated
with lipoteichoic acid followed by rabbit anti-lipoteichoic acid and ferritin-
conjugated goat anti-rabbit globulin. (a) × 60 000, (b) × 120 000. From
Beachey *et al.* (1977).

or its denatured α-1 chain. Deacylated lipoteichoic acid had no inhibitory effect on
platelet aggregation induced by adenosine diphosphate or epinephrine (Beachey
et al., 1977).

Murine and human lymphocyte binding of group A streptococcal lipoteichoic acid
preparations was similarly rapid (Beachey *et al.,* 1979a) and while both T and B
lymphocytes possessed single populations of specific binding sites for the lipoteichoic

acid, mitogenic stimulation of T, but not of B, lymphocytes has been reported by these workers as a consequence of the binding. This contrasts with the B-lymphocyte mitogenicity of lipopolysaccharides and possibly reflects the distinctness of the binding sites for each type of amphophile referred to earlier. Maximal stimulation of T-lymphocytes occurred at concentration of 12.5 μg of lipoteichoic acid per 5×10^6 lymphocytes, at which concentration approximately 12% of the available binding sites would be occupied. It should be noted, however, that the magnitude of the mitogenic stimulation by group A streptococcal lipoteichoic acid was significantly less than for other mitogens, and lipoteichoic acid preparations from other organisms gave weaker mitogenic stimulation to the group A streptococcal lipoteichoic acid when tested in the same systems (Beachey, Ofek and Wicken, unpublished observations). Unequivocal proof of lipoteichoic acid mitogenicity must await the testing of vigorously purified materials.

As far as the binding of streptococcal lipoteichoic acid by human oral mucosal cells is concerned, a postnatal development of binding capacity has been observed (Ofek *et al.*, 1977). Similarly streptococcal cells from which the lipoteichoic acid was derived showed scant adhesive properties to buccal mucosal cells taken soon after birth, adult levels of adhesion only being reached 48–72 h after birth. That there may be a connection between lipoteichoic acid and streptococcal binding to eucaryotic cells is discussed in the next section.

5.3 LIPOTEICHOIC ACID AND ITS POSSIBLE INVOLVEMENT IN THE ATTACHMENT OF BACTERIA TO THE SURFACES OF ANIMAL CELLS

That many bacteria possess adhesive properties has been known for some time but it is only relatively recently that marked research interest has been shown in the phenomenon. While bacteria can be found colonizing a wide variety of surfaces, both animate and inanimate, procaryotic–eucaryotic cell interaction and, in particular, the possible role of lipoteichoic acid in this process, is the area to be considered here. For an excellent general review of bacterial adhesion to the surface of animal cells the reader is referred to Jones (1977).

To date, lipoteichoic acids have only been implicated as adhesins, a term used to describe any substance on the bacterial surface that mediates the attachment of the bacterium to another surface (Duguid, 1959), in the genus *Streptococcus*. Oral streptococci have been the subjects of a long series of investigations by Gibbons *et al.* (for reviews see Gibbons and van Houte, 1975; Gibbons, 1975), which have pointed convincingly to both the ecological importance of adhesion as well as its selectivity. *Streptococcus salivarius* is found preferentially on the dorsum of the tongue, *S. sanguis,* and *S. mutans* on the tooth surface, while *S. mitis* is virtually exclusive to non-keratinized mucosa (Gibbons and van Houte, 1975). This distribution remains true whether the species are introduced into or arise naturally in the oral cavity. The conclusion (Gibbons and van Houte, 1975) that different adhesins or adhesive mechanisms are operating with different oral species and which, in turn, dictate their habitat is plausible. The mechanism of attachment appears to be a fibrillar layer on the surface of the streptococcal cell (Gibbons *et al.*, 1972; Nalbandian *et al.*, 1974). Trypsin treatment, which removes the fibrillar material, destroys adhesive properties (Gibbons *et al.*, 1972), and was associated with the loss of the M protein antigen (Ellen and Gibbons, 1972 and 1974).

The binding of group A streptococci (*Streptococcus pyogenes*) to human skin epithelial cells and buccal mucosal cells has been investigated in some detail by Beachey and co-workers. *S. pyogenes* isolated from skin adhered in greater numbers to human skin epithelial cells than buccal mucosal cells. Streptococcal strains isolated from the throat showed the reverse situation. M-protein-producing strains of group A streptococci did not show significantly greater adherence than M-negative strains (Alkan *et al.*, 1977). More importantly, pre-incubation of human cells with lipoteichoic acid from *S. pyogenes*, inhibited subsequent binding of streptococcal strains (Alkan *et al.*, 1977; Beachey, 1975) and these results were interpreted by these workers as indicating a more central role for lipoteichoic acid than M protein as streptococcal adhesins. It could be argued, of course, that pre-incubation of human cells with lipoteichoic acid resulted in binding of the latter and gain of considerable negative charge by the eucaryotic cell, with subsequent repulsion of similarly charged bacterial cells.

Like the oral streptococci, a fibrillar contact between bacterium and epithelial cell is readily observed in electron micrographs (Fig. 5.5). An early association of M-protein with these fibrillar structures, or fimbriae (Swanson *et al.*, 1969), led to the idea, mentioned above, that the antiphagocytic M protein mediated the adhesion of group A streptococci to oral epithelial cells (Ellen and Gibbons, 1972), but left open the question of whether the fimbriae were composed of M-protein alone. Beachey and Ofek (1976) dissociated the M-protein from group A streptococci by digestion with dilute solutions of pepsin at pH 5.8 without loss of the electron microscopically-observable fimbriae and also without loss of the ability of the cocci to adhere to epithelial cells. Treatment of cells with trypsin or HCl at pH 2.0 removed the fimbriae and all adhesive properties of the streptococci. Hyaluronidase or hydroxylamine at pH 10.0 were, on the other hand, entirely without effect on the appearance of fimbriae or adhesion (Beachey and Ofek, 1976). Loss of M-protein by pepsin treatment was also associated with loss of the antiphagocytic properties of M-positive strains of group A streptococci. Thus, these studies suggested that a fimbrial-associated substance other than M-protein was responsible for the adhesive properties of group A streptococci. Evidence that the adhesin is lipoteichoic acid (Beachey and Ofek, 1976) is compelling, but not conclusive.

(1)   Immunochemical labelling specific to lipoteichoic acid showed a location of at least part of the latter on the surface of M-protein denuded cells and in an orientation that could be equated to the fimbriae.

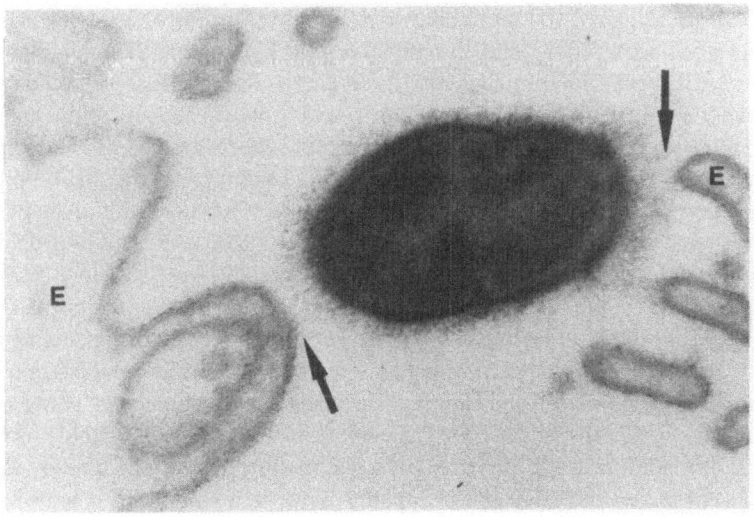

Fig. 5.5 Fimbriae radiating from group A streptococcus to membranes (arrows) of human oral epithelial cells (E). From Beachey and Ofek (1976).

(2)   Adhesion of streptococci to epithelial cells could be blocked by pre-incubating epithelial cells with lipoteichoic acid or pre-incubating the streptococci with anti-lipoteichoic acid antibody.
(3)   Lipoteichoic acid could be extracted from untreated or pepsin-treated strepto-cocci under the mild conditions of acetate buffer at pH 6.0 (Beachey and Ofek, 1976).

(1) and (2) are consistent with an adhesin role for lipoteichoic acid while (3) is assumed to show a surface location of at least part of the lipoteichoic acid. It would need to be proved that the extraction conditions were incapable of removing lipoteichoic acid from deeper within the cell. More recently, excretion of lipoteichoic acid in micellar and deacylated monomer forms by group A streptococci under normal growth conditions has been shown to be greatly enhanced by the presence of penicillin (Alkan and Beachey, 1978). Lipoteichoic acid-depleted cells also lost their adhesive properties for epithelial cells.

While a cellular and extracellular location for the lipoteichoic acid of group A streptococci clearly conforms well with the generalized model discussed earlier in this chapter (Fig. 5.2), two aspects of the proposed role for lipoteichoic acid as an adhesin are puzzling.

(1) If it is assumed that the interaction with epithelial cells is a hydrophobic one, as seems to be implied by Beachey and co-workers, then the lack of effect on adhesion by pretreatment of streptococci with hydroxylamine at pH 10 needs to be explained. Such conditions will readily deacylate lipoteichoic acids in solution ᾱnd one would imagine a similar effect on surface lipoteichoic acid in intact or pepsin-treated streptococci.

(2) It is difficult to conceive of a model in which a molecule such as lipoteichoic acid with only one hydrophobic 'end' can undergo two hydrophobic interactions and bridge streptococcus and epithelial cell. Energy considerations alone would seem to make unlikely a configuration for lipoteichoic acid in which the glycolipid end projects into the aqueous environment from the streptococcal surface. It is perhaps possible that some of the extracellular lipoteichoic acid remains closely associated with the M-protein (or other pepsin-resistant proteins, as it has not been proved that M-protein is the sole protein) of the fimbriae in both a hydrophobic and hydrophilic association that minimizes interaction between glycolipid and the aqueous environment. Close approach to an epithelial cell could result in a rearrangement with consequent hydrophobic interaction of the lipid 'end' of the lipoteichoic acid with the eucaryotic membrane while maintaining an ionic or hydrophilic association of the glycerophosphate chain with the fimbriae. If indeed lipoteichoic acid, while still associated in some way with the streptococcal cell, can undergo hydrophobic interaction with an epithelial cell, it is difficult to explain any specificity of binding to particular types of eucaryotic cells or the apparently limited number of receptor sites (Bartelt and Duncan, 1978). That some bacterial cells can have hydrophobic regions at their surface, however, recently has been demonstrated for strains of porcine enteropathogenic *Escherichia coli* using the new technique of hydrophobic interaction chromatography (Smyth *et al.*, 1978).

At the present time, it must be concluded that the case for streptococcal lipoteichoic acid as an adhesin remains unproven, albeit it is the most attractive candidate to date. One of the problems in this area of research has been aluded to earlier in this chapter, lipoteichoic acids are extraordinarily difficult to purify from contaminating protein and or polysaccharides. The latter may indeed play a role in the adhesion process. A recent report (De Vuono and Panos, 1978) showed that acylated, but not deacylated lipoteichoic acid of Group A streptococci or its L-forms (which differ chemically) can inhibit adhesion of either the streptococcus or its L-form to human kidney cells in tissue culture. These preparations of lipoteichoic acid contained 2–5% protein, but were devoid of nucleic acids.

REFERENCES

Alkan, M.L. and Beachey, E.H. (1978), *J. clin. Invest.*, **61**, 671–677.
Alkan, M., Ofek, I. and Beachey, E.H. (1977), *Infect. Immun.*, **18**, 555–557.
Archibald, A.R. (1974), *Adv. microb. Physiol.*, **11**, 53–95.
Armstrong, J.J., Baddiley, J., Buchanan, J.G., Carss, B. and Greensberg, G.R. (1958), *J. Chem. Soc.*, 4344–4354.
Baddiley, J. (1972), *Essays Biochem.*, **8**, 35–97.
Baddiley, J. and Davison, A.L. (1961), *J. gen. Microbiol.*, **24**, 295–299.
Bartelt, M.A. and Duncan, J.L. (1978), *Infect. Immun.*, **20**, 200–208.
Beachey, E.H. (1975), *Trans. Ass. Am. Phvscns*, **88**, 285–292.
Beachey, E.H., Chiang, T.M., Ofek, I. and Kang, A.H. (1977), *Infect. Immun.*, **16**, 649–654.

Beachey, E.H., Dale, J., Grebe, S., Ahmed, A., Simpson, W.A. and Ofek, I. (1979a), *J. Immunol.*, **122**, 189–195.

Beachey, E.H., Dale, J.B., Simpson, W.A., Evans, J.D., Knox, K.W., Ofek, I. and Wicken, A.J. (1979b), *Infect. Immun*, **23**, 618–625.

Beachey, E.H. and Ofek, I. (1976), *J. exp. Med.*, **143**, 759–771.

Button, D. and Hemmings, N.L. (1976), *J. Bact.*, **128**, 149–156.

Chiang, T.M., Alkan, M.L. and Beachey, E.H. (1979), *Infect. Immun.*, (In press).

Ciznar, I. and Shands, J.W. (1971), *Infect. Immun.*, **4**, 362–367.

Coley, J., Duckworth, M. and Baddiley, J. (1975), *J. gen. Microbiol.*, **73**, 587–591.

Cooper, H.C., Chorpenning, F.W. and Rosen, S. (1978), *Infect. Immun.*, **19**, 462–470.

DeVuono, J. and Panos, C. (1978), *Infect. Immun.*, **22**, 255–265.

Dejter-Suszynski, M., Harpaiz, N., Flowers, H.M. and Sharon, N. (1978), *Eur. J. Biochem.*, **83**, 363–373.

Dickson, M.R. and Wicken, A.J. (1974), *Int. Congr. Electron Microscopy*, 8th, Canberra, **2**, 114–115.

Duguid, J.P. (1959), *J. gen. Microbiol.*, **21**, 271–286.

Ellen, R.P. and Gibbons, R.J. (1972), *Infect. Immun.*, **5**, 826–830.

Ellen, R.P. and Gibbons, R.J. (1974), *Infect. Immun.*, **9**, 85–91.

Ellwood, D.C. and Tempest, D.W. (1972), *Adv. Microbiol. Physiol.*, **7**, 83–117.

Ganfield, M-C.W. and Pieringer, R.A. (1975), *J. biol. Chem.*, **250**, 702–710.

Gibbons, R.J. (1975), In: *Microbiology 1975* (Schlessinger, D., ed.), American Society for Microbiology.

Gibbons, R.J., van Houte, J. and Liljemark, W.F. (1972), *J. Dent. Res.*, **51**, 424–435.

Gibbons, R.J. and van Houte, J. (1975), *A. Rev. Microbiol.*, **29**, 19–44.

Hammerling, U. and Westphal, O, (1967), *Eur. J. Biochem.*, **1**, 46–50.

Hay, J.B., Wicken, A.J. and Baddiley, J. (1963), *Biochim. biophys. Acta*, **71**, 188–190.

Hewett, M.J., Knox, K.W. and Wicken, A.J. (1970), *J. gen. Micrʌbiol.*, **6**, 315–322.

Huff, E., Cole, R.M. and Theodore, T.S. (1974), *J. Bact.*, **120**, 273–281.

Jackson, R.W. and Moskowitz, M. (1966), *J. Bact.*, **91**, 2205–2209.

Jones, G.W. (1977), In: *Microbial Interactions.* (Reissing, J.C., ed.), pp. 140–176. Chapman and Hall, London.

Joseph, R. and Shockman, G.D. (1975), *Infect. Immun.*, **12**, 333–338.

Kessler, R.E. and Shockman, G.D. (1979), *J. Bact.*, **137**, 869–877.

Knox, K.W., Hewett, M.J. and Wicken, A.J. (1970), *J. gen. Microbiol.*, **60**, 303–313.

Knox, K.W. and Wicken, A.J. (1973), *Bact. Rev.*, **37**, 215–257.

Knox, K.W. and Wicken, A.J. (1978), *Secretory Immunity and Infection*, (McGhee, J.R., Mestecky, J. and Babb, J.L., eds.), Plenum Publishing Corp. New York, pp. 629–637.

Markham, J.L., Knox, K.W., Wicken, A.J. and Hewett, M.J. (1975), *Infect. Immun.*, **12**, 378–386.

Melvaer, K.L., Helgeland, K. and Rolla, G. (1974), *Arch. Oral Biol.*, **19**, 589–595.

Moskowitz, M. (1966), *J. Bact.*, **91**, 2200–2204.

Nalbandian, J., Freedman, M.L., Tanzer, J.M. and Lovelace, S.M. (1974), *Infect. Immun.*, **10**, 1170–1179.

Ofek, I., Beachey, E.M., Jefferson, W. and Campbell, G.C. (1975), *J. exp. Med.,* **141**, 990–1033.

Ofek, I., Beachey, E.H., Eyal, F. and Morrison, J.C. (1977), *J. Infect. Dis.,* **135**, 267–274.

Owens, P. and Salton, M.R.J. (1975), *Biochem. biophys. Res. Commun.,* **63**, 875–880.

Pavlovskis O. and Slade, H.D. (1969), *J. Bact.,* **100**, 641–646.

Powell, D.A., Duckworth, M. and Baddiley, J. (1975), *Biochem. J.,* **151**, 387–393.

Rölla, G. (1976), In: *Microbial Aspects of Dental Caries,* Microbial Abst. Special Suppl. 1, 309–324, (Stiles, Loesche and O'Brien, eds),.

Rosan, B. (1978), *Science,* **201**, 918–920.

Roseman, S. (1975), In: *Cell Membranes – Biochemistry, Cell Biology,* (Weissman, G. and Clarborne, R. eds), pp. 55–64, H.P. Publishing Co. Inc.

Shands, J.W. (1973), In: *Bacterial Lipopolysaccharides* (Kass, E.H. and Wolff, M.S. eds), pp. 189–193. University of Chicago Press, Chicago.

Shockman, G.D. and Slade, H.D. (1964), *J. gen. Microbiol.,* **34**, 297–305.

Silvestri, L.J., Craig, R.A., Ingram, L.O., Hoffmann, E.M. and Bleiweis, A.S. (1978), *Infect. Immun.,* **22**, 107–115.

Slade, H.D. and Hammerling, U. (1968), *J. Bact.,* **95**, 1572–1579.

Smith, D.G. and Shattock, P.M.F. (1964), *J. gen. Microbiol.,* **34**, 165–175.

Smyth, C.J., Jonnson, E., Olsson, O., Soderlind, J., Rosengren, S., Hjerten and Wadstrom R. (1978), *Infect. Immun.,* **22**, 462–472.

Spector, A.A. (1975), *J. Lipid Res.,* **16**, 165–179.

Springer, G.F., Adye, J.C., Bezkorovainy, A. and Jurgensons, B. (1974), *Biochemistry,* **13**, 1379–1389.

Steck, T.L. (1974), *J. Cell Biol.,* **62**, 1–19.

Stewart, F.S. and Martin, W.T. (1962), *J. Path. Bact.,* **84**, 251–261.

Swanson, J., Hsu, K.C. and Gotschlich, E.C. (1969), *J. exp. Med.,* **130**, 1063–1073.

Toon, P., Brown, P.E. and Baddiley, J. (1972), *Biochem. J.,* **127**, 399–409.

van Driel, D., Wicken, A.J., Dickson, M.R. and Knox, K.W. (1973), *J. Ultrastruct. Res.,* **43**, 483–497.

Westphal, O., Luderitz, O. and Bister, F. (1952), *Naturforsch.,* **7b**, 148–155.

Wiegandt, H. (1976), *Adv. exp. Med. Biol.,* **71**, 3–14.

Wicken, A.J. and Baddiley, J. (1963), *Biochem. J.,* **87**, 54–62.

Wicken, A.J., Broady, K.W., Evans, J.D. and Knox, K.W. (1978a), *Infect. Immun.,* **22**, 615–616.

Wicken, A.J., Elliott, S.D. and Baddiley, J. (1963), *J. gen. Microbiol.,* **31**, 231–239.

Wicken, A.J., Gibbens, J.W. and Knox, K.W. (1973), *J. Bact.,* **113**, 365–372.

Wicken, A.J. and Knox, K.W. (1970), *J. gen. Microbiol.,* **60**, 293–301.

Wicken, A.J. and Knox, K.W. (1971), *J. gen. Microbiol.,* **67**, 251–254.

Wicken, A.J. and Knox, K.W. (1975a), *Science,* **187**, 1161–1167.

Wicken, A.J. and Knox, K.W. (1975b), *Infect. Immun.,* **11**, 973–981.

Wicken, A.J. and Knox, K.W. (1977a), *Microbiology,* 360–365.

Wicken, A.J. and Knox, K.W. (1977b), *Prog. Immunol.* **III**, 135–143. Aust. Acad. of Sciences.

Wicken, A.J. and Knox, K.W. (1978b), *Secretory Immunity and Infection,* (McGhee, Mes' ɔcky, J. and Babb, J.L., eds), pp. 619–628, Plenum Publishing Corp., New Yoɪ

6 Attachment of *Mycoplasma pneumoniae* to Respiratory Epithelium

ALBERT M. COLLIER

6.1	Introduction	*page*	161
6.2	Biologic properties of mycoplasmas		161
	6.2.1 Definition		161
	6.2.2 Taxonomy		162
	6.2.3 Morphology		163
	6.2.4 Motility		166
	6.2.5 Metabolism		166
	6.2.6 Reproduction		167
6.3	Attachment of *Mycoplasma pneumoniae* to surfaces		167
	6.3.1 Broth culture		167
	6.3.2 Glass and plastic		167
	6.3.3 Red blood cells		168
	6.3.4 White blood cells		170
	6.3.5 Tissue culture cells		170
	6.3.6 Alveolar macrophages		171
	6.3.7 Spermatozoa		171
6.4	Attachment of *Mycoplasma pneumoniae* to natural target cells		172
	6.4.1 Dispersed tracheal epithelial cells		172
	6.4.2 Hamster trachea in organ culture		173
	6.4.3 Human fetal trachea in organ culture		176
	6.4.4 Respiratory tract of chick embryo		177
	6.4.5 Respiratory tract of hamster		178
6.5	*Mycoplasma pneumoniae* disease in man		178
	6.5.1 Clinical features		178
	6.5.2 Pathology		179
	6.5.3 Pathogenesis		180
6.6	Conclusion		181
	References		181

Acknowledgements
I would like to thank my colleagues in the Pediatric Division of Infectious Diseases and the laboratory of Dr Joel B. Baseman for their collaboration in the study of *M. pneumoniae* disease. I would also like to thank my fellow mycoplasmologists for providing me with reprints and preprints of their publications. Work in our laboratory on *M. pneumoniae* has been sponsored by the U.S. Army Medical Research and Development Command, Public Health Services Grant HL-19171 from the National Heart, Lung and Blood Institute, and R-804577 from the Environmental Protection Agency.

Bacterial Adherence
(*Receptors and Recognition,* Series B, Volume 6)
Edited by E.H. Beachey
Published in 1980 by Chapman and Hall, 11 New Fetter Lane, London EC4P 4EE
© 1980 Chapman and Hall

6.1 INTRODUCTION

In the pathogenesis of infectious diseases, the infecting organism may gain entrance into the host by trauma or the bite of a vector on the external skin; however, most infections begin on the much larger internal surface areas of the respiratory, gastro-intestinal or urogenital systems covered by mucous membranes. In order to be successful in establishing an infection, the microbial organism must overcome the defense mechanisms present on mucous membranes and be able to attach and multiply on the lining epithelial cells (Smith, 1977).

The luminal epithelium of the conducting airways of the respiratory tract are covered by a mucous blanket that is propelled by the beating cilia of the ciliated epithelial cells. Respiratory microbial pathogens must be able to penetrate this protective mucous layer and attach to the underlying epithelial cells to prevent from being swept out of the body by the moving mucous blanket. This chapter will be devoted to the attachment of *Mycoplasma pneumoniae*. This microbial pathogen is an extracellular parasite of the mucous membranes of the lower respiratory tract of humans and the most frequent cause of pneumonia in adolescents and young adults.

6.2 BIOLOGIC PROPERTIES OF MYCOPLASMAS

6.2.1 Definition

Mycoplasmas are the smallest free-living organisms able to exist independently in nature and grow on lifeless media. These micro-organisms, smaller than some of the larger viruses, are approximately 300 nm in diameter and filterable through a 450 nm pore size filter. They contain both DNA and RNA and the genetic information necessary to code for a maximum of about 200 functions. The genome size of mycoplasmas ranges from 10^8-10^9 daltons. They therefore lack many of the enzyme systems possessed by more complex micro-organisms and mammalian cells. Most of the mycoplasmas require media supplemented with factors such as nucleic acid precursors, cholesterol, native protein and NAD for growth. Mycoplasmas lack a rigid cell wall and chemical precursors of cell wall peptidoglycan and are therefore resistant to most forms of penicillin and other antibiotics directed at inhibition of cell wall synthesis. These micro-organisms are contained by a 100 Å thick triple-layered unit membrane very similar to the membrane of animal cells. They reproduce at a slower rate than many micro-organisms and, when grown on agar, typically produce microscopic colonies resembling fried eggs that embed in the medium.

6.2.2 Taxonomy

Mycoplasmas have presented problems for the taxonomist. Some have suggested that they must be animal forms since they lack a cell wall; however, they do not possess a nuclear membrane and therefore are procaryotic and more properly classified as plants. The first organism in the mycoplasma group, *Mycoplasma mycoides,* was isolated by Nocard *et al.* in 1898 and later shown to be the etiologic agent of bovine pleuro-pneumonia. Only within the last two decades has it been appreciated that man also suffers from mycoplasma-induced disease. The recent isolation of mycoplasmas from diseased insects and plants, rumens of cattle and sheep, and soil and coal refuse piles illustrates the expanding knowledge of the ecology of these unique organisms. The mycoplasmas have now been elevated to a taxonomic level equivalent to the true bacteria and viruses and placed in a new and separate class Mollicutes (see Table 6.1).

Table 6.1 Taxonomy of the class mollicutes

Class: Mollicutes
       Order: Mycoplasmatales
          Family   I: *Mycoplasmataceae*
                   Genus I: *Mycoplasma* (about 50 species)
                   Genus II: *Ureaplasma* (one species)
          Family II: *Acholeplasmataceae*
                   Genus I: *Acholeplasma* (six species)
          Family III: *Spiroplasmataceae* (one specie)
                   Genus I: *Spiroplasma* (one species)

Genera of uncertain taxonomic position
   Thermoplasma (one species)
   Anaeroplasma (two species)

The genus *mycoplasma* of which *Mycoplasma pneumoniae* is a member requires sterol for growth. Members of this genus have been isolated from almost all types of animal hosts, but only *M. pneumoniae* is a proven pathogen of man. The genus *Ureaplasma* is unique in that it not only requires sterol for growth, but is also able to hydrolyze urea. The *ureaplasmas* have been isolated from man and animals and form small colonies on agar. The *Acholeplasmas* do not require sterol for growth and have been isolated from animals. The *Spiroplasmas* require sterol and are unique in that they are helical filaments. These organisms have been isolated from plants and insects. The *Thermoplasmas* are thermophilic (optimum 59°C) and acidophilic (optimum pH 1.0–2.0) and have been found in burning coal refuse piles. The *Anaeroplasmas* are anaerobic and have been isolated from the rumens of sheep and goats.

The role of members of the class Mollicutes as pathogens is underscored by the fact that approximately half of the named mycoplasma species now placed in this class have been associated with disease states. In contrast, less than 300 of the multitude of bacterial species among the Schizomycetes are pathogenic.

6.2.3 Morphology

The mycoplasmas possess only the minimum set of organelles necessary for cell growth and replication. They have a plasma membrane to separate the cytoplasm from the external environment, ribosomes to assemble the cell proteins, and a double stranded deoxyribonucleic acid (DNA) molecule to provide information for protein synthesis (Razin, 1978).

In that mycoplasmas have no rigid cell wall, it would be expected that they would be amorphous and without any defined shape. Recent studies employing phase contrast (Bredt, 1968) and electron microscopy (Biberfeld, 1970 and Collier *et al.*, 1972) have provided evidence that *M. pneumoniae* range in shape from that of a coccus to a filamentous form. When log phase cultures of *M. pneumoniae* were examined by scanning and transmission electron microscopy they were found to be branching filaments that were asymmetrical because of the presence of an organized terminal structure (see Fig. 6.1). This specialized organelle consisted of a central core, with a dense central filament; the core was surrounded by a lucent space, which was enveloped by an extension of the organism's plasma membrane (see Fig. 6.2).

Wilson and Collier (1976) examined *M. pneumoniae* using specialized staining methods for thin-section electron microscopy. Indium trichloride stained the nuclear material and ribosomes but not the tip structures. When ethanolic phosphotungstic acid stain was employed, the central filament and terminal button of the terminal structure were visualized implicating the presence of basic proteins. The results of ruthenium red and tannic acid staining provided evidence that the entire *M. pneumoniae* organism was enveloped by an extracellular mucoprotein layer that was especially concentrated around its terminal structure. Two other mycoplasma species have been demonstrated to possess a specialized terminal structure. *Mycoplasma pulmonis,* which causes respiratory disease in rodents, has a globular shape with a short protrusion, giving it the appearance of a miniature Florence flask (Richter, 1970). *Mycoplasma gallisepticum,* a respiratory pathogen of poultry, is an ovoid, or pear shaped structure, having polarity established by the presence of a small bleb (Maniloff, 1973).

In that mycoplasmas have no cell wall and no subcellular organelles that are enclosed by membranes, it has been possible to obtain a limiting membrane preparation that was free of contamination with other membrane types and cell wall products. These organisms have therefore become a valuable model for membrane study. The chemical composition and structure of mycoplasma cell membranes have been examined, and it was found that the isolated mycoplasma membranes resemble plasma membranes of other procaryotes in gross chemical composition, being

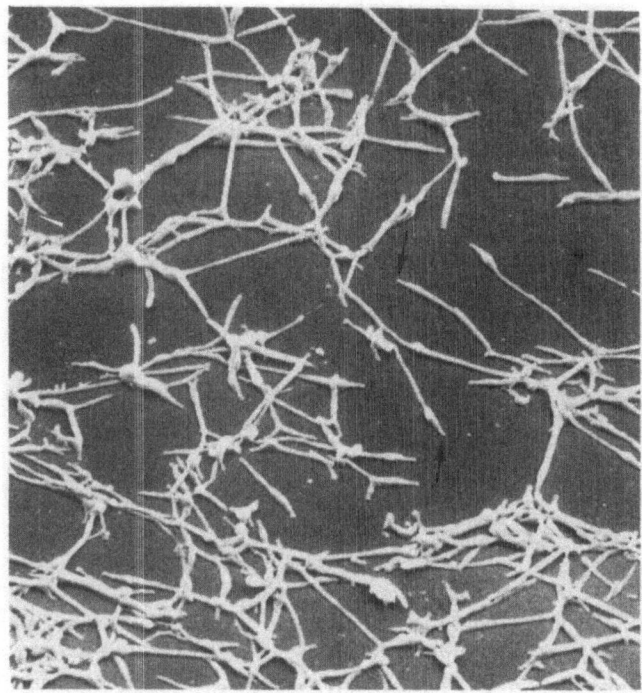

Fig. 6.1 Scanning electron photomicrograph of *M. pneumoniae* attached
to glass demonstrating the filamentous, branching morphology. The
specialized tip structure is seen in close approximation to the glass surface
(arrows). × 7000.

composed mainly of proteins and lipids. Proteins comprise roughly two-thirds of the
mass of the membrane, while the rest is mostly lipids. Carbohydrates are only minor
components of mycoplasma membranes found in glycolipids, lipopolysaccharides,
polysaccharides, and glycoproteins. The location of the carbohydrate containing
components on the cell surface may play a role in the interaction of the parasite
with the cell membrane of its host target cell (Razin, 1978).

Plant lectins have been utilized to search for glycoproteins in *M. pneumoniae*
membranes. This has been done by testing the binding to cells or to isolated membranes of ^{125}I-labeled lectins or by testing for agglutination of cells by lectins.
Schiefer *et al.* (1974), using lectins, found carbohydrate structures presumably bound
to glycolipids, containing both galactose and glucose units, exposed on the surface
of *M. pneumoniae*. The *M. pneumoniae* organisms were agglutinated by the lectins
of *Ricinus communis* and *Canavalia ensiformis*. Agglutination was inhibited by
galactose and glucose respectively and was lost after treatment with periodate.
Pronase treatment of the organisms did not reveal any new sugar determinants.

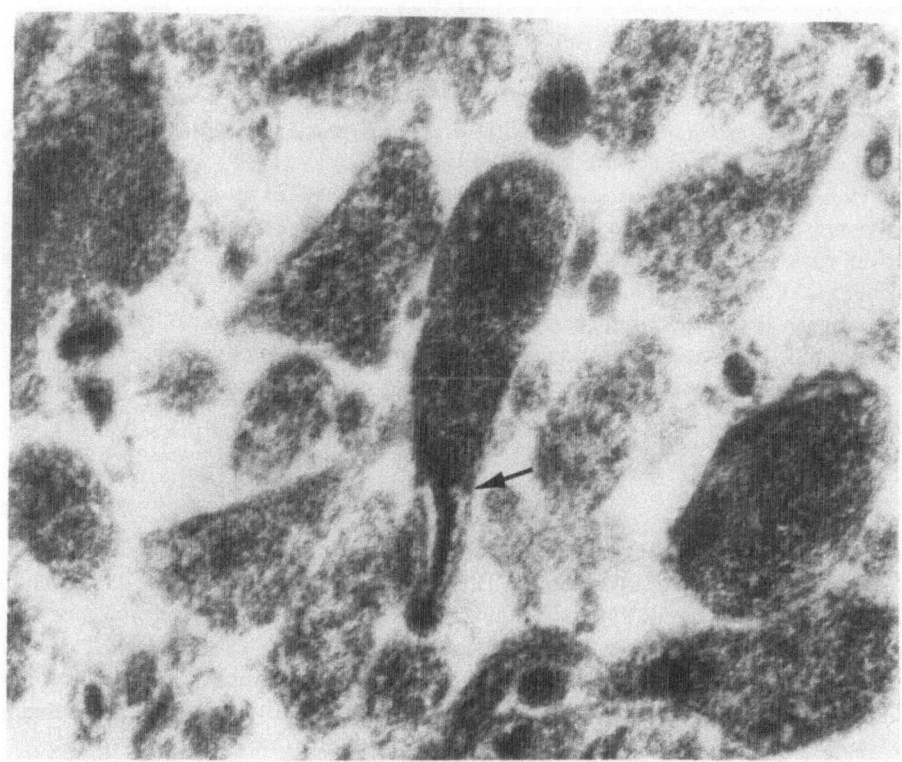

Fig. 6.2 Transmission electron photomicrograph of *M. pneumoniae* grown in broth culture. The internal morphology of the micro-organism is seen with the specialized tip structure (arrow) consisting of a dense central core surrounded by a lucent space enveloped by an extension of the organism's unit membrane. x 72 000.

Kahane and Tully (1976) examined the binding of plant lectins of *Ricinus communis*, and concanavalin A and wheat germ agglutinin to *M. pneumoniae* cells and membranes. The binding of these lectins to *M. pneumoniae* provides evidence for the presence of D-galactose, α-methyl-D-mannoside and N-acetyl-D-glucosamine in the membrane. The binding of lectins to whole cells was similar to that found with isolated membranes providing evidence that significant binding only occurred on the outer surface of the *M. pneumoniae* membrane. Trypsin or pronase treatment of membranes increased the capacity to bind lectins indicating that additional carbohydrate groups on the mycoplasma membrane were masked by a protein layer or protein complexes on the membrane. Kahane and Brunner (1977) isolated a glycoprotein from *M. pneumoniae* membranes with a molecular weight of approximately 60 000. The carbohydrate-containing polypeptide was obtained by extraction with

lithium diiodosalicylate and consisted of 80–90% amino acids and about 7% carbohydrate. The amino acids were mainly glycine and histidine and the carbohydrates
were mainly glucose, galactose and glucosamine.

Evidence for the presence and exposure of glycolipids in the mycoplasma cell
surface has been provided by studies with *M. pneumoniae,* where an antiserum to the
glycolipids was found to agglutinate the cells (Sobeslavsky, 1968). Anionic sites on
M. pneumoniae have been visualized by Schiefer *et al.* (1976) using the electron
microscope and polycationized ferritin. The anionic sites were uniformly distributed
over the entire membrane surface. Chemical and enzymatic treatment of *M. pneumoniae*
indicated that the anionic sites may be lipid phosphate groups.

Mycoplasmas are the only procaryotes depending on cholesterol for growth
Cholesterol is incorporated into the cell membrane of the mycoplasmas and prevents
the membrane from becoming too viscous, therefore playing a role in their survival
and growth (Razin, 1978).

6.2.4 Motility

The motility of both virulent and avirulent *Mycoplasma pneumoniae* cells have been
studied by Radestock and Bredt, 1977. They demonstrated that virulent *M. pneumoniae*
cells possessed a gliding motility and that the avirulent mutant of *M. pneumoniae* was
non-motile and demonstrated reduced adherence. Bredt (1974) also described a
gliding motility by *M. gallisepticum* and *M. pulmonis.* These strains were examined on
liquid-covered surfaces by phase contrast microscopy. Quantitative microcinematography
(30 frames per minute) was also performed. They demonstrated that the virulent
M. pneumoniae strain had a leading end. The leading direction of movement was
always oriented with the specialized tip structure first. The average speed was
constant between 0.2–0.5 μm s^{-1}. Temperature, viscosity, pH and the presence of
yeast extract in the medium influenced the motility significantly, while changes in
glucose, calcium ion, and serum content were less effective. Motility was affected by
iodoacetate, *p*-mercuribenzoate and mitomycin C at inhibitory or sub-inhibitory
concentrations. Sodium fluoride, sodium cyanide, dinitrophenol, chloramphenicol,
puromycin, colchicine and cytochalasin B at minimal inhibitory concentrations did
not affect motility. Movement of the virulent strain was inhibited by *M. pneumoniae*
antisera made against both the virulent and avirulent *M. pneumoniae.* Their studies
with adsorbed antisera suggested that surface components involved in motility were
heat labile. Preliminary evidence has been presented that an actin-like protein is
present in *M. pneumoniae* and this may play some part in motility (Neimark, 1976).

6.2.5 Metabolism

The nutritional requirements of most mycoplasmas have not been delineated due to
the lack of a chemically defined growth medium. The standard basal medium now

in use for the growth of *M pneumoniae* contains yeast extract, serum and glucose. This provides lipoproteins, sterols, nucleic acids and an energy source. *Mycoplasma pneumoniae* is a fermentative organism deriving its carbon and energy from the dissimilation of hexoses such as glucose. Glucose is catabolized by glycolytic pathways with the production of acid (Tully, 1978). *M. pneumoniae* does not possess the ability to metabolize galactose (Hu *et al.,* 1975) and like all other mycoplasmas lacks the orotic acid pathway for pyrimidine synthesis and the enzymatic pathways for the *de novo* synthesis of purine bases. This requires that at least one purine or pyrimidine base be supplemented for mycoplasma growth (Razin, 1978). Hydrogen peroxide is one of the end product of *M. pneumoniae* respiration (Razin, 1969).

6.2.6 Reproduction

The concensus of opinion amongst mycoplasmologists is that mycoplasmas replicate by binary fission like other procaryotes. Phase contrast cinematography of cell division with *Mycoplasma homonis* provided evidence for the transformation of filaments into chains of cocci, which subsequently fragment into single coccoid cells (Bredt *et al.,* 1973).

6.3 ATTACHMENT OF *MYCOPLASMA PNEUMONIAE* TO SURFACES

The role of attachment is a particularly important attribute of *M. pneumoniae*, an extracellular parasite, which produces disease without invasion of the tissue or blood stream of man. The organism must be able to adhere and colonize the epithelial lining of the respiratory tract without being eliminated by the ciliary action and movement of the mucous blanket. In order to examine the attachment process of *M. pneumoniae* many experimental model systems have been employed.

6.3.1 Broth culture

Boatman and Kenny (1971) demonstrated that virulent *Mycoplasma pneumoniae* growing in fluid medium produced free-floating granules which become larger as the culture aged. They called these granules 'spherules'. The spherules were made up of *M. pneumoniae* attaching to each other in a mass. In thin section, spherules were composed of lobulated cells connected together by membranes, and ring-shaped cells. When they treated the spherules with crude porcine pancreatic lipase, large numbers of free organisms were released.

6.3.2 Glass and plastic

Somerson *et al.* (1967) demonstrated that *M. pneumoniae* attached to glass surfaces.

During growth in broth, *M. pneumoniae* formed macroscopic spherical colonies which adhered to the surface of Povitsky bottles used for large-scale cultivation of the organism. Within 48 hours, the colonies formed a confluent layer which covered the glass surface beneath the broth medium. When the broth was removed and the layer of organisms on the glass was washed repeatedly with saline, the colonies adhered tenaciously to the surface. Colonies could be removed by scraping or treatment with 2.5% trypsin. The organisms could not be detached by adding EDTA. This finding suggested that divalent cations were probably not involved in attachment.

Taylor-Robinson and Manchee (1967a) studied the adherence of *M. pneumoniae* to glass and plastic. Adherence occurred more readily to plastic than to pyrex glass, which in turn was superior to soda glass. These authors also found that there was no correlation between hemadsorbtion and adherence to glass or plastic. They found that *M. pneumonia* adhered to five different types of plastic petri dishes.

Biberfeld and Biberfeld (1970) studied the ultrastructural features of *Mycoplasma pneumoniae* by both scanning and transmission electron microscopy when grown in broth on glass and plastic surfaces. They observed that the organisms were filamentous and grew with over-crossing and formed a dense network on the surface with the establishment of colonies composed of rounded and elongated forms. The filaments were usually thinner at the ends and terminated with a knob-like structure. Sectioned organisms were seen to contain ribosome-like structures and a specialized structure at their thinner end, which consisted of a dense rod structure ending with a platelike thickening surrounded by electron-lucent cytoplasm. The strong affinity of the rodlike structure of the terminal end of the organism for uranyl and lead stains suggested that it contained nucleoprotein. They postulated that this condensation of nucleoprotein might reflect a stage in the multiplication of the organism. This peripheral localization of the structure might suggest that it is concerned with the locomotion of the organism. Bredt (1974) observed that *M. pneumoniae* attached to glass could not be removed with ethylene glycol bis (β-aminoethyl ether)-N, N-tetraacetic acid and ethylenediamine tetraacetic acid but could be removed by trypsin and other proteases, suggesting a protein-linked bond.

6.3.3 Red blood cells

Del Giudice and Parvia (1964) first studied the relationship between *Mycoplasma pneumoniae* and erythrocytes by macroscopic techniques. They observed that guinea-pig erythocytes poured over *M. pneumoniae* colonies on agar medium adsorbed to the colonies after a short period of incubation at 37°C. The phenomenon did not occur with other prototypes of human oral mycoplasmas, or with miscellaneous mycoplasma strains. Hemadsorption was inhibited by sera from patients convalescing from primary atypical pneumonia caused by infection with *M. pneumoniae*. Manchee and Taylor-Robinson (1968) studied hemadsorption and hemagglutination by *M. pneumoniae*. All six strains of *M. pneumoniae* tested hemadsorbed human and guinea pig erythrocytes. *M. pneumoniae*, strain FH (the only strain tested), hemagglutinated human and guinea pig erythrocytes. Manchee and Taylor-Robinson (1969) examined erythrocytes treated with neuraminidase and compared them with untreated cells for their ability to adsorb to mycoplasma colonies or to be agglutinated by suspensions of the mycoplasmas. *M. pneumoniae* colonies did not adsorb neuraminidase-treated erythrocytes. The neuraminidase was obtained from *Vibrio cholerae* or A/swine influenza virus and the results obtained by treating erythrocytes with each was closely similar. Neuraminidase treatment of human group O erythrocytes did not block their hemagglutination by *M. pneumoniae*. These authors therefore suggested that the mechanisms of *M. pneumoniae* cell attachment in hemadsorption and hemagglutination tests were not identical.

Sobeslavsky *et al.* (1968) demonstrated that the adsorption sites on *M. pneumoniae* for guinea pig erythrocytes were specifically blocked by homologous but not heterologous antisera. The nature of the *M. pneumoniae* adsorption sites was studied by testing the capacity of different fractions of the organism to block the action of hemadsorption-inhibiting antibodies. The hemadsorption-inhibiting antibodies in rabbit antisera were blocked by the lipid extract of *M. pneumoniae* and by the glycerophospholipid hapten of the organism, but not by the protein or polysaccharide fractions. This indicated that the glycerophospholipid hapten, which was the serologically active component primarily responsible for reaction with complement-fixing antibodies in all antisera and metabolism-inhibiting antibodies in human and rabbit antisera, was also involved in the adsorption reaction with guinea pig erythrocytes. A component of *M. pneumoniae* other than the glycerophospholipid hapten appeared to be involved since the hemadsorption inhibition activity of guinea pig and mule antisera was not blocked by the lipid extract of the organism nor by the hapten. Unlike the glycerophospholipid hapten, the other active component was not heat-stable, since exposure of *M. pneumoniae* to 56°C for 30 min abolished its blocking activity for hemadsorption-inhibiting antibodies in guinea pig and mule sera.

Gorski and Bredt (1977) studied the adherence mechanism of *Mycoplasma pneumoniae* and found evidence that a protein was the responsible substance for attachment to sheep erythrocytes and serum-coated latex particles. In these experiments, adherence was tested on *M. pneumoniae* grown on cover slips. Adherence was observed at 37°C and 22°C but not at 4°C. Variation of the pH over a range from 6–8 and the removal of ions by EDTA or EGTA did not influence attachment. When 1M NaCl was added, no removal of adhered particles was noted. Neuraminidase at a concentration of 25 units ml^{-1} for 1 h destroyed the adherence receptors of the sheep erythrocytes, but did not interfere with the attachment of serum-coated latex particles. They found adherence to be inhibited by pretreatment with aldehydes (glutaraldehyde 0.01% and formaldehyde 3%). Inhibitors of energy metabolism (iodoacetate 5 mM and *p*-chlormercuribenzoate 0.01 mM) inhibited the adherence. Inhibitors of macromolecular synthesis—chloramphenicol 0.1 mg ml^{-1}, mitomycin C 5 mg ml^{-1} ultraviolet light for 4 h—produced no effect. Pretreatment of cell surfaces with gastric mucin (0.3%) and sialic acid (4%) resulted in inhibition of adherence. Proteases (trypsin, chymotrypsin, pronase and proteinase K) in a concentration of

$10 \mu g \, ml^{-1}$ removed the adherence ability of the mycoplasmas. In an attempt to restore adherence, *M. pneumoniae* cells were pretreated with $10 \mu g \, ml^{-1}$ trypsin or pronase and then covered with growth medium. Adherence was restored after 3—4 h if the cells were incubated at 37°C, but not at 4°C. Addition of $10 \mu g \, ml^{-1}$ of chloramphenicol, $5 \mu g \, ml^{-1}$ of mitomycin C or irradiation with UV inhibited the restoration completely. These authors also found that pretreatment with anti-*M. pneumoniae* antisera inhibited attachment. This inhibiting activity was absorbed from the antiserum by native *M. pneumoniae* cells and cells which had been pre-treated with sodium periodate (0.05 M, 60 min at 4°C). Absorption was not effective with cells that had been pretreated with heat (60°C). Digestion with trypsin ($1 \, mg \, ml^{-1}$, 37°C) destroyed the adsorbing material totally after 30 min and gel diffusion tests showed only minimal loss of antigenic material after this treatment.

6.3.4 White blood cells

Zucker-Franklin *et al.* (1966a) studied the interaction of *M. pneumoniae* with mammalian cells. The organisms were seen to attach to neutrophils and eosinophils. The authors found an intimate relationship between *M. pneumonia* and the plasma membrane of mammalian cells. They found the attraction of *M. pneumoniae* to the surface membranes of peripheral blood leukocytes and platelets to be much more pronounced than the reaction which took place between leukocytes and bacteria under these condtions. Evidence of phagocytosis was present in specimens which had been incubated for more than 1 h; polymorphonuclear leucocytes and eosinophils showed large vacuoles in which the organisms could only rarely be recognized. *M. pneumoniae* appeared to proliferate on the plasma membrane of the mammalian cells which had markedly increased their surface area by means of long cytoplasmic processes which extended toward and surrounded the organisms. Zucker-Franklin *et al.* (1966b) examined the uptake of *M. pneumoniae* organisms by monocytes and lymphocytes contained in human peripheral blood buffy coats. *M. pneumoniae* adhered to all leucocytes regardless of whether the cells originated from peripheral blood or from lymph. The incubation of *M. pneumoniae* with cells in human peripheral blood buffy coats resulted in the uptake of the mycoplasmas by more than 50% of the mononuclear cells. *M. pneumoniae* adhered to the plasma membrane of all leukocytes.

6.3.5 Tissue culture cells

Clyde (1961) first demonstrated the attachment of microcolonies of *M. pneumoniae* to rhesus monkey kidney cells in tissue culture by indirect fluorescent antibody staining and an intensified Giemsa stain. The microcolonies on the surfaces of the monkey kidney cells averaged $10 \mu m$ in diameter. The *M. pneumoniae* organisms were shown to be sensitive to tetracycline, but resistant to pencillin, in tissue culture;

growth inhibition was demonstrated by the addition of both rabbit antiserum and serum from patients convalescent from atypical pneumonia (Clyde, 1963). Boatman and Kenny (1971) also studied *M. pneumoniae* in association with HeLa cell cultures. They found individual mycoplasma often aligned in radial apposition to the membranes of the HeLa cells. Manchee and Taylor-Robinson (1969) examined HeLa cells treated with neuraminidase for their ability to adsorb to mycoplasma colonies or be agglutinated by suspensions of the mycoplasmas. *M. pneumoniae* colonies were found to adsorb untreated HeLa cells but not neuraminidase-treated HeLa cells. Zucker-Franklin *et al.* (1966a) examined HeLa cells infected with *M. pneumoniae* by transmission electron microscopy. The limiting membrane of the mycoplasmas were seen in close apposition to the cell membrane of the HeLa cells. Complete obliteration of the space between the mycoplasmas and HeLa cell was frequently seen. No evidence of phagocytosis or active invasion was apparent.

6.3.6 Alveolar macrophages

Powell and Clyde (1975) examined the interaction of *M. pneumoniae* and guinea pig alveolar macrophages by phase contrast and transmission electron microscopy. The alveolar macrophages were washed out of guinea pig lungs and allowed to attach to cover slips. The cover slips were then inoculated with *M. pneumoniae*. After further incubation for 24 hours the cover slips were examined. The macrophages usually had multiple *M. pneumoniae* organisms attached to their surfaces. Transmission electron microscopy demonstrated the micro-organisms to attach by their specialized terminal structure to the macrophage cell membrane; however; there was no evidence that the attached organisms were phagocytosed. After the addition of rabbit anti-mycoplasma serum rapid activation of the infected macrophages was observed and the attached mycoplasmas were endocytosed. Sixty minutes after addition of the antiserum no *M. pneumoniae* organisms were seen attached to the macrophage membranes or the cover slip within reach of the spreading macrophages. Powell and Muse (1977) examined the *in vitro* process of guinea pig alveolar macrophage phayocytosis of *M. pneumoniae* by scanning electron microscopy. They again demonstrated attachment of *M. pneumoniae* to the macrophage membranes without phagocytosis until the organisms had been opsonized by antibody.

6.3.7 Spermatozoa

Taylor-Robinson and Manchee (1967b) studied spermadsorption and spermagglutination by mycoplasmas. Adsorption of bovine and human spermatozoa to colonies of *M. pneumoniae* was investigated and there was increased adsorption at 37°C over 22°C and over 4°C. There was increased spermadsorption with time up to 5.5 h. They also investigated the relationship between motility and viability of spermatozoa and their adsorption to colonies of *M. pneumoniae*. Spermadsorption was rapid and

complete 20 min after collection when 90% of the spermatozoa were actively
motile and viable. Spermatozoa in suspensions incubated at 37°C for 9.5 h or at
56°C for 5 min, were non-motile, but about half the spermatozoa remained viable
and spermadsorption occurred, although less rapidly and less extensively than with
an untreated spermatozoa suspension. Spermatozoa in suspensions subjected to a
temperature of 56°C for 5.5 h were non-viable and adsorbed poorly. Spermadsorption
could be demonstrated best with fresh spermatozoa at 37°C. Spermatozoa adsorbed
strongly to mycoplasma colonies at 22°C and 4°C. Those which adsorbed to colonies
of *M. pneumoniae* did not detach spontaneously even after 2 or 3 days at this temperature. Hemadsorption was inhibited by specific antiserum and therefore inhibition
of adsorption of human spermatozoa to *M. pneumoniae* colonies was attempted with
goat antisera. Pre-inoculation goat serum at a dilution of 1:5 did not inhibit spermadsorption, but a 1:40 dilution of antiserum completely inhibited spermadsorption.

6.4 ATTACHMENT OF *MYCOPLASMA PNEUMONIAE* TO NATURAL TARGET CELLS

In an attempt to examine the role of attachment of *M. pneumoniae* to differentiated
epithelial cells of the respiratory tract both *in vitro* and *in vivo*, experimental models
have been employed.

6.4.1 Dispersed tracheal epithelial cells

Sobeslavsky *et al.* (1968) obtained respiratory epithelial cells by scraping the tracheal
mucous membranes of vervet monkeys, rats and chickens. The respiratory cells were
dispersed by vigorous pipetting and then mycoplasma colonies growing on agar were
overlaid with the tissue cells. After 30 min incubation at room temperature, the
unattached cells were removed by washing and the colonies examined for adsorbed
epithelial cells by low-power microscopy (100 x). The monkey, rat, and chicken
tracheal epithelial cells adsorbed to *M. pneumoniae* colonies. Adsorption to the
colonies was approximately the same whether the test was performed at 37°C, 4°C,
or room temperature. Maximum adsorption of tracheal epithelial cells occurred in
15–30 min and there was no evidence of release of the epithelial cells after 24 h. The
investigators observed that adsorption via the non-ciliated surface of the epithelial
cells occurred as frequently as adsorption via the ciliated surface. When the
epithelial cells were treated with receptor-destroying enzyme (*Vibrio cholerae*
culture filtrate) the ability to adsorb was lost suggesting the *M. pneumoniae* attached
to the epithelial cells by neuraminic acid receptors.

   In order to study the nature of the epithelial cell receptor in more detail, monkey
tracheal epithelial cells were treated with KIO_4, purified neuraminidase from
V. cholerae or influenza B virus. Each of these treatments was successful in removing

the epithelial cell receptors for *M. pneumoniae*. When the *M. pneumoniae* colonies were pretreated with neuraminic acid containing substances—neuraminelactose, gastric mucin or *N*-acetylneuraminic acid—adsorption of tracheal epithelial cells no longer occurred. Pretreatment of the epithelial cells for 30 min with heat, trypsin, parainfluenza virus type 3 on *M. pneumoniae* failed to remove the receptors for *M. pneumoniae* colonies.

The authors also found that pretreatment of the *M. pneumoniae* colonies with the appropriate concentrations of formaldehyde, merthiolate and actinomycin D prevented adsorption of the tracheal cells. The concentrations employed of these substances also caused *M. pneumoniae* death. When the *M. pneumoniae* colonies were treated with appropriate concentrations of KIO_4, trypsin or puromycin, or treatment with heat ($56°C$ for 3 min) or pH 4.5, adsorption of tracheal epithelial cells was inhibited. After these treatments the *M. pneumoniae* were still viable. Eight units of receptor destroying enzyme or 4000 μg ml^{-1} of aminonucleoside did not affect the adsorbtive ability of *M. pneumoniae* colonies.

In these studies, the adsorption sites on *M. pneumoniae* colonies for monkey tracheal epithelial cells were specifically blocked by homologous but not heterologous antisera. These studies also demonstrated that the afinity of *M. pneumonia* for respiratory tract epithelium was unique among the mycoplasma species that infect man. The authors postulated that the ability of attachment to the respiratory epithelium played a role in virulence, since attachment would provide an opportunity for peroxide, secreted by the organism, to attack the epithelial cell membrane without being rapidly destroyed by catalase or peroxidase present in extracellular body fluids.

6.4.2 Hamster trachea in organ culture

Hamster tracheal organ culture was first employed for the study of mycoplasma disease by Collier *et al.* (1969). This *in vitro* model provided a controllable environment for analysis of the interaction between *M. pneumoniae* and the natural target cell of this pathogen, the ciliated respiratory epithelium. In this system, *M. pneumoniae* produced ciliostasis and cytopathology. The time required for these changes to occur were related inversely to the initial number of organisms placed in the culture. The cytopathology consisted of epithelial cell cytoplasmic eosinophilia and vacuolization, nuclear swelling and chromatin margination, loss of cilia and finally desquamation of the epithelial layer. Microcolonies of *M. pneumoniae* were observed on the luminal surface of the epithelium. No ciliostasis or cytopathology was seen in control cultures or cultures inoculated with four other human mycoplasma species (*M. fermentans, M. hominis, M. orale,* type I and *M. salavarium*).

Immunofluorescent and electron microscopy methods were used to investigate the relationship of the *M. pneumoniae* organisms to the cultured epithelium. Immunofluorescence demonstrated the organisms to be localized to the epithelial layer. Electron microscopy permitted visualization of individual organisms which were found to be present on the luminal surface of the epithelium between the cilia and in

Fig. 6.3 Transmission electron photomicrograph of the luminal border of
a hamster trachea ciliated epithelial cell infected in organ culture for 72 h
with *M. pneumoniae*. The filamentous organisms are seen to attach by
their specialized tip (arrows) to the epithelial cell between the cilia.
c = cilium, m = mycoplasma, mv = microvillus. x 40 000.

intercellular spaces. The *M. pneumoniae* were filamentous in morphology and
oriented to the epithelial cell membrane with the specialized terminal structure in
very close approximation (Figs. 6.3 and 6.4). No evidence of intracellular localization was seen (Collier *et al.*, 1971, Collier and Baseman, 1973, Muse *et al.*, 1976).

The hamster tracheal organ culture model has been utilized to examine the role
of attachment in the pathogenesis of *M. pneumoniae* disease. Collier and Baseman
(1973) labeled *M. pneumoniae* organisms with radioisotopes for attachment studies
with tracheal ring epithelium and used radioautography and liquid scintillation
counting as means of quantitation. Using these techniques with hamster tracheal
organ culture, Powell *et al.* (1976) demonstrated that after attachment of labeled
M. pneumoniae, radioautographs revealed emulsion grains on the luminal surface of
the tracheal rings. To confirm that the grains did represent isotope-labeled mycoplasmas, high-resolution radioautography was performed using electron microscopy.

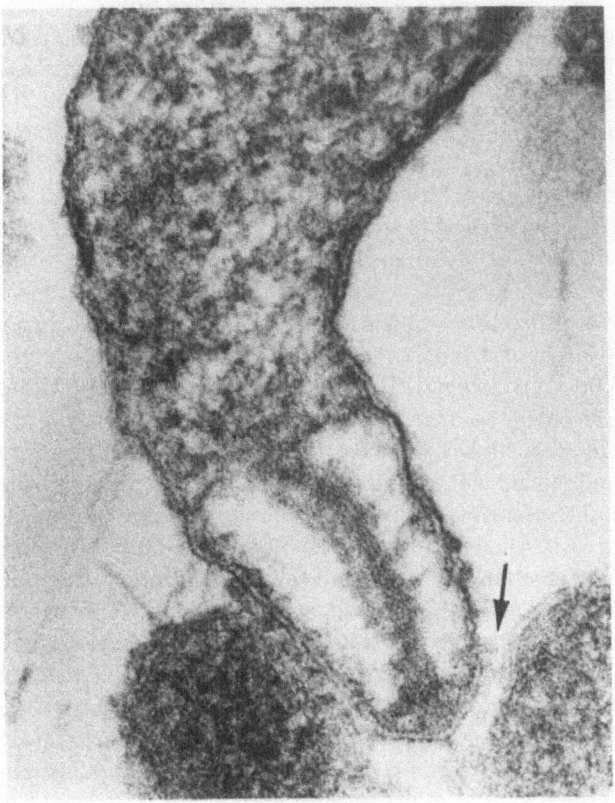

Fig. 6.4 Transmission electron photomicrograph of an *M. pneumoniae* organism attached to a hamster tracheal epithelial cell. The close approximation (arrow) of the mycoplasma cell membrane to the cell membrane of the host epithelial cell is demonstrated. x 250 000.

Individual grains were found to be associated with *M. pneumoniae* organisms. The scintillation technique was employed to accurately quantitate mycoplasma attachment. Attachment of *M. pneumoniae* to hamster tracheal rings in organ culture was found to occur rapidly and to follow first-order kinetics. Powell *et al.* (1976) also examined factors that influenced the attachment of *M. pneumoniae* in this system. Maximum inhibition of attachment occurred when the radiolabeled organisms were heat-killed at 56°C for 30 min prior to incubation with the tracheal rings. There was also a decrease in attachment when Formalin-fixed rings were incubated with normal organisms and when normal rings and normal organisms were incubated at 4°C for 4 h. Tracheal rings were also treated with neuraminidase, glucose oxidase, galactose oxidase or sodium periodate for 30 min prior to incubation with *M. pneumoniae*. The organism attachment was significantly decreased only on rings pretreated with neuraminidase or periodate. Pretreatment of *M. pneumoniae* with the same agents prior to incubation with untreated rings resulted in a significant decrease in attachment only with periodate; however, the authors observed a 1000-fold decrease in organism viability after treatment with periodate. These studies indicated that proteins and carbohydrate moieties may play a role in the attachment of *M. pneumoniae* to respiratory epithelial cells.

The alterations in macromolecular biosynthesis and metabolic activity produced in the epithelial cells of tracheal organ cultures after attachment by *M. pneumoniae* were examined by Hu *et al.* (1975 and 1976) utilizing ^{14}C-labeled galactose, ^{14}C-labeled orotic acid and ^{14}C-labeled amino acids. A decline in protein and RNA synthesis was demonstrated in the tissue prior to the appearance of histological changes suggesting that the primary effect of mycoplasma infection on tracheal organ culture was at the transcription or translational level. The addition of erythromycin within 24 hours after infection prevented the onset of abnormal orotic acid uptake and subsequent cytopathology. In these studies, evidence was presented that *M. pneumoniae* organisms need to attach and remain metabolically active in order to produce epithelial cell injury; however, the mechanism of injury mediation was not known.

Gabridge *et al.* (1974) inoculated *M. pneumoniae* membranes into hamster tracheal organ cultures with resultant ciliostasis and cell cytopathology; however, Hu *et al.* (1976) could not confirm these findings. Gabridge *et al.* (1977a) presented data that the mechanism of attachment of *M. pneumoniae* membranes was different from the sialic acid receptor site-mediated attachment of viable *M. pneumoniae* cells. Gabridge *et al.* (1977b) found that attachment of *M. pneumoniae* to hamster tracheal epithelium was inversely correlated with degree of ciliation. Engelhardt and Gabridge (1977) also examined the effect of vitamin A deficiency-induced metaplasia on the attachment of *M. pneumoniae* to the tracheal epithelium. There was a more than two-fold decrease in the attachment of *M. pneumoniae* to squamous epithelium as compared to ciliated columnar epithelium.

Hu *et al.* (1977) employed organ cultures of hamster trachea in providing evidence that virulent *M. pneumonia* organisms possess a surface protein that plays a role in attachment to the respiratory epithelium. This protein was trypsin-sensitive and when *M. pneumoniae* organisms were mildly treated with a protease, attachment no longer occurred. After incubating the protease-treated organism in mycoplasma medium, the protein was regenerated and the ability to attach returned. Erythromycin inhibited resynthesis of the protein. Lactoperoxidase-catalyzed iodination of intact *M. pneumoniae* organisms further confirmed that the protein was an external surface protein.

6.4.3 Human fetal trachea in organ culture

Tracheal tissue from human fetuses (16–24 weeks gestation) has been utilized to examine the interaction of *M. pneumoniae* with the natural target cell of this respiratory pathogen *in vitro*. Butler (1969) was the first to use human fetal tracheal organ cultures as a means for isolating and growing mycoplasmas. He inoculated

M. pneumoniae at a high titer ($10^{5.7}$ colony form units ml^{-1}) into human fetal tracheal organ cultures and observed cessation of cilia motion and rounding of epithelial cells followed by loss of the epithelial layer. However, no organisms could be recovered from the organ culture at the end of the experiment.

Collier and Clyde (1971) and Collier (1972) examined the interaction between human fetal tracheal organ cultures and five mycoplasma species indigenous to man (*M. pneumoniae, M. hominis, M. fermentans, M. salivarium,* and *M. oral,* type 1). Of the five mycoplasma species studied, only *M. pneumoniae* produced ciliostasis and cytopathology. No attachment of *M. hominis, M. fermentans, M. salivarium* or *M. orale,* type 1 to the respiratory epithelium in organ cultures was observed. *M. pneumoniae* produced complete ciliostasis by four days and a sequence of cytopathological changes that were restricted to the epithelial layer. These changes consisted of cytoplasmic vacuolization, nuclear swelling with chromatin margination, and loss of cilia. The cilia were lost through a process of protrusion of the apical portion of the cell into the lumen with the cilia forming a sunburst configuration followed by loss of the apical portion of the cell into the medium. This process had been observed in sputum specimens collected from patients with respiratory tract infection by Papanicolaou and termed ciliocytophthoria.

When the *M. pneumoniae*-infected tracheal rings were examined by immuno-fluorescent and electron microscopy, *M. pneumoniae* organisms were found to be concentrated on the luminal surface of the epithelium in an extracellular position. Electron microscopy demonstrated that the micro-organisms attached to the epithelial cell by close approximation of the organism's specialized tip with the host cell membrane; but no fusion of the two membranes at the point of contact was observed. The *M. pneumoniae* organisms attached to all cell types making up the luminal epithelium, but no organisms were seen within the epithelial cells.

An avirulent *M. pneumoniae* organism was also examined in the human fetal tracheal organ culture system. This organism had been derived from the parent virulent strain by passage 169 times in broth. The avirulent strain possessed the specialized tip structure but had lost the ability to attach to epithelial and red blood cells and was no longer motile. Lipman and Clyde (1969) compared the avirulent strain with the virulent parent strain and found no difference in their ability to produce hydrogen peroxide. The avirulent strain produced no ciliostasis or cyto-pathology in the tracheal cultures. No attachment to the human fetal respiratory epithelium by avirulent *M. pneumoniae, M. hominis, M. fermentans, M. salivarium,* or *M. orale,* type I was observed.

6.4.4 Respiratory tract of chick embryo

In early studies by Donald and Liu (1959), an attempt to find the virus of primary atypical pneumonia (later proven to be *M. pneumoniae*) in infected chick embryo lungs by electron microscopy was unsuccessful. However, the authors did demonstrate the presence of *M. pneumoniae* on the luminal surface of infected bronchi of the chick embryo lungs by immunofluorescence reactions employing convalescent human serum and fluorescein-labeled anti-human γ-globulin rabbit serum. Marmion and Goodburn (1961) were first to suggest that the agent producing primary atypical pneumonia might be a mycoplasma when they demonstrated extracellular, minute cocco-bacilli on chick bronchial epithelium.

6.4.5 Respiratory tract of hamster

When Syrian hamsters are inoculated intranasally with a broth culture of *M. pneumoniae,* they develop a pneumonia that is histologically similar to that seen in humans with natural disease. Collier and Clyde (1974) examined the lungs of hamsters infected for 2 weeks with *M. pneumoniae.* After being anesthetized, the lungs were fixed *in situ* by perfusion with glutaraldehyde. The lungs were then removed and fixed in osmium tetroxide and prepared for electron microscopy. Areas containing bronchi showing peribronchial infiltrates with mononuclear cells were found by light microscopy of 1 μm sections. Thin sections of corresponding areas were cut and examined by electron microscopy. The *M. pneumoniae* organisms were recognized by their dense staining and characteristic morphological features. They were found only at the respiratory epithelial luminal border where peribronchial infiltrates were present. The organisms were filamentous in shape and attached to the bronchial epithelial lining cells with a constant orientation of the organism's terminal specialized organelle which formed the point of contact with the host epithelial cell. The *M. pneumoniae* were seen only in the conducting airways and at no time were the organisms seen within epithelial cells.

6.5 *MYCOPLASMA PNEUMONIAE* DISEASE IN MAN

Eaton *et al.* (1944) isolated a filterable agent from the sputa of patients with cold agglutinin positive 'primary atypical pneumonia' that produced pneumonias in cotton rats and hamsters. Chanock *et al.* (1962) provided proof that the agent isolated by Eaton was not a virus, but a pleuropneumonia-like micro-organism, when they succeeded in growing the agent on artificial media for the first time. Chanock *et al.* (1961) and Rifkind *et al.* (1962) then fulfilled Koch's postulates by demonstrating experimental disease produced in human volunteers inoculated with organisms isolated in the laboratory. The morbidity of the disease was favorably altered by treatment with demethylchlortetracycline. Chanock *et al.* (1963) subsequently proposed the name *Mycoplasma pneumoniae* for the organism.

6.5.1 Clinical features

Natural *M. pneumoniae* disease is limited to man. Illnesses due to this infectious agent occur worldwide. These infections are most frequent in school-aged children, particularly those 5—19 years old and in closed populations such as military, college students and prisoners. Transmission of the disease is thought to occur by the airborne route and disease has been produced experimentally by aerosol inoculation (Rifkind *et al.*, 1962). The incubation period is from 2—3 weeks, therefore accounting for the slow spread through closed groups.

The disease produced by *M. pneumoniae* is usually mild and self-limiting with infrequent complications and sequelae. The most common manifestation of the disease is cough, which may be paroxysmal. Other general symptoms include: malaise, headache, fever and sore throat. Many clinical studies report the occurrence of tracheobronchitis and pneumonia infiltrates in the lungs. The diagnosis of *M. pneumoniae* disease is confirmed by isolation of this agent from the respiratory tract or a four-fold serologic rise in antibody titer in acute and convalescent serum samples. Following infection a lengthy carrier state of up to 5 months may occur with *M. pneumoniae* being present in the pharynx on culture.

6.5.2 Pathology

In that patients with *M. pneumoniae* disease rarely die, information available concerning pathologic changes in the infected host is limited. Maisel *et al.* (1967) reported a fetal case from which *M. pneumoniae* was isolated directly from the lung in artificial media. The pathologic findings were similar to eight earlier cases reported by Parker *et al.* (1947) without specific etiologic information. These autopsies demonstrated thick exudate in the bronchi and bronchioles with inflamed and ulcerated linings. Microscopic examination revealed evidence of interstitital pneumonitis and desquamative bronchitis and bronchiolitis with infiltration of mononuclear cells in the peribronchial areas. The airways contained necrotic epithelial cells, mononuclear and polymorphonuclear cells and fibrin. There have been no reports of the visualization of *M. pneumoniae* organisms in the lungs of the few reported human cases coming to autopsy.

Collier and Clyde (1974) examined sputum specimens from patients with culture proved natural *M. pneumoniae* disease. Sputum samples from the lower respiratory tract were collected for culture and for light and electron microscopy. Sputum samples from 10 patients were cultured and yielded $10^2 - 10^6$ colony-forming units ml^{-1} of sputum. Light microscopic examination of sputum smears stained with Wright's stain revealed polymorphonuclear leucocytes, mononuclear cells and desquamated ciliated epithelial cells. On the portions of sputum examined by electron microscopy, 1 μm sections were cut using a glass knife and clumps of desquamated epithelial cells located. Thin sections of this area were prepared and examined by transmission electron microscopy. The luminal border of the desquamated epithelial cells was heavily parasitized by *M. pneumoniae* organisms that were easily identified by their dense staining, filamentous appearance, and specialized tip structure. The organisms were attached to the epithelial cells by their differentiated terminal structure.

6.5.3 Pathogenesis

In that little data have been provided by the pathological studies of natural
M. pneumoniae disease, most of the information on pathogenesis has been obtained
by study of experimental models. The ability of the *M. pneumoniae* organisms to
avoid the host defenses and gain access and attach to the surface membranes of the
epithelium lining of the conducting airways has been documented in experimental
models and natural human disease. The attachment of the organism permits the
establishment of infection in a susceptible host. The ability of *M. pneumoniae* to
evade phagocytosis by alveolar macrophages in the absence of opsonic antibody may
also play an important role in the establishment of disease. Attachment to the
epithelium prevents the organism from being swept out of the respiratory tract by the
normal mucociliary clearance mechanism. Studies performed with *M. pneumoniae* in
hamster and human fetal tracheal organ cultures have provided evidence that this
micro-organism produced injury to ciliated cells with resultant slowing of cilia or
ciliostasis. If this process is also present in humans with natural disease, this may
interfere with the normal functioning of the mucociliary escalator thus aiding the
M. pneumoniae organisms in their ability to infect the respiratory tract. The improper
functioning of the mucociliary escalator may lead to stasis of mucous secretions in
the lower respiratory tract resulting in the frequent cough, an attempt to clear
secretions, that is seen in this disease.

The success of *M. pneumoniae* in producing disease may be in part due to
'biological mimicry' by antigenic similarities between its membrane glycerophos-
pholipids and those of the host tissues. Evidence for this is provided by patients with
M. pneumoniae disease who develop antibodies to lung, brain and liver tissue
(Biberfeld, 1970) and a frequent cold hemagglutinin response involving antibody to
the host's erythrocyte I-antigen (Feizi and Taylor-Robinson, 1967). Patients with
M. pneumoniae pneumonia have tuberculin anergy for weeks or months following
illness (Biberfeld and Sterner, 1976) and there are circulating immune complexes
in the serum of some *M. pneumoniae* pneumonia patients during acute phases of
disease (Biberfeld and Norberg, 1974). There is epidemiologic evidence that
repeated infections with *M. pneumoniae* are required to 'sensitize' the patients
before symptomatic disease occurs. This may explain the paucity of disease below
6 years of age (Fernald *et al.*, 1975). Fernald and Clyde (1976) have hypothesized
that the immune response mounted to *M. pneumoniae* may be a substantial contribution to the expression of disease in man.

The mechanism of host cell injury by *M. pneumoniae* after attachment has not
been defined. However, it has been demonstrated that a prerequisite for epithelial
cell injury in the hamster tracheal organ culture model is that *M. pneumoniae*
organisms attach and remain metabolically active. Gabridge and Dee Barden Stahl

(1978) have presented evidence that the addition of 0.01 mM adenine sulfate to hamster tracheal organ cultures exerted a significant protective effect against *M. pneumoniae* infection. In that paper, Gabridge *et al.* postulate that the epithelial cell cytotoxicity produced by *M. pneumoniae* infection is due to nucleic acid depletion.

6.6 CONCLUSION

In summary, when *M. pneumoniae* attach to respiratory epithelial cells, a specialized terminal structure is located immediately adjacent to the membrane of the host cell. The adhesin on the membrane of *M. pneumoniae* is probably a high molecular weight protein. Trypsin or heat treatment of *M. pneumoniae* prevents attachment to tracheal epithelial cells, erythrocytes, and polystyrene beads. The attachment ability can be restored to trypsin-treated *M. pneumoniae* by incubation in growth medium for 4 h at 37°C, but not at 4°C. In that the adhesin is not restored in the presence of chloramphenicol, mitomycin C or ultraviolet irradiation, this protein probably does not accumulate intracellularly, but has to be newly synthesized after removal by trypsin.

One receptor on eucaryotic cells for *M. pneumoniae* has been demonstrated to be a sialic acid moiety. Pretreatment of respiratory epithelial cells with neuraminidase decreased *M. pneumoniae* attachment about 50%. The reason why neuraminidase treatment does not completely block attachment of *M. pneumoniae* to respiratory epithelial cells is unclear, although there is recent evidence that extensive treatment of eucaryotic cells with neuraminidase does not remove all the sialic acid residues (Razin, 1978).

The importance of *M. pneumoniae* attachment in the pathogenesis of this common human respiratory disease has been well documented. More information on the mechanism of attachment should prove helpful in preventing or altering the disease process in favor of the human host.

REFERENCES

Biberfeld, G. (1970), *Clin. exp. Immunol.*, **8**, 319–333.
Biberfeld, G. and Biberfeld, P. (1970), *J. Bact.*, **102**, 855–861.
Biberfeld, G. and Nornerg, R. (1974), *J. Immun.*, **112**, 413–415.
Biberfeld, G. and Sterner, G. (1976), *Scand. J. Infect. Dis.*, **8**, 71–73.
Boatman, E.S. and Kenny, G.E. (1971), *J. Bact.*, **106**, 1005–1015.
Bredt, W. (1968), *Path. Microbiol.*, **32**, 321–326.
Bredt, W., Heunert, H.H., Hofling, K.H. and Milthaler, B. (1973), *J. Bact.*, **113**, 1223–1227.
Bredt, W. (1974), *INSERM*, **33**, 47–54.

Butler, M. (1969), *Nature* (London), **224**, 605–606.

Chanock, R.M., Rufkind, D., Kraveta, H.M., Knight, V. and Johnson, K.M. (1961), *Proc. natn. Acad. Sci. U.S.A.*, **47**, 887–890.

Chanock, R.M., Hayflick, L. and Barile, M.F. (1962), *Proc. natn. Acad. Sci. U.S.A.*, **48**, 41–49.

Chanock, R.M., Dienes, L., Eaton, M.D., Edward, D.G., Freundt, E.A., Hayflick, L., Hers, J.F.Ph., Jensen, K.E., Liu, C., Marmion, B.P., Morton, H.E., Mufson, M.A., Smith, P.F., Somerson, N.L. and Taylor-Robinson, D. (1963), *Science*, **140**, 662.

Clyde, W.A., Jr. (1961), *Proc. Soc. exp. Biol. Med.*, **107**, 715–718.

Clyde, W.A., Jr. (1963), *Am. Rev. Resp. Dis.*, **88**, 212–217.

Collier, A.M., Clyde, W.A., Jr. and Denny, F.W. (1969), *Proc. Soc. exp. Biol. Med.*, **132**, 1153–1158.

Collier, A.M. and Clyde, W.A., Jr. (1971), *Infect. Immun.*, **3**, 694–701.

Collier, A.M., Clyde, W.A., Jr. and Denny, F.W. (1971), *Proc. Soc. exp. Biol. Med.*, **136**, 569–573.

Collier, A.M. (1972), In: *Pathogenic Mycoplasmas*. CIBA Foundation Symposium (Elliot, K. and Birch, J., eds.), pp. 307–327. Associated Scientific Publishers, Amsterdam.

Collier, A.M. and Baseman, J.B. (1973), *Ann. N.Y. Acad. Sci.*, **225**, 277–289.

Collier, A.M. and Clyde, W.A., Jr. (1974), *Am. Rev. Resp. Dis.*, **110**, 765–773.

Del Giudice, R.A. and Pavia, R. (1964), *Bact. Proc.* 71.

Donald, H.B. and Liu, C. (1959), *Virology*, **9**, 20–29.

Eaton, M.D., Meiklejohn, G. and Van Herick, W. (1944), *J. exp. Med.*, **79**, 649–668.

Engelhardt, J.A. and Gabridge, M.G. (1977), *Infect. Immun.*, **15**, 647–655.

Feizi, T. and Taylor-Robinson, D. (1967), *Immunol.*, **13**, 405–409.

Fernald, G.W., Collier, A.M. and Clyde, W.A., Jr. (1975), *Pediatrics*, **55**, 327–335.

Gabridge, M.G., Johnson, C.K. and Cameron, A.M. (1974), *Infect. Immun.*, **10**, 1127–1134.

Gabridge, M.G., Barden-Stahl, Y.D. Polisky, R.G. and Engelhardt, J.A. (1977a), *Infect. Immun.*, **16**, 766–772.

Gabridge, M.G., Agee, C.C. and Cameron, A.M. (1977b), *J. Infect. Dis.*, **135**, 9–19.

Gabridge, M.G. and Dee Barden Stahl, Y. (1978), *Med. Microbiol. Immun.*, **165**, 43–55.

Gorski, F. and Bredt, W. (1977), *FEMS Letters*, **1**, 265–267.

Hu, P.C., Collier, A.M. and Baseman, J.B. (1975), *Infect. Immun.*, **11**, 704–710.

Hu, P.C., Collier, A.M. and Baseman, J.B. (1976), *Infect. Immun.*, **14**, 217–224.

Hu, P.C., Collier, A.M. and Baseman, J.B. (1977), *J. exp. Med.*, **145**, 1328–1343.

Kahane, I. and Tully, J.G. (1976), *J. Bact.*, **128**, 1–7.

Kahane, I. and Brunner, H. (1977), *Infect. Immun.*, **18**, 273–277.

Lipman, R.P. and Clyde, W.A., Jr. (1969), *Proc. Soc. exp. Biol. Med.*, **131**, 1163–1167.

Maisel, J.C., Babbitt, L.H. and John, T.J. (1967), *J. Am. med. Ass.*, **202**, 287–290.

Manchee, R.J. and Taylor-Robinson, D. (1968), *J. gen. Microbiol.*, **50**, 465–478.

Manchee, R.J. and Taylor-Robinson, D. (1969), *J. Bact.*, **98**, 914–919.

Maniloff, J. and Morowitz, H.J. (1972), *Bact. Rev.*, **36**, 263–290.

Marmion, B.P. and Goodburn, G.M. (1961), *Nature*, **189**, 247–248.

Muse, K.E., Powell, D.A. and Collier, A.M. (1976), *Infect. Immun.*, **13**, 229–237.

Neimark, H.C. (1977), *Proc. natn. Acad. Sci. U.S.A.*, **74**, 4041–4045.

Parker, F., Jr., Jolliffe, L.S. and Finland, M. (1947), *Arch. Pathol.*, **44**, 587–608.

Powell, D.A. and Clyde, W.A., Jr. (1975), *Infect. Immun.*, **11**, 540–550.

Powell, D.A., Hu, P.C., Wilson, M.H., Collier, A.M. and Baseman, J.B. (1976), *Infect. Immun.*, **13**, 959–966.

Powell, D.A. and Muse, K.A. (1977), *Lab. Invest.*, **37**, 535–543.

Radestock, U. and Bredt, W. (1977), *J. Bact.*, **129**, 1495–1501.

Razin, S. (1969), *A. Rev. Microbiol.*, **23**, 317–356.

Razin, S. (1978), *Microbiol. Rev.*, **42**, 414–470.

Richter, C.B. (1970), In: *U.S. Atomic Energy Commission Symp.*, **21**, 365–382, Oak Ridge, Tenn.

Rifkind, D., Chanock, R., Mravetz, H., Johnson, K. and Knight, V. (1962), *Am. Rev. Resp. Dis.*, **85**, 479–489.

Schiefer, H.-G., Gerhardt, U., Brunner, H. and Krupe, M. (1974), *J. Bact.*, **120**, 81–88.

Schiefer, H.-G., Kruass, H., Brunner, H. and Gerhardt, U. (1976), *J. Bact.*, **127**, 461–468.

Smith, H. (1977), *Bact., Rev.*, **41**, 475–500.

Sobeslavsky, O., Prescott, B. and Chanock, R.M. (1968), *J. Bact.*, **96**, 695–705.

Somerson, N.L., James, W.D., Walls, B.E. and Chanock, R.M. (1967), *Ann. N.Y. Acad. Sci.*, **143**, 384–389.

Taylor-Robinson, D. and Manchee, R.J. (1967a), *J. Bact.*, **94**, 1781–1782.

Taylor-Robinson, D. and Manchee, R.J. (1967b), *Nature*, **215**, 484–487.

Tully, J.G. (1978), In: *Mycoplasma Infection of Cell Cultures* (McGarrity, G.J., Murphy, D.G. and Nichols, W.W., eds), pp. 1–33, Plenum Publishing Corporation, New York.

Wilson, M.H. and Collier, A.M. (1976), *J. Bact.*, **125**, 332–339.

Zucker-Franklin, D., Davidson, M. and Thomas, L. (1966a), *J. exp. Med.*, **124**, 521–532.

Zucker-Franklin, D., Davidson, M. and Thomas, L. (1966b), *J. exp. Med.*, **124**, 533–542.

7 Adhesive Properties of Enterobacteriaceae

J. P. D U G U I D and D. C. O L D

7.1	Introduction	*page*	187
	7.1.1 Terminology		187
	7.1.2 Haemagglutination, adhesion and fimbriae		188
7.2	MS adhesin and type-1 fimbriae		189
	7.2.1 Haemagglutinating specificity and mannose sensitivity		189
	7.2.2 MS adhesin in different enterobacteria		194
	7.2.3 Association of MS adhesin with fimbriae		194
	7.2.4 Adhesion to cells other than erythrocytes		196
	7.2.5 Phase variation and cultural conditions		197
	7.2.6 Inhibition by D-mannose and its analogues		199
	7.2.7 MS adhesive sites on fimbriae		200
	7.2.8 Function of the MS adhesin		201
7.3	MRE adhesins and type-MRE fimbriae		202
	7.3.1 Specificity, properties and occurrence		202
	7.3.2 Elution at raised temperatures		203
	7.3.3 Multiple adhesins in the same strain		204
	7.3.4 Association of MRE adhesins with fimbriae		205
	7.3.5 Adhesion to epithelium		205
	7.3.6 Function of MRE adhesins		206
7.4	Other adhesins and fimbriae		207
	7.4.1 MR/K adhesin and type-3 fimbriae		207
	7.4.2 MR/P adhesin and type-4 fimbriae		210
	7.4.3 Non-adhesive fimbriae of types 2 and 6		211
7.5	Conclusion		213
	References		215

Bacterial Adherence
(*Receptors and Recognition,* Series B, Volume 6)
Edited by E.H. Beachey
Published in 1980 by Chapman and Hall, 11 New Fetter Lane, London EC4P 4EE
© 1980 Chapman and Hall

7.1 INTRODUCTION

Many bacteria in different genera and species of Enterobacteriaceae form surface substances or appendages that render them adhesive for a variety of substrates. The properties of the commonest of these adhesive factors, or 'adhesins', namely the mannose-sensitive haemagglutinin borne on type-1 fimbriae, have been well defined, but those of most others are still poorly understood. The various adhesins differ in their pattern of affinities for different substrates, the cultural conditions required for their optimal development and the most effective methods of testing for their presence. A careful selection of methods is therefore needed to ensure their detection and, particularly, to distinguish different adhesins which may be present in the same bacterial strain.

7.1.1 Terminology

Whilst most of the known adhesive factors render the bacteria capable of agglutinating red blood cells, and so may be called haemagglutinins, they also enable the bacteria to adhere to other kinds of cells and substrates, and as their natural function is probably concerned with substrates other than red cells, they are more appropriately named 'adhesins' than 'haemagglutinins' (Duguid, 1959).

The numerous, generally adhesive filamentous appendages of enterobacteria are called 'fimbriae' (Duguid *et al.*, 1955) rather than 'pili' (Brinton, 1959), both because the former term has priority (Duguid *et al.*, 1966; Duguid and Anderson, 1967) and because it seems useful to follow Ottow (1975), Jones (1977) and Ørskov *et al.* (1977) in reserving 'pili' for the scanty sex fimbriae that are directly concerned with the conjugative transfer of DNA. These pili may be adhesive for specific types of recipient bacteria, but have no other known adhesive property and will not be considered further in this review.

Various types of common fimbriae differing in dimensions, composition, stability, antigens and adhesive specificity have been described, but because knowledge of their characters is still incomplete, their present classification must be only provisional. A classification based on the easily determined haemagglutinating properties is likely to be most helpful for the present, and that proposed by Duguid *et al.* (1966), whose types 1—4 correspond with Ottow's (1975) group 1, subtypes 1—4, will be used.

Some adhesins, e.g. the K88 antigen of *Escherichia coli* (Stirm *et al.*, 1967a), are associated with the presence on the bacteria of filaments that seem to lack the regular shape and definite points of origin characteristic of typical fimbriae. Until it is determined whether such filaments are a specially delicate type of fimbriae or

187

artefacts produced from a non-filamentous substance during preparation for electron microscopy, it seems wise to follow Jones (1977) in describing them non-committally as 'fibrillae'.

7.1.2 Haemagglutination, adhesion and fimbriae

Haemagglutination was the first observed manifestation of adhesive properties in enterobacteria. Guyot (1908) found that the bacteria of 12 out of 18 strains of *E. coli* rapidly agglutinated the red blood cells of one or more of 13 animal species tested. Each bacterial strain reacted similarly with the red cells from different individual animals of the same species, but different strains reacted with the red cells of different species, showing that they possessed different kinds of haemagglutinin. Rosenthal (1943) later showed that haemagglutinating cultures of *E. coli* also agglutinated leucocytes, sperms, yeasts, fungal spores and pollens, and Kauffmann (1948) found haemagglutinins in 78 of 112 strains of *E. coli* in 0-serogroups 1–25. Griffitts (1948) found that among a variety of enterobacteria, only *E. alkalescens* agglutinated human red cells, but his methods were such that weaker reactions of the other bacteria might have been missed.

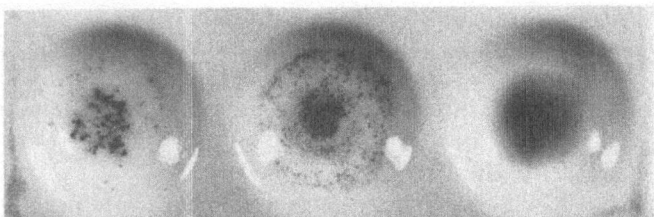

Fig. 7.1 Strong (left), moderately strong (centre) and negative (right) haemagglutination reactions in the rocked tile test. x 1. Reproduced from Gillies and Duguid (1958) with permission, Editors, *Journal of Hygiene, Cambridge.*

Interest in the discoveries that the haemagglutinating properties of the influenza virus (Hirst, 1941) and *Bordetella pertussis* (Keogh *et al.*, 1947) reflected adhesive affinities for receptor substances present on respiratory epithelium as well as on erythrocytes, led Duguid *et al.* (1955) to re-investigate these properties in *E. coli*. They demonstrated haemagglutination by mixing a dense suspension of bacteria centrifuged from an aerobic static culture in broth or peptone water with a suspension of washed red blood cells on a white tile, which was continuously rocked to and fro for about 10 minutes. Usually, the cells were coarsely clumped within a minute or two (Fig. 7.1). In mixtures mounted in wet films, the bacteria were seen to adhere firmly to the red cells, binding them together in clumps, and this observation showed that haemagglutination was due to an adhesive property of the bacteria. Duguid *et al.* (1955) distinguished three groups of *E. coli* strains with different patterns of haemagglutinating activity on the red cells of different animal species, which they termed groups I, II and III, and a fourth group (IV) that was entirely non-haemagglutinating.

With the electron microscope, they showed that haemagglutinating bacteria of groups I and II bore numerous filamentous appendages. These fimbriae were estimated to be about 10 nm in width and were mostly between 0.3 and 1 μm long. They were peritrichously arranged and usually numbered between 100 and 250 per bacterium, but in any pure culture a proportion of the bacteria were non-fimbriate. The few strains in group III, which had haemagglutinating specificities different from those in groups I and II, were invariably non-fimbriate, as were all the non-haemagglutinating strains (group IV). That haemagglutination was associated with the fimbriae in the strains of groups I and II was suggested by the findings that fimbriae were present in all these strains, but absent from all non-haemagglutinating strains, and that certain 'variable' strains became fimbriate and haemagglutinating under some conditions of culture, but non-fimbriate and non-haemagglutinating under others.

7.2 MS ADHESIN AND TYPE-1 FIMBRIAE

7.2.1 Haemagglutinating specificity and mannose sensitivity

A majority of strains of *E. coli* correspond to group I of Duguid *et al.* (1955). All such strains show the same pattern of haemagglutinating specificity for the red blood cells of different animal species. They agglutinate the cells of most species, e.g. guinea-pig, fowl, horse, mouse, pig and rhesus monkey, very strongly, human cells of all blood groups moderately strongly, sheep and goat cells weakly, and ox cells not at all (Table 7.1). The only difference between strains is that some strains are less strongly active than others against the whole range of cells, presumably because they form a smaller amount of the same adhesin.

The haemagglutinating activity of group-I strains, and similar activity in other enterobacteria, is best developed in stationary-phase cultures grown serially for periods of 24 h or longer at 37°C in static liquid media under air, though usually it is also manifest in cultures grown at 20°C and often, but less strongly, in cultures grown on agar (Duguid *et al.*, 1955; Duguid and Gillies, 1957; Duguid, 1959; Duguid *et al.*, 1966).

After Collier and de Miranda (1955) had observed that D-mannose, alone of many sugars, strongly inhibited an *E. coli* haemagglutinin, it was shown that the haemagglutinating activity of all group-I strains of *E. coli*, and that of similarly haemagglutinating strains of *Shigella flexneri*, klebsiellae and salmonellae, was inhibited by small concentrations, e.g. 0.01–0.5% (w/v), of D-mannose, methyl-α-D-mannoside and yeast mannan (Duguid and Gillies, 1957; Duguid, 1959; Duguid *et al.*, 1966) and a very few substances with molecular structures related to D-mannose (Old, 1972). The mannose-sensitive (MS) haemagglutinin in such strains was designated the 'MS adhesin' by Duguid (1959).

Table 7.1 Patterns of activity of the MS, MRE, MR/K and MR/P adhesins of enterobacteria against the red blood cells of different animal species

Red blood cells	Haemagglutinating activity of the adhesin											
	MS (strong culture)	MRE in representative strains of *E. coli* (a–f), *S. salinatis* (g) and *S. sendai* (h)								MR/K	MR/P (weak culture)	MR/P (strong culture)
		(a)	(b)	(c)	(d)	(e)	(f)	(g)	(h)			
Untreated cells												
Ox	–	–	–	++	++	+++	–	++	+++	–	–	++
Sheep	+	++	+++	+++	–	+++	–	++	–	–	++	+++
Man, group 0	++	+++	–	++	++	–	++	–	+++	–	+	++
Rhesus monkey	+++	+++	++	+++	–	–	–	+	++	–	·	·
Pig	+++	+	+++	++	–	–	–	–	+	–	·	·
Dog	+++	–	+	+	–	–	–	++	+	–	·	·
Horse	+++	–	–	+	+	–	–	–	–	–	+	++
Mouse	+++	–	–	–	++	–	–	++	–	–	·	·
Rat	+++	+	–	+	++	–	–	++	+++	–	·	·
Rabbit	+++	+	–	+	–	–	–	–	+++	–	·	·
Guinea-pig	+++	–	–	+	+	–	–	++	–	–	+	+++
Fowl	+++	–	–	+	+++	–	–	–	–	–	++	+++
Toad	+++	++	+	–	+++	–	–	·	·	–	·	·
Frog	+++	–	–	–	++	–	–	·	·	–	·	·

(continued on the next page)

Table 7.1 Patterns of activity of the MS, MRE, MR/K and MR/P adhesins of enterobacteria against the red blood cells of different animal species (*continued from previous page*)

Red blood cells	MS (strong culture)	MRE								MR/K	MR/P (weak culture)	MR/P (strong culture)
		in representative strains of *E. coli* (a–f), *S. salinatis* (g) and *S. sendai* (h)										
		(a)	(b)	(c)	(d)	(e)	(f)	(g)	(h)			
Cells treated with tannic acid												
Ox	−	...	...	...	...	...	...	++	+++	++	...	++
Man, group 0	++	...	...	...	...	...	...	...	...	+++	...	...
Guinea-pig	+++	...	...	...	...	...	...	...	...	+++	...	...

−, No haemagglutination; +, ++ and +++, Increasing strengths of haemagglutination; ..., Not tested.

Data for MS adhesin in *E. coli* and *Sh. flexneri* from Duguid *et al.* (1955) and Duguid and Gillies (1957), for MRE adhesin in *E. coli* from Duguid (1964) and Duguid *et al.* (1979), for MRE adhesin in *S. salinatis* and *S. sendai* from Duguid *et al.* (1966), for MR/K adhesin in *K. aerogenes* from Duguid (1959), and for MR/P adhesin in *Proteus* spp. from Duguid and Gillies (1958), Coetzee *et al.* (1962) and Shedden (1962), but see Section 7.4.2.

Table 7.2 Types of adhesins in strains of enterobacterial species

Species	Number of strains forming adhesins of type(s)									References
	MS only	MS and MRE	MRE only	MS and MR/K	MR/K only	MR/K and MR/P	MS MR/K, MR/P	None		
Escherichia coli	198	95	21	0	0	0	0	73		(a), (b)
E. alkalescens	1	1	7	0	0	0	0	2		(c)
Erwinia atroseptica	0	0	0	0	0	0	0	12		(d)
Erw. carotovora	13	0	0	0	0	0	0	12		(d)
Erw. chrysanthemi	0	0	0	0	0	0	0	2		(d)
Erw. rhapontici	0	0	3	0	0	0	0	0		(d)
Shigella boydii	0	0	0	0	0	0	0	21		(e)
Sh. dysenteriae	0	0	0	0	0	0	0	26		(e)
Sh. flexneri	103	0	0	0	0	0	0	77		(e)
Sh. sonnei	0	0	0	0	0	0	0	20		(e)
Salmonella typhi	122	0	0	0	0	0	0	28		(f)
S. paratyphi A	2	0	0	0	0	0	0	76		(f)
S. paratyphi B	106	0	0	0	0	0	0	29*		(f)
S. typhimurium	1711	0	0	0	0	0	0	319*		(f), (g)
S. gallinarum	0	0	0	0	0	0	0	14*		(f)
S. pullorum	0	0	0	0	0	0	0	30*		(f)
S. salinatis	0	1	0	0	0	0	0	0		(f)
S. sendai	0	0	6	0	0	0	0	0		(f)
Other salmonellae	356	0	0	0	0	0	0	47		(f)
Arizona spp.	6	0	0	0	0	0	0	5		(f)
Citrobacter freundii	8	0	0	0	0	0	0	2		(h)

(continued on the next page)

Table 7.2 Types of adhesins in strains of enterobacterial species (*continued from previous page*)

Species	MS only	MS and MRE	MRE only	MS and MR/K	MR/K only	MR/K and MR/P	MS, MR/K, MR/P	None	References
	Number of strains forming adhesins of type(s)								
C. ballerupensis	8	0	0	0	0	0	0	2	(h)
Enterobacter cloacae	10	0	0	0	0	0	0	3	(i), (j)
Serratia marcescens	0	0	0	8	2	0	0	0	(j)
Klebsiella aerogenes	14	0	0	151†	37	0	0	1	(j), (k)
K. atlantae	0	0	0	0	0	0	0	7	(j), (k)
K. edwardsii	0	0	0	0	0	0	0	7	(j), (k)
K. ozaenae	0	0	0	0	0	0	0	170‡	(j), (k)
K. pneumoniae	11	0	0	0	0	0	0	0	(j), (k)
K. rhinoscleromatis	0	0	0	0	0	0	0	9	(j), (k)
Proteus mirabilis	0	0	0	0	0	3	0	0	(h)
P. vulgaris	0	0	0	0	0	3	0	0	(h)
P. morganii	0	0	0	0	0	2	1	0	(h)
P. rettgeri	1	0	0	0	1	1	0	0	(h)
Providencia spp.	0	0	0	3	3	1	3	0	(h)

* Non-haemagglutinating type-2 fimbriae were formed by 19 strains of *S. paratyphi B*, 8 of *S. typhimurium*, 14 of *S. gallinarum* and 11 of *S. pullorum*.

† One MS⁺ MR/K⁺ strain of *K. aerogenes* also formed an MRE adhesin.

‡ Non-haemagglutinating type-6 fimbriae were formed by 9 strains of *K. ozaenae*.

References: (a) Duguid *at al.* (1955). (b) Duguid *et al.* (1979), includes strains of (a). (c) Miss P.B. Crichton, unpublished. (d) Christofi *et al.* (1979). (e) Duguid and Gillies (1957). (f) Duguid *et al.* (1966). (g) Duguid *et al.* (1975), includes strains of (f). (h) Duguid and Old, unpublished. (i) Constable (1956). (j) Duguid (1959). (k) Duguid (1968), includes strains of (j).

7.2.2 MS adhesin in other enterobacteria

As there may be other, yet undiscovered mannose-sensitive adhesins with different properties in other bacteria, the present use of the term, MS adhesin, must be understood as applying only to such adhesins as have the same properties, e.g. specificity for different cells, as the MS adhesin of *E. coli* group I. MS adhesins of this kind have now been found in many genera and species of enterobacteria. Table 7.2 gives the numbers of strains with MS adhesin found in different species by the investigators listed in the footnote. The first column shows the strains that formed only the MS adhesin and the second and fourth columns the strains that formed both the MS adhesin and another, mannose-resistant (MRE or MR/K) adhesin (see Sections 7.3.1 and 7.4.1). It is in the first (MS only) type of strain that the properties of the MS adhesin are most easily studied.

7.2.3 Association of MS adhesin with fimbriae

The fimbriae of the group-I strains of Duguid *et al.* (1955) were probably the same structures as those described as 'filaments' by Houwink and van Iterson (1950) and Brinton *et al.* (1954). That they bear the MS adhesin was shown by three kinds of observation.

(1) All MS haemagglutinating strains of *E. coli* group I (Duguid *et al.*, 1955), *Enterobacter cloacae* (Constable, 1956), *Sh. flexneri* (Duguid and Gillies, 1957), klebsiellae (Duguid, 1959, 1968) and salmonellae (Duguid *et al.*, 1966) were found to form fimbriae 7—10 nm in width, whereas most strains of these organisms that lacked MS adhesin (MS⁻) were found to be non-fimbriate. The exceptional MS⁻ strains that formed fimbriae 7—10 nm wide were those of *E. coli* group II, in which the fimbriae (type MRE, see Section 7.3.4) had mannose-resistant adhesive properties, and a minority of salmonella and klebsiella strains, in which the fimbriae (types 2 and 6, see Section 7.4.3) were non-adhesive.

(2) The 'variable' strains of *E. coli* group I, and similar strains of *Sh. flexneri* and salmonellae, were found to grow in a mainly fimbriate, and strongly haemagglutinating phase when cultured serially in tubes of broth or peptone water, and to change to a non-fimbriate and non-haemagglutinating phase when cultured serially on agar plates (Duguid *et al.*, 1955; Duguid and Gillies, 1957; Duguid and Wilkinson, 1961; Duguid *et al.*, 1966). During these changes, the strength of the haemagglutinating activity varied directly with the proportion of fimbriate bacteria in the culture.

(3) Bacteria-free preparations of detached and purified fimbriae were shown to agglutinate red blood cells (Brinton, 1959; Old, 1963; Salit and Gotschlich, 1977a).

The fimbriae with MS adhesin were classified as 'type 1' by Duguid *et al.* (1966) and as 'group 1, subtype 1' by Ottow (1975). Their properties include the ability to cause haemagglutination at all temperatures of testing between 0° and 55°C (Duguid and Gillies, 1957; Duguid *et al.*, 1966), their resistance to inactivation by heating at

Table 7.3 Differences between the MS and MRE adhesins of *E. coli* and other enterobacteria

Property	MS adhesin	MRE adhesins
Activity in the presence of 0.5% (w/v) D-mannose	Absent	Present
Activity at raised temperature (e.g. 50°C)	Present	Absent*
Activity after bacteria have been heated at 65°C for 30 min.	Present	Absent
Activity after bacteria have been exposed to 0.5% (w/v) formaldehyde at 37°C for 4 h	Present	Absent
Pattern of activities with erythrocytes of different species of animals	Similar pattern in all strains of bacteria	Different patterns in different strains of bacteria
Optimal conditions for development of the adhesin	Prolonged (e.g. 48 h) culture in aerobic static broth	Culture on buffered agar for 24 h at 37°C. Not developed at 15°–20°C
Organelles associated with the adhesin	Type-1 fimbriae	Type-MRE fimbriae or no visible fimbriae

* Activity present at 3–5°C. The reactions of particular strains of bacteria with particular species of red blood cells may be absent at 10°, 20°, 30°, 40° or 50°C. Those of some strains with some, e.g. fowl, red cells may fail to 'elute' even at 50°C.

65°C for 30 min or exposure to 0.5% (w/v) formaldehyde at 37°C for 4 h, and their susceptibility to removal from the bacteria by heating at 90°C for 1 h (Gillies and Duguid, 1958; Duguid, 1959; Duguid *et al.*, 1966; see also Table 7.3), and their specific antigens, which mediate the agglutination of fimbriate bacteria by anti-fimbrial serum (Gillies and Duguid, 1958; Duguid and Campbell, 1969).

A possible explanation of the adhesive property of type-1 fimbriae is suggested by the observations that the electrophoretic mobility of the fimbriate bacteria of *E. coli* strain B is only two-thirds that of the non-fimbriate bacteria (Brinton *et al.*, 1954) and that the fimbriae are proteins consisting mainly of amino acids with non-polar groups (Brinton and Stone, 1961; Old, 1963; Brinton, 1965). The fimbriate bacteria must therefore have a lower density of negative surface charge and a more hydrophobic character than the non-fimbriate bacteria, and these properties will facilitate their close approach to the negatively charged, hydrophilic surfaces of erythrocytes and other cells (see also Chapter 1).

7.2.4 Adhesion to cells other than erythrocytes

Bacteria with MS adhesin, and no other adhesive factor, adhere strongly to many kinds of cells besides red blood cells; indeed, to most of the animal, plant and fungal cells so far tested as substrates. In the simplest form of adhesion test, a mixture of cells and fimbriate bacteria in 0.85% (w/v) NaCl solution is observed in a wet film under the phase-contrast microscope. Within a minute or so, large numbers of bacteria are seen to adhere to the surfaces of the cells, and their firm attachment is shown by their lack of Brownian movement and failure to shift when disturbed by movement of the film.

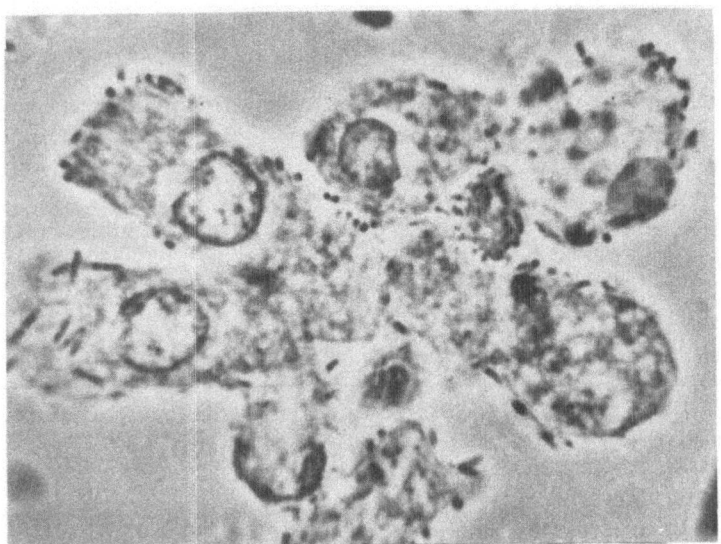

Fig. 7.2 Adhesion of fimbriate *Sh. flexneri* bacilli with MS adhesin to guinea-pig intestinal epithelial cells. Wet film. × 1300. Reproduced from Duguid and Gillies (1957) with permission, Editors, *Journal of Medical Microbiology*.

In this way, Duguid and Gillies (1957) showed that type-1 fimbriate bacteria, but not non-fimbriate bacteria, of strains of *E. coli*, *Ent. cloacae* and *Sh. flexneri* adhered rapidly to human and guinea-pig intestinal epithelial cells (Fig. 7.2) and that the adhesion was inhibited by the addition of D-mannose. Later, type-1 fimbriate bacteria of these species, and of klebsiellae, serratiae and salmonellae, were shown to have a similar mannose-sensitive adhesiveness for ox, human and guinea-pig leucocytes (Fig. 7.3); ox, human, mouse and guinea-pig intestinal epithelium; cells of guinea-pig liver, kidney and tracheal epithelium; the protozoon *Trichomonas*

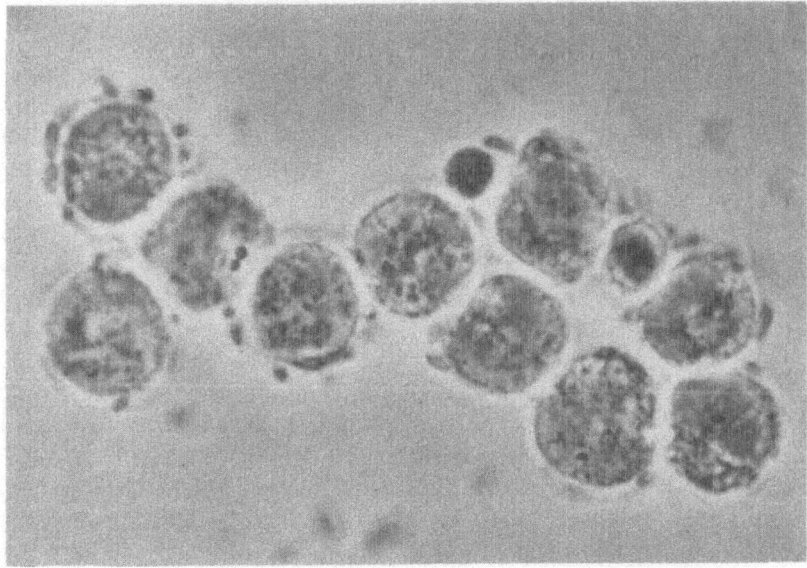

Fig. 7.3 Fimbriate *Klebsiella* bacilli with MS adhesin adhering to, and agglutinating guinea-pig leucocytes. Wet film. x 2200. Reproduced from Duguid (1959) with permission, Editors, *Journal of General Microbiology*.

vaginalis; the yeast cells of *Candida albicans* and the hyphae of *Aspergillus niger* and other moulds; and the root hairs of red clover and cress seedling plants that had germinated under bacteria-free conditions (Duguid, 1959; Duguid *et al.*, 1966). The ability of the bacteria to adhere to ox leucocytes and epithelial cells, though they could not adhere to ox erythrocytes, showed that an adhesin's affinity for erythrocytes does not necessarily correspond with its affinity for other kinds of cells from the same species of animal.

Although the test for adhesion during a few minutes' observation of a wet film detects the presence of the MS and other strong adhesins, it may not detect weaker adhesive properties in non-fimbriate bacteria, which may bring about attachment to, and penetration into epithelial cells during longer periods of exposure *in vitro* (e.g. Ogawa *et al.*, 1968) or *in vivo* (e.g. Labrec *et al.*, 1964; Takeuchi, 1967).

7.2.5 Phase variation and cultural conditions

The formation of fimbriae by an enterobacterium may be subject both to spontaneous variation and to variation determined by the conditions of culture, and this variability must be taken into account when selecting methods for detection of the fimbrial adhesin. Brinton *et al.* (1954) were the first to show that the bacteria of *E. coli* strain B varied reversibly between a fimbriate and a non-fimbriate phase; the change appeared to be spontaneous and took place in either direction at the rate of about once per 1000 bacteria per generation. The predominant fimbrial phase in cultures of *E. coli, Sh. flexneri* and salmonellae was later found to be determined by the conditions of culture (Duguid *et al.,* 1955; Duguid and Gillies, 1957; Duguid *et al.,* 1966). A few serial cultivations for periods of 48 h in static tubes of broth gave cultures with 50–95% of fimbriate bacteria, whilst a few serial cultivations on agar plates gave cultures with few or no fimbriate bacteria. As several successive cultivations under the inducing conditions were required to complete the change of phase, it appeared that the effect of the cultural conditions was to promote the selective outgrowth of spontaneously arising variant bacteria rather than directly to induce or inhibit fimbrial synthesis.

Culture in static broth favours the growth of fimbriate bacteria because the fimbriae enable the bacteria rapidly to establish themselves in a thin pellicle on the surface of the broth, where the supply of atmospheric oxygen allows them to multiply to form a population several times greater than that of non-fimbriate bacteria in a culture without a pellicle (Duguid and Gillies, 1957; Duguid and Wilkinson, 1961; Duguid *et al.,* 1966; Old *et al.,* 1968). Some non-fimbriate bacteria, e.g. *Sh. sonnei, Klebsiella edwardsii* and FIRN strains of *Salmonella typhimurium,* can form non-fimbrial pellicles on static liquid cultures, but do so only at a later stage of growth (e.g. after 24–48 h) than fimbriate bacteria (6–10 h). The enormous growth advantage of the early, fimbriae-dependent pellicle formation has been demonstrated in mixed cultures of related, genotypically fimbriate and non-fimbriate bacteria (Old and Duguid, 1970, 1971).

The reason why type-1 fimbriae promote early pelliculate growth is unknown. If they do so merely by causing the bacteria to adhere to one another, it is difficult to explain why such adhesion should take place at the broth-air interface, but not throughout the broth. It seems more likely that the fimbriae, or their tips, are sufficiently hydrophobic to hold the bacteria at the surface of the broth after they have reached it by passive diffusion or aerotactic locomotion.

As the variation of the fimbrial phase in bacterial cells appears often to take place spontaneously, it may be caused by spontaneous repressions and de-repressions of the *fim* gene, analogous to those of the H1 and H2 genes in the phase variation of the flagellar antigens of salmonellae (Stocker, 1949; Lederberg and Iino, 1956). But sometimes the variation may also be subject to direct induction by particular environmental conditions. Thus, when fimbriate stationary-phase bacteria of *S. typhimurium* were grown logarithmically in a subculture in fresh broth, they changed from the fimbriate to the non-fimbriate phase so quickly, within a few hours, that the conditions of logarithmic growth appeared to be repressing fimbrial synthesis rather than selecting non-fimbriate variant bacteria (Duguid *et al.,* 1966). Another example of a direct environmental influence is seen in the finding that the synthesis of type-1 fimbriae is dependent on cyclic AMP and subject to catabolite repression by many carbohydrates (Saier *et al.,* 1978).

7.2.6 Inhibition by D-mannose and its analogues

The powerful inhibitory action of D-mannose on adhesion by type-1 fimbriate bacteria extends to their reactions with all types of animal, plant and fungal cells tested as substrates (Duguid and Gillies, 1957; Duguid, 1959; Duguid *et al.,* 1966) and to the attachment of purified type-1 fimbriae to monkey kidney cells (Salit and Gotschlich, 1977b). Yet the action of D-mannose is highly specific, for among many substances tested, only D-mannose, methyl-α-D-mannoside, 1, 5-anhydromannitol, D-mannoheptulose and α-D-mannose 1-phosphate were strongly inhibitory; D-fructose was moderately inhibitory and 2-deoxy-D-glucose, 6-deoxy-D-mannose and methyl-β-D-mannopyranosides were weakly inhibitory (Old, 1972). Thus, modification of the hydroxyl groups at the C-2, C-3, C-4 and C-6 positions in the D-mannopyranosyl molecule caused loss of inhibitory activity, suggesting that these were the active groups. The activity of D-fructose was probably due to its assumption of a β-D-fructopyranosyl form, in which the stereochemistry of D-mannose is almost reproduced. This high specificity of the inhibitory action suggests that the sites with which D-mannose reacts are located on the bacterial fimbriae and not on the very diverse types of cells that serve as substrates for the fimbriae.

Duguid and Gillies (1957) found that the binding of D-mannose to its receptor sites was so rapidly reversible that they could not show whether these sites were on the fimbriate bacteria or on the red blood cells by demonstrating an inhibition of haemagglutination by pre-treatment of the bacteria or the red cells. When either the bacteria or the red cells were suspended in 2% (w/v) D-mannose solution and then washed free from the sugar by centrifugation and re-suspension in saline, they remained normally active in giving haemagglutination. Similar reversibility of the inhibition of adhesion of *E. coli* to epithelial cells by methyl-α-D-mannoside was shown by Ofek *et al.* (1977). It must be concluded that the binding of D-mannose to its reaction site is temporary and that the bound D-mannose is in dynamic equilibrium with the free D-mannose in the ambient fluid.

Evidence that D-mannose reacts specifically with type-1 fimbriate bacteria in the absence of other cells was obtained in the finding that the early, fimbriae-dependent formation of a surface pellicle on broth cultures was inhibited by the addition of D-mannose or methyl-α-D-mannoside, but not by that of other fermentable or unfermentable sugars (Old *et al.,* 1968; Old and Duguid, 1970). This observation was originally made with *S. typhimurium,* and we have since repeated it with *E. coli* and different serotypes of *Sh. flexneri.* If pellicle formation is due to a hydrophobic property of fimbriae that holds the bacteria at the broth-air interface, D-mannose presumably acts by adsorbing to mannose-receptor sites on the fimbriae and, perhaps by covering hydrophobic groups, rendering the fimbriae more hydrophilic. If, on the other hand, pellicle formation is due to an adhesion of fimbriate bacteria to one another, the action of D-mannose could be explained either by its reacting with receptor sites on the fimbriae or by its reacting with sites on the bacterial surface to which other fimbriate bacteria would otherwise adhere. The finding that pre-treatment of fimbriate bacteria with yeast mannan, unlike pre-treatment with D-mannose, inhibited adhesion of the bacteria to leucocytes, may be seen as further evidence that the receptor sites for D-mannose are on the fimbriae (Bar-Shavit *et al.*, 1977), but this effect of the mannan might be due to its serving as a substrate for, and so blocking the adhesive sites on the fimbriae, rather than to its reacting with distinct mannose-receptor sites.

The simplest explanation of the inhibitory action of D-mannose is that it serves as a soluble analogue of fixed D-mannose-like residues on the surface of erythrocytes and other cells, and so binds to, and blocks adhesive sites on the fimbriae that otherwise would bind to the residues on the cells (Ofek *et al.*, 1977, 1978). If this is so, D-mannose-like residues must be present on the very wide range of animal, plant and fungal cells that serve as substrates for the MS adhesin. Other possible explanations of the action of D-mannose are

(1) that it binds to, and covers hydrophobic groups on the fimbriae, making the fimbriae more hydrophilic and thus repellent to other cells, and
(2) that fimbriae are allosteric proteins which the binding of D-mannose alters from a hydrophobic and adhesive to a hydrophilic and non-adhesive form.

7.2.7 MS adhesive sites on fimbriae

The relatively low electro-negative charge on bacteria with type-1 fimbriae and the relatively hydrophobic character of the fimbrial protein (Brinton *et al.*, 1954; Brinton and Stone, 1961) are likely to facilitate a close approach of the bacteria to cellular substrates, but by themselves these properties seem insufficient to account for the firm adhesion of the bacteria to the cells. The adhesion is more probably due to the presence on the fimbriae of specific sites that bind to receptor substances, perhaps D-mannose-like, on the substrates. The location and structure of such fimbrial adhesive sites remains to be determined.

The MS-adhesive fimbriae in different enterobacteria are not identical in composition, in as much as that is manifest by their antigenic character. Although there is a partial sharing of type-1 fimbrial antigens among the species, *E. coli, Sh. flexneri* and *K. aerogenes,* and a partial sharing among the genera, *Salmonella, Arizona* and *Citrobacter*, there is no sharing or similarity of type-1 fimbrial antigens between these two groups of organisms (Gillies and Duguid, 1958; Duguid and Campbell, 1969). Still other, different antigenic forms of type-1 fimbriae are present in the genera, *Edwardsiella, Enterobacter, Hafnia, Providencia* and *Serratia* (Nowotarska and Mulczyk, 1977). If, therefore, sites with a single specific molecular configuration are responsible for the MS adhesiveness of type-1 fimbriae, these sites are not sufficiently numerous or antigenically potent to be detectable by agglutination of the bacteria with antifimbrial serum. Further evidence to the same effect is the apparent antigenic identity of the type-1 fimbriae of typical, haemagglutinating strains of *S. paratyphi B* with the variant, non-haemagglutinating (type 2) fimbriae present in a minority of strains (Old and Payne, 1971; see Section 7.4.3).

If, as these observations suggest, the MS adhesive sites occupy only a very small part of the surface of the fimbriae, their likeliest location would seem to be at the peripheral tips. In a watery medium, the fimbriae presumably radiate peritrichously from the bacteria owing to a mutual repulsion of their lateral surfaces, which must be sufficiently hydrophilic to prevent them binding together. The outward facing tips, which may be hydrophobic and bear the adhesive sites, are the parts of the fimbriae that will first come into contact with the surfaces of cells and other substrates.

7.2.8 Function of the MS adhesin

The MS adhesin of parasitic enterobacteria may help them to colonize the intestinal, urinary or biliary tract by attaching them to the mural epithelium and so enabling them to resist removal by the outflowing luminal contents. It may also increase their opportunity to invade the epithelium. Clearly, however, fimbrial adhesion is not *essential* for intestinal infection by salmonellae and shigellae, because some fully pathogenic types of these organisms are invariably non-fimbriate and non-adhesive, e.g. *S. typhi* phage type B1, FIRN strains of *S. typhimurium, Sh. dysenteriae, Sh. boydii* and *Sh. sonnei* (Duguid and Gillies, 1957; Duguid *et al.*, 1966).

The association of salmonella and shigella bacteria with the surface of epithelial cells that induces their intake by the cells, and thus the initiation of infection, must be brought about by non-fimbrial substances, perhaps the O-antigens, which do not cause a rapid, firm adhesion of the bacteria to the cells, like that shown by fimbriate bacteria within a few minutes of observation in a wet film. The finding, however, that most salmonellae infecting mammals form type-1 fimbriae suggests that the MS adhesive property confers at least a marginal advantage on these parasites (Duguid *et al.*, 1966). Some support for this view has been obtained in experimental oral infections of mice with genotypically fimbriate and non-fimbriate lines of *S. typhimurium*, in which the former line gave more numerous infections and longer periods of faecal dissemination than the latter (Duguid *et al.*, 1976).

A significant minority (e.g. 30%) of strains of *E. coli* isolated from the faeces of healthy subjects are non-fimbriate and non-haemagglutinating, as also are smaller minorities (e.g. 10—20%) of strains of enteropathogenic serotypes from outbreaks of infantile diarrhoea and strains from the urines of patients with urinary-tract infections (Duguid, 1964, 1968; Duguid *et al.*, 1979). These findings suggest that the possession of an adhesin is not an essential property in bacteria that colonize the intestine, cause gastroenteritis or infect the urinary tract. In the reported series, however, the possibility was not excluded that the non-adhesive organisms isolated from gastro-intestinal or urinary infections were merely concomitants or secondary invaders in the patients from whom they were isolated. A collection of *E. coli* strains from the urine of patients with acute symptomatic pyelonephritis or cystitis included some that adhered to human urinary-tract epithelial cells much more strongly than strains from the urine of patients with asymptomatic bacteriuria, but both these groups included a proportion of entirely non-adherent strains (Svanborg Edén *et al.*, 1976).

Whatever its role in parasitic enterobacteria, the MS adhesin on type-1 fimbriae must have a different function in free-living saprophytes, such as *K. aerogenes, Ent. cloacae* and *Serratia marcescens*. These saprophytes may benefit from the ability to adhere to plant root hairs, fungal mycelium and other organic bodies that yield nutrient solutes and from which they might otherwise be removed. Fimbriate saprophytes that inhabit stagnant, anaerobic water may also benefit from the effect of their fimbriae in enabling them to grow in a pellicle on the surface of the water where they have access to atmospheric oxygen for energy production. A further possible role of type-1 fimbriae is in the promotion of the conjugational transfer of DNA by non-specifically enhancing, and so stabilizing, the mutual adhesion of mating pairs of bacteria. Although DNA transfer is primarily dependent on the presence of sex pili in the donor bacterium, experiments on the transfer of the *coli* factor have shown that the rate of transfer between the donor and recipient strains is greatly increased if at least one of the strains forms type-1 fimbriae (Mulczyk and Duguid, 1966; Meynell and Lawn, 1967).

If the adhesive property of type-1 fimbriae has a useful function, its exercise may be advantageous only in some circumstances of the life of the bacteria. In other circumstances, the bacteria may benefit from being able to disperse from exhausted to fresh substrates. The value of being able to vary between an adhesive and a dispersive phase may be the reason why most bacteria that form type-1 fimbriae are capable of undergoing fimbrial phase variation (Duguid, 1959).

7.3 MRE ADHESINS AND TYPE-MRE FIMBRIAE

7.3.1 Specificity, properties and occurrence

Some strains of *E. coli* form adhesins different from the MS adhesin formed by the majority. Thus, the strains of group II, which were fimbriate, and those of group III, which were non-fimbriate, showed patterns of haemagglutinating specificity for the red blood cells of 18 animal species that differed from the essentially constant pattern of the MS$^+$ strains of group I and also between the individual strains in groups II and III (Duguid *et al.*, 1955). The specificity patterns of some representative strains from the larger series of Duguid (1964) and Duguid *et al.* (1979) are shown in Table 7.1.

These adhesins differ from the MS adhesin in several properties besides their specificity (Table 7.3). They are well developed in cultures grown on agar plates at 37°C, but poorly developed or absent in those grown at 15–20°C, and they generally show the phenomenon of 'elution', i.e. dispersion of the agglutinated bacteria and red cells when a haemagglutination test is warmed (Duguid *et al.*, 1955). Their agglutinating activity with red cells is resistant to inhibition by D-mannose and its analogues (Duguid and Gillies, 1957), but is destroyed when the bacteria are heated at 65°C for 30 min or exposed to 0.5% (w/v) formaldehyde at 37°C for 4 h, treatments that do not inactivate the MS adhesin (Duguid *et al.*, 1966; Duguid *et al.*, 1979). They were described as mannose-resistant and eluting (MRE) haemagglutinins by Duguid (1964), who found them in a minority of non-fimbriate strains of *E. coli* isolated from infantile diarrhoea.

MRE adhesins have now been found in many strains of *E. coli* from different sources (e.g. Stirm *et al.*, 1967a; Jones and Rutter, 1974; Burrows *et al.*, 1976; Evans *et al.*, 1977; Nagy *et al.*, 1977; Ørskov and Ørskov, 1977; Evans and Evans, 1978; Svanborg Edén and Hansson, 1978; Duguid *et al.*, 1979), in *S. salinatis* and *S. sendai* (Duguid *et al.*, 1966), *Erwinia rhapontici* (Christofi *et al.*, 1979), *E. alkalescens* and an exceptional strain of *K. aerogenes* (Table 7.2).

7.3.2 Elution at raised temperatures

Duguid *et al.* (1955) found that whilst the haemagglutinating activity of MS$^+$ strains of *E. coli* group I was only slightly weaker in tests made at 40–50°C than in those made at 3–5°C, that of the MRE$^+$ strains of groups II and III was strongest at 3–5°C and generally absent at 40–50°C. When the mixtures of bacteria and red blood cells that had agglutinated at 3–5°C were warmed during continued rocking of the tile, the bacteria eluted from the red cells and the clumps dispersed. The exact temperature between 5° and 50°C at which elution took place varied both with the bacterial strain and with the species of red cells. Some reactions demonstrable in the cold were not demonstrable at ambient temperature (*c*. 20°C), whilst others were demonstrable even at body temperature (37°C). Strong reactions of some strains with certain red cells, particularly those of fowl, failed to elute even at 50°C, although the reactions of the same strains with other kinds of red cells eluted at this, or a lower temperature.

Agglutination and elution of the same mixture of MRE$^+$ bacteria and red blood cells could be induced many times in succession by alternate cooling and warming during continued rocking of the test. Elution is therefore not due to the destruction of a receptor substance on the red cells by a temperature-dependent bacterial enzyme analogous to the receptor-destroying enzyme (neuraminidase) of haemagglutinating and eluting influenza viruses. The probable explanation of the elution of MRE$^+$ bacteria is that their adhesive affinity for red cells is so weak that it is readily overcome by the increased molecular agitation at raised temperatures.

The adhesive strength of the MRE haemagglutinin of *E. coli* strains with K88 antigen is even weaker than that of most MRE$^+$ strains, for Jones and Rutter (1972) found that the activity of this haemagglutinin is more readily demonstrated in a static settling test than in a rocked tile test. Apparently it is liable to be overcome by the greater mechanical shearing forces in the rocked test. Duguid *et al.* (1979) have confirmed Jones and Rutter's finding for the K88 antigen, and also their finding that the reactions of strains with other MRE haemagglutinins are better demonstrated in a rocked test than in a settling test; the advantage of the rocked test is probably that it induces more frequent collisions between the bacteria and the red cells.

7.3.3 Multiple adhesins in the same strain

More than one type of adhesin may be present in the same strain. Thus, *E. coli* strain no. 116 of Duguid *et al.* (1955), aberrantly classified in group I, formed an MRE adhesin active against only ox red blood cells as well as the MS adhesin active against 17 other species of red cells. Similarly, the strain of *S. salinatis* of Duguid *et al.* (1966) formed both type-1 fimbriae with MS adhesin and a non-fimbrial MRE adhesin. Each adhesin of the latter strain was subject to a reversible phase variation, the MS adhesin disappearing on serial culture on agar and the MRE disappearing on serial culture in broth. Among 387 strains of *E. coli* in 155 0-serogroups, Duguid *et al.* (1979) found 95 genotypically capable of forming both an MS and an MRE adhesin (MS$^+$/MRE$^+$ strains), 198 forming only an MS adhesin (MS$^+$/MRE$^-$), 21 only an MRE adhesin (MS$^-$/MRE$^+$) and 73 neither adhesin (MS$^-$/MRE$^-$). The eleven strains of *E. coli* from urinary-tract infections found by Svanborg Edén and Hansson (1978) to give mannose-sensitive agglutination of guinea-pig red cells, but mannose-resistant adhesion to urinary-tract epithelial cells, presumably formed both an MS and an MRE adhesin.

Care must be taken to identify separately the MS and MRE adhesins present in the same strain. The presence and properties of the MRE adhesin are easily demonstrated in tests made at $3-5°$C in the presence of D-mannose, which eliminates all the effects of the MS adhesin. If the duration of a test with live bacteria is long enough for D-mannose to be removed by fermentation, an unfermentable analogue such as methyl-α-D-mannoside should be used.

The separate presence of the MS adhesin may be demonstrated by three methods:

(1) by showing that the bacteria agglutinated with guinea-pig red blood cells elute on warming in tests containing D-mannose, but not in tests without D-mannose,
(2) by showing that the bacteria cause a mannose-sensitive agglutination of a species of red cells, e.g. horse, fowl or mouse, against which their MRE adhesin is inactive, and
(3) by showing that bacteria heated at $65°$C or treated with formaldehyde to destroy their MRE adhesin give mannose-sensitive reactions.

It is possible that some strains form two or more kinds of MRE adhesin, active against different species of cells, but the existence of such strains is unknown and it would be difficult to demonstrate the separate presence of different MRE adhesins in a strain unless variant cultures forming only one or other of them could be obtained.

7.3.4 Association of MRE adhesins with fimbriae

The finding by Duguid *et al.* (1955) that some, but not all, MS⁻/MRE⁺ strains form
fimbriae is perplexing. Of 19 such strains examined with the electron microscope by
Duguid *et al.* (1979), 12 were fimbriate and 7 non-fimbriate. These findings might
suggest that the MRE adhesins are not associated with fimbriae, but there is other
evidence to show that in the fimbriate strains they are so associated. The MRE
adhesins of the fimbriate (group II) and non-fimbriate (group III) strains of Duguid
et al. (1955) had different patterns of activity against the different species of red
blood cells, suggesting that their composition might be different. In the group-II
strains, moreover, the association of the MRE adhesin with the fimbriae was
indicated by the finding that whilst both the adhesin and the fimbriae were present
in cultures grown at 37°C, both were absent from those grown at 15–20°C.

The fimbriae of these MS⁻/MRE⁺ strains were not included in the classification
of types of fimbriae proposed by Duguid *et al.* (1966), but were provisionally called
'type MRE fimbriae' by Duguid *et al.* (1979). Further examples of MRE adhesins
associated with such fimbriae in *E. coli* are those associated with the K88 antigen
(Stirm *et al.*, 1967a), the K99 antigen (Burrows *et al.*, 1976), the colonization
factor antigens CFA/I and CFA/II (Evans *et al.*, 1977; Evans and Evans, 1978) and
the antigen 987 (Nagy *et al.*, 1977). Strains forming the adhesin, fimbriae and
antigen in cultures grown at 37°C, failed to form them in cultures grown at 18°C,
and the adhesin and fimbrial antigen were gained or lost simultaneously in cultures
gaining or losing a plasmid that determined their joint production.

The structure of the MRE adhesins in the strains so far not found to be fimbriate
remains to be discovered. Possibly, these adhesins are diffuse, non-filamentous
surface substances or, possibly, filaments too fragile or thin to be seen by the
electron-microscope methods used in the search for them.

Chemical analysis has shown that purified preparations of the MRE antigens of
K88 (Stirm *et al.*, 1967b; Jones and Rutter, 1974) and K99 (Morris *et al.*, 1977;
Isaacson, 1977) strains of *E. coli* are proteins with traces of, probably contaminating,
carbohydrates; that of CFA/I is probably also protein (Evans *et al.*, 1978a). No
serological cross-reactions are known between these antigens and other recognized
K-antigens of *E. coli*, which are polysaccharides (Ørskov *et al.*, 1977).

7.3.5 Adhesion to epithelium

An association between MRE haemagglutinating activity and an adhesiveness for
small intestinal epithelium has been shown most clearly by comparative observations
on MRE⁺ and MRE⁻ variants from porcine strains of *E. coli* with K88 antigen (Jones
and Rutter, 1972), calf strains with K99 antigen (Burrows *et al.*, 1976) and human
strains with CFA/I and CFA/II antigens (Evans *et al.*, 1978a; Evans and Evans, 1978).

The adhesiveness of MS⁻/MRE⁺ bacteria for epithelium does not necessarily
correspond with the affinity of the bacteria for red blood cells from the same animal

species. Thus, an MRE⁺ strain of *E. coli* that agglutinated the red cells of ox and
sheep, but not those of man and guinea-pig, nevertheless adhered to human amnion
and guinea-pig intestinal epithelium (Duguid, 1964) and MRE⁺ strains of *S. sendai*
that agglutinated guinea-pig, human and ox, but not mouse red cells, adhered to the
intestinal epithelium of mice as well as to that of guinea-pigs (Duguid *et al.*, 1966).
Moreover, it should not be assumed that because a bacterium elutes from red cells
at 37°C, it will fail to adhere to epithelium at the same temperature *in vivo*.

7.3.6 Function of MRE adhesins

As noted above (Section 7.2.8), the finding of a minority of *E. coli* strains possessing
neither an MS nor an MRE adhesin in collections from infantile diarrhoea and urinary
infections suggested, but did not prove, that the possession of an adhesin was un-
necessary for infectivity in these strains. Clear evidence, on the other hand, has been
obtained for a role of MRE adhesins in promoting intestinal infection by strains
containing the antigens K88, K99 and 987, for which suitable experimental models,
in piglets, calves and lambs, are available.

The survival value of the MRE adhesin associated with the K88 antigen was
established in experiments showing that its presence enabled the bacteria to colonize
the small intestine in neonatal piglets (Smith and Linggood, 1971; Jones and Rutter,
1972; Rutter and Jones, 1973). Removal of the plasmid controlling K88 synthesis
was accompanied by losses of the abilities to form K88 antigen, to colonize the
intestinal mucosa and to cause diarrhoea, and re-introduction of the plasmid into
K88-negative strains led to the simultaneous gain of these three properties.

Similar experiments with the MRE-haemagglutinating K99 antigen in calves and
lambs (Ørskov *et al.*, 1975) and piglets (Moon *et al.*, 1977) and the MRE antigen
987 in piglets (Nagy *et al.*, 1977) have shown that these adhesins are essential for
colonization of the small intestine and the production of diarrhoea in the specific
hosts. Moreover, the vaccination of pregnant animals with purified MRE fimbriae
has prevented diarrhoea in the offspring when they were later challenged by
inoculation of the MRE⁺ bacteria (Rutter and Jones, 1972; Nagy *et al.*, 1978).

The observation that 98% of a series of enterotoxigenic strains of *E. coli* from
adult human patients with diarrhoea contained one or other of the fimbrial MRE
adhesins, CFA/I and CFA/II, suggests that these adhesins are the principal colonization factors in such human strains (Evans and Evans, 1978). Loss of the plasmid
controlling CFA/I or CFA/II caused the strain to lose its ability to colonize the
small intestine in infant rabbits (Evans *et al.*, 1975; Evans and Evans, 1978). The
pathogenic role of CFA/I was further shown by the observation that, when human
volunteers were infected orally with 10⁸ CFA⁺ or CFA⁻ bacteria, diarrhoea and
prolonged excretion of the bacteria were caused only in those receiving the CFA⁺
inoculum (Evans *et al.*, 1978b).

Thus there is good evidence that the MRE adhesins play an important part in the
causation of *E. coli* diarrhoea in man, but it is still unclear why these adhesins have been

found in only a small minority of strains of enteropathogenic serotypes, including O55 and O111, isolated from outbreaks of infantile diarrhoea (Duguid, 1964).

7.4 OTHER ADHESINS AND FIMBRIAE

7.4.1 MR/K adhesin and type-3 fimbriae

Most strains of *K. aerogenes,* as defined by Cowan *et al.* (1960), and *S. marcescens* form both type-1 fimbriae with MS adhesin and thinner, type-3 fimbriae, which have mannose-resistant (MR) adhesive properties different from those of MRE adhesins (Duguid, 1959). Of 203 strains of *K. aerogenes* in capsule serotypes 1—72, 151 formed both types of fimbriae and adhesin, 14 formed only type-1 fimbriae and MS adhesin, 37 only type-3 fimbriae and MR adhesin, and one formed neither type of fimbriae or adhesin (Duguid, 1968). Of 10 strains of *S. marcescens,* 8 formed both type-1 and type-3 fimbriae, and 2 only type 3. We have recently observed a similar MR adhesin in 11 of 12 strains of *Proteus* and all of 10 strains of *Providencia* (Table 7.2).

The type-3 fimbriae were thinner and more numerous than type 1 in the metal-shadowed films of Duguid (1959) and this difference was shown more clearly in films negatively stained with phosphotungstic acid by Thornley and Horne (1962), who estimated the diameters of the type-1 and type-3 fimbriae as, respectively, 7 and 5 nm, and their numbers per bacterium as 200—400 and 400—700.

The adhesive properties of the type-3 fimbriae were most easily demonstrated in the klebsiella strains that formed only this type of fimbriae, but were confirmed in the strains also forming type-1 fimbriae by tests made in the presence of D-mannose to prevent the effects of the MS adhesin (Duguid, 1959). Bacteria with only type-3 fimbriae did not agglutinate the fresh, untreated red blood cells of any animal species, nor, in wet films, adhere to such red cells, to leucocytes or to intestinal epithelial cells. They did, however, agglutinate the red cells of ox, man, guinea-pig and other species, if these cells were first treated with tannic acid (Table 7.1). In wet films, moreover, they adhered strongly to *untreated* hyphae and conidia of aspergillus and other moulds (Fig. 7.4), yeast cells of 'rough', but not 'smooth' strains of candida, root hairs of clover and cress plant seedlings (Fig. 7.5), and the surfaces of glass slides and cellulose fibres. In all cases the adhesion was unaffected by the addition of D-mannose to the test.

This mannose-resistant adhesin of klebsiella and serratia differs from the MRE adhesins of *E. coli* and the MR/P adhesins of *Proteus* and *Providencia* (see Section 7.4.2), because it is inactive against red blood cells not treated with tannic acid, and it also differs from the MRE adhesins because the bacteria do not elute from their substrates, e.g. tanned red cells, on warming to 50°C. To distinguish it from these other adhesins, it will be called the 'MR/K adhesin'.

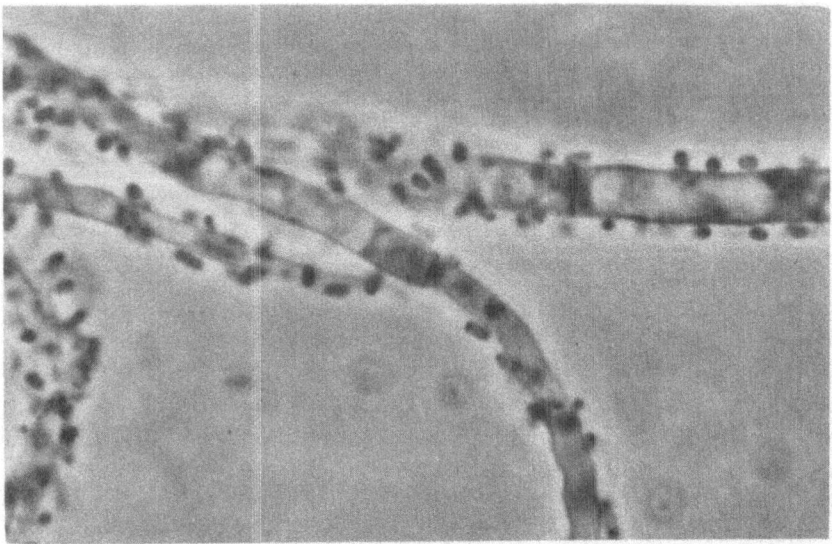

Fig. 7.4 Adhesion of *Klebsiella* bacilli with MR/K adhesin to fungal hyphae. Wet film. x 2000. Reproduced from Duguid (1959) with permission, Editors, *Journal of General Microbiology*.

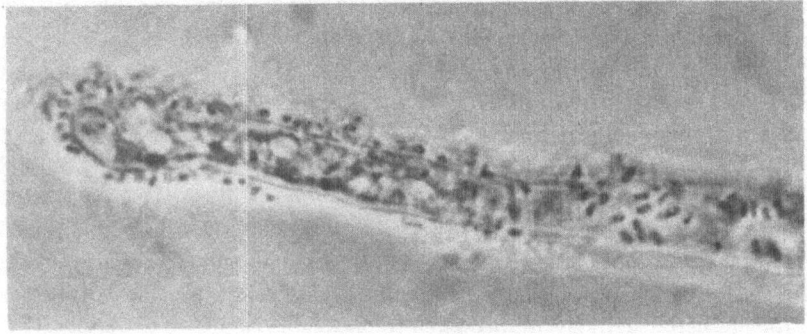

Fig. 7.5 Adhesion of *Klebsiella* bacilli with MS and MR/K adhesins to a root hair of a cress plant seedling. Wet film. x 1300. Reproduced from Duguid (1959) with permission, Editors, *Journal of General Microbiology*.

The strains of *K. aerogenes* and *S. marcescens* that formed both type-1 and type-3 fimbriae showed a combination of the adhesive properties of the MS and MR/K adhesins, e.g. MS agglutination of untreated guinea-pig red blood cells and MR agglutination of tanned ox red blood cells. Generally, a pure culture seeded from a single colony and grown for 48 h came to contain some bacteria with type-1

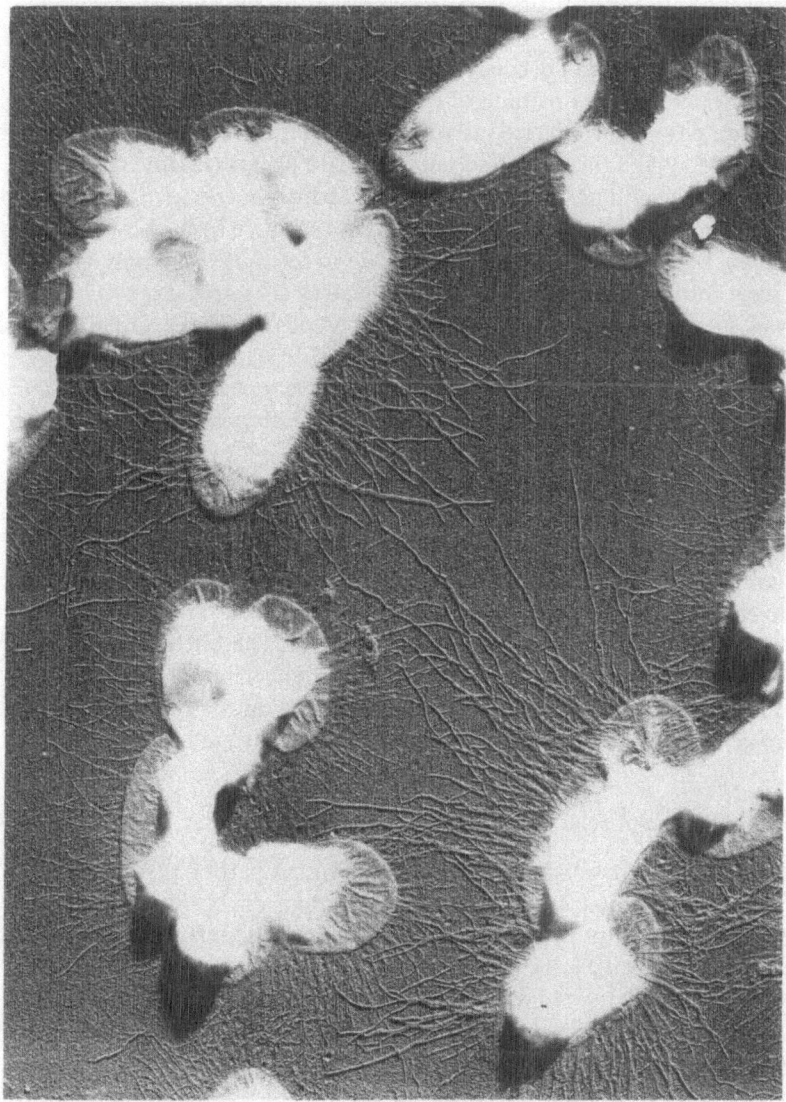

Fig. 7.6 Bacilli with thick, type-1 fimbriae and other bacilli with thin, type-3 fimbriae in a pure culture of *K. aerogenes* capsule serotype 54. Metal-shadowed film. x 20 000.

fimbriae, some with type-3 fimbriae and some without fimbriae. That of the strain illustrated in Fig. 7.6 contained about 20% of bacteria with the thicker type-1 fimbriae, 80% with the thinner type-3 fimbriae and very few without fimbriae. It was unusual for both types of fimbriae to be present on the same bacterial cell. In some strains the proportions of the two types of fimbriate bacteria were increased by serial culture in broth and decreased by serial culture on agar, but in other strains they were unaffected by the method of culture. Thus, some strains of *K. aerogenes* appear to undergo reversible phase variations by which the bacteria change between type-1 fimbriate, type-3 fimbriate and non-fimbriate phases (Duguid, 1959).

Type-3 fimbriae and MR/K adhesin were not formed in species of *Klebsiella* other than *K. aerogenes*. Neither type-1 nor type-3 fimbriae were formed by any strain of *K. atlantae, K. edwardsii, K. ozaenae* and *K. rhinoscleromatis*, and only type-1 fimbriae were formed by strains of *K. pneumoniae* as defined *sensu stricto* by Cowan *et al.* (1960) and Cowan (1974) (Duguid, 1968). The broader definition of *K. pneumoniae*, as given by Ørskov (1974) in Bergey's Manual, which includes *K. aerogenes, K. atlantae* and *K. edwardsii* as well as *K. pneumoniae (sensu stricto)*, seems unsatisfactory, because these species are clearly disparate both in their ecology and in the type of fimbriae they form.

An ecological distinction may be drawn between, on the one hand, *K. aerogenes* and *S. marcescens,* which live saprophytically in soil and water, are rarely pathogenic, and typically form both type-1 and type-3 fimbriae, and, on the other hand, *K. pneumoniae (sensu stricto), K. atlantae, K. edwardsii, K. ozaenae* and *K. rhinoscleromatis,* which commonly infect the respiratory tract, are not known to inhabit soil and water, and never form type-3 fimbriae. The MR/K adhesin of type-3 fimbriae may have a useful function in holding saprophytic bacteria on the surfaces of plant and fungal cells which exude nutrient solutes and on the surfaces of organic and inorganic materials where nutrient solutes are concentrated by adsorption.

7.4.2 MR/P adhesin and type-4 fimbriae

A mannose-resistant adhesin with haemagglutinating properties different from those in other enterobacteria was found in all of many strains of *Proteus mirabilis, P. vulgaris, P. morganii* and *P. rettgeri* by Duguid and Gillies (1958), Coetzee *et al.* (1962) and Shedden (1962). Bacteria from cultures in broth or peptone water commonly agglutinated the red blood cells of sheep and fowl strongly, and those of ox, man, horse, mouse, rabbit and guinea-pig moderately strongly, though bacteria from less active cultures failed to agglutinate the ox, human and rabbit cells (Table 7.1, cf. i and j). Whether the different patterns of reactions given by different cultures reflected merely a variation in the strength of the same adhesive factor or a variation in the relative contents of two or more adhesins with different affinities for different species of red cells, cannot be deduced from the published findings. *P. rettgeri,* however, was distinguished from the other species of *Proteus* because it agglutinated all species of red cells to the same degree and because its haemagglutinating activity was stronger in cultures grown at ambient temperature than in those at 37°C (Coetzee *et al.,* 1962).

In rocked tile tests, the reactions of *Proteus* strains were similar, whether the tests were made with or without D-mannose, and whether at 4°, 22°, 37° or 55°C; i.e., they were mannose-resistant and non-eluting. In wet films, the bacteria adhered to the yeast cells of *Candida* and the surface of glass slides (Shedden, 1962). Since the *Proteus* adhesin differs from the MR/K adhesin of *Klebsiella* by its activity on red blood cells not treated with tannic acid, it will be distinguished as the 'MR/P adhesin'.

All haemagglutinating cultures of *Proteus* contained bacteria showing fimbriae in metal- shadowed films, and the association of the fimbriae with the MR/P adhesin was confirmed by the finding that the bacteria became both non-fimbriate and non-haemagglutinating when cultured serially 2—5 times on agar plates, and again both fimbriate and haemagglutinating when cultured serially 2—5 times in tubes of broth (Duguid and Gillies, 1958; Coetzee *et al.*, 1962; Shedden, 1962). The fimbriae bearing MR/P adhesin were classified as 'type 4' by Duguid *et al.* (1966).

The width of the type-4 fimbriae is uncertain. In the metal-shadowed films of Coetzee *et al.* (1962) and Shedden (1962) it seemed to be about the same as that of type-1 fimbriae, i.e. 7—10 nm, but the fimbriae of *P. rettgeri* demonstrated by Hashimoto *et al.* (1963) were thinner, and those of one strain of *P. mirabilis* examined in a film with phosphotungstic acid by Hoeniger (1965) were about 4 nm wide. Recently, Silverblatt (1974) described the formation of two types of fimbriae by a strain of *P. mirabilis*, fimbriae about 4 nm wide in the early stages of growth in shaken broth or on agar, and fimbriae about 7 nm wide after one to six serial cultures for 48 h in stationary tubes of broth. The haemagglutinating properties of the two types were not determined, so it cannot be said which of them, if either, bore the MR/P adhesin or whether the MR/P adhesive properties might be due to their combined activities. Both types appeared to mediate adhesion of the bacteria to renal pelvic epithelium in rats.

Recently, we have found that *Proteus* and *Providencia* strains may have multiple adhesins, e.g. combinations of MS, MR/K and MR/P adhesins (Table 7.2). Moreover, different types of MR/P adhesin acting mainly on different species of red blood cells, particularly on those of fowl, guinea-pig or horse, were present in different strains, and varied in amount with the conditions of culture. These observations may explain apparent anomalies in previously reported studies.

7.4.3 Non-adhesive fimbriae of types 2 and 6

Fimbriae resembling type-1 fimbriae in morphology, but differing from them in not conferring haemagglutinating, adhesive and pellicle-forming properties, and designated 'type 2' (Fig. 7.7), have been described in some strains of a few serotypes of *Salmonella*, namely *S. gallinarum, S. pullorum* and *S. paratyphi B* (Duguid *et al.*, 1966) and *S. typhimurium* (Duguid *et al.*, 1975). Serological analysis showed that the type-1 and type-2 fimbriae of *S. paratyphi B* were antigenically similar, suggesting that the type-2 fimbriae were only slightly altered, mutant forms of the type-1 fimbriae (Old and Payne, 1971). A mutational origin of the type-2 fimbriae was further suggested by the observation that some cultures of *S. paratyphi B* strains that formed only type-2

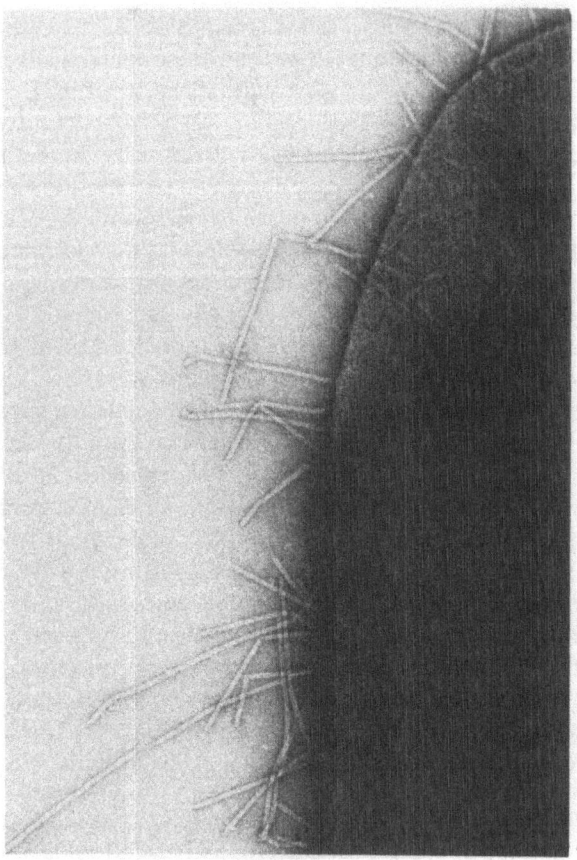

Fig. 7.7 Non-adhesive type-2 fimbriae on *S. paratyphi B* strain S66. Negative staining with uranyl acetate. × 108 000.

fimbriae gave rise spontaneously to mutant bacteria with type-1 fimbriae and MS adhesin. If the chemical difference between these two types of fimbriae can be determined, it should reveal the chemical basis of the MS adhesive property.

Non-haemagglutinating fimbriae, designated 'type-6', have been reported in about 5% of strains of *K. ozaenae* (Duguid *et al.,* 1966; Duguid, 1968). They were peritrichously arranged, scanty (e.g. 10–20 per bacterium), long (up to 10 μm) and about 10 nm in width (Fig. 7.8), and possibly represent sex fimbriae (pili). Other, thinner non-haemagglutinating fimbriae have been described in exceptional strains of *S. enteritidis* isolated from pasta (Rohde *et al.,* 1975) and strains of *Yersinia enterocolitica* 0-group 10 (Aleksić *et al.,* 1976).

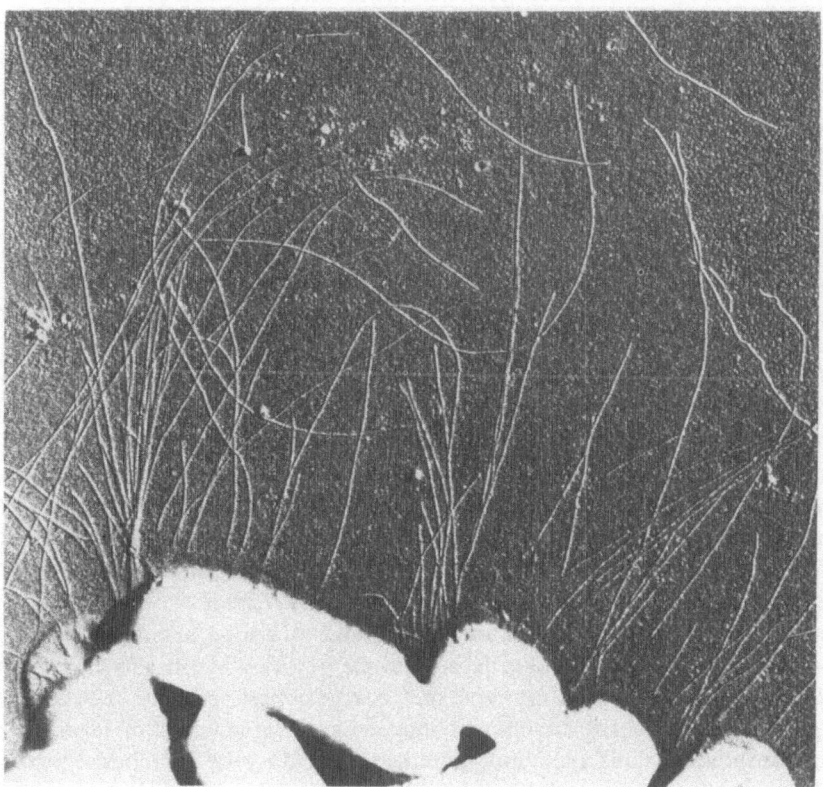

Fig. 7.8 Non-adhesive type-6 fimbriae on *K. ozaenae* capsule serotype 4. Metal-shadowed film. x 18 000.

7.5 CONCLUSION

A main purpose of this review has been to illustrate the diversity of types and the wide distribution of the adhesive factors in Enterobacteriaceae. Attention has been concentrated on the adhesive properties of these factors, and not on other, e.g. biochemical and genetical, aspects of the fimbrial and non-fimbrial adhesins, which have recently been reviewed by Ottow (1975) and Jones (1977).

Until much further information has been obtained about the ultra-structure, chemical composition and genetic control of the different adhesins, it seems best to classify them by their readily observed properties, such as their specificity for different substrates, their sensitivity to inhibition by D-mannose and to elution at raised temperatures, and their stability on exposure to heat and formaldehyde. By these criteria, four main types of adhesin, MS (mannose-sensitive), MRE (mannose-resistant and eluting), MR/K (mannose-resistant/klebsiella) and MR/P (mannose-resistant/proteus) have so far been identified in enterobacteria, and several subtypes of MRE and MR/P adhesins with different affinities for different species of red blood cells. But it is quite possible that there are further types of adhesin still to be discovered, or that the properties now attributed to a single adhesin represent the combined properties of two or more adhesins present in the same bacterium.

That bacteria commonly form more than one type of adhesin is shown by the finding that many strains of *E. coli* form both an MS and an MRE adhesin and many strains of *K. aerogenes,* both an MS and an MR/K adhesin. The use of D-mannose to abolish the effects of the MS adhesin in such strains facilitates the separate demonstration of the mannose-resistant adhesin, but the distinction of two different mannose-resistant adhesins in the same bacterium is more difficult. In an exceptional strain of *K. aerogenes* that formed three types of adhesins, MS, MRE and MR/K, it was possible to distinguish the MRE adhesin, which acted on untreated erythrocytes, from the MR/K adhesin, which acted only on erythrocytes treated with tannic acid, because the former, but not the latter, was absent in cultures grown at 20°C. Other mannose-resistant adhesins may not be so easily distinguishable from one another. Before the full range of enterobacterial adhesins can be defined, much further work will be required, particularly on the less studied genera such as *Proteus, Providencia, Enterobacter, Erwinia, Hafnia* and *Yersinia.*

In some studies of the pathogenic role of adhesive properties in enterobacteria, the reported observations of these properties in the experimental strains of bacteria have been insufficient to define the type, or types, of adhesin present, or to prove that supposedly non-adhesive strains were indeed devoid of any kind of adhesin. Such studies should include a very full investigation of the possible adhesive properties. A wide variety of species of erythrocytes and other cells should be tested as substrates; haemagglutination should be observed by a variety of methods, including tests made with very dense suspensions of bacteria, tests with and without D-mannose and tests at different temperatures from 4–50°C; and the bacteria to be tested should be grown under a variety of conditions, including serial and prolonged culture in static broth and overnight culture on plates of buffered nutrient agar. In the case of adhesins that may be formed only under certain cultural conditions, or after spontaneous phase variation, the attempt should be made to demonstrate whether they are, in fact, formed *in vivo*.

Although medical interest centres on the possible role of adhesins in the causation of infection by pathogenic bacteria, the adhesins are widely distributed in commensal and free-living saprophytic species, in which no doubt they originally evolved. Hypotheses have been advanced about the possible functions of the adhesins in such organisms, but little experimental work has been done to prove their value.

REFERENCES

Aleksić, S., Rohde, R., Müller, G. and Wohlers, B. (1976), *Zbt. Bakt. Hyg., I. Abt. Orig.* A, **234**, 513–520.

Bar-Shavit, Z., Ofek, I., Goldman, R., Mirelman, D. and Sharon, N. (1977), *Biochem. biophys. Res. Comm.*, **78**, 455–460.

Brinton, C.C. (1959), *Nature*, **183**, 782–786.

Brinton, C.C. (1965), *Trans. N.Y. Acad. Sci.*, **27**, 1003–1054.

Brinton, C.C. and Stone, M.J. (1961), *Bact. Proc.*, G96.

Brinton, C.C., Buzzell, A. and Lauffer, M.A. (1954), *Biochim. biophys. Acta*, **15**, 533–542.

Burrows, M.R., Sellwood, R. and Gibbons, R.A. (1976), *J. gen. Microbiol.*, **96**, 269–275.

Christofi, N., Wilson, M.I. and Old, D.C. (1979), *J. appl. Bact.*, **46**, 179–183.

Coetzee, J.N., Pernet, G. and Theron, J.J. (1962), *Nature*, **196**, 497–498.

Collier, W.A. and de Miranda, J.C. (1955), *Antonie van Leeuwenhoek, J. Microbiol. Serol.*, **21**, 135–140.

Constable, F.L. (1956), *J. Path. Bact.*, **72**, 133–136.

Cowan, S.T. (1974), *Cowan and Steel's 'Manual for the identification of medical bacteria'.* Cambridge University Press, Cambridge, p.111.

Cowan, S.T., Steel, K., Shaw, C. and Duguid, J.P. (1960), *J. gen. Microbiol.*, **23**, 601–612.

Duguid, J.P. (1959), *J. gen. Microbiol.*, **21**, 271–286.

Duguid, J.P. (1964), *Rev. lat-amer. Microbiol.*, **7**, Suppl. 13–14, 1–16.

Duguid, J.P. (1968), *Arch. Immun. Ther. Exp.*, **16**, 173–188.

Duguid, J.P. and Anderson, E.S. (1967), *Nature*, **215**, 89–90.

Duguid, J.P. and Campbell, I. (1969), *J. med. Microbiol.*, **2**, 535–553.

Duguid, J.P. and Gillies, R.R. (1957), *J. Path. Bact.*, **74**, 397–411.

Duguid, J.P. and Gillies, R.R. (1958), *J. Path. Bact.*, **75**, 519–520.

Duguid, J.P. and Wilkinson, J.F. (1961), In: *Microbial Reaction to Environment.* (Meynell, G.G. and Gooder, H. eds), Symposia of the Society for General Microbiology, **11**, University of London Press, London, pp. 69–99.

Duguid, J.P., Anderson, E.S. and Campbell, I. (1966), *J. Path. Bact.*, **92**, 107–138.

Duguid, J.P., Clegg, S. and Wilson, M.I. (1979), *J. med. Microbiol.*, **12**, 213–227.

Duguid, J.P., Darekar, M.R. and Wheater, D.W.F. (1976), *J. med. Microbiol.*, **9**, 459–473.

Duguid, J.P., Smith, I.W., Dempster, G. and Edmunds, P.N. (1955), *J. Path. Bact.*, **70**, 335–348.

Duguid, J.P., Anderson, E.S., Alfredsson, G.A., Barker, R. and Old., D.C. (1975), *J. med. Microbiol.*, **8**, 149–166.

Evans, D.G. and Evans, D.J. (1978), *Infect. Immun.*, **21**, 638–647.

Evans, D.G., Evans, D.J. and Tjoa, W.S. (1977), *Infect. Immun.*, **18**, 330–337.

Evans, D.G., Evans, D.J., Tjoa, W.S. and DuPont, H.L. (1978a), *Infect. Immun.*, **19**, 727–736.

Evans, D.G., Satterwhite, T.K., Evans, D.J. and DuPont, H.L. (1978b), *Infect. Immun.*, **19**, 883–888.

Evans, D.G., Silver, R.P., Evans, D.J., Chase, D.G. and Gorbach, S.L. (1975),
    Infect. Immun., **12**, 656–667.
Gillies, R.R. and Duguid, J.P. (1958), *J. Hyg., Camb.*, **56**, 303–318.
Griffitts, J.J. (1948), *Proc. Soc. exp. Biol. Med.*, **67**, 358–362.
Guyot, G. (1908), *Zbt. Bakt. Abt. I Orig.*, **47**, 640–653.
Hashimoto, M., Tsuchimoto, S., Kato, A., Takemoto, T., Yoshino, T., Ikuta, M. and
    Nakajima, M. (1963), *Bull. Tokyo med. dent. Univ.*, **10**, 181–203.
Hirst, G.K. (1941), *Science*, **94**, 22–23.
Hoeniger, J.F.M. (1965), *J. gen. Microbiol.*, **40**, 29–42.
Houwink, A.L. and van Iterson, W. (1950), *Biochim. biophys. Acta*, **5**, 10–44.
Isaacson, R.E. (1977), *Infect. Immun.*, **15**, 272–279.
Jones, G.W. (1977), In: *Microbial Interactions*. (Reissig, J.L. ed.), Receptors and
    Recognition (Series B), **3**, Chapman and Hall, London, pp. 141–176.
Jones, G.W. and Rutter, J.M. (1972), *Infect. Immun.*, **6**, 918–927.
Jones, G.W. and Rutter, J.M. (1974), *J. gen. Microbiol.*, **84**, 135–144.
Kauffmann, F. (1948), *Acta path. microbiol. scand.*, **25**, 502–506.
Keogh, E.V., North, E.A. and Warburton, M.F. (1947), *Nature*, **160**, 63.
Labrec, E.H., Schneider, H., Magnani, T.J. and Formal, S.B. (1964), *J. Bact.*, **88**,
    1503–1518.
Lederberg, J. and Iino, T. (1956), *Genetics*, **41**, 743–757.
Meynell, G.G. and Lawn, A.M. (1967), *Genet. Res., Camb.*, **9**, 359–367.
Moon, H.W., Nagy, B., Isaacson, R.E. and Ørskov, I. (1977), *Infect. Immun.*, **15**,
    614–620.
Morris, J.A., Stevens, A.E. and Sojka, W.J. (1977), *J. gen. Microbiol.*, **99**, 353–357.
Mulczyk, M. and Duguid, J.P. (1966), *J. gen. Microbiol.*, **45**, 459–477.
Nagy, B., Moon, H.W. and Isaacson, R.E. (1977), *Infect. Immun.*, **16**, 344–352.
Nagy, B., Moon, H.W., Isaacson, R.E., To, C.C. and Brinton, C.C. (1978), *Infect.
    Immun.*, **21**, 269–274.
Nowotarska, M. and Mulczyk, M. (1977), *Arch. Immun. Ther. Exp.*, **25**, 7–16.
Ofek, I., Beachey, E.H. and Sharon, N. (1978), *T.I.B.S.*, **3**, 159–160.
Ofek, I., Mirelman, D. and Sharon, N. (1977), *Nature*, **265**, 623–625.
Ogawa, H., Nakamura, A. and Nakaya, R. (1968), *Jap. J. med. Sci. Biol.*, **21**,
    259–273.
Old, D.C. (1963), Ph. D. Thesis, University of Edinburgh, Scotland.
Old, D.C. (1972), *J. gen. Microbiol.*, **71**, 149–157.
Old, D.C. and Duguid, J.P. (1970), *J. Bact.*, **103**, 447–456.
Old, D.C. and Duguid, J.P. (1971), *J. Bact.*, **107**, 655–658.
Old, D.C. and Payne, S.B. (1971), *J. med. Microbiol.*, **4**, 215–225.
Old, D.C., Corneil, I., Gibson, L.F., Thomson, A.D. and Duguid, J.P. (1968),
    J. gen. Microbiol., **51**, 1–16.
Ørskov, I. (1974), Bergeys *Manual of Determinative Bacteriology*. (Buchanan, R.E.
    and Gibbons, N.E. eds), The Williams and Wilkins Co., Baltimore, pp. 321–324.
Ørskov, I. and Ørskov, F. (1977), *Med. Microbiol. Immun.*, **163**, 99–110.
Ørskov, I., Ørskov, F., Jann, B. and Jann, K. (1977), *Bact. Rev.*, **41**, 667–710.
Ørskov, F., Ørskov, I., Smith, H.W. and Sojka, W.J. (1975), *Acta path. microbiol.
    scand.*, B, **83**, 31–36.

Ottow, J.C.G. (1975), *Ann. Rev. Microbiol.*, **29**, 79–108.

Rohde, R., Aleksić, S., Müller, G., Plavsić, S. and Aleksić, V. (1975), *Zbl. Bakt. Hyg.*, *I. Abt.* Orig. A, **230**, 38–50.

Rosenthal, L. (1943), *J. Bact.*, **45**, 545–550.

Rutter, J.M. and Jones, G.W. (1973), *Nature*, **242**, 531–532.

Saier, M.H., Schmidt, M.R. and Leibowitz, M. (1978), *J. Bact.*, **134**, 356–358.

Salit, I.E. and Gotschlich, E.C. (1977a), *J. exp. Med.*, **146**, 1169–1181.

Salit, I.E. and Gotschlich, E.C. (1977b), *J. exp. Med.*, **146**, 1182–1194.

Shedden, W.I.H. (1962), *J. gen. Microbiol.*, **28**, 1–7.

Silverblatt, F.J. (1974), *J. exp. Med.*, **140**, 1696–1711.

Smith, H.W. and Linggood, M.A. (1971), *J. med. Microbiol.*, **4**, 467–485.

Stirm, S., Ørskov, F., Ørskov, I. and Birch-Andersen, A. (1967a), *J Bact.*, **93**, 740–748.

Stirm, S., Ørskov, F., Ørskov, I. and Mansa, B.(1967b), *J. Bact.*, **93**, 731–739.

Stocker, B.A.D. (1949), *J. Hyg., Camb.*, **47**, 398–413.

Svanborg Edén, C. and Hansson, H.A. (1978), *Infect. Immun.*, **21**, 229–237.

Svanborg Edén, C., Hanson, L. Å., Jodal, U., Lindberg, U. and Sohl Akerland, A. (1976), *Lancet*, **ii**, 490–492.

Takeuchi, A. (1967), *Am. J. Path.*, **50**, 109–136.

Thornley, M.J. and Horne, R.W. (1962), *J. gen. Microbiol.*, **28**, 51–56.

8 The Adhesive Properties of *Vibrio cholerae* and other *Vibrio* Species

GARTH W. JONES

8.1	Introduction	*page*	221
8.2	Taxonomy and habitats of *V. cholerae*		222
	8.2.1 Taxonomy		222
	8.2.2 Ecology		223
8.3	Adhesive properties		225
	8.3.1 Haemagglutinating properties of *V. cholerae*		226
	8.3.2 The association with, and the adhesion to, intestinal mucosae by *V. cholerae*	.	235
	8.3.3 *V. cholerae* adhesins		239
8.4	Ecological significance of adhesion and association		243
8.5	Conclusions		245
	References		245

Bacterial Adherence
(*Receptors and Recognition,* Series B, Volume 6)
Edited by E.H. Beachey
Published in 1980 by Chapman and Hall, 11 New Fetter Lane, London EC4P 4EE
© 1980 Chapman and Hall

8.1 INTRODUCTION

The study of the adhesive properties of micro-organisms should be concerned with the location of the organism within the habitat and with the nature of that habitat. From such observations, evidence of the production of adhesive substances that materially assist the organism in its colonization of the habitat may be obtained. For example, if the microbial population is found to be in intimate association with a substratum then it is valid to assume that the microbes remain permanently associated with the surface because the population elaborates adhesive substances. Alternatively, the existence of adhesive properties may be presumed if it is found that the hydro-dynamic flow and shear within the habitat (e.g. the gastro-intestinal tract) would tend to remove unattached bacteria (Gibbons and van Houte, 1975) or that because of the low nutriment content of the environment (e.g. many natural surface waters) the bacteria would benefit from localization at interfaces where nutriments may accumulate (ZoBell, 1972). Seldom is it possible, however, to study *in situ* the details of the events that bring about the adhesion of bacteria to surfaces and recourse has to be made to artificial *in vitro* systems (see also Chapter 1). This is particularly true of apparently strict human pathogens such as *Vibrio cholerae* (Finkelstein, 1973) for which the evidence of adhesion within the human intestine is non-existent and the synthesis of adhesive substances by *V. cholerae in vivo* can be deduced only from a multitude of diverse *in vitro* studies. Unfortunately, adhesive properties can be investigated with *in vitro* models without regard to the many other microbial activities that determine an organism's fitness to survive in nature and under conditions when growth and even continued viability may be of little importance. Indeed, it is very likely that most organisms can be made to attach to some suitable surface if the appropriate surface conditions are induced in the organism and/or suitable physico-chemical conditions found for the test system. Thus, the question is not does the organism possess adhesive properties but rather do those properties that are demonstrable contribute significantly to the organism's ability to colonize the habitat under study?

Some species of bacteria such as *V. cholerae* are not restricted to single habitats. Established populations of this organism can be found in the human intestine and in a variety of surface waters. It seems reasonable to assume that colonization of these different environments does not occur in identical fashions but that the establishment of viable populations in either would require rather different bacterial attributes. If these essential attributes include adhesive properties then these are likely to be different also. That is, several adhesins may be synthesized by bacteria of the species *V. cholerae,* the specificity and nature of which will depend upon the niche occupied by the particular strain of vibrio. Accordingly, the following may be

221

hypothesized. Strains of *V. cholerae* may exist that colonize both the human intestine and water. Such strains may produce different adhesins according to the habitat occupied by the vibrio at the time. *Escherichia coli* provides an excellent example of a bacterium, strains of which can produce at least two adhesive substances (Chapter 7). Alternatively, races of *V. cholerae* may exist that colonize only the intestine whereas other races colonize only surface waters. Such races may produce adhesins with properties that uniquely suit the vibrios to these habitats. If one entertains the possibility that water is the most likely habitat of ancestral vibrios from which pathogenic vibrios that colonize the human intestine have evolved, then it it necessary to consider the possibility that evolution has resulted in the appearance of new adhesive properties in pathogenic vibrios. These newer adhesins may be present in addition to or may have replaced pre-existing adhesins. Furthermore, it can be envisaged that the evolution of new adhesive activities may have occurred upon several distinct occasions. Consequently, vibrios that now colonize the intestine may elaborate several strain specific adhesins. For example, it appears from geographical evidence that classical and eltor strains of *V. cholerae* developed as human pathogens in different areas and perhaps at different times (e.g. Pollitzer, 1959; Finkelstein, 1973) and hence may produce quite distinct adhesins. Enteropathogenic *E. coli* again appears to provide examples of this phenomenon (Chapter 7). Finally, the adhesins of water vibrios may be identical to those of vibrios that colonize the intestine but the latter may have evolved in other ways that allow them to proliferate in the human gut.

These suggestions may be considered to be premature at this time. However, it must be emphasized that the investigation of the adhesive properties of vibrios is in its infancy and is confined entirely to the study of vibrios in model systems that may in no way portray the adhesive behavior of vibrios in the human intestine nor indeed in any other habitat. I do not intend to imply that model systems consisting of various forms of intestinal tissues of laboratory animals and of erythrocytes are worthless. Indeed, they are extremely useful. However, the results obtained with such systems need to be carefully and intelligently interpreted; this general problem has been addressed (Freter, 1978). From the foregoing, it would seem that the ecology and taxonomy of *V. cholerae* are subjects pertinent to a discussion of the adhesive properties of these organisms and accordingly a brief summary of these topics is included.

8.2 TAXONOMY AND HABITATS OF *V. CHOLERAE*

8.2.1 Taxonomy

Many years ago it was found that those strains of *V. cholerae* most adept at coloniz-ing the human intestine and giving rise to outbreaks of cholera could be distinguished from vibrios present in water and other habitats because they produced a cell wall antigen designated O1 and fermented certain carbohydrates (reviewed by Pollitzer,

1959). Commonly called cholera vibrios, the group is composed of two biotypes (the cholera biotype, also called the classical biotype, and the eltor biotype) and three serotypes (Shewan and Véron, 1974). Other vibrios commonly known as non-cholera, non-agglutinable or water vibrios differ in that they produce cell wall antigens other than the O1 antigen (Sakazaki *et al.,* 1970) and commonly demonstrate different fermentation patterns (Pollitzer, 1959; Shewan and Véron, 1974). For the purpose of this chapter, the term cholera vibrio will be used for all serotypes of the overtly pathogenic forms (i.e., classical and eltor strains) and the term non-cholera vibrios for all others that do not produce the O1 antigen. This does not mean, however, that the latter are not pathogenic. It is becoming increasingly apparent that these vibrios cause a cholera-like disease (Lindenbaum *et al.,* 1965; McIntyre and Feeley, 1965; McIntyre *et al.,* 1965; Smith and Goodner, 1966; Aldova *et al.,* 1968; Finkelstein, 1973; Hughes *et al.,* 1978; Năcescu and Ciufecu, 1978).

In contrast to the early system of classification, which was one devoted to differentiating vibrios according to prevailing views of pathogenicity, more recent studies show that cholera and non-cholera vibrios comprise a single distinct species (Sakazaki *et al.,* 1967; Colwell, 1970) which on the basis of DNA homology show no evidence of a marked evolutionary divergence (Citarella and Colwell, 1970) that may account for the different degrees of pathogenicity and distribution in nature of these bacteria. Indeed, substances considered perhaps more characteristic of pathogenic bacteria such as haemolysins (Freter, 1951; Sakazaki *et al.,* 1967; Colwell, 1970) mucinases (Freter, 1955) and enterotoxins (McIntyre *et al.,* 1965; Sakazaki *et al.,* 1967; Finkelstein, 1973; Colwell *et al.,* 1977) are produced by cholera and non-cholera vibrios as are many other enzymes including, perhaps significantly, chitinase (Sakazaki *et al.,* 1963) which is to be discussed later. Moreover, those traits originally thought to differentiate cholera vibrios from non-cholera vibrios such as serotype and fermentation group (Bhattacharji and Bose, 1964; Smith and Goodner, 1965; Pesigan *et al.,* 1967; Sen *et al.,* 1967; El-Shawi and Thewaini, 1969) and enterotoxin production (Dutta *et al.,* 1963; Bhattacharji and Bose, 1964; Bhattacharya *et al.,* 1971) may be inducible or mutable characteristics.

8.2.2 Ecology

(a) *Water*

Aquatic bacteria such as *Vibrio natriegens* (*Baneckea natriegens*) and *Vibrio anguillarium* display adhesive properties (Meadows, 1971; Zachary *et al.,* submitted for publication), as does *Vibro parahaemolyticus* which is found attached to the external surfaces of esturine copepods (Kaneko and Colwell, 1973; Kaneko and Colwell, 1975b) and which can also adsorb to chitin particles when the salinity of the water is at an appropriate level (Kaneko and Colwell, 1975a). Although there is no reason to suppose that *V. cholerae* is associated with copepods, recently it has been shown that *V. cholerae* adsorbs to chitin particles also (Nalin *et al.,* 1978).

V. cholerae, as do many aquatic bacteria (Chapter 11), attaches to glass surfaces
(Lankford and Legsomburana, 1965; Freter and Jones, 1976) but this property is
independent of those that enable vibrios to attach to animal cells (Freter and Jones,
1976). Historically non-cholera vibrios have been associated with water habitats as is
evidenced by their synonym water vibrios. Recent ecological studies have established
that non-cholera vibrios do, in fact, constitute part of the autochthonous flora of
aquatic habitats (Colwell and Kaper, 1977; Colwell *et al.,* 1977). Population densities
may be exceedingly low and confined to particular salinity zones. Although non-
cholera vibrios may be isolated from fresh water sources, e.g. rivers (Müller, 1977),
their survival and the survival of cholera vibrios in fresh water is enhanced by increased
salinity (Felsenfeld, 1966; Pesigan *et al.,* 1967; Müller, 1977). However, the persistance
of cholera vibrios in water has been the subject of epidemiological rather than ecological
investigations. The principal concern has been with the existance of potential sources
of infection. Consequently, water has seldom been considered as a natural habitat
of cholera vibrios, but rather as a vector of disease. The general tendency has been
to conclude that the presence of cholera vibrios in water is contingent upon continual
faecal pollution (Read and Pandit, 1941; Abou-Gareeb, 1960; Benenson *et al.,* 1965)
(although it should be noted that the rate at which vibrios are killed in water is
increased in the presence of sewage (Geldreich, 1972)). Laboratory experiments
support this position; cholera vibrios do not in general survive long in water, although
survival is very variable and depends upon several factors (Pollitzer, 1959; Felsenfeld,
1963; Neogy, 1965; Pesigan, 1965; Pandit *et al.,* 1967; Pesigan *et al.,* 1967; Müller,
1977). However, in most of these studies little or no account has been taken of the
small numbers of vibrios that may constitute a viable population in water (Colwell
and Kaper, 1977; Colwell *et al.,* 1977), no special precautions have been taken to
facilitate the isolation of stressed bacteria (Colwell and Kaper, 1977) and no regard
has been paid to the niche vibrios may occupy in aquatic habitats. Consequently,
few conclusions on the colonization of aquatic habitats by cholera vibrios can be
drawn from such studies. The results of one investigation (Pandit *et al.,* 1967),
however, indicate that cholera vibrios, after a marked decrease in their initial
numbers, may establish stable populations more commensurate with population
levels of non-cholera vibrios in natural habitats (Colwell and Kaper, 1977; Colwell
et al., 1977).

(b) *Intestine*
V. cholerae is found throughout the gastro-intestinal tract of cholera patients
(Gorbach *et al.,* 1970) where they appear to be confined to the lumen (Gangarosa
et al., 1960; Fresh *et al.,* 1964). It is likely that *V. cholerae,* like so many other
non-invasive bacteria of the gastro-intestinal tract (Chapter 2 and 3), colonize the
mucosal surfaces and most probably the mucosal surface of the small intestine; such
is indicated by the following information. Cholera toxin acts on the epithelial cells
of the small intestine and causes a fluid loss from this organ (Finkelstein, 1973).
Accordingly, it can be argued that close association of the vibrios with the tissue

would provide the most effective means of delivering the toxin to its site of action. This possibility is supported by the very different observations of Freter *et al.* (1961) who found that cholera vibrios excreted in the stools of cholera patients were unable to grow in this medium. They concluded, therefore, that because of the short transit time of fluid through the intestine of cholera patients and the inability of the contents of the lumen to support the growth of vibrios, that the vibrios most probably colonized and grew on the intestinal mucosa. Furthermore, *V. cholerae* cultures are able to associate with the mucosa of the small intestine of laboratory animals or with components thereof (Le Brec *et al.*, 1965; Patnaik and Ghosh, 1966; Freter, 1969; Elliot *et al.*, 1970; Guentzel and Berry, 1975; Jones and Freter, 1975; Freter and Jones, 1976; Jones *et al.*, 1976; Jones and Freter, 1976; Nelson *et al.*, 1976; Schrank and Verwey, 1976; Nelson *et al.*, 1977; Freter *et al.*, 1978b). Finally, it should be noted that Lankford (1960) almost two decades ago advanced the argument that cholera vibrios colonized the mucosa of the small intestine; an hypothesis which was consistent with the established antibacterial role of peristalsis in this organ (Dixon, 1960) and the subsequent demonstration of the presence of adhesive properties in cholera vibrio cultures by Lankford and Legsomburana (1965) and by others. If valid comparisons can be made between cholera vibrios and apparently similar enteropathogens such as *E. coli* (Chapter 7), then the conclusion that *V. cholerae* colonizes the mucosa of the small intestine of man is difficult to avoid.

In summary, cholera and non-cholera vibrios are clearly members of a single species although the species appears to include types that have evolved or are evolving to occupy different habitats (i.e., the human intestine and/or water). On the one hand are the non-cholera vibrios that form part of the autochthonous flora of particular aquatic habitats. Some of these bacteria cause cholera-like disease but whether or not these are types identical to those found in water is unknown. Whether or not pathogenic non-cholera vibrios and cholera vibrios will be found to colonize the human intestine in identical manners is an imponderable question at this time. The situation is complicated by the claim that cholera vibrio-like characteristics are accentuated in non-cholera vibrios by animal passage (Dutta *et al.*, 1963; Bhattacharya *et al.*, 1971). On the other hand are the cholera vibrios, the agents of classical cholera. Nothing is known about the aquatic niche these organisms may occupy nor has any serious attempt been made to identify such a niche. It has been suggested, however, that cholera vibrios in water change and assume the characteristics of non-cholera vibrios (Bhattacharji and Bose, 1964). This was not confirmed (El-Shawi and Thewaini, 1969) and the original observation may have been the result of an unfortunate choice of culture (Sakazaki *et al.*, 1970).

8.3 ADHESIVE PROPERTIES

The terms haemaglutinin, adhesin and receptor are used in the ways suggested previously (Jones, 1977). The term 'association' (Freter, 1978) is used to describe the interaction of vibrios with mucosal surfaces, a process that appears to involve the motility and the chemotaxis, as well as the adhesive activities of the vibrio cells.

Arguments of the sort that support the supposition that the adhesion of bacteria such as *V. cholerae* to surfaces in habitats such as the gastro-intestinal tract are well established (Chapters 2, 3, and 7) and need not be reiterated. In contrast, the contribution of motility and chemotaxis to the complex process of colonization has gone unnoticed until recently. Both these activities, however, have a profound influence on the adhesion of vibrios in model systems and will be considered here as part of the overall process of association that may eventually lead to the adhesion of vibrios to mucosal surfaces.

8.3.1 Haemagglutinating properties of *V. cholerae*

Although haemagglutination tests can provide a simple means with which adhesive properties may be investigated and differentiated, the model is not without drawbacks and results can be too easily misinterpreted (Jones, 1977). Perhaps the most serious misinterpretations which are pertinent here and which may account for some of the anomalies to be discussed are as follows: Qualitative and quantitative differences in haemagglutinating activity are not always distinguished; such can lead to erroneous assumptions about the number of haemagglutinins produced by a strain. The presence of several haemagglutinins is easily overlooked and only the major haemagglutinin which is produced by the particular conditions of culture and which is active on the particular species of erythrocyte in a particular haemagglutination test is observed; this is also the case when possible inhibitors are titrated against single dosages of haemagglutinin. The presence of cell-free haemagglutinins is often incorrectly interpreted to mean that the bacterial cells are adhesive also, or conversely, that the presence of cell-free haemagglutinins indicates a lack of stable adhesive activity on the bacterial cell (see Jones, 1977).

Haemagglutinating activity in *V. cholerae* cultures has been known for many years (see Table 8.1). Over a decade ago Lankford proposed that these properties reflected the adhesive activities of vibrios in the intestine (Lankford, 1960). However, it is only recently that interest in this subject has been renewed following the realization that the vibrio's ability to colonize the human intestine probably involves adhesion to the mucosa. Earlier studies were concerned with the haemagglutination test as a means of differentiating eltor and classical types for epidemiological purposes. Thus, from the results of some of these investigations (Table 8.1) it is apparent that almost all eltor vibrios produce haemagglutinins whereas only a small percentage of classical strains do so. This difference is in part a reflection of the success of the differential tests (Barua and Mukherjee, 1963; Finkelstein and Mukherjee, 1963) which were designed to differentiate between plate-grown cultures of these cholera vibrio. The frequent absence of haemagglutinating activities in agar grown cultures of classical strains may be due to the release of the haemagglutinins from the vibrio cell surface into the agar medium (Lankford and Legsomburana, 1965).

Table 8.1 Haemagglutinating activity of *V. cholerae*

V. cholerae	Number positive/ number tested	Medium[1]	Erythrocytes[2]	Remarks	Reference
Eltor vibrios	—[4]	—	Sheep Guinea-pig Goat	HA[5] active over range $0-37°C^3$	Doorenbos, 1932
	349/349	Solid	Chicken	—	Finkelstein and Mukherjee, 1963
	25/25	Solid	Sheep Human Rabbit Chicken	—	Barua and Mukherjee, 1963
	503/503	Solid	Chicken	—	Swanson and Gillmore, 1964
	121/121	Solid	Chicken Goat Human	—	Zinnaka *et al.*, 1964
	111/115	Solid	Chicken	HA-negative eltor strains isolated 1905–1933 and probably altered by extensive passage	Feeley, 1965

1 Variety of media and growth conditions used; distinguished as liquid (broth) or solid (agar) media only.
2 Unless otherwise stated it is assumed that all erythrocytes were reactive.
3 Unless otherwise stated tests were done at room temperature.
4 Figures estimated or no figures available; estimates do not noticeably alter the overall picture of the distribution of haemagglutinins among *V. cholerae* strains.
5 HA: haemagglutinin or haemagglutination.

(*continued on the next page*)

Table 8.1 Haemagglutinating activity of *V. cholerae* (continued from previous page)

V. cholera	Number positive/number tested	Medium[1]	Erythrocytes[2]	Remarks	Reference
Eltor vibrios	349/349[4]	Solid	Chicken		Rizvi et al., 1965
	1/1	Solid	Chicken	HA active at 4°C	Ghosh et al., 1965
	6/6[4]	Liquid	Chicken		Barua and Mukherjee, 1965
	123/123	Solid	Sheep	Sheep erythrocyte agglutinins not active at 4°C, but active at 30 and 37°C;	
	82/82	Solid	Rabbit		
	52/52	Solid	Human	HA of 5 strains inhibited to some degree by D-mannose	
	110/110	Solid	Chicken		
	9/9	Solid	Guinea-pig		
	1/1	Solid	Guinea-pig	HA partially inhibited by D-mannose	Tweedy et al., 1968
		Liquid	Human		
	26/26	Solid	Sheep Human Rabbit Chicken	Including 8 strains isolated from water	Neogy and Sanyal, 1969
	1/1	Liquid	Human	—	Jones et al., 1976
	1/1	Liquid	Chicken	Cell-bound HA; not inhibited by D-mannose or L-fucose	Finkelstein et al., 1977

1 Variety of media and growth conditions used; distinguished as liquid (broth) or solid (agar) media only.
2 Unless otherwise stated it is assumed that all erythrocytes were reactive.
3 Unless otherwise stated tests were done at room temperature.
4 Figures estimated or no figures available; estimates do not noticeably alter the overall picture of the distribution of haemagglutinins among *V. cholerae* strains.
5 HA: haemagglutinin or haemagglutination.

(continued on the next page)

Table 8.1 Haemagglutinating activity of *V. cholerae* (*continued from previous page*)

V. cholerae	Number positive/ number tested	Medium[1]	Erythrocytes[2]	Remarks	Reference
Eltor vibrios	1/1	Solid Liquid	Human	HA of only agar-grown cultures inhibited by D-mannose; non-motile mutants HA-positive	Bhattacharjee and Srivastava, 1978
Classical vibrios	7/10	Solid	Human	HA active at 35°C	Griffitts, 1948
	17/17	Liquid	Rabbit	–	Bales and Lankford, 1961
	0/287	Solid	Chicken	HA-positive forms arise during broth passage	Finkelstein and Mukherjee, 1963
	0/53	Solid	Sheep Human Rabbit Chicken	Weak HA reactions	Barua and Mukherjee, 1963
	8/11	Liquid		–	
	22/50	Solid	Chicken Human Goat	Incidence of HA-positive cultures related to age and passage of stock	Zinnaka *et al.*, 1964
	–[4]	Solid Liquid	Rabbit	HA active at 4°C; diffusable in agar cultures	Lankford and Legsomburana, 1965

1 Variety of media and growth conditions used; distinguished as liquid (broth) or solid (agar) media only.
2 Unless otherwise stated it is assumed that all erythrocytes were reactive.
3 Unless otherwise stated tests were done at room temperature.
4 Figures estimated or no figures available; estimates do not noticeably alter the overall picture of the distribution of haemagglutinins among *V. cholerae* strains.
5 HA: haemagglutinin or haemagglutination.

(*continued on the next page*)

Table 8.1 Haemagglutinating activity of *V. cholerae* (*continued from previous page*)

V. cholerae	Number positive/ number tested	Medium[1]	Erythrocytes[2]	Remarks	Reference
Classical vibrios	19/105	Solid	Chicken	—	Feeley, 1965
	8/1300[4]	Solid	Chicken	Passaged in broth before testing	Rizvi *et al.*, 1965
	3/3[4]	Solid	Chicken	HA active at 4°C	Ghosh *et al.*, 1965
	20/22[4]	Liquid			
	2/10	Solid	Guinea-pig	Agar cultures inactive on sheep, rabbit, human and chicken erythocytes; both cultures active on sheep, rabbit, human and chicken	Barua and Mukherjee, 1965
	8/11	Liquid	Sheep Rabbit Human Chicken		
	18/24	Solid	Sheep	HA develops in cultures upon serial transfer in broth; clonal variants arise in agar cultures and rabbit intestine	Neogy *et al.*, 1966
	1/4	Liquid	—	Subculture in liquid medium increases the incidence of HA-positive strains	Neogy and Mukherjee, 1966

1 Variety of media and growth conditions used; distinguished as liquid (broth) or solid (agar) media only.
2 Unless otherwise stated it is assumed that all erythrocytes were reactive.
3 Unless otherwise stated tests were done at room temperature.
4 Figures estimated or no figures available; estimates do not noticeably alter the overall picture of the distribution of haemagglutinins among *V. cholerae* strains.
5 HA: haemagglutinin or haemagglutination.

(*continued on the next page*)

Table 8.1 Haemagglutinating activity of *V. cholerae* (continued from previous page)

V. cholerae	Number positive/ number tested	Medium[1]	Erythrocytes[2]	Remarks	Reference
Classical vibrios	1/1	Liquid	Guinea-pig Human	Not present in agar cultures. Partially sensitive to D-mannose	Tweedy et al., 1968
	0/11	Solid	Sheep Human Rabbit Chicken	Two rough strains HA-positive	Neogy and Sanyal, 1969
	1/1	Liquid	Human	L-fucose sensitive requiring Ca^{2+} for activity. Non-motile mutants HA-negative	Jones and Freter, 1976
	10/10	Liquid	Human		Jones et al., 1976
	3/3	Liquid	Chicken	Mainly cell-free HA	Finkelstein et al., 1977
	0/80	Solid	Chicken	—	Swanson and Gilmore, 1964
Non-cholera vibrios	4/4	Liquid	Rabbit	—	Bales and Lankford, 1961
	15/19	Solid	Chicken	—	Finkelstein and Mukherjee, 1963

[1] Variety of media and growth conditions used; distinguished as liquid (broth) or solid (agar) media only.
[2] Unless otherwise stated it is assumed that all erythrocytes were reactive.
[3] Unless otherwise stated tests were done at room temperature.
[4] Figures estimated or no figures available; estimates do not noticeably alter the overall picture of the distribution of haemagglutinins among *V. cholerae* strains.
[5] HA: haemagglutinin or haemagglutination.

(continued on the next page)

Table 8.1 Haemagglutinating activity of *V. cholerae* (continued from previous page)

V. cholerae	Number positive/ number tested	Medium[1]	Erythrocytes[2]	Remarks	Reference
Non-cholera vibrios	1/8	Solid	Chicken Human	Sheep and rabbit-inactive	Barua and Mukherjee, 1963
	3/3	Solid	Chicken	HA most active at 4°C	Ghosh et al., 1965
	5/19 5/18 4/15 7/16	Solid	Sheep Rabbit Human Chicken	—	Barua and Mukherjee, 1965
	1/1	Liquid	Guinea-pig Human	HA not produced by agar cultures; partially inhibited by D-mannose	Tweedy et al., 1968
	6/10 5/10	Solid	Sheep Human Rabbit Chicken	Ten strains from cholera patients and ten from water	Neogy and Sanyal, 1969
	24/38	Solid	Chicken	—	Swanson and Gillmore, 1964

1 Variety of media and growth conditions used; distinguished as liquid (broth) or solid (agar) media only.
2 Unless otherwise stated it is assumed that all erythrocytes were reactive.
3 Unless otherwise stated tests were done at room temperature.
4 Figures estimated or no figures available; estimates do not noticeably alter the overall picture of the distribution of haemagglutinins among *V. cholerae* strains.
5 HA: haemagglutinin or haemagglutination.

The results should not be interpreted to mean, therefore, that the majority of classical strains lack adhesive properties. Indeed, classical strains frequently produce haemagglutinins after passage in broth (Finkelstein and Mukherjee, 1963; Zinnaka *et al.*, 1964; Neogy and Mukherjee, 1966; Neogy *et al.*, 1966) and haemagglutinating variants are not uncommon in stock cultures (Zinnaka *et al.*, 1964). Perhaps of greatest interest is the appearance of haemagglutinating clones of non-haemagglutinating vibrios upon passage in the ligated loops of rabbits (Neogy *et al.*, 1966). Whether or not those haemagglutinins that develop in broth cultures of classical strains are the same as those produced by stock cultures passaged on agar remains uninvestigated, as does the significance of the apparent induction of haemagglutinating activity in vibrios growing in the rabbit intestine (Neogy *et al.*, 1966). About half of the strains of non-cholera vibrios produce haemagglutinins. It should be noted, however, that with few exceptions there is little or no evidence that the cultures designated as non-cholera vibrios, non-agglutinable vibrios or water vibrios are indeed *V. cholerae*. Unfortunately, many of the earlier publications lack important detail and the most that can be concluded from such studies is that *V. cholerae* cultures produce haemagglutinins, that the haemagglutinins of different strains are probably not identical, that haemagglutinating activity is not a unique feature of cholera vibrios and that haemagglutinins are produced by organisms isolated from diseased persons and from water (Neogy and Sanyal, 1969).

Although a lack of detail makes the task of comparing the haemagglutinins of *V. cholerae* a difficult and imprecise one, it is possible to cite examples that support the position that all the haemagglutinins of *V. cholerae* are not identical.

There seems little evidence that the species of erythrocyte influence the activities of the major haemagglutinins of eltor vibrio cultures, whereas the converse may be true of classical strains and non-cholera vibrios (Table 8.1). For example, Barua and Mukherjee (1965) demonstrated that a proportion of classical strains grown on agar agglutinated guinea-pig erythrocytes, but not erythrocytes of four other species. However, this might be due to differences in the amount of haemagglutinin rather than differences in the specificity of the haemagglutinins produced by these strains. In this respect, the observations of Majer and Aminoff (1952) and of Finkelstein *et al.* (1977) that erythrocytes from only some individuals of a species are suitable for the demonstration of the haemagglutinins of *V. cholerae* strains is of importance. Similar observations have been made with *Escherichia coli* haemagglutinins (Jones and Rutter, 1974).

The growth medium (i.e. broth or agar) has a marked influence on the haemagglutinating properties of *V. cholerae*. Whereas eltor vibrios are active when grown on agar or in broth media, this is not the case with other types (Table 8.1). Classical types and some non-cholera vibrios (Finkelstein and Mukherjee, 1963; Zinnaka *et al.*, 1964; Neogy and Mukherjee, 1966; Neogy *et al.*, 1966; Tweedy *et al.*, 1968) are more active after growth in broth culture, a state similar to the production of type 1 fimbriae by *E. coli* (Chapter 7). The haemagglutinins of classical strains, in contrast to those of eltor vibrios, are more readily released from the cell surface. In broth cultures during the early logarithmic phase of growth, the haemagglutinins of classical types are cell-bound, but subsequently disappear to be replaced some 10–15 h later by a cell-free haemagglutinin (Lankford and Legsomburana, 1965). The delay between the disappearance of the first haemagglutinin and the appearance of the second haemagglutinin indicates that either distinct haemagglutinins are produced by different phases of the culture or that there is present in the broth a sufficient concentration of haemagglutinin inhibitor (Finkelstein *et al.*, 1977) to mask the activity of the haemagglutinin upon its initial release from the cell. Eltor vibrios retain the haemagglutinin on the cell and little appears in the cell-free state (Lankford and Legsomburana, 1965; Finkelstein *et al.*, 1977). Similar patterns of release occur in agar cultures (Lankford and Legsomburana, 1965). Although it has been claimed that the cell-bound haemagglutinins of both eltor and classical types are more heat-sensitive than the cell-free form (Lankford and Legsomburana, 1965) this has not been confirmed (Finkelstein *et al.*, 1977).

Some vibrio haemagglutinins have a higher activation temperature (Barua and Mukherjee, 1965; Jones *et al.*, 1976) than others (Ghosh *et al.*, 1965). Some haemagglutinins require calcium ions before their activities become apparent (Jones *et al.*, 1976), whereas other vibrio haemagglutinins are active in calcium-free solutions (Lankford and Legsomburana, 1965; Tweedy *et al.*, 1968; Finkelstein *et al.*, 1977; Guentzel *et al.*, 1978). Furthermore, it has been found that non-motile mutants of one strain of the classical type lack haemagglutinating activity (Jones and Freter, 1976), but haemagglutinins are present in cultures of non-motile mutants of other strains (Lankford and Legsomburana, 1965; Guentzel *et al.*, 1977; Bhattacharjee and Srivastava, 1978).

D-mannose may inhibit the haemagglutinins of *V. cholerae* totally (Barua and Mukherjee, 1965; partially (Barua and Mukherjee, 1965; Tweedy *et al.*, 1968) c: not at all (Jones and Freter, 1976; Bhattacharjee and Srivastava, 1978). Partial inhibition defined as a reduction in the sizes of the aggregates formed by the erythrocytes (Tweedy *et al.*, 1968) is difficult to interpret and may signify any of several possibilities: D-mannose may

(1) alter the physicochemical properties of the test system making the environment unsuitable for the activity of the haemagglutinin,
(2) specifically act at sub-optimal concentrations to inhibit a proportion of the haemagglutinating activity, or
(3) inactivate only mannose-sensitive haemagglutinins to disclose mannose-resistant haemagglutinins.

The latter appears to be the most likely explanation in the case of the eltor vibrio studied by Bhattacharjee and Srivastava (1978). Agar cultures of this strain produced a mannose-sensitive haemagglutinin, whereas the major activity in broth cultures not noticeably inhibited by either L-fucose or D-mannose (Finkelstein *et al.*, 1977; Guentzel *et al.*, 1977; Freter, personal communication).

In summary, therefore, there is reason to believe that the species *V. cholerae* produces several haemagglutinins. The best indication of this is perhaps the carbo-hydrate inhibition studies. However, as stated above, when only a single standard dose of haemagglutinin is used, it is the sensitivity of the major haemagglutinin alone which is detected. In other cases the existence of apparently distinct *V. cholerae* haemag-glutinins may be the result of variations in the test systems, quantitative differences in the requirements for haemagglutinin synthesis and/or activity or even pleiotropic effects. Most disconcerting, however, is the possibility that minor haemagglutinins which may contribute greatest to the colonization of the intestine, have remained unnoticed.

8.3.2 The association with, and the adhesion to, intestinal mucosae by *V. cholerae*

The complex nature of the interaction of cholera vibrios with the mucosal surfaces of animal intestines may be a reflection of the complexity of these surfaces. In its simplest form the mucosa consists of two habitats, the inner habitat of the brush-border surfaces of the epithelial cells and the outer habitat of the mucous gel which provides a mantle of varying thicknesses over the epithelium. The most obvious feature of the brush-border surface is the glycocalyx (Ito, 1969), the oligosaccharide components of which may differ in composition, not only at different positions along the intestine, but may also change as the cells migrate from the crypts to the tips of the villi (Etzler and Branstrator, 1974). The mucus of the small intestine is poorly characterized, although that of the pig appears to be typical of epithelial mucus (Gibbons *et al.*, 1975). The oligosaccharide moieties of glycoproteins are character-istically heterogeneic (Montgomery, 1972) and this is probably as true of the intestinal mucus as it appears to be of the glycocalyx. Such factors as the chance inclusion of foreign elements, and the action of degrading enzymes will contribute to the mucus and brush-borders forming diverse and variable habitats that are neither chemically nor physically uniform. On the one hand, this probably presents the vibrios with many sites of attachment within the mucus and on the epithelial cell surface. On the other hand, the thickness of the mucus mantle probably determines the degree and rate of penetration of the vibrios to the epithelium.

The association of vibrios with the mucosae of animal models results from a series of interactions that involve the motility and chemotaxis of vibrios in addition to their ability to adhere to receptors on the epithelial cells. Three phases of asso-ciation are apparent in animal models. Firstly, vibrios have the ability to detect and move towards the mucosa by following gradients of chemo-attractants (Allweiss *et al.*, 1977). This response appears to be but the manifestation of the general chemotactic tendencies of vibrios rather than a specific response to tissue compon-ents. Secondly, vibrios penetrate into and through the mucus gel for considerable distances in short periods of time (Freter *et al.*, 1978b; Freter *et al.*, 1978c). Thirdly,

vibrios may, after penetration of the mucus, adhere to the brush-border surfaces of
the epithelial cells (Nelson *et al.*, 1976; Nelson *et al.*, 1977). Interference with any
one of these steps possibly disrupts the process of association.

(a) *Motility and chemotaxis*

Non-motile cholera vibrios associate less with the mucosal surface of the mouse and
the rabbit intestine compared to their motile parent strains (Freter *et al.*, 1977;
Freter *et al.*, 1978c). (Similar, although clearly not comparable, results are obtained
when motile vibrios are examined in the presence of agglutinating antibody (Bellamy
et al., 1975; Schrank and Verwey, 1976)). The difference between motile and non-
motile vibrios is not only that fewer non-motile bacteria are present in the mucus,
but that motile vibrios are more widely distributed and penetrate more deeply
between the villi (Freter *et al.*, 1977; Freter *et al.*, 1978c). Although motility provides
the motor force necessary for the vibrios to move through the mucus gel, the movement is random unless given direction by chemotaxis. Thus, non-chemotactic but
straight swimming motile vibrios, although more adept than non-motile vibrios at
penetrating into the mucus, nevertheless, are far less effective at this than are
chemotactic vibrios (Freter *et al.*, 1978b; Freter *et al.*, 1978c). However, if dense
populations of vibrios (i.e. bacterial pastes) are placed in direct contact with the
mucosal surface neither motility nor chemotaxis is of import, and chemotactic, non-
chemotactic and non-motile vibrios associate equally well with the surface (Freter
et al., 1977; Freter *et al.*, 1978a). Chemotactic and non-chemotactic vibrios once
located within the mucus remain there in the absence of an external gradient of
attractants (Freter *et al.*, 1977). The taxin gradients not only increase the efficiency
of the initial contact between the vibrios and the mucosal surface but, in addition,
appear to extend some distance towards the crypts where the relative numbers of
chemotactic vibrios far exceed the numbers of non-chemotactic vibrios (Freter
et al., 1978b; Freter *et al.*, 1978c).

The mucous gel does not present an impenetratable barrier to vibrios (Freter *et al.*,
1977) and indeed may provide a suitable habitat (Freter and Jones, 1976; Freter
et al., 1977). The physical structure of the mucus, furthermore, may assist the
migration of vibrios towards the tissue surface in a manner analogous to that proposed
for the migration of spermatozoa within cervical mucus (Gibbons and Sellwood, 1973).
The properties of mucus are such that, if the gel is stretched or if it is caused to flow,
the glycoprotein moieties align parallel to the force of stress. Vibrios readily
penetrate stretched gels along tracts that parallel the alignment, but less readily at
right angles to the introduced forces (Jones *et al.*, 1976). It can be imagined, therefore, that vibrios most readily move through the mucus along tracts created within
the gel as it flows from the epithelial surface towards the lumen. Indeed, mucus flow
may not only facilitate penetration to the tissue surface in this manner, but may be
responsible for the creation of the taxin gradients.

(b) *Association of vibrios with mucosal surfaces*
In both mouse and rabbit models, *V. cholerae* populations are found within the
mucous gel (Williams *et al.*, 1973; Schrank and Verwey, 1976; Freter *et al.*, 1977).
The apparent ease with which vibrios migrate through mucus (Freter *et al.*, 1977,
Freter *et al.*, 1978b; Freter *et al.*, 1978c) suggests that little permanent attachment
occurs in this region, although some evidence to the contrary does exist (Freter, 1978).
   Cholera vibrios can attach to the brush-border surfaces of intestinal epithelial cells
of the rabbit and may accumulate there to form layers several cells deep (Nelson
et al., 1976). The initial absence of vibrios on the cell surfaces (Nelson *et al.*, 1976),
at a time when it would seem that large numbers are present in the adjacent mucus
(Freter *et al.*, 1977), suggests that the mucous layer is lost during tissue preparation
(Freter *et al.*, 1977) and that the vibrio populations of the brush-border surfaces
alone are being observed in this study. Simple systems consisting of suspensions of
cholera vibrios and brush-borders, provide direct evidence that vibrios do attach to
the brush-border membranes of both rabbit (Jones *et al.*, 1976; Jones and Freter,
1976) and mouse intestinal epithelial cells (Freter, personal communication).
However, *V. cholerae* show a degree of selectivity in their attachment to the
epithelial cells of the guinea-pig and rabbit intestine. In the former, adherent vibrios
are confined to the small intestine but are absent from the large intestine (Le Brec
et al., 1965); the fluorescent antibody technique used probably detects both brush-
border and mucus populations of vibrios and such are probably formed as the result
of not only adhesion, but also of chemotaxis. The unequal distribution of vibrios on
adjacent villi of the rabbit intestine (Nelson *et al.*, 1976) is most probably due to an
unequal distribution of the mucus over the epithelium providing the vibrios with
easier access to some areas compared to others. However, the absence of vibrios
from the tips of the villi (Nelson *et al.*, 1976), even though this site is readily
accessible to the vibrios (Freter *et al.*, 1977), is probably due to a lack of suitable
receptors in this region. In contrast to the rabbit (Nelson *et al.*, 1976), the attachment of vibrios to tissue fragments obtained from the rat intestine (Thaller,
personal communication) is markedly dependent on the age of the rat and is some
60-fold greater with material obtained from preweaned rats compared with weaned
rats (Hirschberger *et al.*, 1978). Thus, it appears that vibrios cannot attach to the
surfaces of all epithelial cells and are perhaps selective in their attachment to others.
Indeed, in any preparation of brush-borders isolated from the rabbit small intestine,
vibrios are found to attach to some brush-border membranes more readily than
they do to others (unpublished observations). All of this and the results of studies
on haemagglutination discussed previously implies that the attachment of vibrios
to eucaryotic cells depends upon the presence of appropriate receptors. Adhesion
to the intact mucosal surface, however, is probably modified further during the
initial phase of interaction by the extent of the overlying mucus and the presence
of taxin gradients.
   The number of available sites within the mucosa of the small intestine that
vibrios can occupy is not unlimited and can become saturated when large numbers

of vibrios are placed in contact with intestinal tissue (Freter and Jones, 1976). However, layers of vibrios several cells deep can attach to brush-borders *in situ* (Nelson
et al., 1976) and to the isolated membranes *in vitro* (Jones and Freter, 1976).
Consequently, limited adhesion sites on the brush-borders is unlikely to account for
the observed competition and a more plausible reason is that negative chemotaxis
which results from oxygen depletion at the tissue surface or the production of
negative attractants by the bacteria or the tissue (Allweiss *et al.,* 1977) reduces the
efficiency with which vibrios usually invade the mucus and arrive at the epithelial
surface.

(c) *Adhesion of vibrios to brush-border surfaces*
The extent of the adhesion of one vibrio strain to brush-border membranes *in vitro*
is directly related to the vibrio concentration and inversely related to the numbers
of brush-border membranes present (Jones *et al.,* 1976). Furthermore, it is both
temperature- and time-dependent. At 4°C, little or no adhesion occurs whereas
marked adhesion occurs at 25 and 37°C. (The haemagglutinating activity of this
vibrio has a similar temperature dependence (Jones *et al.,* 1976). At 37°C, however,
adhesion is not permanent (Jones *et al.,* 1976) and the bacteria eventually dissociate.
The membranes, after the dissociation of the vibrios, remain receptive to the attachment of fresh suspensions of vibrios. Consequently, marked enzymic destruction of
the brush-border receptors is an unlikely, but not excluded, reason for dissociation.
Haemagglutination by other vibrio strains appears to be remarkably similar in these
respects (Lankford and Legsomburana, 1965) as does, in part, the dissociation of
vibrios from mucosal surfaces (Nelson *et al.,* 1976). Dissociation from brush-borders
in situ is accompanied by morphological changes in the microvilli and perhaps
increased mucus secretion (Nelson *et al.,* 1976). There is no evidence in the latter
study that the vibrio population actually leaves the mucosa (i.e., vibrios may be in
the mucus, which is subsequently removed by tissue preparation), however, another
study (Freter *et al.,* 1978c) suggests that this is the case. Presumably the movement
of the vibrios is the result of the creation of a reversed gradient of chemotaxins or
the dissolution of the mucous gel by mucinases that act rapidly to depolymerize
rabbit mucus (unpublished results). Calcium ions promote the activities of the
adhesins of some (Jones *et al.,* 1976), but not all (Guentzel *et al.,* 1977; Guentzel
et al., 1978), strains of cholera vibrio; strontium may be substituted for calcium in
the case of one strain (Jones *et al.,* 1976). The possible function of the calcium ions
has been discussed (Jones *et al.,* 1976) and the role of ions summarized elsewhere
(Jones, 1977). Calcium, however, is not required for association with mucosae
(Freter and Jones, 1976). This anomaly may be caused by the release of adequate
calcium by the tissue, but is more probably due to the fact that the over-riding
mechanisms studied in this system is chemotaxis. Non-motile aflagellate mutants of
some (Jones and Freter, 1976), but not all (Guentzel *et al.,* 1977), cholera vibrio
do not adhere to brush-borders. (This and the influence of calcium ions on adhesion
closely parallels the similar observations made on the haemagglutinating activities

of cholera vibrios.) Reduced contact between brush-borders and non-motile compared with motile vibrios is not the cuase of the absence of adhesion because non-motile vibrios impacted onto the brush-borders by centrifugation still failed to adhere whereas the adhesion of motile vibrios was extensive (Jones and Freter, 1976). Motility and adhesive properties, including haemagglutinating activity, are not invariably linked characters of this strain; motile but non-adhesive and non-haemagglutinating vibrios can be obtained from agar cultures (Jones *et al.*, 1976).

Besides the possible similar effect of cell wall modification on both flagellum and adhesin production, other explanations for the apparent inconsistent relation between motility and adhesion are several: Firstly, some strains produce more than one active adhesin, but only one adhesin ceases to be produced when the flagellum is not synthesized. Secondly, adhesins are not distributed similarly on the cell surfaces of all vibrio strains. For example, some adhesins may be confined to the flagella sheath and such strains would re-acquire adhesive activities upon reversion to a flagellated state (Jones and Freter, 1976). Thirdly, mutation or the condition of culture may increase the loss of adhesins to the medium (i.e. as cell-free adhesins) without affecting motility; this appears to be the reason for the existence of non-adhesive, agar grown cultures (Jones *et al.*, 1976).

8.3.3 *V. cholerae* adhesins

Unlike haemagglutination, which can be brought about by cell-free components, adhesion requires that the adhesin remains on the surface of the bacteria. An intimate contact between the vibrio and the brush-border membrane, however, does not appear to be a requisite for attachment. Layers of bacteria accumulate on brush-borders *in situ* (Nelson *et al.*, 1976) and on the isolated membrane if brush-borders and vibrios are impacted together by centrifugation (Jones and Freter, 1976). The distances that separate the vibrios from the membrane surface are of the dimensions (Nelson *et al.*, 1976) consistent with that proposed for adhesion rather than that thought of as adsorption (Jones, 1977). Indeed, the actual separation is probably much less than that portrayed in the published photographs from which it is apparent that the glycocalyx (at least 300 Å deep (Ito, 1965) remains unstained. Those vibrios lying immediately adjacent to the tissue may be attached via specific adhesins, whereas those positioned some distance away may be attached to neighboring bacteria by interbacterial sites of adhesion (Jones, 1977), mediated by other adhesive systems. However, direct attachment to the tissue of even the most distant bacteria is possible if the adhesive organelles are of sufficient length and of sufficiently small cross-sectional area, i.e., are morphologically like fimbriae (Jones, 1977), to penetrate between the vibrios that compose the layers nearest the tissue. Fimbriae are indeed produced by vibrios and are of sufficient length to anchor the distal bacteria to the epithelial surfaces if the tissue receptors are not occluded by bacteria more proximal to the surface. However, fimbriae have not been observed on vibrios attached to brush-borders (Nelson *et al.*, 1976; Nelson *et al.*, 1977). An alternative explanation is that the layers and microcolonies of vibrios are embedded in a commonly synthesized matrix of adhesive material similar in function to that proposed for the K88 antigen of *E. coli* (Jones, 1975).

The forms taken by bacterial adhesins have been reviewed (Jones, 1977). One such form is that of fimbriae which both cholera and non-cholera vibrios produce (Barua and Chatterjee, 1964; Tweedy *et al.*, 1968). Unlike the straight fimbriae of *E. coli*, the fimbriae of *V. cholerae* are curved and up to 2.5 μm in length. The diameters of the fimbriae range from 60–80 Å (Barua and Chatterjee, 1964) or approximately 100 Å (Tweedy *et al.*, 1968) in cultures of cholera vibrios, and about 130 Å in diameter in cultures of non-cholera vibrios (Tweedy *et al.*, 1968). The single strains of classical and non-cholera vibrios examined (Tweedy *et al.*, 1968) produced few fimbriae (an average of less than 10 per cell) and only about 10% of cells were fimbriate. Vibrios of a strain of the eltor biotype, in contrast, produced up to 50 fimbriae per cell and about half the vibrios were fimbriate (Tweedy *et al.*, 1968). It seems that eltor types produce fimbriae more readily than both classical cholera vibrios and non-cholera vibrios and this may account for the greater adhesiveness of eltor strains. Fimbriation is increased after several passages in broth (Tweedy *et al.*, 1968), as are the haemagglutinating properties of *V. cholerae* (see above). However, it is by no means certain that these fimbriae are the adhesive organelles responsible for the attachment of vibrios to brush-border surfaces. Several investigators have failed to find fimbriae on vibrio cells (Finkelstein and Mukherjee, 1963; Lankford and Legsomburana, 1965), although this might be due to the ease with which fimbriae are lost from the surface of vibrios. Furthermore, it is disconcerting that the sex pili of *V. cholerae* (Bhaskaran *et al.*, 1969) are similar in diameter and numbers per cell to the fimbriae described above. Although sex pili are shorter, this again may be an artifact introduced during preparation.

Substances often called slime or slime envelopes are produced by cholera (Barua and Chatterjee, 1964; Lankford and Legsomburana, 1965; Tweedy *et al.*, 1968) and non-cholera vibrios (Tweedy *et al.*, 1968). Under the electron microscope the material produced by a non-cholera vibrio appeared as a dense network of strands quite unlike fimbriae (Tweedy *et al.*, 1968). Extracts believed to contain the slime material produced by cholera vibrios were found to be essentially protein in composition (Lankford and Legsomburana, 1965). There is no evidence, however, that the slime envelopes possess adhesive properties. Indeed, slime production and haemagglutinating activity of a non-cholera vibrio were found to be inversely related (Tweedy *et al.*, 1968).

Recently, Finkelstein and associates (1977), have characterized a cell-free haemagglutinin of a non-toxigenic mutant (M13) of a classical biotype of *V. cholerae*, which they found had a molecular weight of only approximately 68 000 even in the absence of depolymerizing agents. It can be predicted from physicochemical considerations that adhesins are likely to be in the form of long filaments that allow attachment to occur when the bacterial and animal cell surfaces remain separated by distances at which the forces of repulsion between the surfaces are at a minimum (Jones, 1977). In general this is true of the majority of known adhesins which appear to be composed of many subunits each of which is often less than 100 000 daltons (Jones, 1977) which may aggregate spontaneously under appropriate conditions (Brinton, 1965; Gilboa-Garber *et al.*, 1977). Consequently, the small haemagglutinin of *V. cholerae* (Finkelstein *et al.*, 1977) is unusual (the crude haemagglutinin fraction investigated by Lankford and Legsomburana (1965) may be estimated to have a somewhat similar size, however) and its organization on the surface of the bacterial cell poses an intriguing problem.

Good evidence that the haemagglutinin of the classical strain of cholera vibrio investigated by Finkelstein *et al.*, (1977) is involved in the adhesion of the vibrios to tissue cells is provided by the finding that the partially purified haemagglutinin inhibits the attachment of an eltor vibrio strain to both slices of rabbit liver and to the intestinal epithelium of infant rabbits (Finkelstein *et al.*, 1977). The question of immediate concern is whether inhibition results from direct competition between the cell-free and cell-bound haemagglutinins for the same receptors, or whether inhibition is due to the steric hinderance caused by the attachment of the cell-free haemagglutinin to sites adjacent to those occupied by the vibrio adhesin (Jones, 1977). (In similar experiments with *E. coli* adhesins, competition between only identical adhesins occurs (Isaacson *et al.*, 1978)). If direct competition between the haemagglutinins of the classical and the eltor strain does occur, then the haemagglutinins at least react with the same receptor complex or contrary to the evidence cited earlier, are similar or identical adhesive substances.

(a) *Inhibition of adhesion and association, and the possible nature of receptors*
Inhibition studies besides providing more precise characterization of interactions may also provide some insight of the nature of the interaction. On the one ʰand this might be the nature of the tissue receptor to which the vibrios attach or, on the other hand, the taxins produced by the tissue to which the vibrios respond. Accordingly, chemotaxis (i.e. the association of vibrios with intact tissue slices measured after short periods of contact (Freter *et al.*, 1977)) and adhesion (i.e. the attachment of vibrios to brush-border membranes (Jones *et al.*, 1976; Jones and Freter, 1976) which do not attract vibrios (Freter *et al.*, 1977)) must be clearly differentiated. Inhibition of adhesion, however, does not necessarily mean that the inhibitor constitutes part of the tissue receptor (Jones, 1977), nor does inhibition of association mean that the inhibitor is the taxin produced by the tissue (for an account of chemotaxis see Adler, 1975); the latter is certainly not the case if the inhibitor reduces motility.

D-mannose not only inhibits the haemagglutinins of some *V. cholerae*, but can inhibit the attachment of a strain of cholera vibrio (strain P) to the brush-border membranes isolated from the rabbit (Jones and Freter, 1976) and the mouse (Freter, personal communication) intestine. Moreover, Hirshberger *et al.* (1978) found that the adhesion of a different strain of classical cholera vibrio (strain VRL8) to rabbit and preweaned rat, but to not weaned rat, mucosal scrapings was inhibited

by α-methyl-D-mannoside and mannan. D-mannose, however, does not inhibit the
association of the vibrio strain P to rabbit intestinal tissue (Freter and Jones, 1976)
although the sugar, in addition to being an inhibitor of adhesion, is also a positive
attractant for this strain (Freter *et al.*, 1978c). Bhattacharjee and Srivastava (1978),
in contrast, found that D-mannose did inhibit the association of an eltor vibrio strain
with intact slices of rabbit intestinal tissue and also, in constrast to other studies
(Jones and Freter, 1976) inhibited haemagglutination. It is possible that the inhibition
of association observed in this study (Bhattacharjee and Srivastava, 1978) is a consequence of the inhibition of the adhesion of vibrios to brush-borders, but it is equally
possible that the reduced association with the rabbit mucosa is a result of D-mannose
interfering with normal chemotaxis. Perhaps an important difference between the
adhesins of these classical and eltor vibrio strains that may invalidate comparison, is
that the mannose-sensitive adhesin of the eltor strain was produced only by agar
cultures (Bhattacharjee and Srivastava, 1978), whereas cell-bound adhesins of the
classical strain were present only in broth culture (Jones *et al.*, 1976).

L-fucose and L-fucosides also inhibit the haemagglutinating and adhesive
properties of cholera vibrio strain P (Jones and Freter, 1976). Only the L-isomer
is inhibitory and inhibition increases with increasing molecular size of the fucoside
(Jones and Freter, 1976). This has been confirmed in studies on the interaction of
classical strain VRL8 with rat as well as rabbit intestinal tissue (Hirschberger *et al.*,
1978). It is possible that some *V. cholerae* strains produce adhesins that react with
at least two different tissue receptors (Jones and Freter, 1976; Bhattacharjee and
Srivastava, 1978). The adhesion of *V. cholerae* strain P, for example, is inhibited by
a mixture of L-fucose and D-mannose, but this is no greater than the inhibition
caused by L-fucose alone (Jones and Freter, 1976). Besides the fact that there is an
absence of co-operative inhibition by these two dissimilar carbohydrates, the
existence of some adhesive activity even in the presence of both sugars (Jones and
Freter, 1976) suggests that another adhesin, which is inhibited by neither carbohydrate, is also produced. It is also noteworthy, that like D-mannose, L-fucose does
not reduce the association of the vibrios with mucosal slices (Freter and Jones, 1976),
but unlike D-mannose, L-fucose is not a chemo-attractant for this strain (Freter
et al., 1978c).

As noted previously, the association of vibrios with mucosae involves mechanisms
quite distinct from that of adhesion. Likewise, the inhibition of association is, at
least in part, distinct from the inhibition of adhesion. Pepsinized mucosal scrapings
inhibit both adhesion to brush-borders and association with mucosal slices (Freter
and Jones, 1976). Fractionation of this crude mixture of substances revealed the
presence of two activities, one of which inhibited association (Freter *et al.*, 1977),
possibly by acting as a chemo-attractant, while the other inhibited adhesion and
most probably contained L-fucose.

Only preliminary incursions on the subject of the nature of the tissue receptors
for *V. cholerae* adhesins have been made. Although there is no direct evidence that
specific receptors exist, the evidence of the selectivity with which vibrios attach to

animal cells refered to earlier and the inhibition of adhesion by some monosaccharides suggests that this is so. Indeed inhibition of the adhesion of some *V. cholerae* strains by carbohydrates suggests that the receptors for these particular organisms may be glycoproteins. This is consistent with the supposed nature of the receptors of other bacteria (Jones, 1977) although one should be cautious about ascribing to a receptor the composition of the inhibitor (Jones, 1977). If one assumes that carbohydrates are the receptors for some vibrio adhesins, then the adhesins of other vibrios, the activities of which are not reduced by carbohydrates, may react with receptors of quite different chemical composition or may require receptors with particular stereochemical configurations which are to be found as surface components of some eucaryotic cells (Jones, 1977).

8.4 ECOLOGICAL SIGNIFICANCE OF ADHESION AND ASSOCIATION

The association of vibrios with the intestinal mucosa, for the most part, can be accounted for by the motility and chemotaxis of the bacteria. Toxigenic non-motile classical and eltor vibrios are less virulent for the suckling mouse than are the motile parental vibrios (Guentzel and Berry, 1975), from which it may be surmized that non-motile vibrios are less well-equipped to colonize the intestine than are chemotactic vibrios. Likewise, vibrios treated with agglutinating antibody, i.e., their motility is reduced or eliminated, are less virulent in the rabbit ileal loop (Schrank and Verwey, 1976) and in the infant mouse (Bellamy *et al.*, 1975). Although both non-motile and motile non-chemotactic vibrios have a reduced capacity to associate with mucosal surfaces, they differ in that the latter can adhere to brush-border membranes (Freter *et al.*, 1977). However, even in stagnant ileal loops of the rabbit, where removal by peristalsis is of little consequence, non-chemotactic vibrios grow significantly less well than chemotactic vibrios (Freter *et al.*, 1978c). From this it may be supposed that association with the mucosa besides reducing removal by peristalsis, also provides the vibrios with a habitat that is more favorable for growth than the contents of the lumen. It is no surprise, therefore, to find that chemotactic vibrios have an ecological advantage over non-chemotactic vibrios and outgrow and eliminate the latter strain from the gnotobiotic mouse intestine (Freter *et al.*, 1977). What is quite unexpected, however, is that in the infant mouse the roles are reversed and the non-chemotactic vibrio predominates (Freter *et al.*, 1977). The most likely explanation of this phenomenon is that the contents of the intestine of the baby, but not the adult, mouse contains attractants that reverse the migration of the chemotactic vibrio towards the mucosa leaving the distribution of the non-chemotactic vibrio unaffected.

The reversion of non-motile vibrios to motile forms which occurs more readily in the intestine than in culture (Guentzel and Berry, 1975), suggests that, in the environment of the intestine, motility confers a selective advantage on the vibrio cell.

Surprisingly, however, vibrio populations observed microscopically immediately after removal from the mouse intestine under anaerobic conditions are largely non-motile unless provided with energy sources or oxygen (Freter *et al.,* 1977); the significance of this has been discussed (Freter *et al.,* 1977). The expenditure of energy on motility in the absence of a functional chemotactic system appears to be such a liability to vibrios in the mouse intestine, that selection of non-motile types occurs (Freter *et al.,* 1977). Presumably in the strain studied, mutation to a non-motile state is a more likely event than reversion to the ecologically more advantageous chemotactic state.

Although much of the behavior of vibrios in the intestines of animal models can be indeed explained by the studies on motility and chemotaxis, the additional ability to adhere to the epithelial cells may provide the vibrios with a means of establishing more permanent viable populations in the mucosa that are relatively unaffected by the vagaries of chemical gradients. Chemotaxis, it may be surmized, increases the efficiency and effectiveness of such adhesion by bringing the vibrios into contact with the tissue surface. It is probable that only part of the vibrio population attaches or remains in an attached state and thereby theoretically provides some ecological advantage to the species (Gibbons and van Houte, 1975). In this respect, the relevance of the observation that not all vibrios in a culture are adhesive (Jones *et al.,* 1976) is uncertain. The few other studies which show that vibrios dissociate from surfaces, including mucosal surfaces (Nelson *et al.,* 1976; Freter *et al.,* 1977), after a period of attachment may be significant, and that this may be perhaps a phase of colonization during which vibrios repopulate the environment. The mechanisms by which vibrios may be released from the attached state is possibly enzymatic. A likely candidate for such a role is the receptor-destroying enzyme of *V. cholerae*; this sialidase does not seem to be responsible for haemagglutinating activity (Majer and Aminoff, 1952) and clearly would not act directly on the receptors if these are found to be oligosaccharides, the principal receptor determinants of which are L-fucose and D-mannose. Furthermore, mucinase might be responsible not only for the invasion of the mucus but also for the release of vibrio populations from the mucosa.

The ability of *V. cholerae* to colonize aquatic habitats may depend upon the ability of the bacteria to elaborate adhesive substances, some of which may be identified with those activities that allow the vibrios to associate with and to adhere to the intestinal mucosa. Like *V. parahaemolyticus, V. cholerae* is confined to aquatic regions of particular salinity. The restricted areas in which *V. parahaemolyticus* is found is probably related to the influence of salt concentrations on the adhesion of the vibrios to chitin and hence to copepods; chitinase production by *V. parahaemolyticus* (Sakazaki *et al.,* 1963) plays some role in this association (Kaneko and Colwell, 1975a). *V. cholerae,* which also adsorbs to chitin particles (Nalin *et al.,* 1978) and produces chitinase (Sakazaki *et al.,* 1963), may be of restricted distribution in water for a similar reason. In this respect, it is of some interest to note that the surface waters in those areas where classical cholera is endemic and the waterways along which cholera apparently speads most readily are frequently of a brackish nature (Felsenfeld, 1966; Stock, 1976).

At this point, it is worth noting certain other perhaps striking similarities between *V. parahaemolyticus* and *V. cholerae*. Both species are aquatic microbes but, like *V. cholerae*, only particular types of *V. parahaemolyticus* cause enteric disease (Sakazaki *et al.,* 1968). The types of *V. parahaemolyticus* that are most often associated with disease comprise but a small proportion of the vibrio populations that constitute some of the sources of infection for man (Fujino *et al.,* 1974); likewise, cholera vibrios may constitute a small proportion of aquatic *V. cholerae* populations and this may account for the infrequency with which the pathogenic vibrios are isolated from water when cholera is not prevalent in the neighbourhood. Like *V. cholerae*, *V. parahaemolyticus* also adheres to mammalian cells and such adhesive properties are found in both pathogenic types (Carruthers, 1977) and in the supposed non-pathogenic forms (Joseph *et al.,* 1978).

8.5 CONCLUSIONS

Adhesive properties are not a unique feature of cholera vibrios but are exhibited by cultures of cholera and non-cholera vibrios alike and by *V. cholerae* strains isolated from both cholera patients and from water. There is evidence that these properties are not the same in all cultures and that some cultures elaborate at least two adhesins. There is no easily arrived at explanation for the widespread occurance of dissimilar adhesins, nor for the function of these adhesins at present. Not only must a determined effort be made to recognize and differentiate these substances, but a better understanding of the habitats occupied by bacteria of this ubiquitous species must be attempted and a correlation sought between the existence of a particular adhesin in culture and the ability of the organism to colonize the intestinal mucosa of man and/or to occupy a particular aquatic niche. What must be avoided is the tendency to ascribe to these properties an adhesive function in the intestine of man because it is at this time both appropriate and desirable to do so.

REFERENCES

Abou-Gareeb, A.H. (1960), *J. Hyg. (Cantab),* **58**, 21–33.

Adler, J. (1975), *A. Rev. Biochem.,* **44**, 341–356.

Aldova, E., Laznickova, K., Stepankova, E. and Lietava, J. (1968), *J. Inf. Dis.,* **118**, 25–31.

Allweiss, B., Dostal, J., Carey, K.E., Edwards, T.F. and Freter, R. (1977), *Nature,* **226**, 448–450.

Bales, G.L. and Lankford, C.E. (1961), *Bact. Proc.,* 118.

Barua, D. and Chatterjee, S.N. (1964), *Ind. J. med. Res.,* **52**, 828–830.

Barua, D. and Mukherjee, A.C. (1963), *Bull. Calcutta Sch. Trop. Med.,* 11, 85–86.

Barua, D. and Mukherjee, A.C. (1965), *Ind. J. med. Res.*, **53**, 399–404.

Bellamy, J.E.C., Knop, J., Steele, E.J., Chaicumpa, W. and Rowley, D. (1975), *J. Inf. Dis.*, **132**, 181–188.

Benenson, A.S., Ahmad, S.Z. and Oseasohn, R.O. (1965), *Proc. Cholera Res. Symp.*, Honolulu, Hawaii, Jan. 24–29, 1965, pp. 332–336, U. S. Government Printing Office, Washington, D.C.

Bhaskaran, K., Dyer, P.Y. and Rogers, G.E. (1969), *Aust. J. exp. Biol. med. Sci.*, **47**, 647–650.

Bhattacharjee, J.W. and Srivastava, B.S. (1978), *J. gen. Microbiol.*, **107**, 407–410.

Bhattacharji, L.M. and Bose, B. (1964), *Ind. J. med. Res.*, **52**, 777–786.

Bhattacharya, S., Bose, A.K. and Gosh, A.K. (1971), *Appl. Microbiol.*, **22**, 1159–1161.

Brinton, C.C. (1965), *Trans. N.Y. Acad. Sci.*, **27**, 1003–1054.

Carruthers, M.M. (1977), *J. inf. Dis.*, **136**, 588–592.

Citarella, R.V. and Colwell, R.R. (1970), *J. Bact.*, **104**, 434–442.

Colwell, R.R. (1970), *J. Bact.*, **104**, 410–433.

Colwell, R.R. and Kaper, J. (1977), In: *Bacterial Indicators: Health Hazards Associated with Water.* (Hoadly, A.W. and Dutka, B.J., eds), pp. 115–125, American Society for Testing Materials, Philadelphia.

Colwell, R.R., Kaper, J. and Joseph, S.W. (1977), *Science*, **198**, 394–396.

Dixon, J.M.S. (1960), *J. Path. Bact.*, **79**, 131–140.

Doorenbos, W. (1932), *An. Inst. Pasteur*, **48**, 457–469 (in French).

Dutta, N.K., Panse, M.V. and Jhala, H.I. (1963), *Br. med. J.*, **1**, 1200–1203.

Elliot, H.L., Carpenter, C.C.J., Sack, R.B. and Yardley, J.H. (1970), *Lab. Invest.*, **22**, 112–120.

El-Shawi, N. and Thewaini, A.J. (1969), *Bull WHO*, **40**, 163–166.

Etzler, M.E. and Branstrator, M.L. (1974), *J. Cell Biol.*, **62**, 329–343.

Feeley, J.C. (1965), *J. Bact.*, **89**, 665–670.

Felsenfeld, O. (1963), *Bull WHO*, **28**, 289–296.

Felsenfeld, O. (1966), *Bull WHO*, **34**, 161–195.

Finkelstein, R.A. (1973), *CRC Crit. Rev. Microbiol.*, **2**, 553–623.

Finkelstein, R.A., Arita, M., Clements, J.D. and Nelson, E.T. (1977), *Proc. 13th Joint Cholera Res. Conf. Atlanta, Georgia*, Sept. 19–21, 1977, pp. 137–151, U.S. Government Printing Office, Washington, D.C.

Finkelstein, R.A. and Mukherjee, S. (1963), *Soc. exp. Biol. Med.*, **112**, 355–359.

Fresh, J.W., Versage, P.M. and Reyes, V. (1964), *Arch. Path.*, 77, 529–537.

Freter, R. (1951), *Zabl. Bakt. Orig.*, **156**, 523–536 (in German).

Freter, R. (1955), *J. Inf. Dis.*, **97**, 238–245.

Freter, R. (1969), *Texas Report Biol. Med.*, **27**, 299–316.

Freter, R. (1978), Cholera and Related Diarrheas–Molecular Aspects of a Global Health Problem. *Proc. Nobel Symp. 43.*, S. Karger, Basle, in press.

Freter, R., Allweiss, B., O'Brien, P.C.M. and Halstead, S.A. (1977), *Proc. 13th Joint Cholera Res. Conf., Atlanta, Georgia*, Sept. 19–22, 1977, pp. 152–181. Government Printing Office, Washington, D.C.

Freter, R. and Jones, G.W. (1976), *Inf. Immun.*, **14**, 246–256.

Freter, R., O'Brien, P.C.M. and Halstead, S.A. (1978a), In: *Secretory Immunity and Infection.* (McGhee, J.R., Mestecky, J. and Babb, J.L., eds), pp. 429–437. Plenum Publishing Corporation, New York and London.

Freter, R., O'Brien, P.C.M. and Macsai, M.S. (1978b), 5th Int. Symp. Intest. Microecol. Columbia, Mo. May 31-June 1, 1978. *J. clin. Nutr.*, (1979), in press.

Freter, R., O'Brien, P.C.M. and Macsai, M.S. (1978c), *Proc. 14th Joint Cholera Res. Conf.*, Karatsu City, Japan. Sept. 27–29, 1978. National Institutes of Health, Tokyo, in press.

Freter, R., Smith, H.L. and Sweeney, F.J. (1961), *J. Inf. Dis.*, **109**, 35–42.

Fujino, T., Sakaguchi, G., Sakazaka, R. and Takeda, Y. (eds), (1974), *Proc. Int. Symp. Vibrio parahaemolyticus*, Saikon Publ. Co. Ltd. Tokyo.

Gangarosa, E.J., Beisel, W.R., Benyajati, C., Sprinz, H. and Piyaratn, P. (1960), *Am. J. trop. Med.*, **9**, 125–135.

Geldreich, E.E. (1972), In: *Water Pollution Microbiology*. (Mitchell, R., ed.), pp. 207–241, Wiley–Interscience, New York, London, Sydney and Toronto.

Ghosh, A.K., Ganguly, R. and Shrivastava, D.L. (1965), *Ind. J. Med. Res.*, **53**, 1–7.

Gibbons, R.A., Jones, G.W. and Sellwood, R. (1975), *J. gen. Microbiol.*, **86**, 228–240.

Gibbons, R.A. and Sellwood, R. (1973), In: *The Biology of the Cervix*. (Blandan, R.J. and Moghissi, K., eds), pp. 251–265. University of Chicago Press, Chicago, Illinois.

Gibbons, R.J. and van Houte, J. (1975), *A. Rev. Microbiol.*, **29**, 19–44.

Gilboa-Garber, N., Mizrahi, L. and Garber, N. (1977), *Can. J. Biochem.*, **55**, 975–981.

Gorbach, S.L., Banwell, J.G., Jacobs, B., Chatterjee, B.D., Mitra, R., Brigham, K.L. and Neogy, K.N. (1970), *J. Inf. Dis.*, **121**, 38–45.

Griffitts, J.J. (1948), *Proc. Soc. exp. Biol. Med.*, **67**, 358–362.

Guentzel, M.N. and Berry, L.J. (1975), *Inf. Immun.*, **11**, 890–897.

Guentzel, M.N., Jocz, R.J., Wong, G.G., Gay, T.V. and Guerrero, D.L. (1978), *Abst. A. Meeting Am. Soc. Microbiol.*, Las Vegas, Nevada, May 14–19, 1978. p. 14.

Guentzel, M.N., Wong, G.G. and Gay, T.V. (1977), *Abst. 13th Joint Cholera Res. Conf.*, p. 77, U.S. Government Printing Office, Washington, D.C.

Hirschberger, M., Thaler, M.M. and Mirelman, D. (1978), *Pediat. Res.*, **12**, 436.

Hsieh, H. and Liu, P.V. (1970), *J. Inf. Dis.*, **121**, 251–259.

Hughes, J.M., Hollis, D.G., Gangarosa, E.J. and Weaver, R.E. (1978), *A. Int. Med.*, **88**, 602–606.

Isaacson, R.E., Fusco, P.C., Brinton, C.C. and Moon, H.W. (1978), *Inf. Immun.*, **21**, 392–397.

Ito, S. (1965), *J. Cell Biol.*, **27**, 475–491.

Ito, S. (1969), *Fedn. Proc. fedn. Am. Socs. exp. Biol.*, **28**, 12–25.

Jones, G.W. (1975), In: *Microbiology 1975*, (Schlessinger, D., ed.), pp. 137–142, American Society for Microbiology, Washington, D.C.

Jones, G.W. (1977), In: *Microbial Interactions*, (Reissig, J.L., ed.), Receptors and Recognition Series B, Vol. 3, pp. 139–176, Chapman and Hall, London.

Jones, G.W., Abrams, G.D. and Freter, R. (1976), *Inf. Immun.*, **14**, 232–239.

Jones, G.W. and Freter, R. (1975), *Proc. 11th Joint Cholera Res. Conf.*, New Orleans, Louisiana, Nov. 4–6, 1975, pp. 186–200, U.S. Government Printing Office, Washington, D.C.

Jones, G.W. and Freter, R. (1976), *Inf. Immun.*, **14**, 240–245.

Jones, G.W. and Rutter, J.M. (1974), *J. gen. Microbiol.*, **84**, 135–144.

Joseph, S.W., Merrell, B.R., Sochard, M. and Brown, W.P. (1978), *Scanning Electron Microscopy*, **2**, 727–732.

Kaneko, T. and Colwell, R.R. (1973), *J. Bact.*, **113**, 24–32.

Kaneko, T. and Colwell, R.R. (1975a), *Appl. Microbiol.*, **29**, 269–274.

Kaneko, T. and Colwell, R.R. (1975b), *Appl. Microbiol.*, **30**, 251–257.

Lankford, C.E. (1960), *Ann. N.Y. Acad. Sci.*, **88**, 1203–1212.

Lankford, C.E. and Legsomburana, U. (1965), *Proc. Cholera Res. Symp.*, Honolulu, Hawaii, Jan. 24–29, 1965, pp. 109–120, U.S. Government Printing Office, Washington, D.C.

Le Brec, E.H., Sprinz, H., Schneider, H. and Formal, S.B. (1965), *Proc. Cholera Res. Symp.*, Honolulu, Hawaii, Jan. 24–29, 1965, pp. 272–276, U.S. Government Printing Office, Washington, D.C.

Lindenbaum, J., Greenough, W.B., Benenson, A.S., Oseasohn, R.O., Rizvi, S. and Saad, A. (1965), *Proc. Cholera Res. Symp.*, Honolulu, Hawaii, Jan. 24–29, 1965, pp. 200–204, U.S. Government Printing Office, Washington, D.C.

Mayer, J. and Aminoff, D. (1952), *Bull. Res. Council Israel*, **2**, 212.

McIntyre, O.R. and Feeley, J.C. (1965), *Bull WHO*, **32**, 627–632.

McIntyre, O.R., Feeley, J.C., Greenough, W.B., Benenson, A.S., Hassan, S.I. and Saad, A. (1965), *Am. J. Trop. Med. Hyg.*, **14**, 412–418.

Meadows, P.S. (1971), *Arch. Microbiol.*, **75**, 374–381.

Montgomery, R. (1972), In: *Glycoproteins, their composition, structure and function*, 2nd Edn. (Gottschalk, A. ed.), pp. 518–528, Elsevier Pub. Co., Amsterdam, London, New York.

Müller, G. (1977), *Zbl. Bakt. Hyg. 1 Abt. Orig. B* **165**, 487–497 (in German).

Năcescu, N. and Ciufecu, C. (1978), *Zbl. Bakt. Hyg. Abt. Orig. A*, **240**, 334–338.

Nalin, D.R., Reid, A. and Levine, M.M. (1978), *18th Intersc. Conf. Antimicrob. Agents and Chemother.*, Atlanta, Georgia, Oct. 1–4, 1978, Abst. 119, American Soc. Microbiology, Washington, D.C.

Nelson, E.T., Clements, J.D. and Finkelstein, R.A. (1976), *Inf. Immun.*, **14**, 527–547.

Nelson, E.T., Höchli, M., Hackenbrock, C.R. and Finkelstein, R.A. (1977), *Proc. 12th Joint Cholera Res. Conf.*, U.S.-Japan Cooperative Medical Science Program. Soppora, Japan, 1976, pp. 81–87, National Institutes of Health, Tokyo, Japan.

Neogy, K.N. (1965), *Bull. Calcutta Sch. trop. Med.*, **13**, 10–11.

Neogy, K.N. and Mukherjee, A.C. (1966), *Nature*, **212**, 303.

Neogy, K.N. and Sanyal, S.N. (1969), *Bull. WHO*, **40**, 329–330.

Neogy, K.N., Sanyal, S.N., Mukherjee, M.K. and Nandy, P.K. (1966), *Bull. Calcutta Sch. trop. Med.*, **14**, 1–3.

Pandit, C.G., Pal, S.C., Murti, G.V.S., Misra, B.S., Murty, D.K. and Shrivastav, J.B. (1967), *Bull. WHO*, **37**, 681–685.

Patnaik, B.K. and Ghosh, H.K. (1966), *Br. J. exp. Path.*, **47**, 210–214.

Pesigan, T.P. (1965), *Proc. Cholera Res. Symp.*, Honolulu, Hawaii, Jan. 24–29, 1965, pp. 317–321, U.S. Government Printing Office, Washington, D.C.

Pesigan, T.P., Gomez, C.Z., Gaetos, D. and Barua, D. (1967a), *Bull. WHO*, **37**, 795–797.

Pesigan, T.P., Plantilla, J. and Rolda, M. (1967b), *Bull. WHO.*, **37**, 779–786.

Pollitzer, R. (1959), *Cholera.* World Health Organization, Geneva.

Read, W.B.B. and Pandit, S.R. (1941), *Ind. J. med. Res.*, **29**, 403–415.

Rizvi, S., Huq, M.I. and Benenson, A.S. (1965), *J. Bact.*, **89**, 910–912.

Sakazaki, R., Gomez, C.Z. and Sebald, M. (1967), *Jap. J. Med. Sci. Biol.*, **20**, 265–280.

Sakazaki, R., Iwanami, S. and Fukumi, H. (1963), *Jap. J. Med. Sci. Biol.*, **16**, 161–188.

Sakazaki, R., Tamura, K., Gomez, C.Z. and Sen, R. (1970), *Jap. J. Med. Sci. Biol.*, **23**, 13–20.

Sakazaki, R., Tamura, K., Kato, T., Obara, Y., Yamai, S. and Hobo, K. (1968), *Jap. J. Med. Sci. Biol.*, **21**, 325–331.

Schrank, G.D. and Verwey, W.F. (1976), *Inf. Immun.*, **13**, 195–203.

Sen, R., Vaishnav, P. and Majumder, J. (1967), *Bull. WHO.*, **37**, 491–492.

Shewan, J.M. and Véron, M. (1974), In: *Bergey's Manual of Determinative Bacteriology.* (Buchanan, R.E. and Gibbins, N.E. eds), pp. 340–345, Williams and Wilkin Co., Baltimore, Maryland.

Smith, H.L. and Goodner, K. (1965), *Proc. Cholera Res. Symp.*, Honolulu, Hawaii, Jan. 24–29, 1965, pp. 4–8, U.S. Government Printing Office, Washington, D.C.

Smith, H.L. and Goodner, K. (1966), *Far East med. J.*, **2**, 301–302.

Stock, R.F. (1976), *Cholera in Africa.* International African Institute, London.

Swanson, R.W. and Gillmore, J.D. (1964), *Bull. WHO.*, **31**, 422–425.

Tweedy, J.M., Park, R.W.A. and Hodgkiss, W. (1968), *J. gen. Microbiol.*, **51**, 235–244.

Williams, H.R., Verwey, W.F., Schrank, G.D. and Hurry, E.K. (1973), *Proc. 9th Joint Cholera Res. Conf.*, Grand Canyon, Arizona, Oct. 1–3, 1973, pp. 161–173. Dept. State Publications, Washington, D.C.

Zachary, A., Taylor, M.E., Scott, F.E. and Colwell, R.R. Intern. *Biodeteriol. Bull.* Submitted for publication.

Zinnaka, Y., Shimodori, S. and Takeya, K. (1964), *Jap. J. Microbiol.*, **8**, 97–103.

ZoBell, C.E. (1972), In: *Marine Ecology.* (Kinne, O. ed.), 1 (3), pp. 1251–1270, Wiley–Interscience, London, New York, Sydney, Toronto.

9 Adherence of *Neisseria gonorrhoeae* and other *Neisseria* Species to Mammalian Cells

PETER J. WATT and MICHAEL E. WARD

9.1	Introduction	*page*	253
9.2	Clinical aspects of neisserial infections		253
	9.2.1 Pathobiology of neisserial infections		254
9.3	Surface structures of *Neisseria*		257
	9.3.1 Fimbriae		258
	9.3.2 Capsules		259
	9.3.3 Outer envelope proteins		259
	9.3.4 Lipopolysaccharides		261
9.4	Experimental models in the study of neisserial adhesion		262
	9.4.1 Organ culture models		263
	9.4.2 Isolated cell models		263
	9.4.3 Leukocyte interactions		
9.5	Physico-chemical factors in host cell–neisserial interactions		270
	9.5.1 Electrostatic interactions		270
	9.5.2 Long-range attraction (the DLVO theory)		273
	9.5.3 The role of ionic bridging		275
	9.5.4 The effect of pH		276
	9.5.5 Hydrophobic interactions		276
9.6	Surface carbohydrates in host cell–neisserial interactions		279
9.7	Conclusions		282
	Note		283
	References		284

Acknowledgements

Our thanks to the many people with whom we have discussed ideas concerning neisserial attachment and who gave us access to unpublished data. We are grateful to Dr van Heynigen for samples of purified gangliosides; Mrs Sue Davies who patiently typed the manuscript and the Medical Research Council for a programme grant in support of our work.

Bacterial Adherence
(*Receptors and Recognition*, Series B, Volume 6)
Edited by E.H. Beachey
Published in 1980 by Chapman and Hall, 11 New Fetter Lane, London EC4P 4EE
© 1980 Chapman and Hall

9.1 INTRODUCTION

The natural habitats of bacteria of the genus *Neisseria* are the mucosal surfaces of man and other mammals. A critical factor in their adaptation for life in such an hostile environment is the development of a mechanism for anchorage to the mucosal cells whose surfaces are continually washed by flows of mucus and other secretions. Clearly, then, both the common commensal neisseria and the major pathogens *N. gonorrhoeae* and *N. meningitidis* must possess surface components which function as adhesins. If we are to explain the attachment processes in molecular terms we need detailed information on the architecture of the neisserial surface. Equally critical to any understanding of the role of adhesion in the pathogenesis of neisserial infections are clinico-pathological studies of the natural disease. These topics will be discussed before analysing our current understanding of the cohesive mechanisms.

9.2 CLINICAL ASPECTS OF NEISSERIAL INFECTIONS

The text-book image of meningococci as invasive organisms causing life-threatening septicaemia or meningitis contrasts to that of gonococci, which tend to remain localized to the genital tract producing relatively trivial disease. This superficial analysis is misleading. In a 3 year non-epidemic period the prevalence of naso-pharyngeal carriage ranged from 4.9–10% (Greenfield *et al.*, 1971) but the number of reported cases of clinical disease in the United States suggests a prevalence of only 0.00023% (Meningococcal Disease Surveillance Group, 1976). Thus, the usual course of meningococcal disease is asymptomatic infection of the posterior nasopharynx and both meningococci and gonococci must be considered as primary pathogens of epithelial surfaces. Further, both infections are unique to man indicating a high degree of adaptation to the human mucosal surface.

Like meningococcal carriage in the nasopharynx, gonococcal infections of the pharynx (Bro-Jørgensen and Jensen, 1973) and rectum (Bhattacharyya and Jephcott, 1974) are rarely symptomatic. The prevalence of asymptomatic genital infection with gonococci depends to some extent on the population being studied (Watt and Lambden, 1978) suggesting that gonococcal strains vary in virulence. Certainly, gonococci causing asymptomatic cervical infection in women tend to produce asymptomatic infection in their male sexual partners (Blount, 1972). Further work is required to determine if strains from carriers have only a limited ability to grow on mucosal surfaces or whether they multiply freely, but are less invasive. A third possibility is that invasion occurs but these strains do not invoke an inflammatory response. What is clear is that gonococcal strains associated with asymptomatic infection are virulent, being those strains most commonly isolated from cases of disseminated gonococcal infection (Schoolnik *et al.*, 1976).

The prevalence of gonorrhoea is tribute to the infectiousness of the disease (the risk of males acquiring gonorrhoea during intercourse with asymptomatic women may exceed 50% (Holmes *et al.*, 1970)). Yet relatively few gonococci can be recovered from infected patients. In men with acute gonorrhoea of around 24 hours' duration some 10 000 gonococci were recovered by scraping the urethral mucosa (Ward *et al.*, 1970) whilst in women, vaginal washout techniques recovered some 4×10^2 – 1.8×10^7 gonococci (Lowe and Kraus, 1976). This suggests that the infectious dose is small. Studies on human volunteers have confirmed this supposition: the introduction of only 1×10^3 gonococci into the urethra can initiate gonorrhoea (Brinton *et al.*, 1978). The observation that micturition immediately after intercourse may not prevent naturally acquired gonorrhoea (Bernfeld, 1972) suggests that the small numbers of infecting organisms rapidly adhere to the genital mucosa to avoid being flushed from the surface.

Meningococci are less infectious than gonococci. In instances where meningococci were naturally introduced into households the rate of acquisition of meningococcal carriage by other members of that family is such that less than 50% of susceptible individuals became carriers even when continually exposed for 5 years (Greenfield *et al.*, 1971). This low infectivity is difficult to explain when the median duration of pharyngeal carriage of 9.6 months (Greenfield *et al.*, 1971) indicates that meningococci are well adapted for survival on mucosal surfaces. No data are available for commensal neisseria.

9.2.1 Pathobiology of neisserial infections

Electron microscopic studies of mucosal cells obtained from the urethra of men with early gonorrhoea showed gonococci attached to and partially embedded in the surface of epithelial and mucous secreting cells (Ward and Watt, 1972; Ovčinnikov and Delektorski, 1971; Novotny *et al.*, 1975 (Fig. 9.1). In the laboratory, gonococci readily adhered to human cells in tissue culture (Swanson, 1973; Ward and Watt, 1975) to human fallopian tube organ cultures (Carney and Taylor-Robinson, 1973; Ward *et al.*, 1974) and to red cells (Punsalung and Sawyer, 1973; Buchanan and Pearce, 1976). Gonococci were more efficient than commensal organisms in adhering to vaginal epithelial cells (Mårdh and Weström, 1976) while the attachment of gonococci to human sperm *in vitro* has led to the concept of the 'bacterial hitch-hiker' riding on motile sperms into the upper female genital tract (James *et al.*, 1976).

Electron micrographs of gonococcal attachment in both the natural infection and in human fallopian tube organ cultures suggest that the initial interaction is with microvillous projections from the host cell surface (Ward and Watt, 1975). These microvilli frequently appear twisted towards the invading bacteria (Fig. 9.2) and are particularly well-developed on the cell surface nearest to the bacteria. This suggests that their formation might be stimulated by the proximity of gonococci; an effect

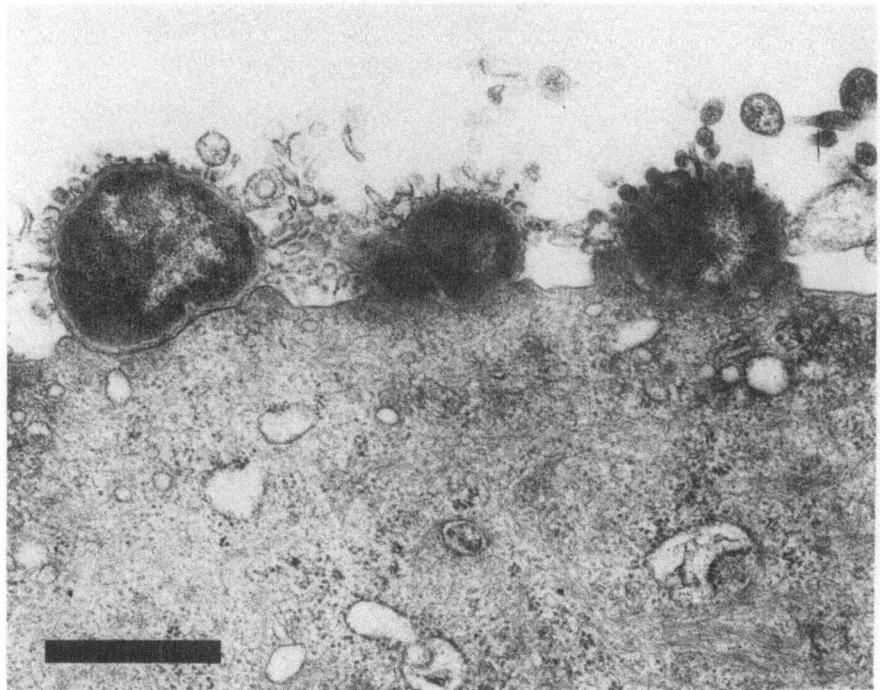

Fig. 9.1 Electron micrograph showing gonococci closely adherent to a urethral epithelial cell from a male patient with early symptomatic gonorrhoea. The membrane of the host cell appears pushed up around the gonococcus to form a 'cushion-like' structure. The bar represents 500 nm. (From Ward and Watt, 1972.)

comparable to that reported for the mucosal pathogen *Bordetella pertussis* (de Bault and Yoo, 1974). Some gonococci become enfolded by microvillous processes; resorption of these processes would bring gonococci into contact with the host cell surface (Ward and Watt, 1975).

To understand the biology of neisserial infections we must explain how the pathogenic neisseria cause disease whilst the commensals do not. Our working hypothesis is that bacterial adherence to mucous secreting columnar epithelia requires high avidity attachment possessed by pathogenic neisseria, but not by commensal organisms. Certainly, gonococci infect the columnar epithelium in the genital tract at sites like the endocervix and posterior male urethra (Harkness, 1948) while meningococci infect the columnar epithelium of the posterior nasopharynx, sites which are only transiently colonized with low numbers of commensal organisms (Sparkes *et al.*, 1977). Commensal neisseria essentially adhere to squamous epithelia whose surface consists of dead cells forming a passive substrate for bacterial attachment

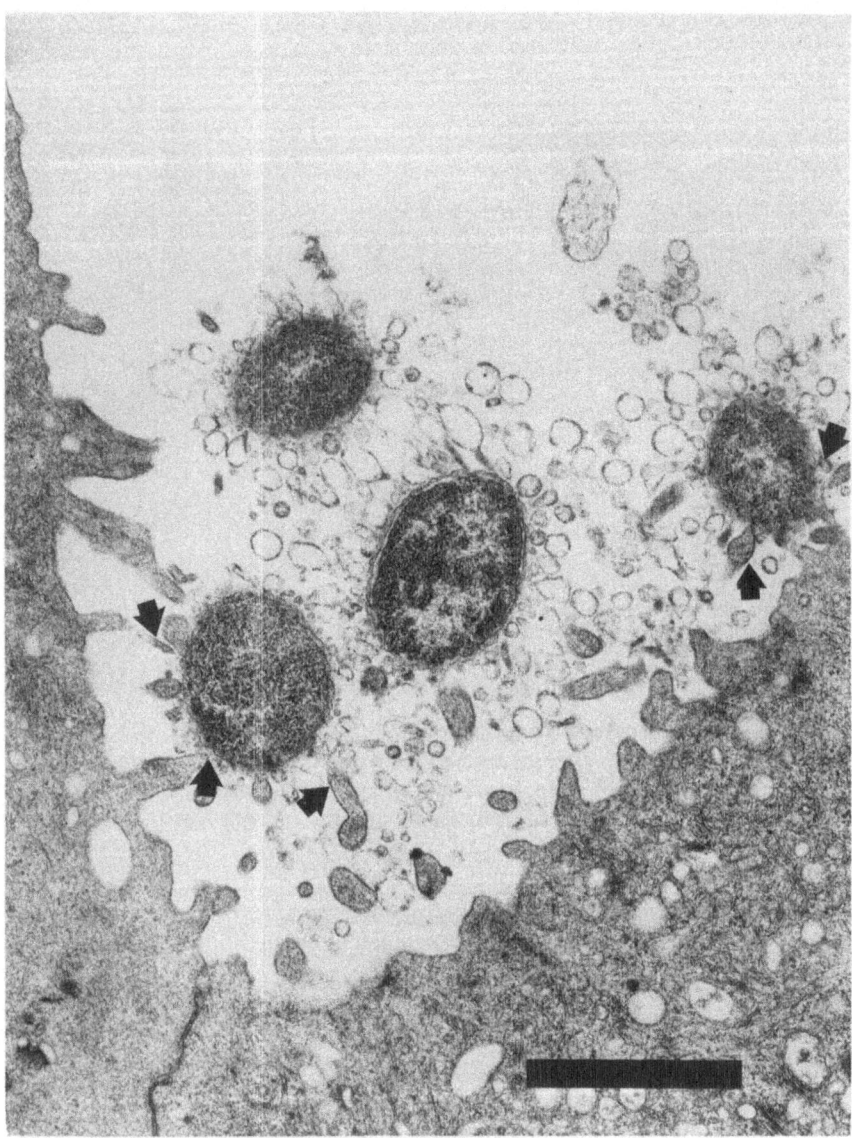

Fig. 9.2 Electron micrograph of gonococci close to the mucosal surface of a
male patient with early symptomatic gonorrhoea. Some microvilli (arrows)
from the host cell have made contact with the bacteria. The vesicles surrounding the gonococci may be blebs derived from the gonococcal outer envelope
or remnants from the inflammatory exudate. The bar represents 1 μm. (From
Ward and Watt, 1975.)

By contrast, the surface membrane of columnar epithelia is in active movement necessitating a high avidity mechanism if the organisms are to adhere. Interestingly, gonococcal attachment to an MRC 5 fibroblast monolayer is markedly enhanced when the epithelial cell surface is immobilized by cytochalasin B (Watt and Ward, 1977), suggesting that cell surface movement may hinder attachment. Furthermore, unlike gonococci commensal *N. subflava* did not attach to the mucous-secreting columnar epithelial surface of human fallopian tube organ cultures (McGee *et al.*, 1977).

Once the gonococcus is firmly bound to the epithelial surface by high avidity binding, interiorization of a proportion of the organisms is inevitable (Watt *et al.*, 1978). Evidence from other systems suggests that the initial interaction of surface ligands on a particle with receptors on the surface of professional phagocytes is insufficient to trigger ingestion. Griffin *et al.* (1975) produced evidence suggesting that the ingestion of red cells by macrophages required sequential circumferential interaction of particle-bound ligand with specific plasma membrane receptors not involved in the original attachment process. We do not know if this 'zipper-mechanism' of particle uptake occurs with inefficient, non-professional phagocytes such as columnar epithelial cells. However, as would be predicted, shortly after interiorization gonococci lie within tight membrane-bound vesicles (Ward *et al.*, 1975) rather than in large phagosomes. Moreover, this invasion process is impaired by cytochalasin B (Watt *et al.*, 1976).

Thus, gonococci penetrate mucosal surfaces as a result of uptake by mucosal cells, in marked contrast to salmonellae which erode their way through the brush border of the intestine to lie free within the host cell's cytoplasm (Takeuchi, 1967).

9.3 SURFACE STRUCTURES OF *NEISSERIA*

The mechanism of neisserial adhesion to host cell surfaces must, of necessity, be mediated by macromolecules associated with the bacterial outer membrane. The function of this outer membrane in gram-negative bacteria is complex (Di Rienzo *et al.*, 1978) with different components involved in structural integrity, the formation of non-specific diffusion pores as well as uptake systems for iron, sugar, vitamins, etc. Clearly, such molecules will contribute to the physical characteristics of the bacterial surface and by their effects on charge density, hydrophobicity and polymer interactions must influence the cohesive properties of any specific mediators of neisserial adhesion. Pathogenic neisseria possess surface structures which act as a critical defence against host phagocytes and which form a diffusion barrier protecting the vulnerable cytoplasmic membrane from the lytic action of antibodies and complement. This induces a conflict of requirements in that bacteria with markedly hydrophilic surfaces show enhanced resistance to phagocytosis by leukocytes (van Oss, 1978), but impaired adhesion to mucosal cell membranes (Perers *et al.*, 1977). The implication is that the adhesive properties of neisseria (avidity, pH optima, ionic

Bacterial Adherence

requirements, etc.) will be complex functions resulting from the interaction of dominant surface components; in particular fimbriae, capsular antigen, lipopolysaccharide and the different outer-membrane proteins.

9.3.1 Fimbriae

Fimbriae, protein filaments extending from the bacterial surface, are universally present on gonococci and meningococci when primarily isolated from the patient (Jephcott *et al.*, 1971; Swanson *et al.*, 1971; DeVoe and Gilchrist, 1975). After subculture non-fimbriated variants soon outgrow the fimbriated gonococci indicating that growth within the host exerts a selective pressure in favour of the fimbriated state. This may also apply to commensal neisseria since fimbriae have been reported to occur on stock cultures of several *Neisseria* species (Wistreich and Baker, 1971), but are not universally present (McGee *et al.*, 1977).

Gonococcal fimbriae have been purified by several groups of workers and many of their properties established (Buchanan *et al.*, 1973; Robertson *et al.*, 1977; Brinton *et al.*, 1978). The fimbriae, which are composed of multiple units of a single protein, have a diameter of 7 nm and extend up to 2—4 μm from the gonococcal surface. The fimbrial protein contains some 200 amino-acid residues with a molecular weight from different isolates in the range 19 000 ± 2500 (Buchanan *et al.*, 1978). Although such differences may account for the known antigenic variations in fimbriae the remarkable similarity in the proportions of different amino-acids (Table 9.1) is indicative of a common function. This table emphasizes the high percentage of amino-acids with hydrophobic properties; indeed NH_2^- terminal analysis has demonstrated that the first 24 residues in the in the NH_2^- terminal region are

Table 9.1 Comparison of the % composition of amino acids for purified gonococcal fimbriae isolated from various strains

Amino acid character	Gonococcal strains			
	P9[1]	201[1]	33[2]	CDC-B[3]
Basic*	13	10.8	13.9	15.3
Acidic†	25	23.6	22.3	23
Essentially hydrophobic‡	26	24	27.4	26
Total non-polar§	45.7	45.3	45.7	45.8

* Lys. Arg. His
† Glu. Asp
‡ Val. Leu. Met. Phe. Ile. Tyr. Trp.
§ Val. Leu. Met. Phe. Ile. Tyr. Trp. Gly. Ala.
1. Data calculated from Robertson *et al.* (1977)
2. Data calculated from Buchanan *et al.* (1978)
3. Data calculated from Brinton *et al.* (1978)

all hydrophobic except for threonine and glutamic acid in positions 2 and 5 (Hermodson, *et al.*, 1978). The fact that fimbriae have an isolectric point of p*I* 5.3 suggests that the Asx and Glx residues chiefly represent the acidic amino-acids aspartic and glutamic acids (Robertson *et al.*, 1977). These authors also suggest that the fimbrial protein of strain P9 contains 1−2 hexose groups per protein subunit.

Although little information is available on the properties of fimbriae from other neisseria there is evidence to suggest that different types of fimbriae occur in the same species. DeVoe and Gilchrist (1975) report that a small colony variant of Group B meningococci had short, small-diameter (2.0 nm) fimbriae while a large colony variant had long, large-diameter (4.5 nm) fimbriae. Other workers (McGee *et al.*, 1977) suggest that these large fimbriae have dimensions identical to those of gonococci (7 nm diameter) while a single report suggests a comparable sub-unit molecular weight (Brinton *et al.*, 1978). Commensal neisseria possess numerous short fimbriae and a few long fimbriae, but the true dimensions of these fimbriae is in dispute (McGee *et al.*, 1977: Wistreich and Baker, 1971).

9.3.2 Capsules

Capsules are important virulence factors present in meningococci freshly isolated from patients. The failure of non-fimbriated capsulated meningococci to adhere to human buccal cells can be explained by the capsule blocking binding-sites present on non-capsulated variants (Craven and Frasch, 1978). This inhibition could be due to charge effects since the capsules are acidic linear polymers of sialic acid (Groups B and C) or 2 acetamido-2 deoxy-D-mannopyranosyl phosphate (Group A). Further, the capsule polymers are rigid rods with chain lengths varying from 76−150 sugar units (Liu *et al.*, 1977) and must affect binding by steric interactions. In carrier strains fimbriae permit attachment to host cells, presumably by penetrating the capsule barrier. However, 8 of 12 fresh patient isolates bound poorly to cells suggesting that the inhibitory effect of large capsules is not easily overcome (Craven and Frasch, 1978). Similarly, the 'capsulated' variant of gonococcus P9 shows 57% inhibition of binding to buccal cells when compared to the non-capsulated prototype (Watt *et al.*, 1978). Thus, although capsules are critical in the resistance of pathogens such as meningococci to host defence systems, the penalty is a reduction in the ability of the capsulated organisms to adhere to mucosal cell surfaces (see also Section 1.2).

9.3.3 Outer envelope proteins

Early work on outer membrane proteins suggested that gonococci possess a major protein and lesser amounts of a characteristic secondary protein (Johnston *et al.*, 1976). The major protein, termed Protein I, was responsible for serotype specificity, varied in molecular weight from 32 000−39 000 and was expressed at the gonococcal

Table 9.2 Outer-membrane proteins present in opacity variants of *N. gonorrhoeae* P9

OM Protein	Molecular weight	Opacity variant						
		P9-1	P9-2	P9-6	P9-11	P9-13	P9-16	P9-19
Protein I	36 000	+	+	+	+	+	+	+
Protein II*	29 000	–	–	+	–	–	–	+
Protein IId*	28 850	–	–	–	+	–	–	–
Protein IIa*	28 500	–	–	–	+	+	–	–
Protein IIb*	28 000	–	–	–	–	–	+	–
Protein IIc*	27 500	–	–	–	–	–	–	+
Fimbrial protein	19 500	–	+	–	–	–	–	–

surface, as shown by intense labelling with ^{125}I using the lactoperoxidase system (Heckels, 1978). The comparable protein in meningococci, with a 41 000 mol. wt., may exist in the membrane as trimers and tetramers (Frasch and Mocca, 1978), suggesting that it is analogous to the 'porin' of *E. coli* (Di Rienzo *et al.*, 1978). Treatment with proteases removes only a small fragment ($\sim$ 10%) of this protein; presumably only this small component is accessible at the surface.

By contrast, proteases completely digest the secondary protein (designated Protein II) in both meningococci (Frasch and Mocca, 1978) and gonococci (Swanson, 1978). Protein II is characterized by its behaviour in SDS-PAGE, the molecular weight increasing from 24 000 when reacted with SDS at 37°C to 28 000–29 000 when reacted at 100°C (Heckels, 1977). Recent studies on gonococci with differing colonial opacities (Swanson, 1978; Lambden and Heckles, 1979) demonstrated that Protein II was present in comparable amounts to Protein I. Protein II cannot be considered a single entity since selected opacity variants of strain P9 possess distinct proteins termed II*, IIa*, IIb*, IIc*, IId* (Table 9.2). The presence of different proteins was correlated with important virulence properties such as adhesion to human cells and resistance to killing mediated by antibody and complement (Lambden *et al.*, 1979a).

Studies of outer envelope vehicles isolated from a wide range of commensal neisseria suggest a comparable structure with a major protein (mol. wt. 38 000) equivalent to Protein I and a heat-modifiable protein (mol. wt. 30 000) equivalent to Protein II (Russell *et al.*, 1975).

9.3.4 Lipopolysaccharides

A characteristic feature of the genus *Neisseria* is the propensity to bud off large amounts of lipopolysaccharide-rich outer envelope in the form of vesicles or blebs (Stead *et al.*, 1975; Perry *et al.*, 1978). In the natural infection, comparable structures surround gonococci adherent to mucosal surfaces, suggesting the LPS may be significant in the pathogenesis of this disease (Novotny *et al.*, 1975). Analyses of pure lipopolysaccharide obtained from six commensal *Neisseria* species (Johnson *et al.*, 1975), gonococci (Stead *et al.*, 1975) and meningococci (Jennings *et al.*, 1973) show remarkable similarities. The hydrophobic component Lipid A is embedded in the outer envelope and consists of a D-glucosamine oligosaccharide which is both *O*- and *N*-substituted with long-chain fatty acids. Lipid A is linked through an 8-carbon sugar acid 3-deoxy-D-*manno* octulosonic acid (KDO) to the core oligosaccharide. The core oligosaccharide of *N. gonorrhoeae* contains D-glucose, D-galactose, D-glucosamine, KDO and L-glycero-D-*manno*-heptose with two D-galactopyranose and one D-glucosamine units as terminal non-reducing groups (Perry *et al.*, 1978). These external sugars could serve as bridging molecules linking to lectin-like proteins on the host cell surface (Yamada *et al.*, 1975) or the specific surface glycosyltransferases postulated by Roseman (1970). The support for this concept comes from the work of Freimer *et al.* (1978) who found that the binding of S-type *S. typhimurium* to mouse macrophages was inhibited by the sugars present in the O-polysaccharide (glucose, galactose, rhamnose and glucosamine) while the R-type was inhibited by core sugars. However, the binding of non-fimbriated gonococci to human buccal cells was not inhibited by 0.4M galactose or 0.4M glucosamine (Watt *et al.*, 1979). Further evidence against the specific attachment of gonococci to host cells via LPS terminal sugar groups comes from our finding that purified core polysaccharide coupled via KDO to ^{125}I-methylalbumen does not bind to human buccal or red cells (Watt *et al.*, 1979).

Meningococcal LPS core contains the same constituent sugars as gonococci, but the LPS from Groups A, B, X and Y are serologically distinct suggesting different sugar linkages. Smooth(S)-type LPS has not been satisfactorily demonstrated in the pathogenic neisseria (Perry *et al.*, 1978; Frasch, 1979). Amongst the commensal neisseria *N. canis* and *N. subflava* are of the S type, the O-chains being identical linear polysaccharides having repeating tetrasaccharide units of three rhamnoses linked 1–6 to glucose. Other commensal neisseria have Rough (R)-type LPS, although the core compositions are distinct from that found in the R-type LPS of gonococci and meningococci.

The presence of only R-type LPS may confer certain advantages in the adhesiveness of pathogenic neisseria. Thus S–R mutation in *S. typhimurium* results in increased association with and uptake by phagocytic cells (Stendahl *et al.*, 1973). One possibility is that the long O-chains, which may consist of some 60 repeating sugar units, could give rise to steric repulsive forces as a consequence of mutually excluded volumes of polymer chains between the interacting bacteria and the host cell glycocalyx. In mammalian cell systems, this force is very large relative to that arising from the overlap of electrical double layers (Grieg and Jones, 1976).

9.4 EXPERIMENTAL MODELS IN THE STUDY OF NEISSERIAL ADHESION

Fimbriated gonococci rapidly adhere to human cells in tissue culture (Swanson, 1973) to human sperm (James *et al.*, 1976) and to red blood cells (Buchanan and Pearce, 1976). Indeed they would seem to be more effective at attaching to human vaginal epithelial cells than the normal commensal flora of the vagina (Mårdh and Weström, 1976). The evidence on the ability of gonococci to attach to non-human tissues is conflicting. Taylor-Robinson *et al.* (1975) showed that strains of gonococci, which were able to invade the human fallopian tube in organ culture, failed to adhere to the mucosal surface of the rabbit oviduct. In contrast, gonococci were found to attach to the cervix, uterus and male urethra of the guinea-pig, although the animal is immune to infection (Tebbutt *et al.*, 1976). These workers equated attachment to a tissue with the number of adherent gonococci which could be removed from the tissue using a standard washing procedure.

Obviously, these bacteria were only weakly attached to guinea-pig cell surfaces.

If such weak adhesion is non-specific, investigation into the molecular basis of specific adhesion requires careful choice of a model. Investigations into the adhesion of gonococci to tissue cultures of animal cells, human embryonic or transformed cells certainly requires cautious interpretation; the physiology and membrane structure of these cells differs significantly from human mucosal cells. The red cell is equally unsatisfactory, lacking, for example, HLA antigens and the microtubule—microfilament system essential for membrane movement.

9.4.1 Organ culture models

The human fallopian tube organ culture model devised by Taylor-Robinson *et al.* (1974) is highly relevant for the study of gonococcal adhesion to and invasion of mucosal surfaces since infection of the fallopian tube (salpingitis) is a major complication of gonorrhoea. Using this model, we have demonstrated (Watt *et al.*, 1976) that fimbriated gonococci show four-fold greater adhesion to the mucosal surface than the non-fimbriated variant of the same strain and that fimbriated gonococci appear attached to the surface by fimbrial bundles (Ward *et al.*, 1975). However, fimbria-mediated adhesion was not essential for invasion, since the proportion of gonococci penetrating the mucosal surface was 0.7% for fimbriated gonococci and 0.5% for non-fimbriated gonococci. The invasive process was the result of phagocytosis of gonococci by mucosal cells, as demonstrated by blocking experiments using cytochalasin B to inhibit membrane movement of host cells (Watt *et al.*, 1976). A further advantage is that the responsiveness of the fallopian tube mucosa to cyclic hormone changes occurring during menstruation makes this model ideal for studying the indirect effects of sex hormones on gonococcal attachment and invasion.

9.4.2 Isolated cell models

The disadvantages of organ culture models are the limitation in the numbers of experiments possible with a scarce material and the problem that the mucosal surface is susceptible to the chemical treatments designed to elucidate the molecular mechanisms of adhesion. The two widely used test systems using readily available isolated cells are fimbria-mediated gonococcal haemagglutination (Punsalang and Sawyer, 1973) and attachment to buccal epithelial cells (Tramont, 1977). Haemagglutination inhibition is, at best, only semi-quantitive, while the method of quantitating adhesion by staining the buccal cells and counting the attached gonococci by microscopy is tedious and leads to day-by-day variations (Tramont and Wilson, 1977). To overcome the problem, a simple system for separating non-adherent from cell-associated gonococci has been devised (Lambden *et al.*, 1979a). ^{3}H-labelled gonococci suspended in tissue culture medium at a concentration of 2.8×10^7 colony-forming units (cfu ml^{-1}) were added to an equal volume of 10% (v/v) buccal cell suspension

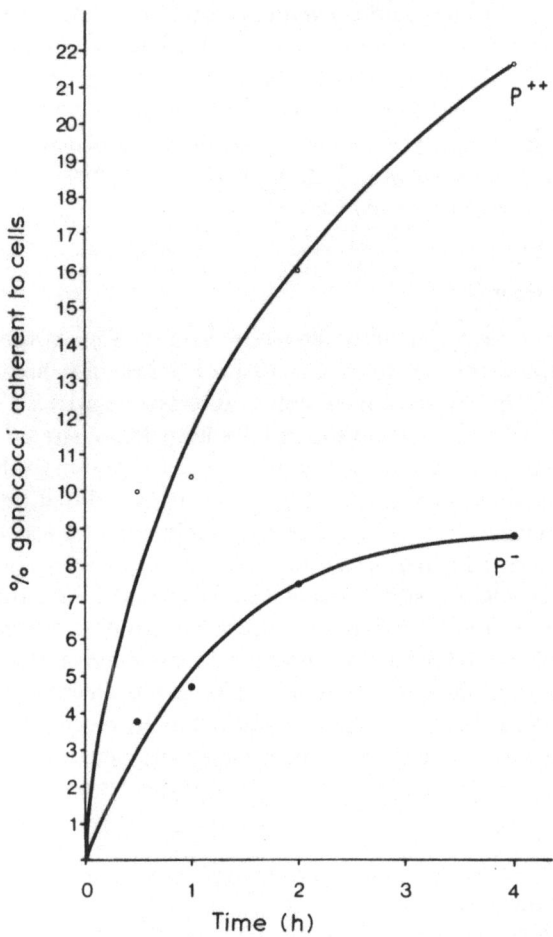

Fig. 9.3 Comparison of the ability of fimbriated and non-fimbriated gonococci to bind to buccal cells. Tritium-labelled fimbriated (P[++]) or non-fimbriated (P[-]) gonococci at a concentration of 2.8 x 10^7 cfu ml[-1] were gently mixed with ar equal volume of 10% (v/v) human buccal cells at 37°C. At various times samples of the suspension were centrifuged on a dextran cushion and the radioactivity in the buccal cell pellet due to the attached gonococci was measured. The graph shows that fimbriation greatly increases the ability of gonococci to attach to the buccal epithelia.

and then incubated with gentle mixing for 1 h at 37°C. Samples of the suspension were layered onto a cushion of 6% (w/v) dextran (Dextraven 110, Fisons) and centrifuged in a swingout head at 500 g for 2 min. The number of gonococci adherent to the buccal cell pellet were determined by scintillation counting. The method is

Table 9.3 Effect of modifying the surface carbohydrates of buccal cells on the attachment of fimbriated and non-fimbriated gonococci

Buccal cell	Fimbriated	Non-fimbriated
No treatment	100*	35
Neuraminidase†	92	30
Neuraminidase plus exoglycosidase 1 mg ml^{-1} ‡	73	34
Neuraminidase plus exoglycosidase 10 mg ml^{-1} ‡	22	35
Periodate/NaBH$_4$ §	100	35
NaBH$_4$ control	100	31

* To facilitate comparison of results the actual percentage binding (26.6%) of fimbriated gonococci to buccal cells is expressed as 100.
† Neuraminidase $100\,\mu$ ml^{-1} in 50mM acetate, pH 5.5, plus 1 mg ml^{-1} Ca^{2+} at 37°C for 30 min.
‡ Contains α-L-fucosidase, α-galactosidase, α-N-acetyl-galactosaminidase, β-N-acetylhexosaminidase + α-mannosidase.
§ 10 mM NaIO$_4$ in 0.1M acetate, pH 5.5, on ice then NaBH$_4$ in 0.1M tris/maleate pH 7.4 − both treatments 30 min.

Table 9.4 The effect of modifying the red cell surface on the binding of ^{3}H-labelled fimbriated and non-fimbriated gonococci

Red cell	Fimbriated	Non-fimbriated
Untreated	*100	64
Trypsin†	114	114
Neuraminidase‡	107	105
Periodate/BH$_4$ §	56	60
BH$_4$ control	104	69

* To facilitate comparison of results the actual percentage binding (37%) of fimbriated gonococci to red blood cells is expressed as 100.
† Trypsin 1 mg ml^{-1} in PBS for 1 h.
‡ Neuraminidase $100\,\mu$ ml^{-1} in 50mM acetate, pH 5.5, plus 1 mg ml^{-1} Ca^{2+}.
§ 10 mM NaIO$_4$ in 0.1M acetate, pH 5.5, on ice then 5 mM NaBH$_4$ in 0.1M tris/maleate, pH 7.4 − both treatments 30 min.

equally applicable to studies of gonococcal binding to red cells or modified Sepharose gels.

Fimbriated gonococci not only bind in greater numbers to human cells, but also show an increased rate of attachment over non-fimbriated organisms (Fig. 9.3). It is conceivable that rapid adhesion is critical in the transmission of the natural infection

Table 9.5 Association of opacity variants of *N. gonorrhoeae* P9 with human PMN leukocytes

	Additional surface protein†						
	NIL	Fimbrial protein	II*	IIa* IId*	IIa*	IIb*	II* IIc*
Association	10.5	8.0	21.5	20.2	32.0	4.6	21.0

Results are expressed as the % of polymorphs with associated gonococci, and are the mean values from three separate experiments.
† After Lambden *et al.* (1979).

where flows of mucus or urine could instantly remove the bacterium from the mucosal surface. The finding that fimbriae enhance the attachment of gonococci to buccal epithelial cells (Table 9.3) and red cells (Table 9.4), but not leukocytes (Table 9.5), suggests that fimbriae bind to a specific component of cell membranes. Furthermore, fimbrial binding to red cells differs from the cohesive process with buccal cells, since fimbria-mediated attachment to red cells is not pH-dependent (Fig. 9.2), while binding to buccal cells by contrast is considerably enhanced between pH 5–7. Again, periodate/borohydride treatment of red cells but not buccal cells (Tables 9.3 and 9.4) impairs fimbria-mediated gonococcal attachment.

Pearce and Buchanan (1978) have shown that the binding of purified gonococcal fimbriae to cells is temperature-dependent. We have confirmed this finding for whole gonococci; at 37°C the binding of fimbriated gonococci to buccal cells is double the attachment at 4°C, while the binding of non-fimbriated variants is reduced by 25% at 4°C. One suggestion (Pearce and Buchanan, 1978) is that, at lower temperatures, the reduced host membrane fluidity affects the mobility of fimbrial binding sites. Impaired mobility would reduce any fimbria-mediated re-alignment of binding sites, thus decreasing the efficiency of attachment. An alternative suggestion is that low temperatures (4°C) induce conformational changes in the fimbria molecule; support for this idea comes from the finding that fimbria binding to polystyrene is markedly reduced at 4°C (Buchanan, 1978).

These data emphasize the difficulty in interpreting fimbria modifications which impair binding such as UV irradiation, heating at 85°C for 1 h or Koshlands reagent (Buchanan *et al.*, 1978). Clearly, the problem is to distinguish specific effects on individual amino-acid residues from conformational changes induced in the fimbrial protein molecule. We have found that periodate/borohydride treatment of fimbriated gonococci markedly reduces binding to buccal cells but not to red cells suggesting that the small carbohydrate component in fimbriae (Section 9.3.1) is critical for adhesion. However, periodate oxidation may cross-link membrane proteins. For example, 2 mM periodate produces extensive cross-linking of red cell spectrin (Gahmberg *et al.*, 1978). Because spectrin lacks carbohydrates, possible mechanisms of cross-linking include the formation of interchain disulphide bonds or tyrosine

Table 9.6 The effect of modifying gonococcal surface components on the binding of ^{3}H-labelled gonococci to buccal cells

Treatment	Relative adherence of gonococci	
	Fimbriated	Non-fimbriated
No treatment	*100	32
UV irradiation†	48	33
Periodate/NaBH$_4$ ‡	18	42
NaBH$_4$ control	100	39
β-galactosidase §	74	33
β-N-acetylglucosaminidase¶	71	NT
Ricin I**	200	NT

* To facilitate comparison of results the % binding of fimbriated gonococci to buccal cells (31%) is expressed as 100.

† Exposed to UV source (254 nm, intensity 1240 μW cm^{-2}) at 5 cm for 1 h.

‡ 10 mM NaIO$_4$ in 0.1M acetate, pH 5.5, on ice then 5 mM NaBH$_4$ in 0.1M tris/maleate, pH 7.4 – both treatments 30 min.

§ β-galactosidase 10 μ ml^{-1} in PBS 1 h.

¶ β-N-acetylglucosaminidase 1 μ ml^{-1} in PBS 1 h.

** Ricin I 100 μg ml^{-1} in PBS 1 h.

NT, not tested.

dimers, it seemed important, therefore, to confirm the finding using other techniques. Treatment of fimbriated gonococci with ricin to block presumptive galactose residues on fimbriae doubles the binding of gonococci to buccal cells (Table 9.6). The latter finding illustrates the difficulties of establishing specific molecular mechanisms; presumably ricin binds to fimbriae either non-specifically by hydrophobic interaction or specifically to galactose residues and then cross-links the gonococci to galactose units on the host cell surfaces. Treatment of fimbriated gonococci with β-galactosidase or N-acetyl-D-glucosaminidase impaired binding by 30% (Table 9.6), but we cannot exclude the possibility of non-specific effects resulting from enzyme adsorption to the fimbriae.

Thus, the ability to quantitate gonococcal attachment to cells permits investigation of the molecular basis of fimbria-mediated adhesion, but the available probes do not give hard answers and the data must be viewed with suspicion until confirmed by several different approaches.

With the exception of Swanson's work on the leukocyte association factor (Section 9.4.3), the adhesion of non-fimbriated gonococci to cells has been little investigated. Since non-fimbriated gonococci can invade the intact mucosal surface (Watt and Ward, 1977) the structural determinants of this interaction warrant attention. Clearly, the mechanism of attachment to different cell types is mediated by different surface components (Table 9.5 and 9.7). Thus, gonococci possessing

Table 9.7 Attachment of *N. gonorrhoeae* P9 opacity variants to buccal epithelial cells, erythrocytes and hydrophobic phenyl-Sepharose

Substrate	Attachment ‡ of gonococci with additional surface protein						
	NIL	Fimbrial protein	II*	IIa* IId*	IIa*	IIb*	II* IIc*
Buccal epithelial cells†	12.0	38.0	24.0	27.3	26.1	17.7	20.6
Erythrocytes†	24.5	45.4	19.0	15.1	15.5	11.0	11.4
Phenyl-Sepharose †	8.9	10.0	3.9	8.2	8.5	8.6	8.1

† Average values from three separate experiments.
‡ Attachment was expressed as the % of ^{3}H-labelled gonococci sedimenting with the buccal cells, erythrocytes or gel.

only LPS and Protein I in the outer-envelope bind best to red cells, while the possession of a second major protein, particularly IIa*, enhances binding to buccal epithelial cells and leukocytes. The nature of these interactions is not known. The only published information is the work of King and Swanson (1978) who showed that periodate oxidation of gonococci (without subsequent borohydride reduction) destroyed the leukocyte association factor. Again, the specificity of periodate oxidation must be questioned.

9.4.3 Leukocyte interactions

The ability of pathogenic bacteria to avoid destruction by phagocytes is an essential prerequisite of life within the host. Invasive *N. meningitidis* are invariably capsulated (Craven and Frasch, 1978) and resistant to phagocytosis by human polymorpho-nuclear leukocytes (PMN). Immunization with pure capsule polysaccharide induces opsonic antibodies which stimulate PMN uptake and destruction of meningococci (Roberts, 1970). Unlike this single determinant of meningococcal phagocytosis, several surface components influence the interaction of gonococci and PMN. Many groups (Thongthai and Sawyer, 1973; Ofek *et al.*, 1974; Dilworth *et al.*, 1975) have reported that non-fimbriated gonococci were readily phagocytosed by PMN but fimbriated variants resist ingestion. The fact that fimbriae are antiphagocytic has been confirmed by demonstrating the opsonic properties of specific anti-fimbrial serum (Buchanan *et al.*, 1978). Microscopic studies (Densen and Mandel, 1978; Buchanan *et al.*, 1978) show fimbriated gonococci adherent to the PMN surface with their fimbriae radiating across the host cell membrane. The mechanism by which such fimbriae block ingestion is unknown. One possibility is that the mesh of fimbriae adherent to the PMN surface seriously disrupts membrane motility; certainly, it is difficult to envisage how a 'zipper-mechanism' of phagocytosis could function in this situation. Another suggestion is that adherent fimbriae decrease the fluidity of the host cell membrane and this impairs particle ingestion. Preliminary evidence supporting this concept comes from spin-label studies of red cells (Senff *et al.*, 1977).

Swanson and his colleagues (Swanson *et al.*, 1975; King and Swanson, 1978) have demonstrated that non-fimbriated gonococci can exhibit either higher or lower levels of association with human PMN than fimbriated organisms of the same strain. These findings led to the hypothesis that a non-fimbria factor, termed leukocyte association factor, was the primary determinant of the interaction of gonococci and PMN. Recent studies (King and Swanson, 1978) show that leukocyte association factor is a protein exhibiting variable molecular weight from 29 000 for strain MS 111 to 28 000 for C109. That leukocyte association protein (LAP) is expressed on the gonococcal surface was confirmed by demonstrating its sensitivity to trypsin and its ready labelling in the ^{125}I-lactoperoxidase system. Studies from our group (Lambden *et al.*, 1979a) on isogenic mutants of gonococcus strain P9 confirm that outer-membrane proteins in the molecular weight range 27 000–29 000 have a greater influence on leukocyte association than fimbriae (Table 9.5). However, unlike the results presented by Swanson, our study demonstrates that one protein (IIb*) of molecular weight 28 000 is associated with decreased leukocyte association, whereas all the other variants show increased association. Thus, we were not able to identify a single LA factor, although it may be relevant that the variant which shows greatest leukocyte association contains a single extra protein (IIa*), with a molecular weight 28 000 which is close to that reported for the LA factor of King and Swanson (1978). Although this variant with 30% leukocyte association does not reach the 50% association defined as LA+ cultures by these workers (1978) we cannot exclude the possibility that this is due to inadvertant technical differences in the assay systems used.

If, as reported by Swanson's group (Swanson *et al.*, 1975), LAP does not facilitate gonococcal attachment to mucosal cells it is difficult to envisage a role for LAP in the pathogenesis of gonorrhoea. In contrast, we find that the additional outer-membrane proteins II*, IIa*, IIb*, IIc* and IId* markedly enhance the binding of non-fimbriated gonococci to buccal cells (Table 9.7) and suggest that the true function of LAP may be in attaching gonococci to mucosal surfaces.

9.5 PHYSICO-CHEMICAL FACTORS IN HOST CELL–NEISSERIAL INTERACTIONS

9.5.1 Electrostatic interactions

Fimbriated gonococci have a pI of 5.3, indicating an excess of ionized acidic groups on their surfaces (Heckels *et al.*, 1976). This net negative charge on the gonococcal surface is present over the pH range of mucosal secretions because the percentage of radio-labelled fimbriated or non-fimbriated gonococci binding to positively charged DEAE Sephadex (18%) or negatively charged CM Sephadex (0.8%) gels remained constant over the pH range 5.0–7.5 (Watt *et al.*, 1979). Eukaryotic cells also carry a net negative charge which can largely be attributed to sialic acid groups in cell surface glycoproteins (Greig and Jones, 1977). The presence of ions adsorbed to the neisserial and host cell surfaces will produce a potential field in the neighbouring medium. Inevitably, as the bacteria approach the host cell surface these potential fields will overlap producing an electrostatic repulsive force acting against attachment. Positively charged groups are also present on the surfaces of eukaryotic cells and presumably the neisserial surface. These cationic groups can be attributed to such components as the basic side chain amino groups of lysine and hydroxylysine, terminal protein amino groups and phospholipid and glycolipid amines (Gasic *et al.*, 1968). In the experiment described, the enhanced binding of the negatively charged gonococci to the DEAE Sephadex when compared with CM Sephadex must be the result of electrostatic attraction. In natural neisserial–host cell interactions, there is no experimental evidence of complementary charge mosaics of opposite sign on the opposing surfaces giving rise to adhesion. The major difficulty in exploring this

Table 9.8 The effect of modifying gonococcal surface charge on the binding of P9 to WISH tissue culture cells ‡

Organism	Modification	p*I*†	Mean no. of bacteria attached per cell
Fimbriated	None	5.3	14.5
Fimbriated	—COOH groups blocked §	8.2	27.6
Fimbriated	—NH$_2$ groups blocked ¶	4.0	2.6
Fimbriated	—NH$_2$ and —COOH groups blocked	*	35.0
Non-fimbriated	None	5.6	4.7

* p*I* not determined as the organisms were insufficiently charged.
† Measured by post-pH equilibrium isoelectric focussing in a 110 ml column using 1% w/v ampholine on a glycerol-stabilized gradient.
‡ From Heckels *et al.*, (1976).
§ 1-Ethyl-3 (3-dimethylaminopropyl) carbodiimide/methylamine-treated.
¶ Formaldehyde-treated.

possibility is that the tool available for pin-pointing regions of differing charge density (electron-microscopic localization of bound colloids) is too coarse to separate charge differences at the molecular level.

The importance of electrostatic repulsion in the interaction of gonococci with host cells has recently been demonstrated (Heckels *et al.*, 1976). Free amino groups on the surface of gonococci were blocked using formaldehyde, and carboxyl groups blocked with 1-ethyl-3 (dimethylaminopropyl) carbodiimide and methylamine. The effect of this treatment on the surface charge of the organisms was determined by measuring the p*I* by equilibrium isoelectric focussing. The mean number of untreated or chemically modified gonococci firmly adherent to the surface of monolayer cultures of WISH cells after extensive rinsing was then determined microscopically. The results (summarized in Table 9.8) show that blocking amino groups, and thus increasing the surface net negative charge, reduced binding by 67%, whilst blocking carboxyl groups effectively reversed the usual negative surface charge on the gonococcus and doubled the mean number of gonococci adherent to the cells. This increased attachment was not simply due to electrostatic attraction to the now positively charged gonococcal surface, because when both the amino and carboxyl groups on the gonococcus were blocked so that the organism failed to migrate in an electrical field, the enhanced adherence of the gonococcus to the cells was retained. Thus, the critical factor was the removal of the electrostatic barrier to the negatively charged gonococcal surface.

In a complementary series of experiments, the ability of 35 nm diameter, radio-labelled, model membrane vesicles (liposomes) of varying surface charge to attach to unmodified gonococci was investigated (Ward and Watt, 1977). ^{3}H-labelled liposomes were mixed with gonococci and the percentage of adherent liposomes

Table 9.9 The influence of the composition and surface charge of model membrane vesicles on their attachment to gonococci (P9II* fimbriated)

Membrane composition‡	Published* zeta potential (mV)	Attachment of membrane vesicles
Lecithin + cholesterol (LC)	0	17.2
LC + cerebroside	0	10.2
LC + stearylamine	+ 14.6	34.0
LC + phosphatidylserine	− 14.2	5.1
LC + mixed gangliosides	− 16.0	3.0

* From Hayward (1974).
† Attachment was expressed as the percentage of the 35 nm ^{3}H-cholesterol-labelled membrane vesicels sedimenting with the gonococci after spinning through Ficoll 400 (Pharmacia, Upsala).
‡ Egg lecithin, 5 μM; cholesterol, 5 μM; cerebroside, 0.61 mg; stearylamine, 0.5 μM; phosphatidylserine, 0.375 μM; mixed bovine gangliosides, 1.8 mg.

determined after centrifuging the gonococci through a cushion of 7% (w/v) Ficoll 400 (Pharmacia). The results (summarized in Table 9.9) showed that, at optimum conditions, attachment of the liposomes was doubled when a positive charge due to stearyl amine was introduced into the membranes and was greatly decreased if a negative charge due to ganglioside or phosphatidyl serine was introduced, again demonstrating the importance of the electrostatic repulsive barrier. Nevertheless, gonococci must be able to overcome this barrier, and the work of Jan and Chien (1973a, b) suggests a mechanism.

Jan and Chien (1973a,b) showed that dextrans of molecular weights 20 000 and 40 000 daltons were incapable of inducing erythrocyte aggregation probably because their molecular lengths were too short (25 and 32 nm) to effect bridging. Dextrans of 80 000 daltons with a molecular length of 55 nm readily induced aggregation. When the electrostatic charge barrier on the erythrocytes was abolished by neuraminidase treatment all the dextrans were capable of effecting aggregation, with electron microscopically determined intercellular distances of 16, 19 and 22 nm respectively for dextrans of 20 000, 40 000 and 80 000 daltons. Thus, it was reasonable to speculate that the electrostatic repulsive forces deriving from untreated erythrocytes operated over a distance of some 20 nm. In the case of neisseria, fimbriae can be expected to serve a similar bridging function between the host and bacterial surfaces; this would be faciliated by their low radius of curvature and by the paucity of charged amino acids in their composition (Table 9.1). Interestingly, in an analogous situation to the work described on dextran-mediated erythrocyte aggregation, Heckels *et al.* (1976) found that fimbriae no longer promoted the adhesion of fimbriated gonococci to human tissue culture cells when the electrostatic repulsive barrier was reduced by chemically blocking negatively charged carboxyl groups on the gonococcal surface. Arguably, this effect might have been due to chemical destruction of the attachment function of the fimbria. However, in recent experiments in this laboratory (Watt *et al.*, 1979) in which the attachment of radio-labelled, fimbriated (P^{++}) and non-fimbriated (P^-) gonococci to human erythrocytes was quantitated (summarized in Table 9.4) it was found that fimbriation doubled the attachment of gonococci to untreated red cells unless the negatively charged groups were removed from the red cell surface by treatment with neuraminidase or trypsin; after these enzymatic treatments fimbriae no longer enhanced adhesion. Thus, one function of fimbriae is to bridge the energy barrier due to electrostatic repulsion between the host and neisserial cell surfaces.

9.5.2 Long-range attraction (the DLVO theory)

Electron micrographs of gonococci sectioned at their point of contact with the mucosal surface after they have been exposed to colloidal thorium before dehydration and embedding (Fig. 9.4) show that thorium particles must have a diameter of less than 13 nm to penetrate the gap between the outer envelope and the host cell membrane (Ward and Watt, 1975). Strong forces of attachment involving direct bonding cannot operate over such long distances. However, the size of the gap is consistent with a secondary minimum adhesion system as defined by Curtis (1973). This concept has arisen from physical studies on the interactions between negatively charged lyophobic colloid particles which became formulated as the DLVO theory (Derjaguin and Landau, 1941; Verwey and Overbeek, 1948). The theory considers that the energy of interaction of two charged particles of like sign and magnitude is the sum of the electrostatic energy of repulsion and the energy of attraction provided by London — van der Waal's forces. This London force arises as a result of atomic and molecular vibrations producing fluctuating dipoles; electromagnetic interactions between atoms and molecules with a similar fluctuation frequency produces an attractive force. Inspection of the relevant equations (Curtis, 1973) reveals that the repulsive energy decreases more rapidly with distance than the attractive energy (Grieg and Jones, 1977). Thus, over a range of separation distances a total potential energy curve displaying a maximum and two minima is obtained. Surface—surface cohesion is favoured at the close approach of the bacteria (primary minimum) and at a greater separation of some 1—10 nm, the secondary minimum. At distances between these minima the interacting forces cause repulsion. Primary minimum adhesion can only occur when the potential energy barrier arising from electrostatic repulsion between the diffuse electrical double layers of the contacting surfaces has been overcome. Curtis (1973) discusses the characteristics of this secondary minimum adhesion system:

(1) The opposed surfaces will be separated by a gap whose size will be determined by the nature and ionic environment of the contacting surfaces. The estimate of a 13 nm gap between the outer envelope of the gonococcus and the host cell membrane

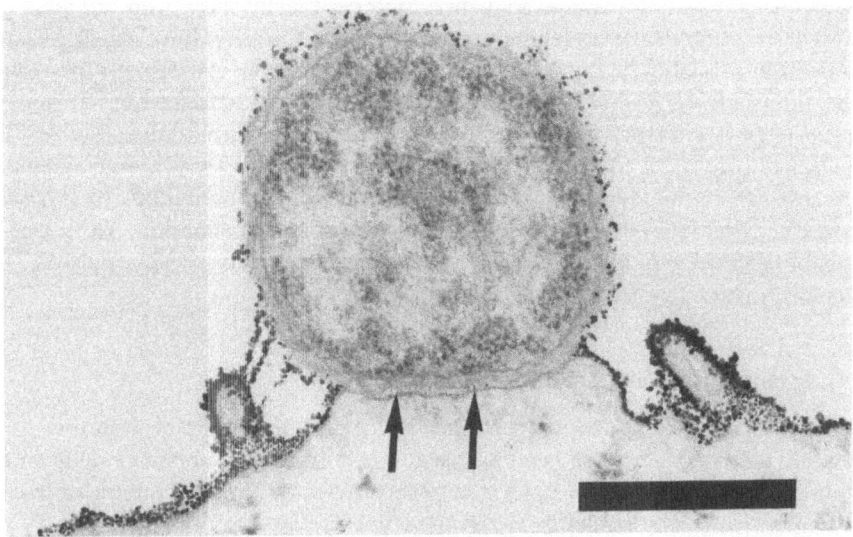

Fig. 9.4 A gonococcus attached to the mucosal surface of a human fallopian tube organ culture. The cell coat has been delineated using colloidal thorium hydroxide prior to dehydration and embedding. Only particles of colloidal thorium less than some 13 nm can penetrate the small gap between the gonococcus and host cell membrane. Tissue glutaraldehyde and osmium-fixed, stained *en bloc* with 1% colloidal thorium hydroxide in 3% v/v acetic acid, pH 2.6. The bar represents 500 μm. (From Ward and Watt 1975.)

is appropriate for secondary minimum adhesion (Ward and Watt, 1975). Increasing ionic strength, by contracting the size of the electrical double layer on the cell surfaces, decreases the electrostatic repulsive force and thus may reduce the size of this gap. In practice, gonococci are exposed to a wide range of ionic strengths since urine osmolarity may range from 50–1400 mosm l^{-1} with associated changes in ionic composition, whilst in prostatic secretions the levels of Zn^{2+} can approach 1 mg ml^{-1} (Fair and Wehner, 1976). Clearly, such large changes in ionic strength must have a considerable effect on bacterial adhesion (Jones, 1977).

(2) Secondary minimum adhesion involves a much weaker attractive force than primary minimum adhesion (Good, 1972). Thus, because of the low energy involved, the adhesion will be redispersible, a fact taken advantage of by Tebbutt *et al.* (1976) who assessed the numbers of gonococci adherent to a mucosal surface by applying a fluid shearing force and counting the number of organisms which detached. A dispersible system of attachment would have the advantage of permitting gonococci to migrate over or detach from the epithelial surface to infect new hosts or a new site.

(3) The attachment system would lack the specificity of adhesion systems which

require a close fit between the contributing surfaces. Again, the studies of Tebbutt *et al.* (1976) support this concept, since gonococci weakly attach to the uterus, cervix and male urethra of guinea pigs even though this animal is resistant to gonococcal infection. Moreover, fimbriated gonococci will readily stick to themselves and such non-specific substrates as glass or polystyrene.

Clearly, the DLVO theory alone cannot explain gonococcal adhesion and other factors such as polymer interactions, electrostatic repulsion, hydrophobic interactions and binding to specific receptors may be of greater significance (see also Section 2.2). Nevertheless, secondary minimum adhesion may be a preliminary stage in gonococcal attachment permitting subsequent primary minimal adhesion. This specific adhesion may be the critical determinant of gonococcal infection of mucosal surfaces. Thus, gonococci adhere to and invade the moving surface of the human fallopian tube, but not the rabbit oviduct, which is resistant to experimental infection (Taylor-Robinson *et al.*, 1975).

9.5.3 The role of ionic bridging

The topographically irregular nature of the gonococcal surface implies that point-to-point contacts can be made with the host cell membrane with gaps of much less than 13 nm. Thus, higher energy attachment mechanisms involving a close contact between the appropriate surfaces cannot be excluded. One such possibility is the direct bridging of surface ligands on the gonococcal and host cell surface by di or trivalent cations such as Ca^{2+} and Fe^{3+}; such bridging requires the contacting surfaces be separated by a gap of some 1 nm.

Experimental investigation of the role of ionic factors in gonococcal adhesion is fraught with problems. Gonococci are aggregated by moderate concentrations of di and trivalent cations and rapidly lyse in non-physiological ionic environments or when deprived of divalent cations. Thus, there is relatively little experimental data. James *et al.* (1976) quantitated the adhesion of gonococci to human sperm by visual counting and showed that the adhesion is markedly enhanced by the presence of 10 mM iron salts. A probable explanation based on DLVO theory is that Stern layer adsorption of the Fe^{3+} counter ion non-specifically enhances adhesion by reducing the magnitude and operating distance of the electrostatic repulsive forces (Watt and Ward, 1977). However, Buchanan *et al.* (1978) have implicated a more specific mechanism. These workers showed that Fe^{3+} enhanced the binding of ^{125}I-labelled fimbriae to human buccal epithelial cells between 1.2 (0.01 mM Fe^{3+})–2.4 times (0.1 mM Fe^{3+}). This enhancement by ferric ion did not occur at pH 4.5 (the average pH of vaginal secretions) and its effect at pH 7.4 was blocked by the prior photo-oxidation of fimbriae with methylene blue, a treatment which selectively destroys histidine residues. It was suggested that ferric ion binds to the histidine residues of fimbriae when they are in an unprotonated state (pH 7.4) bridging between the fimbriae and the buccal epithelial surfaces. Fe^{2+} ions and 10-fold higher concentration of Mg^{2+} or Ca^{2+} did not significantly enhance ^{125}I–fimbria binding suggesting that the role of Fe^{3+} was specific (Buchanan *et al.* 1978). In this laboratory, using similar techniques, the percentage binding of purified ^{125}I-labelled gonococcal fimbriae to buccal epithelia was increased from 7.5–73.2% by the presence of 10 mM Fe^{3+}, but 10 mM Ca^{2+}, Mg^{2+}, or Zn^{2+} had no effect. However, the fimbriae were clearly aggregated by the presence of Fe^{3+}, thus it is not possible to distinguish between enhanced binding of fimbriae to the epithelial surface and enhanced binding of fimbriae to themselves. In an attempt to overcome this problem, buccal epithelial cells were treated with chelating agents to reduce the amount of surface adsorbed divalent cations, and the washed cells were then exposed to 1, 10 and 100 μM of Fe^{3+}. Unadsorbed Fe^{3+} was then removed by washing and the binding of ^{125}I-labelled fimbriae was tested as before. With this system and 100 μM of Fe^{3+} only a 1.5-fold increase in fimbria attachment was attained even though the cells were visibly coated in Fe^{3+} and must have been available to bind to the fimbrial histidine residues (data to be published). In view of this and the fact that in human plasma, the amount of free ionic iron available is only some 10^{-14} M (Weinberg, 1978), it is doubtful whether bridging through Fe^{3+} plays a significant role in gonococcal attachment in the natural disease. Similarly, it has not been possible to obtain experimental evidence that the high concentrations of Zn^{2+} and Mg^{2+} in human prostatic secretions potentiate gonococcal attachment.

9.5.4 The effect of pH

Neisseria colonizing mucosal surfaces exist in a complex environment rich in mucins and varying in ionic composition. In particular, gonococci are exposed to extremes of acidity varying from semen (pH 7.19) and prostatic secretions (pH 6.45) to urine whose pH ranges from 4.8–8. In women, during the menstrual cycle the pH of endocervical mucus varies from 5.9 to 7.3; in the ectocervical mucus from 4.0 to 7.4; in the lateral fornix of the vagina from 3.5 to 5.8 and in the vaginal entrance from 3.5 to 5.3 (Kroeks and Kremer, 1977). However, after coitus the ejaculate exerts a powerful buffering effect perhaps permitting initial attachment at less extremes of pH.

Gonococcal adhesion is influenced by pH; Mårdh and Weström (1976) reported that gonococcal adhesion to vaginal epithelial cells was enhanced threefold at pH 4.5 compared to pH 7.5 while Pearce and Buchanan (1978) have shown comparable increases in the binding of purified fimbriae to buccal epithelial cells. Our data (Fig. 9.5) suggests that fimbria mediated binding to buccal cells is optimum between pH 5.5–pH 7; this is appropriate for the preferred sites of infection of the endocervix and male urethra.

9.5.5 Hydrophobic interactions

Studies using partition between aqueous polymer two-phase systems to assess the

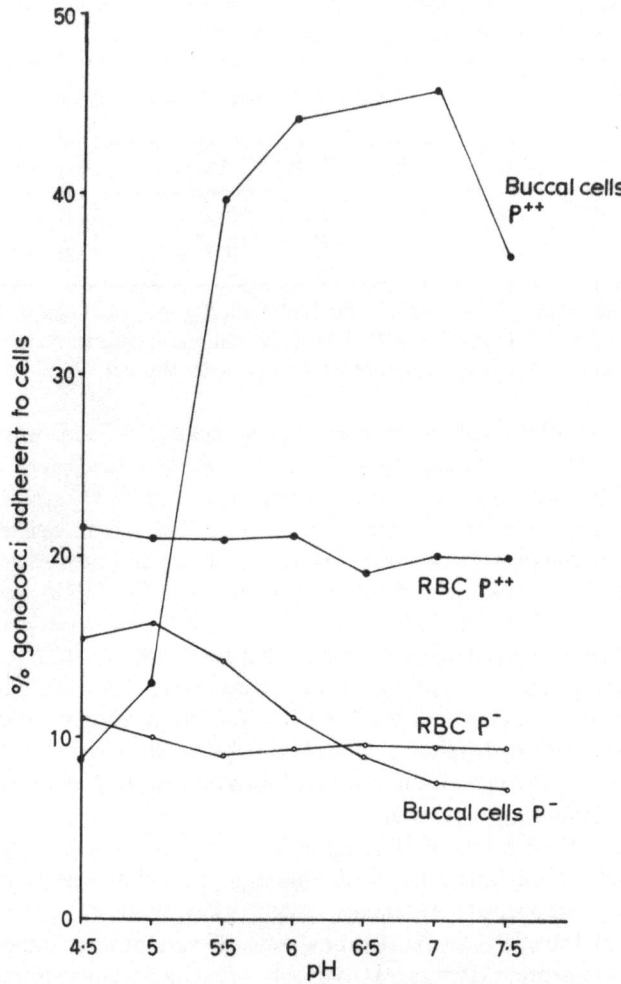

Fig. 9.5 Effect of pH on the attachment of fimbriated and non-fimbriated gonococci to human cells. Radiolabelled fimbriated or nonfimbriated gonococci were incubated for 1 h at 37°C with an equal volume of 10% v/v packed human buccal or group 0 red cells in 100 mM Tris-maleate buffer ranging over the physiological pH range 4.5–7.5. The graph shows that pH change over this range had little effect on the attachment of gonococci to red blood cells. However, the attachment of the fimbriated organism to buccal cells was greatly affected. Thus, the fimbrial receptor on the buccal cells, but not the red cells, was highly susceptible to pH change.

hydrophobic properties of smooth and rough *S. typhimurium* and *E. coli* have provided valuable insights into the determinants of bacterial adhesion to mucosal

Table 9.10 Comparative adsorption to hydrophobic gels of fimbriated and non-fimbriated variants of *N. gonorrhoeae* strain P9

		Adsorption of gonococci to Sepharose gel coupled with				
Gonococci	Nil	Ethyl	Butyl	Hexyl	Octyl	Phenyl
Non fimbriated*	1.2	4.4	4.1	7.9	17.8	10.7
Fimbriated*	1.1	4.1	3.7	9.7	23.6	14.3

* Data are the mean of 3 experiments. ^{3}H-labelled gonococci (2.8×10^7 cfu ml^{-1} were reacted at $37°$C for 1 h with 10% (v/v) gel suspensions. Attachment was expressed as the % of gonococci sedimenting with the gel.

surfaces. Variants of *S. typhimurium* and *E. coli* lacking *O*-chains were lipophilic and demonstrated greater association with intestinal mucosae (Perers *et al.*, 1977) and increased liability to phagocytosis (Stjernström *et al.*, 1977) than hydrophilic strains. Thus, the lack of *O*-polysaccharide chains in pathogenic neisseria may permit hydrophobic interactions between the bacterial surface and the host cell membrane. Unfortunately little experimental data is available. Van Oss (1978) measured the contact angle of a sessile liquid drop on a monolayer of gonococci and found that this value $\zeta = 26.7°$, considerably exceeded that for PMM ($\zeta = 18.0°$). This was also true for typical intracellular pathogens such as brucellae ($\zeta = 27°$), but not capsulated organisms which resist phagocytosis ($\zeta = 17°$). The implication of these findings is that, in bacteria with hydrophobic surfaces, the bacteria–water interfacial tension exceeds the phagocyte/bacterium interfacial tension resulting in an overall free energy change favouring engulfment.

We have used the binding of ^{3}H glucose-labelled gonococci to hydrophobic ligands on Sepharose gels as a means of comparing the relative hydrophobicity of variants of gonococcus P9 (Lambden *et al.*, 1979a; Watt *et al.*, 1979). The results (summarized in Tables 9.7 and 9.10) show that the variant containing the additional outer membrane protein II* was less hydrophobic than variants containing proteins IIa*, IIb*, IIc* or no additional proteins. Clearly, in the absence of *O*-polysaccharide chains, outer membrane proteins become major determinants of gonococcal surface hydrophobicity. The adhesion of non-fimbriated gonococci to host cell surfaces was not solely determined by hydrophobic interactions since the attachment of the relatively hydrophobic variant II* to buccal epithelial cells was double that of a hydrophobic variant lacking this outer membrane protein (Table 9.7). Again, gonococcal leukocyte interactions were found to be more complex than predicated from the van Oss (1978) theory of relative hydrophobicities. Outer membrane variants possessing proteins IIa* and IIb* were equally hydrophobic, as shown by binding to phenyl-Sepharose (Table 9.7), but the IIa* variant had a seven-fold advantage in leukocyte association (Table 9.5).

Given that 46% of the constituent amino acids of gonococcal fimbriae are non-polar (Table 9.1) it is not surprising that fimbriae avidly bind to hydrophobic gels. When 45 μg ^{125}I fimbriae were mixed with 1 ml of 10% (v/v) octyl- or phenyl-Sepharose all the fimbriae sedimented with the gel. Since the area of the gonococcal outer membrane and the area of the fimbriae are of comparable magnitude (about 1 x 10^{-11} m^2), we might expect fimbriae to dominate gonococcal surface hydrophobicity. However, when whole gonococci were interacted with amphipathic gels, fimbriae only conferred a 20% advantage in binding to hexyl-, octyl- and phenyl-substituted Sepharose (Table 9.10). This result emphasizes the facility with which the surface of the gonococcal outer membrane undertakes hydrophobic binding and is in marked contrast to the report that the fimbria-like K88 antigen was solely responsible for the binding of smooth enteropathogenic *E. coli* to octyl- and phenyl-Sepharose gels (Smyth *et al.*, 1978).

Adhesion at mucosal surfaces exists in an aqueous environment so that the exclusion of hydrophobic groups from the water lattice will assist the close approximation of hydrophobic surfaces and facilitate non-specific van der Waals binding. Further hydrophobic interaction may assist binding to specific receptors. Noting that concanavalin A has a hydrophobic region adjacent to the sugar binding site Smyth *et al.* (1978) suggest that lectin-like binding of the K88 antigen to mannose residues may be facilitated by the hydrophobic nature of the protein.

9.6 SURFACE CARBOHYDRATES IN HOST CELL–NEISSERIAL INTERACTIONS

The significance of bacterial adhesins binding to carbohydrate receptors on host-cell membranes has been reviewed recently (Jones, 1977). Clear cut examples are the L-fucose susceptible binding of *Vibrio cholerae* to brush border membranes and the mannose-sensitive type fimbria haemagglutination. In contrast, the binding of fimbriated gonococci to human buccal cells and gonococcal haemagglutination were resistant to a wide range of simple sugars (Punsalang and Sawyer, 1973).

We have confirmed (Watt *et al.*, 1979) that representatives of the constituent sugars of surface glycolipids and glycoproteins including: D-galactose, β-methyl-D-galactoside, α-methyl-D-galactoside, lactose (Gal-β-[1 › 4] -Glc), melibiose (Gal-α-[1 → 6] -Glc) mannose, α-methyl-D-mannoside, α-L-fucose, N-acetyl-D-galactosamine and N-acetyl-D-glucosamine were without effect on the binding of either fimbriated or non-fimbriated gonococci to human buccal cells or erythrocytes. A second approach has been to block surface carbohydrates by pretreatment of buccal cells with lectins. In these experiments (to be published) 50 μl of packed buccal cells were suspended in 15 ml of PBS containing 0.08 μM lectins specific for: β-D-galactose (Ricin), α-L-fucose (Lotus), α-D-mannose (ConA), N-acetyl-D-glucosamine (wheatgerm) and N-acetyl-D-galactosamine (soyabean). When the treated buccal cells were reacted with 200 μg ^{125}I-labelled fimbriae, the percentage of fimbriae adhering to the cells (40%) was the same in the untreated and lectin-treated cells. Taken together, the results of these experiments suggest that gonococcal fimbriae do not act as simple lectins binding to single sugars on the host-cell membrane. However, they cannot exclude the possibility that fimbriae bind, perhaps non-specifically, to a sequence of sugars on cell surface carbohydrates. Evidence supporting this concept comes from the work of Buchanaan *et al.* (1978), who showed that the binding of ^{125}I fimbriae to cells was inhibited by 60% when the buccal cells were treated with an exoglycosidase mixture at 2 mg ml^{-1}. Further fimbria binding to buccal cells was inhibited 20% in the presence of 1 mg ml^{-1} chitin oligosaccharides (β[1 → 4]-linked polymer of *N*-acetyl-D-glucosamine) and 860 μg ml^{-1} yeast mannan (a branched polymer with a backbone of α[1 → 6]-linked D-mannose and side-chains of 2−5 mannose residues linked α [1 → 2] or α [1 → 3]), but not by 10 μM α-methyl-D-mannoside. We can, in part, confirm these findings using whole gonococci. Treatment of buccal cells with 1 mg ml^{-1} exoglycosidases reduced the binding of ^{3}H-labelled fimbriated gonococci by 27% and 10 mg ml^{-1} enzyme by 78% (Table 9.3); the binding of non-fimbriated gonococci was not affected. Comparable results were obtained using red cells where treatment with exoglycosidases reduced the binding of fimbriated gonococci to the level of non-fimbriated gonococci (Table 9.4). In an attempt to confirm the significance of cell-surface carbohydrates, we have treated cells by periodate oxidation followed by borohydride reduction of the resulting aldehydes to polyalcohols. The binding of ^{3}H-labelled gonococci to periodate/borohydride-treated red cells was inhibited by 44% (Table 9.4), but was unaffected in the case of buccal cells (Table 9.3).

Candida albicans with a surface largely composed of mannan has been used in studies on mannose-sensitive fimbriae (Ofek and Beachey, 1978). We have measured the binding of ^{3}H-labelled fimbriated gonococci to 10% v/v suspensions of *Candida*, buccal cells and red cells (unpublished observation). The attachment of fimbriated gonococci to *Candida* (18.5%) was significant and, although only half the binding seen with buccal and red cells, this does not take into account the relative differences in surface area for a given packed cell volume. Although mannans dominate the surface of *Candida* the underlying chitin is exposed, since yeasts are aggregated by lectins specific for *N*-acetyl-D-glucosamine (G. Bull, personal communication). Thus, there are 2 polysaccharides accessible for fimbrial binding.

A further suggestion arising from the purified gonococcal fimbria studies of Buchanan *et al.* (1978) is that cell surfaces possess 'ganglioside-like' fimbrial receptors. The experimental evidence supporting this concept arises from the finding that fimbriae treated with gangliosides (GM1, GD1a, GD1b, GT) at a concentration of 10−20 μM show impaired binding to buccal cells. At concentrations of 0.3 μM GD1a and GD1b inhibited attachment by 20%−30%, but significant binding (40%) persisted even at concentrations of 74 μM.

One difficulty in analysing these experiments is illustrated by our study (Watt *et al.*, 1979) on the effect of gangliosides on the binding of ^{3}H-labelled gonococci to buccal cells (Table 9.11). These results show that micelles of gangliosides

Table 9.11 The effect of gangliosides on the binding of ^{3}H-labelled gonococci to buccal cells

| Treatment of gonococci | Fimbriated | | Non-fimbriated | |
	% gonococci sedimenting with buccal cells	% gonococci sedimenting minus buccal cells	% gonococci sedimenting with buccal cells	% gonococci sedimenting minus buccal cells
No treatment	21.1	5.7	8.8	0.6
GMI*	23.8	4.1	8.4	0.6
GD1a*	23.3	9.1	7.2	0.8
GT1*	28.0	21.5	7.5	0.9

* Gonococci exposed to ganglioside micelles at a contentration of 300 μg ml^{-1}. ^{3}H-labelled gonococci incubated alone (control) or with buccal cells and then layered onto a cushion of 6% Dextran. After centrifugation at 500 g for 2 min the pellet was counted.

GT1 > GD1a cross-link fimbriated gonococci to produce large aggregates which sediment in the absence of buccal cells. Specific cross-linking of fimbriated gonococci by the multivalent ganglioside micelles is a possible explanation for these findings. However, in the same study, we failed to demonstrate the specific binding to gonococci of gangliosides incorporated into ^{3}H-cholesterol/lecithin liposomes. A possible explanation is that ganglioside micelles exert a non-specific effect on fimbriae.

In summary, the evidence supporting gonococcal fimbria adhesion to carbohydrate receptors on the host cell surface is of two types. First, treatment of the cell with a mixture of α-mannosidase, α-L-fucosidase, α-galactosidase, α-N-acetyl-galactosaminidase and β-N-acetyl hexosaminidase markedly impairs fimbria-mediated adhesion to the cell. Secondly, fimbriated gonococci adhere to the surface mannans on yeasts and these mannans, together with chitin and gangliosides inhibit fimbria attachment. Although we cannot exclude fimbria-mediated binding to a single sugar with unique conformation, this is unlikely because adhesion is unaffected by high concentrations of constituent sugars of cell surface carbohydrates or by pretreating cells with lectins specific for these sugars. A simple explanation would be that gonococcal fimbriae bind, perhaps by hydrogen bonding, along the length of the sugar chain to unsubstituted hydroxyl groups. Under these circumstances, specificity would be low and modification of the oligosaccharide sugars to polyalcohols need not inhibit binding, as was observed with periodate/borohydride-treated buccal cells. In support of this concept, we have observed (unpublished) that sugars; tetrasaccharides > trisaccharides > disaccharides > monosaccharides, disperse fimbria-fimbria aggregates while long-chain carbohydrates such as dextrans and poly-sucrose cross-link fimbriae.

9.7 CONCLUSIONS

Experimental studies of neisserial adhesion to host cells have concentrated on the gonococcus. Extension of these studies to meningococci and commensal neisseria is important in order to understand how a group of related bacteria all adapted for life on mucosal surfaces cause such clinically distinct disorders.

Neisserial attachment is clearly a complex, multifactorial process and great care is necessary in interpreting experiments designed to elucidate the molecular basis of adhesion, because attempts to manipulate one determinant may have multiple consequences. Thus, the use of lectins to block sugar groups on mucosal cells will modify surface charge and hydrophobicity, induce steric hindrance of adjacent groups, alter membrane fluidity, affect the microfilament–microtubule system and cyclic nucleotide levels. Nevertheless, analysis of the different experimental approaches used reveals some generally accepted features. The physics of neisserial interaction with eukaryotic cells can simplistically, but usefully, be compared to the behaviour of negatively charged colloid particles. The DLVO theory reasonably explains the long-range forces in attachment; however, the *in vivo* interaction is much more complex involving multiple factors such as steric repulsion from interacting surface polymers, the shear forces of mucosal fluid flows and the effects of host cell surface mobility. All workers seem agreed that fimbriation facilitates gonococcal attachment to a wide range of eukaryotic cells. The exception is polymorphonuclear leukocytes where fimbriae serve an antiphagocytic function. The role of fimbriae in attachment is complex. Fimbriae penetrate the electrostatic barrier between the gonococcus and the host cell membrane initiating attachment. The membrane adhesion may be non-specific involving, for example, hydrophobic interactions or specific binding to a receptor. Evidence is accumulating that fimbriae bind to sugar chains on host-cell membrane glycolipids or glycoproteins. The apparent lack of specificity in this binding will be the subject of further research. In our view, the concept that Fe^{3+} or cell surface gangliosides play a specific role in gonococcal fimbria adhesion to the cell membrane is unproven because the concentration of Fe^{3+} involved grossly exceeds physiological levels and non-specific aggregation due to trivalent ions has not been adequately controlled. Identification of the attachment moiety on gonococcal fimbriae should permit more precise experiments (see Chapter 10).

Outer membrane proteins play an important role in neisserial adhesion. The pioneering work of Swanson (1978) on outer membrane protein variants of gonococci and their relation to colony type has opened up this field of investigation by providing a simple means of selecting the variants. Clearly, different proteins are involved in adhesion to host cells; in particular variants of protein II* enhance the binding of gonococci to leukocytes or buccal epithelial cells, but not to red blood cells. Detailed knowledge of the topographical relationships of the components in the gonococcal outer envelope is the subject of current studies. Undoubtedly, non-fimbria-mediated attachment will be an expanding research area.

The ideal approach to the control of gonococcal and meningococcal infections would be to block their initial attachment to the mucosal surface thereby preventing colonization. Local genital tract antibodies are produced against infecting strains of *N. gonorrhoeae* and are capable of inhibiting the attachment of gonococci to human buccal or vaginal epithelial cells (Tramont, 1977). Parenteral immunization with gonococcal products would be expected to give rise to mucosal antibodies because genital tract secretions are rich in both IgG and IgA (Chipperfield and Evans, 1975; Tramont, 1977). The promise of this approach has been indicated by the work of Brinton *et al.* (1978), who immunized four human volunteers parenterally with purified gonococcal fimbriae and increased the mean number of gonococci required to produce symptomatic gonorrhoea 900-fold. Thus, a detailed understanding of the molecular mechanisms of gonococcal and meningococcal attachment offers the hope of better control of these important infections.

NOTE

Isogenic variants from *Neisseria gonorrhoeae* strain P9 have been shown to produce two distinct types of fimbriae designated α and β. The apparent subunit molecular weights were found to be 19 500 for the α fimbrial protein and 20 500 for the β (Lambden, Robertson and Watt, 1979b). Striking differences were seen in the adhesive properties of α and β fimbriae. The attachment of α fimbriae to human buccal cells was markedly pH dependent with a maximum binding of 44% at pH 6.5. By contrast, β fimbriae showed a steady decline in binding ability over the pH range tested from a maximum of 13% attachment at pH 4.5 to 4% binding at pH 8.5. These findings for β fimbriae are in close agreement with the results of Pearce and Buchanan (1978) for F62 fimbriae which showed a pH optimum of 4.5 and a three to fourfold reduction in attachment at pH 7.5. Further differences in the cohesive properties of α and β fimbriae were demonstrated using host cells modified by enzyme treatments. Removal of sialic acid residues from buccal cell surface carbohydrates by neuraminidase treatment markedly inhibited the binding of α fimbriae but had little effect on β fimbriae. Treatment of the neuraminidase modified buccal cells with a mixture of exoglycosidases further reduced the binding of α fimbriae (6% attachment) to a level comparable to that of β fimbriae (8% attachment). Interestingly, there was no difference in the ability of α and β fimbriae to bind to the human erythrocytes (7% attachment) and this adhesion was unaffected by pH over the range pH 5.0 to 8.5.

These results demonstrate clear differences in the cohesive properties of α and β fimbriae. A possible explanation is that α fimbriae specifically bind to a receptor, involving sialic acid plus other sugar residues, present on the surface of human buccal cells and that β fimbriae lack such receptor recognition.

REFERENCES

Bernfeld, W.K. (1972), *Brt. med. J.,* **4**, 173.
Bhattacharyya, M.N. and Jephcott, A.E. (1974), *Br. J. Vener. Dis.,* **50**, 109—112.
Blount, J.H. (1972), *Am. J. pub. Health,* **62**, 710—712.
Brinton, C.C., Bryan, J., Dillon, J-A., Guerina, N., Jacobson, L.J., Labik, A., Lee, S., Levine, A., Lim, S., McMichael, J., Polen, S., Rogers, K., To, A.C-C. and To, S. C-M. (1978), In: *Immunobiology of Neisseria gonorrhoeae,* (Brooks, G.F., Gotschlich, E.C., Holmes, K.K., Sawyer, W.D. and Young, F.E., eds.), pp. 155—178, American Society of Microbiology, Washington D.C.
Bro-Jørgensen, A. and Jensen, T. (1973), *Br. J. Vener. Dis.,* **49**, 491—499.
Buchanan, T.M. (1978), *J. infect. Dis.,* **138**, 319—325.
Buchanan, T.M. and Pearce, W.A. (1976), *Infect. Immun.,* **13**, 1483—1489.
Buchanan, T.M., Pearce, W.A. and Chen, K.C.S. (1978), In: *Immunobiology of Neisseria gonorrhoeae,* (Brooks, G.F., Gotschlich, E.C., Holmes, K.K., Sawyer, W.D. and Young, F.E., eds.), pp. 242—249, American Society of Microbiology, Washington D.C.
Buchanan, T.M., Swanson, J., Holmes, K.K., Kraus, S.J. and Gotschlich, E.C. (1973), *J. clin. Invest.,* **52**, 2896—2909.
Carney, F.E. and Taylor-Robinson, D. (1973), *Br. J. Vener. Dis.,* **49**, 435—440.

Chipperfield, E.J. and Evans, B.A. (1975), *Infect. Immun.*, **11**, 215–221.
Craven, D.E. and Frasch, C.E. (1978), In: *Immunobiology of Neisseria gonorrhoeae*, (Brooks, G.F., Gotschlich, E.C., Holmes, K.K., Sawyer, W.D. and Young, F.E., eds.), pp. 250–252, American Society of Microbiology, Washington D.C.
Curtis, A.S.G. (1973), *Prog. Biophys. mol. Biol.*, **27**, 317–386.
de Bault, LE. and Yoo, T.J. (1974), *J. Cell Biol.*, **63**, 79a.
Densen, P. and Mandell, G.L. (1978), *J. clin. Invest.*, **62**, 1161-1171.
Derjaguin, B.V. and Landau, L. (1941), *Acta Physiochem. U.S.S.R.*, **14**, 633–656.
DeVoe, I.W. and Gilchrist, J.E. (1975), *J. exp. Med.*, **141**, 297–305.
Dilworth, J.A., Hendley, J.O. and Mandell, G.L. (1975), *Infect. Immun.*, **11**, 512–516.
DiRienzo, J.M., Nakamura, K. and Inouye, M. (1978), *A. Rev. Biochem.*, **47**, 481–532.
Fair, W.R. and Wehner, N. (1976), *Prog. clin. biol. Res.*, **6**, 383–403.
Frasch, C.E. (1979), *Seminar Infect. Dis.*, **2**, 1–47.
Frasch, C.E. and Mocca, L.F. (1978), *J. Bact.*, **136**, 1127–1134.
Freimer, N.B., Ogmundsdottir, H.M., Blackwell, C.C., Sutherland, I.W., Graham L. and Weir, D.M. (1978), *Acta path. microbiol. scand.*, **86B**, 53–57.
Gahmberg, C.G., Virtanen, I. and Wartiovaara, J. (1978), *Biochem. J.*, **171**, 683–686.
Gasic, G.J., Berwick, L. and Sarentino, M. (1968), *Lab. Invest.*, **18**, 63–71.
Good, R.S. (1972), *J. theor. Biol.*, **37**, 413–434.
Greenfield, S., Sheehe, P.R. and Feldman, H.A. (1971), *J. infect. Dis.*, **123**, 67–73.
Greig, R.G. and Jones, M.N. (1976), *J. theor. Biol.*, **63**, 405–419.
Greig, R.G. and Jones, M.N. (1977), *Biosystems*, **9**, 43–55.
Griffin, F.M. Jr., Griffin, J. A., Leider, J. E. and Silverstein, S.C. (1975), *J. exp. Med.*, **142**, 1263–1282.
Harkness, A.H. (1948), *Br. J. Vener. Dis.*, **24**, 137–147.
Hayward, A.M. (1974), *J. mol. Biol.*, **83**, 427–436.
Heckels, J.E. (1977), *J. gen. Microbiol.*, **99**, 333–341.
Heckels, J.E. (1978), *J. gen. Microbiol.*, **108**, 213–219.
Heckels, J.E., Blackett, B., Everson, J.S. and Ward, M.E. (1976), *J. gen. Microbiol.*, **96**, 359–364.
Hermodson, M.A., Chen, K.C.S. and Buchanan, T.M. (1978), *Biochemistry*, **17**, 442–445.
Hoeprich, P.D. (1970), *Calif. Med.*, **112**, 1–9.
Holmes, K.K., Johnson, D.W. and Throstle, H.J. (1970), *Am. J. Epidemiol.*, **91**, 170–174.
James, A.N., Knox, J.M. and Williams, R.P. (1976), *Br. J. Vener. Dis.*, **52**, 128–135.
Jan, K.N. and Chien, S. (1973a), *J. gen. Physiol.*, **61**, 638–654.
Jan, K.N. and Chien, S. (1973b), *J. gen. Physiol.*, **61**, 655–668.
Jennings, H.J., Hawes, G.B., Adams, G.A. and Kenny, C.P. (1973), *Can. J. Biochem.*, **51**, 1347–1354.
Jephcott, A.E., Reyn, A., Birch-Anderson, A. (1971), *Acta path. microbiol. scand.*, **79B**, 437–439.
Johnson, K.G., Perry, M.B., McDonald, I.J. and Russel, R.R.B. (1975), *Can. J. Microbiol.*, **21**, 1969–1980.

Johnston, K.H., Holmes, K.K. and Gotschlich, E.C. (1976), *J. exp. Med.*, **143**, 741–758.

Jones, G.W. (1977), In: *Microbial Interactions* (Receptors and Recognition, Series B, Volume 3) (Reissig, J.L., ed.), pp. 139–176, Chapman and Hall, London.

King, G.J. and Swanson, J. (1978), *Infect. Immun.*, **21**, 575–584.

Kroeks, M.V.A.M. and Kremer, J. (1977), In: *The Uterine Cervix in Reproduction*, (Insler V. and Bettendorf, G., eds.), pp. 109–117, Thieme, Stuttgart.

Lambden, P.R. and Heckels, J.E. (1979), *FEMS Microbiol.*, **5**, 263–265.

Lambden, P.R., Heckels, J.E., James, L.T. and Watt, P.J. (1979a), *J. gen. Microbiol.*, in press.

Lambden, P.R., Robertson, J.N. and Watt, P.J. (1979b), Submitted for publication.

Liu, T-Y., Gotschlich, E.C., Egan, W. and Robbins, J.B. (1977), *J. Infect. Dis.*, **136S**, 71–77.

Lowe, T.L. and Kraus, S.J. (1976), *J. Infect. Dis.*, **133**, 621–626.

McGee, Z.A., Dourmashkin, R.R., Gross, J.G., Clark, J.B. and Taylor-Robinson, D. (1977), *Infect. Immun.*, **15**, 594–600.

Mårdh, P.-A. and Weström, L. (1976), *Infect. Immun.*, **13**, 661–666.

Meningococcal Disease Surveillance Group (1976), *J. Am. Med. Ass.*, **235**, 261–265.

Novotny, P., Short, J.A. and Walker, P.D. (1975), *J. med. Microbiol.*, **8**, 413–427.

Ofek, I. and Beachey, E.H. (1978), *Infect. Immun.*, **22**, 247–254.

Ofek, I., Beachey, E.H. and Bisno, A.L. (1974), *J. Infect. Dis.*, **129**, 310–316.

van Oss, C.J. (1978), *A. Rev. Microbiol.*, **32**, 19–39.

Ovčinnikov, N.M. and Delektorskij, V.V. (1971), *Brt. J. Vener. Dis.*, **47**, 419–439.

Pearce, W.A. and Buchanan, T.M. (1978), *J. clin. Invest.*, **61**, 931–943.

Perers, L., Andaker, L., Edebo, L., Stendahl, P.O. and Tagesson, C. (1977), *Acta path. microbiol. scand.*, Sect. B, **85**, 308–316.

Perry, M.B., Daoust, V., Johnson, K.G., Diena, B.B. and Ashton, F.E. (1978), In: *Immunobiology of Neisseria gonorrhoeae*, (Brooks, G.F., Gotschlich, E.C., Holmes, K.K., Sawyer, W.D. and Young, F.E., eds.), pp. 101–107, American Society of Microbiology, Washington D.C.

Punsalang, A.P., Jr. and Sawyer, W.D. (1973), *Infect. Immun.*, **8**, 255–263.

Roberts, R.B. (1970), *J. exp. Med.*, **131**, 499–513.

Robertson, J.N., Vincent, P. and Ward, M.E. (1977), *J. gen. Microbiol.*, **102**, 169–177.

Roseman, S. (1970), *Chem. phys. lipids*, **5**, 270–297.

Russell, R.R.B., Johnson, K.G. and McDonald, I.J. (1975), *Can. J. Microbiol.*, **21**, 1519–1534.

Schoolnik, G.K., Buchanan, T.M. and Holmes, K.K. (1976), *J. clin. Invest.*, **58**, 1163–1173.

Senff, L.M., Sawyer, W.D. and Haak, R.A. (1977), In: *Abstracts of the Annual Meeting of the American Society of Microbiology*, American Society of Microbiology, Washington, D.C.

Smyth, C.J., Jonsson, P., Olsson, E., Sonderlind, O., Rosengren, J., Hjerten, S. and Wadström, T. (1978), *Infect. Immun.*, **22**, 462–472.

Sparkes, R.A., Purrier, B.G.A., Watt, P.J. and Elstein, M. (1977), In: *The Uterine Cervix in Reproduction*, (Inster, V. and Bettendorf, G., eds.), pp. 271–277, George Thieme, Stuttgart.

Stead, A., Main, J.S., Ward, M.E. and Watt, P.J. (1975), *J. gen. Microbiol.*, **88**, 123–131.

Stendahl, O., Tagesson, C. and Edebo, M. (1973), *Infect. Immun.*, **8**, 36–41.

Stjernström, I., Magnusson, K.-E., Stendahl, O. and Tagesson, C. (1977), *Infect. Immun.*, **18**, 261–265.

Swanson, J. (1973), *J. exp. Med.*, **137**, 571–589.

Swanson, J. (1978), *Infect. Immun.*, **21**, 292–302.

Swanson, J., Kraus, S.J. and Gotschlich, E.C. (1971), *J. exp. Med.*, **134**, 886–906.

Swanson, J., Sparks, E., Young, D. and King, G. (1975), *Infect. Immun.*, **11**, 1352–1361.

Takeuchi, A. (1967), *Am. J. Path.*, **50**, 109–136.

Taylor-Robinson, D., Whytock, S., Green, C.J. and Carney, F.E., Jr. (1974), *Br. J. Vener. Dis.*, **50**, 279–288.

Taylor-Robinson, D., Johnson, A.P. and McGee, Z.A. (1975), In: *Genital Infections and their Complications*, (Danielsson, D., Juhlin, L. and Mårdh, P.-A., eds.), pp. 243–250, Almqvist and Wiksell, Stockholm.

Tebbutt, G.M., Veale, D.R., Hutchison, J.G.P. and Smith, H. (1976), *J. med. Microbiol.*, **9**, 263–273.

Thongthai, C. and Sawyer, W.D. (1973), *Infect. Immun.*, **7**, 373–379.

Tramont, E.C. (1977), *J. clin. Invest.*, **59**, 117–124.

Tramont, E.C. and Wilson, C. (1977), *Infect. Immun.*, **16**, 709–711.

Verwey, E.J.W. and Overbeek, J.T.G. (1948), In: *Theory of the Stability of Lyophobic Colloids*, Elsevier, London

Ward, M.E., Robertson, J.N., Englefield, P.M. and Watt, P.J. (1975), In: *Microbiology 1975* (Schlessinger, D., ed.), pp. 188–199, American Society for Microbiology, Washington, D.C.

Ward, M.E. and Watt, P.J. (1972), *J. infect. Dis.*, **126**, 601–605.

Ward, M.E. and Watt, P.J. (1975), In: *Genital Infections and their Complications* (Danielsson, D., Juhlin, L. and Mårdh, P.-A., eds.), pp. 229–241, Almqvist and Wiksell, Stockholm.

Ward, M.E. and Watt, P.J. (1977), In: *Gonorrhoea, Epidemiology and Pathogenesis* (Skinner, F.A., Walker, P.D. and Smith, H., eds.), pp. 83–93, Academic Press, London.

Ward, M.E., Watt, P.J. and Glynn, A.A. (1970), *Nature*, **227**, 382–384.

Ward, M.E., Watt, P.J. and Robertson, J.N. (1974), *J. Infect. Dis.*, **129**, 650–659.

Watt, P.J. and Lambden, P.R. (1978), In: *Modern Topics in Infection*, (Williams, J.D., ed.), pp. 134–147, William Heinemann, London.

Watt, P.J., Lambden, P.R. and Trust, T.J. (1979) submitted for publication.

Watt, P.J. and Ward, M.E. (1977), In: *The Gonococcus*, (Roberts, R.B., ed.), pp. 652–655, John Wiley and Sons, Inc., New York.

Watt, P.J., Ward, M.E., Heckels, J.E. and Trust, T.J. (1978), In: *Immunobiology of Neisseria gonorrhoeae*, (Brooks, G.F., Gotschlich, E.C., Holmes, K.K., Sawyer, W.D. and Young, F.E., eds.), pp. 253–257, American Society of Microbiology, Washington, D.C.

Watt, P.J., Ward, M.E. and Robertson, J.N. (1976), In: *Sexually Transmitted Diseases*, (Catterall, R.D. and Nicol, C.S., eds.), pp. 89–105, Academic Press, London.

Weinberg, E.D. (1978), *Microbiol. Rev.*, **42**, 45–66.
Wistreich, G.A. and Baker, R.F. (1971), *J. gen. Microbiol.*, **65**, 167–173.
Yamada, K.M., Yamada, S.S. and Pastan, I. (1975), *Proc. natn. Acad. Sci., U.S.A.*, **72**, 3158–3162.

10 Structure and Cell Membrane-Binding Properties of Bacterial Fimbriae

WILLIAM A. PEARCE
and
THOMAS M. BUCHANAN

10.1	Introduction	*page*	291
	10.1.1 Scope, terminology and ecological proposal		291
10.2	Occurrence and morphology of bacterial fimbriae		293
	10.2.1 Gram-negative bacteria		293
	10.2.2 Gram-positive bacteria		304
10.3	Cell membrane-binding properties conferred by fimbriae		305
	10.3.1 Bacterial hemagglutination		305
	10.3.2 Bacterial adherence to host cells		307
	10.3.3 Characteristics of the binding of isolated fimbriae to host cells		315
	10.3.4 Chemical nature of receptors (binding sites) for bacterial fimbriae		321
	10.3.5 Non-specific fimbrial binding		329
10.4	Structure of bacterial fimbriae in relation to adherence		329
	10.4.1 Antigenic structure		329
10.5	Conclusions		339
	References		340

Acknowledgements
We wish to thank all of our colleagues who have provided us with unpublished information. We are particularly grateful to Dr. Kirk C.S. Chen and Mr. Duane A. Olsen, who helped inspire and perform many of the studies on gonococcal fimbriae structure and adherence, to John C. Newland for his generous contribution of the electron micrograph in Fig. 10.1 and to Patra Leaming whose valuable help included typing the manuscript.

Bacterial Adherence
(*Receptors and Recognition,* Series B, Volume 6)
Edited by E.H. Beachey
Published in 1980 by Chapman and Hall, 11 New Fetter Lane, London EC4P 4EE
© 1980 Chapman and Hall

10.1 INTRODUCTION

10.1.1 Scope, terminology and ecological proposal

It has been thirty years since bacteria were recognized to sometimes possess protein-aceous, non-flagellar surface appendages that radiate outwards in a fairly rigid, fila-mentous fashion (Houwink and van Iterson, 1950). At the same time, it was first suggested that they function as organs of attachment. Such appendages were initially named 'fimbriae'(plural of Latin for thread, fiber or fringe) in the same report in which they were recognized to mediate the hemagglutination reactions frequently found for members of the family Enterobacteriaceae (Duguid *et al.,* 1955). Subse-quently, Brinton (1959) introduced the name 'pili' (Latin for hairs or fur) and urged its correctness and, therefore, preference in usage (Brinton, 1965). In deference to the temporal priority and linguistic suitability of the term 'fimbriae' (Duguid *et al.,* 1966), it, as well as the singular (fimbria) and adjectival (fimbrial, fimbriate) forms, will be used throughout this review.

Bacterial fimbriae, which are smaller and more numerous than flagella, are now known to occur widely among, but nearly restricted to, many gram-negative genera, a gram-positive exception being Corynebacteria. Fimbriation of strains is an unstable reflection of an ecologically advantageous expression found in natural environments for these bacteria. Therefore, it can best be studied in strains freshly isolated from natural sources (e.g. intestinal or genito-urinary tract).

Fimbriae, as will be discussed, are heterogeneous and, apparently, multi-functional. Possible functions include attachment between bacteria growing in liquids for surface pellicle formation, twitching motility and adherence. Another function, performed by a special class of fimbriae, such as F-pili (Britnon, 1965), is the conduction of bacterial or viral nucleic acids. This review is only concerned with the functional role of fimbriae as attachment organelles for adherence to animal host cell surfaces. It considers only those fimbriae on bacteria that have been associated with virulence, attachment and/or hemagglutination. Other fimbriae have been reviewed by Brinton (1967), Duguid *et al.* (1966), Duguid (1968), Ottow (1975), Jones (1977) and Frøholm (1978). This article will first review the occurrence and morphology of fimbriae that function in bacterial adherence. Second, it will address the character-istics of fimbrial binding with attention to the ecological, molecular and chemical nature of both fimbriae and their binding sites. Emphasis has been placed on detail-ing those fimbria/host-cell interactions that have been examined in greatest detail to stress the molecular nature of fimbria-mediated adherence. Some of these data include work performed in our own laboratory on *Neisseria gonorrhoeae* fimbrial attachment.

291

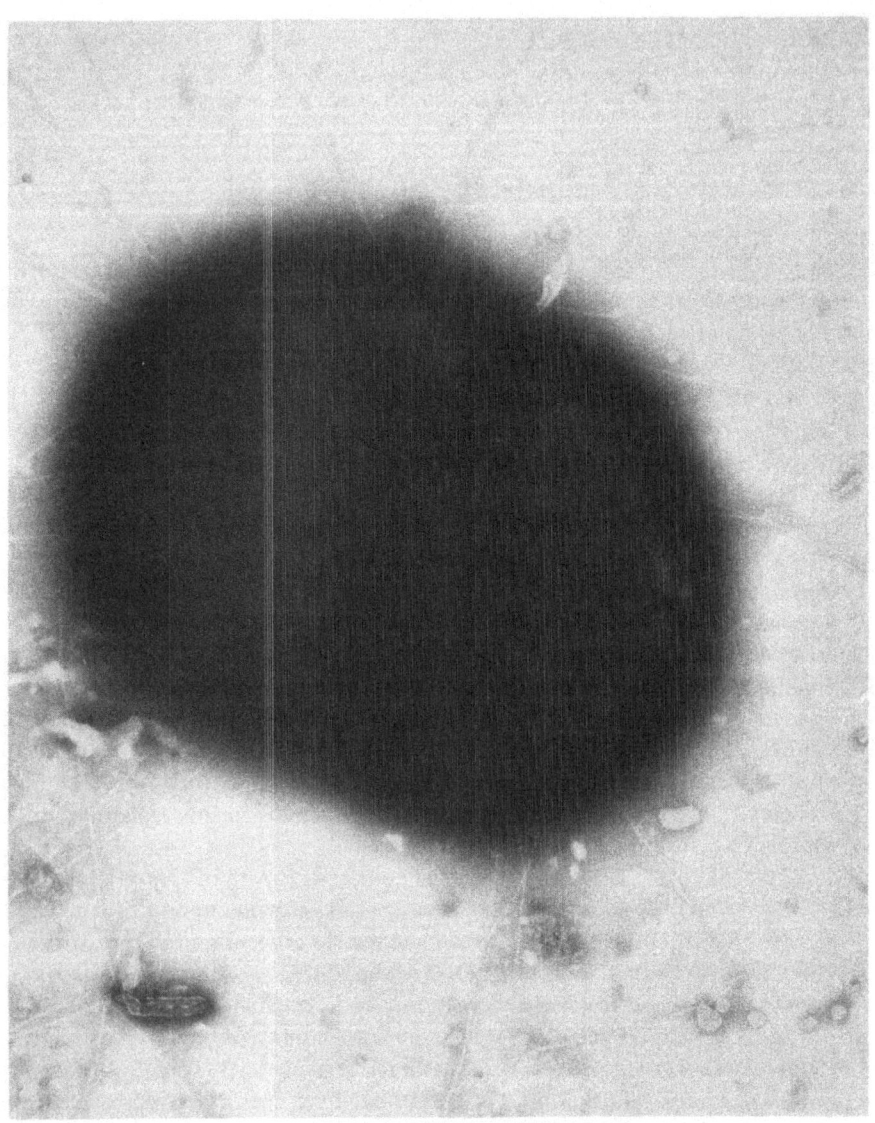

Fig. 10.1 Electron micrograph of fimbriated *N. gonorrhoeae* from strain F62. Bacteria from colony type T_2 (Kellogg *et al.*, 1963) were negatively stained with 1% potassium phosphotungstic acid, pH 7.0. Note the fairly rigid fimbriae radiating uniformly from the surface of the diplococcus. Several membrane blebs are visible on the grid surface. Magnification x 103 250. Photograph courtesy of John Newland.

10.2 OCCURRENCE AND MORPHOLOGY OF BACTERIAL FIMBRIAE

10.2.1 Gram-negative bacteria

Bacterial fimbriae are non-flagellar filamentous appendages approximately 4.0—10.0 nm in width and 0.5—4.0 μm long that are usually distributed evenly and project outwards from the outer membrane (Figs. 10.1 and 10.2). Table 10.1 summarizes the characteristics of fimbriae known or expected to function in adherence. A single bacterium may possess from 100—700 fimbriae. Houwink and van Iterson (1950) first observed and recognized fimbriae as possible attachment organelles in strains of *E. coli*, *Pseudomonas aeruginosa* (pyocyanea) and *Proteus mirabilis*. The ecology, occurrence, classification, genetics and functions of fimbriae have been recently reviewed (Ottow, 1975; Jones, 1977). Fimbriation is a reversible trait in most cases and, for fimbriae considered by this review, is most likely determined by a chromosomal gene(s) (Brinton, 1965, 1967; Ottow, 1975). Brinton (1965, 1967) has found that the stability of fimbriation is mutational in nature, with loss being sudden, complete, random and reversible. The rate of fimbriate to non-fimbriate conversion can approach 10^{-2}/cell division, but the reverse has a much lower frequency (Brinton, 1967). Further, the rate of loss of fimbriation can be strongly influenced by the environment. For example, an 18°C change in temperature can cause a 100-fold change in fimbriation (Brinton, 1967). Brinton proposed that a mutator gene function influenced by environmental control is responsible. The optimal cultural conditions for fimbriation among enterobacteria is serial cultivation in static broth for 24—48 h each subculture (Duguid *et al.*, 1955, 1966; Duguid and Gillies, 1956, 1957, 1958; Duguid, 1959, 1964, 1968; Brinton, 1965). Rich fimbriation is promoted by this hypoxic state, whereas brief exponential growth in liquid media or non-selective growth on solid media greatly reduces or prevents expression of fimbriation. This is also true for *Vibrio* species which have 6—10 times more numerous fimbriae in liquid culture (Tweedy *et al.*, 1968) and *Neisseria* species (Wistreich and Baker, 1971; McGee *et al.*, 1977) except the meningococcus, where solid is equivalent to broth cultures (Devoe and Gilchrist, 1975). Pseudomonads, however, are generally more fimbriated in log phase growth (Weiss, 1971) with a decline occurring during static growth and the production of 'colonization factor' fimbriae by enterotoxigenic *E. coli* is maximal on agar (Evans *et al.*, 1977, 1978). The occurrence and recognition of characteristic colony morphologies on solid media that correlate with fimbriation can allow the selective passage of bacterial cultures that express, albeit often reduced, fimbriation on agar. For example, fimbriated *E. coli* form smaller, smoother colonies than the often rough, non-fimbriated variants (Brinton, 1959). Fimbriation in *P. aeruginosa* has been found to correlate with granular, compact and dry colony morphology (Weiss, 1971). For gonococci, the colony types most common in primary cultures (small, opaque, convex, adherent) have been correlated with virulence (Kellogg *et al.*, 1963, 1968) and fimbriation (Jephcott *et al.*, 1971;

Swanson *et al.*, 1971). No other *Neisseria* species that are fimbriated have colonial variants (Devoe and Gilchrist, 1975; McGee *et al.*, 1977). Fimbriae which function as adhesive organelles in gram-negative bacteria can be classified by their morphology especially their diameter, which remains constant while length is often variable within a given fimbrial type or bacterial strain. Other useful properties include hemagglutinating and attachment ability, host cell species specificity, and whether the attachment mediated by fimbriae is inhibited or not by various carbohydrates, especially those containing D-mannose (Table 10.2). Two classification schemes exist which have often served as reference point for describing fimbriae found subsequently in other bacteria. They are summarized and compared in Table 10.1. Duguid *et al.*, (1966) described different types of fimbriae based upon morphology and adhesive properties in the approximate order of their discovery, usually in enterobacteria. A similar system has been proposed by Brinton (1965, 1967). Only those fimbriae known or thought to function in adherence are described.

Type 1 fimbriae, also referred to as 'common pili (fimbriae)', are found in entero-bacteria (see Chapter 7). They average 7.0 nm in width and 2.0 μm in length with as many as 400 per cell peritrichously located (Table 10.1). These fimbriae are res-ponsible for adhesive properties, especially hemagglutination. Members of the family Enterobacteriaceae with type 1 fimbriae all adhere to fungal, plant and animal cells, including erythrocytes (Duguid *et al.*, 1955, 1966) (with strongest hemagglutination being produced with guinea pig, fowl and horse cells). The most distinguishing pro-perty of type 1 fimbriae is the inhibition of their adhesive activities by D-mannose or α-methyl-D-mannoside (mannose-sensitive) (Duguid and Gillies, 1958; Duguid, 1959, 1964; Old, 1972). Type 1 fimbriae also confer the ability to form pellicles of adherent bacteria on the surface of a static broth (Duguid and Gillies, 1957; Duguid *et al.*, 1966; Old *et al.*, 1968) which is also mannose-sensitive (Old *et al.*, 1968). The occurrence of type 1 fimbriae on *E. coli* and the hemagglutination they mediate were defined in 1955 by Duguid *et al.* For example, 31 of 47 strains examined were richly fimbriated having 100–250 peritrichous fimbriae/bacillus for 30–100% of cells (Table 10.1). Hemagglutination was strictly correlated to the presence of fimbriae and strongest for horse, guinea pig and fowl, weaker for human, and lowest for sheep, goat and ox cells (Table 10.2). Further, antisera which contained some anti-bodies to fimbriae could inhibit these reactions. Such hemagglutination tests led to the future ability to recognize the occurrence of fimbriated bacteria. Among enterobacteria (Table 10.1), similar type 1 fimbriae which facilitated hemagglutin-ation were reported for *Shigella* (Duguid and Gillies, 1957), *Salmonella* (Duguid and Gillies, 1958, Duguid *et al.*, 1966), *Klebsiella* and *Serratia* (Duguid and Gillies, 1958; Duguid, 1959; Cowan *et al.*, 1960; Thornley and Horne, 1962) and *Enterobacter* (Constable, 1956; Duguid, 1959). Duguid (1968) and Jones (1977) point out that the function of these fimbriae remains an enigma, as they are produced by saprophytes, commensals and pathogens alike. Although evidence that their adhesive properties promote pathogenesis is difficult to obtain, it seems clear that such properties allow colonization, an essential step in either mutualism or parasitism. Further, Duguid and

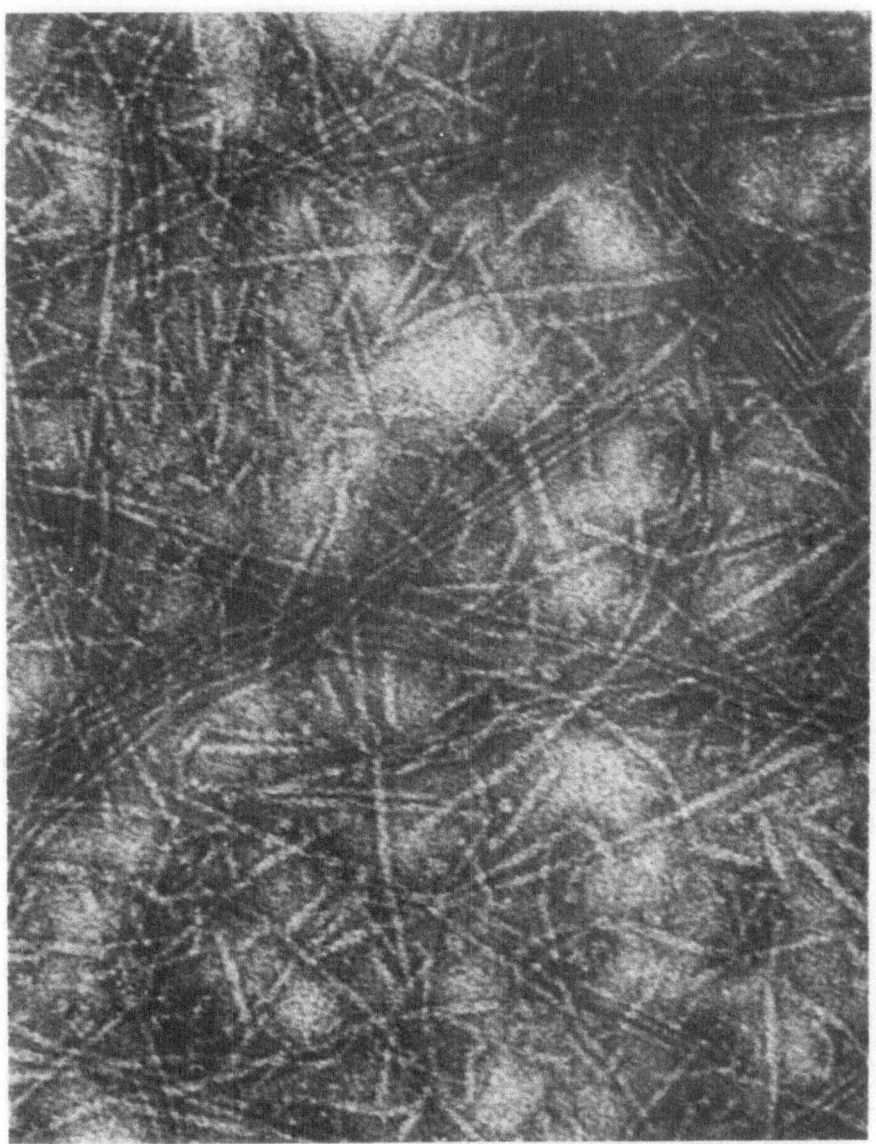

Fig. 10.2 Electron micrograph of purified fimbriae from *N. gonorrhoeae* strain F62. The fimbriae preparation (27:1 purity by SDS-polyacrylamide gel electrophoresis of radiolabeled fimbriae compared to background radioactivity) was negatively stained with 1% potassium phosphotungstic acid, pH 7.0. Magnification x 280 000.

co-workers (1976) have recently demonstrated that fimbriation in *S. typhimurium* significantly increases both the number of infections (26% increase) and deaths (40% increase) in mice inoculated orally as compared with non-fimbriated organisms from the same parent strain.

An interesting variant of common fimbriae may be type 2 fimbriae (Duguid classification), which are identical to type 1 in morphology and closely related antigenically (Old and Payne, 1971), but lack adhesive ability (Duguid, 1958; 1968; Duguid *et al.*, 1966). These were recognized in some strains of *Salmonella gallinarum, pullorum* and *paratyphi B* (Duguid and Gillies, 1958; Duguid *et al.*, 1966; Duguid, 1968) and may be a mutational form of type 1. In addition, many enterotoxigenic strains of *E. coli* isolated from humans, piglets, calves and lambs have been recognized to possess surface structures ('colonization factors') morphologically indistinguishable from common fimbriae (CFA, K88, K99 respectively; Table 10.1) which mediate mannose-resistant adherence (Table 10.2).

Type 3 fimbriae ('thin type') carry a mannose-resistant hemagglutinin reported by Duguid (1959) for *Klebsiella aerogenes* and *Serratia marcescens* strains, which also possessed type 1 fimbriae, but individual bacilli were homogeneous as to fimbrial type. These fimbriae are distinct in being only 4.8 nm in diameter (Thornley and Horne, 1962), 0.5–2 μm long and mediating mannose-resistant adherence generally limited to tannic acid treated erythrocytes (Duguid, 1959; Cowan *et al.*, 1960; Duguid, 1968).

Type 4 fimbriae were recognized in *Proteus* strains (Duguid and Gillies, 1958), are near 4 nm in width (Hoeniger, 1965), up to 2–6 μm long (Brinton, 1965) and facilitate mannose-resistant adherence to untreated cells of a different spectrum from type 1 fimbriae, being most active for sheep and fowl erythrocytes (Duguid and Gillies, 1958; Sheddon, 1962; Duguid *et al.*, 1966; Duguid, 1968). These fimbriae predominate in 4 h, exponential broth cultures or on agar and may contribute to the ability of *P. mirabilis* to cause pyelonephritis (Silverblatt, 1974).

Similar fimbriae of this small diameter and mannose-resistance have been reported for non-pathogenic *Neisseria* (Wistreich and Baker, 1971). No characteristic colony morphology exists which correlates with the presence of these fimbriae (McGee *et al.*, 1977). Also, their ability to agglutinate untreated erythrocytes, with the exception of human red cells, indicates a close similarity to the type 4 fimbriae of *Proteus*. However, no evidence exists to implicate their function in adherence to human mucosal epithelium *in vivo*. Brinton (1965, 1967) and Silverblatt (1974) have reported the additional occurrence on *Proteus mirabilis* grown under hypoxic conditions, such as serial subcultures of 48 hours each in static broth, of 7.0 nm diameter fimbriae (type IV) which are indistinguishable in appearance from common (type 1) fimbriae. Significantly, these larger fimbriae predominate duiing *in vivo* growth and appear to be the principal mediators of adherence during infection as their presence significantly enhances virulence (Silverblatt, 1974).

In contrast to enterobacteria, the fimbriae of the *Pseudomonadaceae* often have a polar or bipolar distribution, are flexible, narrower and have retractile capabilities

(Table 10.1). Type 5 fimbriae are the mannose-sensitive, monopolar contractile fimbriae of the star (cluster)-former *Pseudomonas echinoides* (Marx and Heumann, 1962; Heumann and Marx, 1964). They are approximately 5.0 nm wide, 2–10 µm long, few in number and can adhere to sheep erythrocytes. The fimbriae of other *Pseudomonas* species are also thinner than common fimbriae, such as the polar (PSA) fimbriae of *P. aeruginosa* (c.6.0 nm), (Fuerst and Hayward, 1969; Weiss, 1971; Bradley, 1972). These fimbriae are retractile receptors for RNA phages, but other functions are unknown (Bradley, 1972; Bradley and Pitt, 1974). They are not conductors of bacterial DNA or coded for by drug-resistance plasmids. Fuerst and Hayward (1969) observed that all six *Pseudomonas* species with polar fimbriae have ≤ 10/cell, whereas *P. multivorans* and *P. fragi*, which are peritrichously fimbriate, have approximately 100 and 30/cell respectively. Fimbriae have also been observed on some *Vibrio* species (Tweedy *et al.*, 1968), but their role in adherence in this genus is less well-characterized. These fimbriae are 7.2 nm in average diameter, being similar but not identical to type 1 fimbriae, and can facilitate agglutination of human erythrocytes (Tweedy *et al.*, 1968).

A filamentous hemagglutinin that is 2 nm wide and roughly 40–70 nm long has been purified from *Bordetella pertussis* (Morse and Morse, 1970; Sato *et al.*, 1973, 1974; Arai and Sato, 1976). Although it appears quite different from type 1 fimbriae and its contribution to virulence remains to be demonstrated, this outer membrane structure (126–133 000 subunit molecular weight) can agglutinate a broad spectrum of erythrocytes, HeLa, spleen, L and sarcoma cells (Sato *et al.*, 1973).

Moraxella bovis, M. nonliquefaciens and *M. kingii* all form colonies on solid media which are spreading and agar-corroding and have been correlated to the presence of rich fimbriation (Pedersen *et al.*, 1972; Bøvre and Frøholm, 1972; Bøvre *et al.*, 1970; Frøholm and Bøvre, 1972). These fimbriae are between 6.0–8.5 nm in diameter, similar in length to common fimbriae and most often peritrichously located (Table 10.1).

Devoe and Gilchrist (1974) first reported fimbriae on *Neisseria meningitidis*. They found only 5% of the laboratory strains examined to be fimbriated, but one strain, ATCC 13090, had numerous fimbriae (c. 7.0 nm diameter, McGee *et al.*, 1977) and greater than 80% of the organisms were fimbriated, despite growth on artificial laboratory media. Subsequently, Devoe and Gilchrist (1975) examined primary cultures from more than 30 meningococcal carriers and from a similar number of patients with meningococcal disease. More than 80% of primary cultures were heavily fimbriated, but this was lost on serial subculturing. Further, no colony morphology variants exist which could allow correlation with fimbriation (Devoe and Gilchrist, 1975; McGee *et al.*, 1977). They suggested that artificial media lacked a component necessary for fimbriae expression or that selective pressures in the human host favored persistence of fimbriation. No studies of meningococcal fimbriae function or antigenicity have been reported (see Section 10.4.2).

Before fimbriae were known to occur on *N. gonorrhoeae*, Kellogg *et al.* (1963) had described four distinct colony types on agar medium. Only two of these are seen

Table 10.1 Characteristics of bacterial fimbriae that function in adherence

Code No. (see Table 10.2)	Fimbriae classification		Genus and species	Colonial morphology charateristic of fimbriation	Frequency
	Duguid	Brinton			
A	1	I	*Escherichia coli*	Small, smooth, transparent	31/47
B	1	I	*Klebsiella* species	(encapsulated)	109/140 125/154
C	1	I	*Serratia marcescens*	–	–
D	1	I	*Shigella flexneri*	–	66/98 103/180
E	1	I	*Enterobacter cloacae*	–	10/13
F	1	I	*Salmonella* species	–	65/81 1184/1453
G	3	II	*Klebsiella aerogenes* *Serratia marcescens*	–	Nearly all
H	4	III	*Proteus* species	–	35/35
I	–	IV	*Proteus mirabilis* *Proteus vulgaris*	–	–
J	5	VI	*Pseudomonas echinoides*	–	–
K	–	–	*Pseudomonas* species	–	8/15 Species

(*continued on the next page*)

Table 10.1 Characteristics of bacterial fimbriae that function in adherence (*continued*)

% Bacteria fimbriated	Fimbriae size Width (nm)	Length (μm)	Number per bacterium (location*)	Reference
30–100	7.0	0.2–2.0	50–400 (PE)	Duguid *et al.*, 1955; Duguid, 1964, 1968; Brinton, 1965, 1967; Salit and Gotschlich, 1977a
50–90	6.5–7.0	0.2–1.5	100–400 (PE)	Duguid and Gillies, 1958; Duguid, 1959; 1968; Thornley and Horne, 1962
50–90	6.5–7.0	0.2–1.5	100–400 (PE)	Duguid and Gillies, 1958; Duguid, 1959, 1968
–	7.0	0.3–2.0	100–250 (PE)	Duguid and Gillies, 1956, 1957
–	(7.0)	(0.5–2.0)	100 (PE)	Constable, 1956
5–90	7.0	0.1–1.5	10–300 (PE)	Duguid and Gillies, 1958; Duguid *et al.*, 1966
–	4.8	0.2–1.5	400–700 (PE)	Cowan *et al.*, 1960; Thornley and Horne, 1962; Duguid, 1959, 1968
80% of 4 h cultures	3.0–4.0	0.2–6.0	50–500 (PE)	Duguid and Gillies, 1958; Hoeniger, 1965; Brinton, 1965; Silverblatt, 1974
98% of 96 h cultures	7.0 (helical)	1.0–2.0	100–200 (PE)	Brinton, 1965; Silverblatt, 1974; Weibull and Hedvall, 1953
–	5.0	1.0–10.0	1–50 (MP)	Marx and Heumann, 1962; Heumann and Marx, 1964; Duguid, 1968; Brinton, 1965
–	4.0–6.5	≤ 1.7	≤ 10 (PO,6) 30–100 (PE,2)	Fuerst and Hayward, 1969

* PE = peritrichous; PO = polar; MP = monopolar; BP = bipolar.

(*continued on the next page*)

Table 10.1 Characteristics of bacterial fimbriae that function in adherence (*continued*)

Code No. (see Table 10.2)	Fimbriae classification Duguid	Brinton	Genus and species	Colonial morphology characteristic of fimbriation	Frequency
L	−	−	*Pseudomonas aeruginosa* (PSA)	Compact, dry, granular	10/11
M	1	I	*Pseudomonas multivorans*	−	−
N	1	I	*Aeromonas liquefaciens*	−	−
O	−	−	*Vibrio cholerae*	−	−
P	−	−	*Vibrio eltor*	−	−
Q	4 (?)	III (?)	*Neisseria* species	None	−
R	−	−	*Neisseria meningitidis*	None	All primary isolates
S	−	−	*Neisseria gonorrhoeae*	Small, convex, opaque, adherent	All primary isolates
T	−	−	*Moraxella* species	Spreading, agar corroding (SC)	−

(*continued on the next page*)

Table 10.1 Characteristics of bacterial fimbriae that function in adherence (*continued*)

% Bacteria fimbriated	Fimbriae size Width (nm)	Length (μm)	Number per bacterium (location*)	Reference
4–100	6.0 ± 2.8	1.0–5.0	1–25 (MP or BP)	Houwink and van Iterson, 1950; Bradley, 1972; Fuerst and Hayward, 1969; Weiss, 1971; Bradley and Pitt, 1974
–	6.0 ± .25	≤ 1.7	≥ 100 (PE)	Tweedy et al., 1968; Fuerst and Hayward, 1969
10	7.9 ± 0.3	≤ 1.3	– (PE)	Tweedy et al., 1968
10	7.2 ± 0.3	≤ 1.0	≤ 9 (PE)	Tweedy et al., 1968
50	7.9 ± 0.2	≤ 3.3	≤ 50 (PE)	Tweedy et al., 1968
–	2.0–3.0 4.0–6.0 4.0–4.5 7.0	7.0 0.2–0.3 or 2.3–3.4 0.5–0.7 or 12.5–14.1 0.5–4.3	25–100 (PE)	Wistreich and Baker, 1971; McGee et al., 1977
80	7.0	0.5–4.3	> 50 (PE)	Devoe and Gilchrist, 1974, 1975; McGee et al., 1977
> 90	7.0	0.5–4.0	20–200 (PE)	Kellogg et al., 1963; Jephcott et al., 1971; Swanson et al. 1971; Robertson et al. 1977; McGee et al. 1977; W.A. Pearce, unpublished data
–	6.0–8.5	Several microns	1–200 (PE or PO)	Bøvre et al., 1970; Frøholm and Bøvre, 1972; Bøvre and Frøholm, 1972

* PE = peritrichous; PO = polar; MP = monopolar; BP = bipolar.

(*continued on the next page*)

Table 10.1 Characteristics of bacterial fimbriae that function in adherence (*continued*)

| Code No. (see Table 10.2) | Fimbriae classification | | Genus and species | Colonial morphology characteristic of fimbriation | Frequency |
	Duguid	Brinton			
U	–	–	*E. coli* CFA (enterotoxigenic)	–	98%: Serotypes 015, 025, 063, 078 (CFA/I); 06, 08 (CFA/II)
V	–	–	*E. coli* K88 (enterotoxigenic)	–	Limited serotypes
W	–	–	*E. coli* K99 (enterotoxigenic)	Transparent	9/11
X	4 (?)	III (?)	*Coryne-bacterium renale*	–	–
Y	4 (?)	III (?)	*Coryne-bacterium species*	–	11/11 Species

Table 10.1 Characteristics of bacterial fimbriae that function in adherence (*continued*)

% Bacteria fimbriated	Fimbriae size Width (nm)	Length (μm)	Number per bacterium (location*)	Reference
Most	8.0–9.0	Several microns	⩾ 100 (PE)	Evans *et al.*, 1975, 1978; Ørskov, *et al.* 1976; Evans and Evans, 1978
Most	4.0–8.0	0.1–1.5	⩾ 100 (PE)	Stirm *et al.*, 1967b; Hohmann and Wilson 1975; Nagy *et al.*, 1977
Most	7.0–9.8	Several microns	⩾ 100 (PE)	Ørskov *et al.*, 1975 Burrows *et al.*, 1976; Isaacson, 1977
–	2.5–3.0	⩽ 10	Numerous (PE)	Yanagawa and Otsuki, 1970
91–100(3) 10–37 (5) 0.5–3 (3)	2.0–6.0	0.2–3.0	10–>100(3) < 10 (5) < 10 (3) (PE)	Yanagawa and Honda, 1976

* PE = peritrichous; PO = polar; MP = monopolar; BP = bipolar.

on primary isolation and cultures from both were more virulent in experimental
intra-urethral inoculation of humans (Kellogg *et al.*, 1963, 1968). These colonial
variants were found to be fimbriated in 1971 by Jephcott *et al.* and independently
by Swanson *et al.* Cultures from the other colonial variants lacked fimbriae.
Gonococcal fimbriae resemble type 1 fimbriae (Figs. 10.1 and 10.2), being approximately 7 nm in diameter, 0.5–4 µm in length, with between 100–200 per heavily
fimbriated organism peritrichously located (Swanson *et al.*, 1971; Jephcott *et al.*,
1971; Robertson *et al.*, 1977; Brinton *et al.*, 1978; W.A. Pearce, unpublished data).
Also, fimbriated gonococci were quickly recognized to agglutinate many different
types of erythrocytes, including human (Punsalang and Sawyer, 1973), but unlike
type 1 fimbrial-mediated hemagglutination, it is mannose-resistant. Soon afterwards,
it was demonstrated that fimbriated gonococci were more virulent than non-fimbriated
organisms for the chimpanzee (Brown *et al.*, 1972) and the chick embryo (Buchanan
and Gotschlich, 1973; Bumgarner and Finkelstein, 1973). Further, this enhanced
virulence correlated with greater attachment of fimbriated than non-fimbriated
gonococci to human cells.

10.2.2 Gram-positive bacteria

Among gram-positive bacteria, the only genus reported to possess fimbriae is
Corynebacterium. Here too an adhesive function for fimbriae exists. First reported
for strains of *C. renale* (Yanagawa *et al.*, 1968), which have been classified on a
serological and biochemical basis into three types, those strains of type II possessed
the most fimbriae (Yanagawa and Otsuki, 1970). Fimbriation was stable under
various cultural and subculturing conditions. Bundles of fimbriae (0.4 µm × 10 µm)
are common, but the size of isolated fimbriae is 2.5–3.0 nm in diameter and up to
5–10 µm in length (Yanagawa and Otsuki, 1970). This is similar to type 4 fimbriae
(Duguid classification). Hemagglutination mediated by these fimbriae is limited to
trypsinized sheep erythrocytes (Honda and Yanagawa, 1974) and is not inhibited by
D-mannose. The contribution of these fimbriae to virulence has been questioned
as heavily fimbriated strains of *C. renale* (type II) are parasitic rather than pathogenic
whereas types I and III, less fimbriated strains, are pathogenic (Yanagawa and
Otsuki, 1970). However, *in vivo* experiments with a heavily fimbriated type II strain
demonstrated a clearly enhanced virulence and attachment over less fimbriated strains
(Shimono and Yanagawa, 1977; Honda and Yanagawa, 1978).

Other *Corynebacteria* which parasitize humans and animals also possess fimbriae.
For example, isolates of *C. kutscheri, C. diphtheriae* and *C. pseudodiphtheriticum*
are nearly all fimbriated (Yanagawa and Honda, 1976). Organisms possess up to
100 fimbriae, which are similar to those of *C. renale* in size, being mostly 2–4 nm
in diameter (2–6 nm range) and 0.2–3.0 µm long (Yanagawa and Honda, 1976).
Other species of *Corynebacteria* in this study were less fimbriated but similar.

10.3 CELL MEMBRANE-BINDING PROPERTIES CONFERRED BY FIMBRIAE

10.3.1 Bacterial hemagglutination

The presence of fimbriae on members of Enterobacteriaceae (e.g. *E. coli*) correlates with an ability to hemagglutinate (Duguid *et al.*, 1955). Usually, the strength of hemagglutination reactions reflects the degree of fimbriation. Subsequently, similar fimbrial hemagglutination reactions were reported for *Enterobacter* (Constable, 1956), *Shigella* (Duguid and Gillies, 1957), *Salmonella* (Duguid and Gillies, 1958; Duguid *et al.*, 1966), *Klebsiella* and *Serratia* (Duguid and Gillies, 1958; Duguid, 1959). These and the hemagglutinating properties of other fimbriae described in Section 10.2 are outlined in Table 10.2. These type 1 fimbrial agglutinations exhibit cellular specificity being strongest for erythrocytes of the guinea pig, horse, and fowl > human > sheep, goat > ox (Duguid *et al.*, 1955; Duguid and Gillies, 1957, 1958; Duguid, 1959, 1964, 1968). Confirmation of the role of fimbriae in these attachment reactions was provided by the inhibition produced by antisera partially specific for fimbriae (Duguid *et al.*, 1955; Duguid and Gillies, 1956; Gillies and Duguid, 1958). As mentioned, these reactions could also be inhibited by D-mannose or carbohydrate derivatives containing D-mannose such as α-methyl-D-mannoside.

Mannose-resistant fimbrial hemagglutination may have different red blood cell specificities than those seen with type 1 fimbriae (Table 10.2). Type 3 ('thin type') fimbriae of *Klebsiella* and *Serratia* strains bear a mannose-resistant hemagglutinin, but are limited to tannic acid-treated cells (Duguid, 1959). *Proteus* fimbriae (Duguid and Gillies, 1958; Sheddon, 1962; Duguid *et al.*, 1966) are mannose-resistant, type 4, and have strongest hemagglutination for sheep and fowl cells compared to the weaker agglutination of human erythrocytes (Table 10.2).

The only reported fimbrial hemagglutination by *Pseudomonas* species is the mannose-sensitive reaction displayed by *P. multivorans*, which is peritrichously fimbriate unlike most pseudomonads, for guinea pig cells (Tweedy *et al.*, 1968; Fuerst and Hayward, 1969). The monopolar type 5 fimbriae of *P. echinoides* mediate mannose-sensitive adherence to sheep erythrocytes (Heumann and Marx, 1964), but no agglutination was observed. *P. aeruginosa* strains are able to agglutinate human erythrocytes (Gilboa-Garber, 1972a), but the morphological nature of this hemagglutinin has not been considered. The fimbriae of *V. cholerae* and *V. eltor* facilitate agglutination of human and guinea pig erythrocytes that is partially inhibited by D-mannose (Tweedy *et al.*, 1968).

Most human-associated enterotoxigenic *E. coli* produce one of two, but not both, 'colonization factors', CFA/I or CFA/II (Evans *et al.*, 1977, 1979; Evans and Evans, 1978; Table 10.1). CFA/I mediates the mannose-resistant hemagglutination of human, bovine, chicken and ox cells (Evans *et al.*, 1977, 1979; Ørskov and Ørskov, 1977) while CFA/II positive strains only agglutinate bovine and chicken erythrocytes in the presence of D-mannose (Evans and Evans, 1978; Evans *et al.*, 1979). These 'colonization factor' fimbriae appear morphologically similar to fimbriae structures that are found on *E. coli* strains selectively causing neonatal diarrhea in piglets (K88 antigen: Jones and Rutter, 1972; Wilson and Hohmann, 1974; Jones and Rutter, 1974; Hohmann and Wilson, 1975; Nagy *et al.*, 1977) or in calves and lambs (K99 antigen: Ørskov *et al.*, 1975; Burrows *et al.*, 1976; Isaacson, 1977). All of these colonization-promoting antigens are mannose-resistant hemagglutinins (Jones and Rutter, 1974; Ørskov *et al.*, 1975; Burrows *et al.*, 1976) which can be inhibited by antisera containing antibodies to the homologous colonization factor (Burrows *et al.*, 1976; Ørskov and Ørskov, 1977).

E. coli isolates from human extra-intestinal infections have been shown to possess a mannose-resistant hemagglutinin which is strongly active for human cells (Minshew *et al.*, 1978a,b). This species specificity and mannose resistance is more similar to the 'colonization fimbriae' of enterotoxigenic *E. coli* in contrast to normal intestinal isolates (Table 10.2; Salit and Gotschlich, 1977a). Between 59–64% of 44 extra-intestinal *E. coli* isolates, but only 12–15% of intestinal *E. coli* isolated from normal individuals, could agglutinate human erythrocytes (Minshew *et al.*, 1978a,b). This was shown to be mannose-resistant (Minshew *et al.*, 1978b) and closely correlated with virulence for the chick embryo in the former isolates (Minshew *et al.*, 1978a). In contrast, 31–33% of the extra-intestinal isolates and 75% of normal intestinal *E. coli* could agglutinate chicken erythrocytes and this was mannose-sensitive. While the mediator of these reactions is anticipated to be fimbriae, it remains under investigation. Svanborg-Edén and Hansson (1978) have presented conflicting evidence for a mannose-sensitive fimbrial hemagglutinin on extra-intestinal isolates. They found that fimbriae-mediated agglutination of guinea pig erythrocytes by 91% (11/12) *E. coli* isolates from urinary tract infections was inhibited by D-mannose and α-methyl-D-mannoside.

Keogh and North (1948) reported that cultures or supernatants from *B. pertussis* could hemagglutinate human, mouse and fowl red cells. Antibody to *B. pertussis* could neutralize this agglutination, the hemagglutinin was lost on subculture and there appeared to be a correlation between hemagglutinin content, virulence for mice, and the protective potency of *B. pertussis* vaccines (Keogh and North, 1948). Masry (1952) found this hemagglutinin on most freshly isolated strains, but he disputed its role as a virulence factor. This hemagglutinin was later purified, separated from leukocytosis factor which also carried some hemagglutinating ability, and demonstrated to be responsible for a very broad species reactivity in hemagglutination reactions (Morse and Morse, 1970; Sato *et al.*, 1973; Arai and Sato, 1976).

Many investigators have observed that fimbriated gonococci agglutinate erythrocytes better than non-fimbriated bacteria (see Table 10.2). Punsalang and Sawyer (1973) reported enhanced mannose-resistant hemagglutination by fimbriated gonococci of cells from rabbit, guinea pig, sheep, chicken and human blood group 0, Rh$^+$. This could be inhibited by antiserum to partially purified homologous fimbriae and the inhibition was removed only by adsorption of the antiserum with fimbriated gonococci. The hemagglutination observations of Punsalang and Sawyer were confirmed, modified and extended by Waitkins (1974), Koransky *et al.* (1975) and Buchanan and Pearce (1976). The last group could inhibit the hemagglutination of human A or O, Rh$^+$ cells with antisera to purified fimbriae and the best inhibition was obtained with antisera to the homologous fimbriae among antigenically distinct types.

Strains of *Corynebacterium renale* possessing numerous fimbriae can hemagglutinate trypsinized sheep cells (Honda and Yanagawa, 1974). In the same study fimbriae were demonstrated by electron microscopy to adhere to red cell ghosts. This and hemagglutination were inhibited by homologous antisera, but not by D-mannose. The only other fimbriated *Corynebacteria* that can hemagglutinate is *C. diphtheriae* (Yanagawa and Honda, 1976), which weakly agglutinates trypsinized or untreated sheep cells.

10.3.2 Bacterial adherence to host cells

Table 10.2 summarizes fimbriae-mediated bacterial attachment to host cells. Most studies of type 1 fimbriae function in Enterobacteriaceae have indicated that fimbriae may facilitate adherence of these organisms to host cell surfaces. Duguid and Gillies (1957) noted that the fimbriae of *Shigella flexneri* facilitated this organism's attachment to human colonic epithelium. Similarly, *E. coli* strain K12 adherence to monkey kidney cells in tissue culture is enhanced by the presence of fimbriae (Salit and Gotschlich, 1977b). Further, antibody to purified type 1 fimbriae from these *E. coli* blocked this attachment. Brinton (1967) points out that most gram-negative bacteria possess fimbriae when freshly isolated from natural sources. For an example, he chose to study the infected human urinary tract (niche) where 65 of 72 isolates yielded fimbriated bacteria on culturing (*Proteus,* 12/12; *Klebsiella,* 17/17; *E. coli,* 29/33; *Pseudomonas,* 10/12). Twenty could be seen to be fimbriate in the urine itself. In another study, *E. coli* isolated from patients with symptomatic urinary tract infection adhered in larger numbers to epithelial cells from the urinary tract of humans than did organisms from patients with asymptomatic infection (Svanborg-Edén *et al.,* 1976, 1977, 1978). The presence of fimbriae was correlated to this attachment ability (Svanborg-Edén and Hansson, 1978). Ofek *et al.* (1977) demonstrated that these fimbriated organisms adhere much more readily to human buccal mucosal cells. Other fimbriae of *E. coli,* termed coloniza-tion factor, K88 and K99, appear important for enteropathogenic *E. coli* host specificity. They seem to selectively attach to the small intestinal mucosal epithelium and thus promote diarrheal disease in humans (colonization factor: McNeish *et al.,* 1975; Evans *et al.,* 1977, 1978), piglets (K88: Wilson and Hohmann, 1974) and calves or lambs (K99: Ørskov *et al.,* 1975). Homologous antibodies to these fimbriae can neutralize the activity of these colonization factors *in vivo* and are protective (Evans *et al.,* 1975; Evans and Evans, 1978) as well as block adhesion of bacteria carrying these fimbriae to host tissues or cells *in vitro* (Jones and Rutter, 1972; Wilson and Hohmann, 1974; Burrows *et al.,* 1976). In the rat model of

Table 10.2 Functional properties of fimbriae

Code No. (see Table 10.1)	Genus and species	Presence of fimbriae correlated with virulence	Species specificity of fimbriae-mediated hemagglutination
A	*Escherichia coli*	Yes	Guinea pig, horse, fowl ⩾ swine, monkey, rabbit > human > sheep, goat
B	*Klebsiella* species	Yes	Guinea pig, horse, fowl > human, rabbit > sheep; sheep (trypsinized)
C	*Serratia marcescens*	No	Guinea pig, horse, fowl > human, rabbit > sheep, ox
D	*Shigella flexneri*	No	Guinea pig ⩾ horse, fowl > rabbit, mouse > human, sheep
E	*Enterobacter cloacae*	No	Guinea pig, fowl > human, sheep, ox
F	*Salmonella* species	Yes	Guinea pig > horse, fowl, dog, rhesus monkey > swine, rabbit, mouse, rat > human > sheep
G	*Klebsiella aerogenes* *Serratia marcescens*	No	Human, guinea pig and ox (tannic acid-treated)

(*continued on the next page*)

Table 10.2 Functional properties of fimbriae (*continued*)

Mannose sensitivity	Fimbriae-mediated bacterial attachment to other cells	Mannose sensitivity	Reference
Yes	Guinea pig, rabbit and swine small intestine mucosa, human buccal mucosal cells, Vero monkey kidney cells	Yes	Duguid et al., 1955; Collier and DeMiranda, 1955; Duguid and Gillies, 1957; Duguid, 1959, 1964, 1968; Brinton, 1965, 1967; Tweedy et al., 1968;
	Human uroepithelial cells	No	McNeish et al., 1975; Salit and Gotschlich, 1977a, 1977b; Ørskov and Ørskov, 1977; Nagy et al., 1977; Ofek et al., 1977; Svanborg-Edén and Hansson, 1978; Evans et al., 1979
Yes	Human, guinea pig and ox intestinal epithelial cells and leukocytes, BHK-21 cells	Yes	Duguid and Gillies, 1958; Duguid, 1959; Cowan et al., 1960; Thornley and Horne, 1962; Brinton, 1967; Honda and Yanagawa, 1975
Yes	—	—	Duguid and Gillies, 1958; Duguid, 1959; Duguid et al., 1966
Yes	Human, guinea pig and ox intestinal epithelial cells and leukocytes	Yes	Duguid and Gillies, 1956; 1957; Duguid, 1959; Old, 1972
Yes	Human, guinea pig and ox intestinal epithelial cells and leukocytes	Yes	Constable, 1956; Duguid and Gillies, 1957; Duguid, 1959
Yes	Human, mouse, guinea pig and ox intestinal epithelial cells; human buccal and amnion cells; guinea pig tracheal epithelium	Yes	Duguid and Gillies, 1958; Duguid et al., 1966, 1976; Duguid, 1968; Old, 1972
No	Guinea pig and ox intestinal epithelial cells	No	Duguid and Gillies, 1958; Duguid, 1959; Cowan et al., 1960; Thornley and Horne, 1962

(*continued on the next page*)

Table 10.2 Functional properties of fimbriae

Code No. (see Table 10.1)	Genus and species	Presence of fimbriae correlated with virulence	Species specificity of fimbriae-mediated hemagglutination
H	*Proteus* species	No (Possible)	Sheep, fowl > guinea pig, horse > human, rabbit, ox
I	*Proteus mirabilis*	Yes	−
J	*Pseudomonas echinoides*	No	Sheep (adherence by fimbriated pole only)
M	*Pseudomonas multivorans*	No	Guinea pig
N	*Aeromonas liquefaciens*	No	Human, guinea pig
O	*Vibrio cholerae*	No	Human, guinea pig
P	*Vibrio eltor*	No	Human, guinea pig
Q	*Neisseria* species (non-pathogenic)	No	Mouse, rabbit; human (±)
S	*Neisseria gonorrhoeae*	Yes	Human, rabbit, guinea pig, sheep, fowl

(*continued on the next page*)

Table 10.2 Functional properties of fimbriae (*continued*)

Mannose sensitivity	Fimbriae-mediated bacterial attachment to other cells	Mannose sensitivity	Reference
No	Rat renal pelvis epithelium	—	Duguid and Gillies, 1958; Sheddon, 1962; Duguid et al., 1966; Duguid, 1968; Wistreich and Baker, 1971; Silverblatt, 1974
—	Rat renal pelvis epithelium, rabbit bladder epithelial cells	—	Brinton, 1967; Silverblatt, 1974; Silverblatt and Ofek, 1976
Yes	—	—	Heumann and Marx, 1964
Yes	—	—	Tweedy et al., 1968; Fuerst and Hayward, 1969
Yes	—	—	Tweedy et al., 1968
Partial	Adult and infant rabbit small intestinal mucosa—mediator of adherence is unknown	Partial	Tweedy et al., 1968; Jones and Freter, 1976; Nelson et al., 1976
Partial	—	—	Tweedy et al., 1968
No	—	—	Wistreich and Baker, 1971
No	Human buccal, cervical—vaginal and uroepithelial mucosal cells, amnion and foreskin tissue culture cells, HeLa cells, sperm, fallopian tube epithelium	No	Kellogg et al., 1963, 1968; Punsalang and Sawyer, 1973; Swanson, 1973; Waitkins, 1974; James-Holmquest et al., 1974; Ward et al., 1974, Swanson et al., 1975; Koransky et al., 1975; Buchanan and Pearce, 1976; James et al., 1976; Mårdh and Weström, 1976; Tebbutt et al., 1976; Johnson et al., 1977; Pearce and Buchanan, 1978

(*continued on the next page*)

Table 10.2 Functional properties of fimbriae

Code No. (see Table 10.1)	Genus and species	Presence of fimbriae correlated with virulence	Species specificity of fimbriae-mediated hemagglutination
T	*Moraxella bovis*	Yes	—
U	*E. coli* CFA/I	Yes	Human > bovine, fowl, ox
	E. coli CFA/II	Yes	Bovine, fowl
V	*E. coli* K88	Yes	Guinea pig > fowl ≫ human, sheep, bovine, goat
W	*E. coli* K99	Yes	Guinea pig, sheep
X	*Corynebacterium renale*	Yes	Sheep (trypsinized)
Y	*Corynebacterium diphtheriae*	No	Sheep

Table 10.2 Functional properties of fimbriae (*continued*)

Mannose sensitivity	Fimbriae-mediated bacterial attachment to other cells	Mannose sensitivity	Reference
–	Calf conjunctival mucosa	–	Pedersen *et al.*, 1972
No	Human fetal and infant rabbit small intestine mucosa	No	McNeish *et al.*, 1975; Evans *et al.*, 1977, 1978, 1979; Ørskov and Ørskov, 1977; Evans and Evans, 1978
No	Swine small intestine; isolated piglet intestinal epithelial cells	No	Stirm *et al.*, 1967b; Jones and Rutter, 1972, 1974; Wilson and Hohmann, 1974; Hohmann and Wilson, 1975; McNeish *et al.*, 1975; Nagy *et al.*, 1977
No	Calf small intestine mucosa	–	Ørskov *et al.*, 1975; Burrows *et al.*, 1976; Isaacson, 1977
No	Primary dog and rabbit kidney cells; BHK-21 cells; mouse urinary bladder	–	Honda and Yanagawa, 1974; 1975; 1978; Shimono and Yanagawa, 1977
–	–	–	Yanagawa and Honda, 1976

pyelonephritis, Silverblatt (1974) determined that *Proteus mirabilis* organisms with
7 nm diameter fimbriae more frequently produced infection than non-fimbriated
bacilli, and electron micrographs demonstrated organisms that appeared to attach to
epithelial cells of the renal pelvis by their fimbriae. These larger fimbriae of *Proteus*
predominate during *in vivo* growth (Silverblatt, 1974). The role of these fimbriae
in determining the site of localization in renal infection is further supported by
evidence that they mediate attachment to rabbit urinary tract epithelial cells *in vitro*
(Silverblatt and Ofek, 1976). It is not clear whether the smaller *Proteus* fimbriae
(type 4) also contribute to the ability of *Proteus* to cause pyelonephritis. Other
fimbriae of gram-negative bacteria with a smaller diameter (4—5 nm) have not been
associated with enhanced attachment to human cells (Duguid, 1959). Fimbriae of
Moraxella and *Corynebacteria* also appear to be the principal mediators of adherence
during infection and determine the site of localization. Only heavily fimbriated
cultures of *M. bovis* are able to cause experimental conjunctival infection in calves
and this is attributed to the ability to colonize the epithelial surfaces (Pedersen *et al.*,
1972). Honda and Yanagawa (1975) demonstrated the greater than twofold attachment ability of *C. renale* strains that possessed numerous fimbriae for primary dog
or rabbit kidney and BHK21 cells. This adherence was demonstrated by electron
microscopy to occur by fimbriae and could be inhibited with antifimbriae serum
(Honda and Yanagawa, 1975). This finding of an attachment advantage has been
carried further by *in vivo* experimental models of infection in mice. Shimono and
Yanagawa (1977) could produce pyelonephritis, ureteritis and cystitis significantly
more often by inoculating the urinary bladder of mice with heavily fimbriated
strains of *C. renale* than with less fimbriated ones. Similarly, fimbriae were shown
to mediate *C. renale* attachment to the epithelial mucosa of mouse urinary bladder
in vivo (Honda and Yanagawa, 1978). Attachment could be visualized by scanning
electron microscopy and inhibited by antibody to fimbriae. Presence of numerous
fimbriae conferred a 10- to 30-fold attachment advantage *in vivo* to bladder
epithelium. Table 10.2 points out that the mediator of *V. cholerae* adherence to
rabbit intestinal epithelium is unknown. Nelson *et al.* (1976), however, could not
rule out fimbriae as possible contributors (see Chapter 8).

Many investigators have observed that fimbriated gonococci attach better to
human cells than non-fimbriated organisms (Table 10.2) (see Chapter 9). In
addition to erythrocytes, Punsalang and Sawyer (1973) reported enhancement of
attachment to human buccal mucosal cells for fimbriated gonococci. This epithelial
adherence could be inhibited by antiserum to partially purified fimbriae and
inhibition was removed only by adsorption of the antiserum with fimbriated
gonococci suggesting a role for fimbriae in facilitating attachment to epithelial cells.
Swanson (1973) and Swanson *et al.* (1975) noted that fimbriated gonococci
attached more readily than non-fimbriated organisms to human amnion tissue
culture cells, human foreskin cells and HeLa cells. James-Holmquest *et al.* (1974)
and James *et al.* (1976) demonstrated enhanced attachment of fimbriated
gonococci to human sperm. This increased attachment was blocked by antiserum to

purified fimbriae, suggesting that fimbriae were responsible for the enhanced attachment. Mårdh and Weström (1976) have demonstrated that fimbriated organisms adhered more readily to human vaginal epithelial cells than non-fimbriate gonococci, and subsequently Mårdh and co-workers have noted enhanced attachment to uro-epithelial cells by fimbriated gonocci (P.-A. Mårdh, personal communication). Ward *et al.* (1974) demonstrated the greater adherence of fimbriate gonococci to fallopian tube epithelium and, using scanning and transmission electron microscopy, they also were able to visualize fimbriae anchoring gonococci to the membrane of the fallopian tube epithelial cell layer within three hours after perfusing the tube with gonococci. Thus, it was suggested that fimbriae were involved in the initial adherence of gonococci to human fallopian tube epithelial cells, a natural site of infection (Ward *et al.*, 1974). Tebbutt *et al.* (1976) have confirmed the adherence advantage conferred by fimbriae for gonococci using fallopian tube mucosa, human endocervix and ectocervix. Interestingly, fimbriated gonococci did not have the same adherence advantage for guinea pig epithelial surfaces, or for human bronchial mucosa, perhaps suggesting species and tissue specificity for the presence of the receptor for gonococcal fimbriae. This species specificity was again recently confirmed by Johnson *et al.* (1977) who found rapid attachment of fimbriated gonococci to human fallopian tube mucosa, but little if any attachment to oviducts from the rabbit, pig or cow. Evidence that such fimbriae-mediated adherence functions *in vivo* for gonorrhea has been provided by Tramont (1977). Antibodies to fimbriae, detected by their ability to inhibit attachment of fimbriated gonococci to human buccal cells, were quantitated in human genital secretions following gonorrhea. This antibody was IgG and 11S IgA in immunoglobulin class, was quite specific for the infecting strains and persisted for at least 3–4 weeks.

10.3.3 Characteristics of the binding of isolated fimbriae to host cells

Direct evidence for the adherence role of bacterial fimbriae can best be obtained through the use of isolated, purified fimbriae. This allows sensitive attachment assays to be performed, which can be characterized, quantitated and studied for specificity or examined at a molecular level. Binding site or receptor information can be obtained and, using antisera prepared against purified fimbriae, fimbria antigenic determinants can be studied for proximity to binding moieties or cross-reactivity in a functional assay of inhibiting attachment.

Studies of the binding characteristics of isolated fimbriae for *N. gonorrhoeae* have been conducted. Punsalang and Sawyer (1973) found that a partially purified fimbriae preparation could agglutinate rabbit erythrocytes and bind to buccal epithelial cells in an analagous fashion to fimbriated bacteria. Buchanan and Pearce (1976) used highly purified fimbriae to hemagglutinate human A and 0, Rh^+ cells. This was inhibitable by antibody to purified fimbriae with homologous being better than heterologous antisera at blocking. Pearce and Buchanan (1978) expanded this type of study by utilizing these isolated gonococcal fimbriae that had been

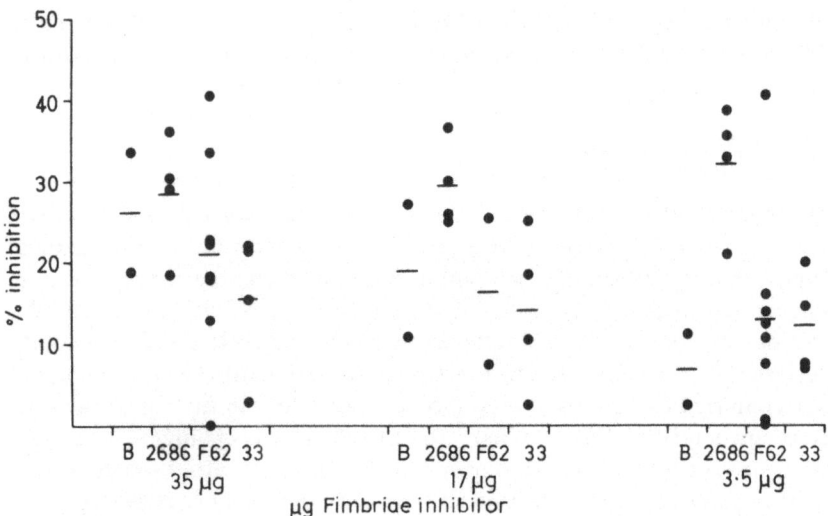

Fig. 10.3 Inhibition of attachment of ^{125}I-labeled *N. gonorrhoeae* fimbriae
from strain B to human buccal mucosal cells at 37°C by unlabeled fimbriae
from 4 antigenically distinct strains: B, 2686, F62 and 33. Labeled fimbriae
and cells were first mixed and incubated at 37°C for 5 min. Unlabeled fimbriae
were then added in 10(3.5 μg), 50(17 μg) and 100(35 μg)-fold excess by weight
relative to labeled fimbriae. % Inhibition was calculated with respect to a
phosphate buffered saline control. ● = individual experiment at either 20 or
40 min incubation at 37°C after addition of inhibitor. — = arithmetic
mean.

radioactively labeled to investigate their attachment to human cells (Table 10.3).
They found that the binding reaction to epithelial cells follows pseudo-first order
kinetics with saturation being achieved within 20–60 minutes at 37°C. Fimbriae of
four antigenically distinct types attached equally to a given cell type implying that
the attachment moiety of each fimbriae was similar. Further antigenically heterologous
fimbriae produce nearly equivalent inhibition of fimbrial attachment in a hapten
blockade (Figs. 10.3 and 10.4) as compared to equal weights of homologous fimbriae.
This suggests that the attachment portion of the gonococcal fimbria that interacts
with the binding site is equivalent for antigenically distinct fimbriae.

Isolated gonococcal fimbriae adherence exhibits a cellular specificity which has
been quantitated to yield the binding site density on various human cells (Table 10.3).
For example, fimbriae attached in a nearly 50-fold greater density, based on cell
surface area, to buccal mucosal epithelium as compared to human erythrocytes. In
general, fimbriae attached in 10- to 40-fold higher amounts to human cervical-
vaginal or buccal mucosal cells, sperm and fallopian tube mucosa (1–10 fimbriae μm^{-2},
10^4 fimbriae binding sites per cervical-vaginal cell) than to human red blood cells

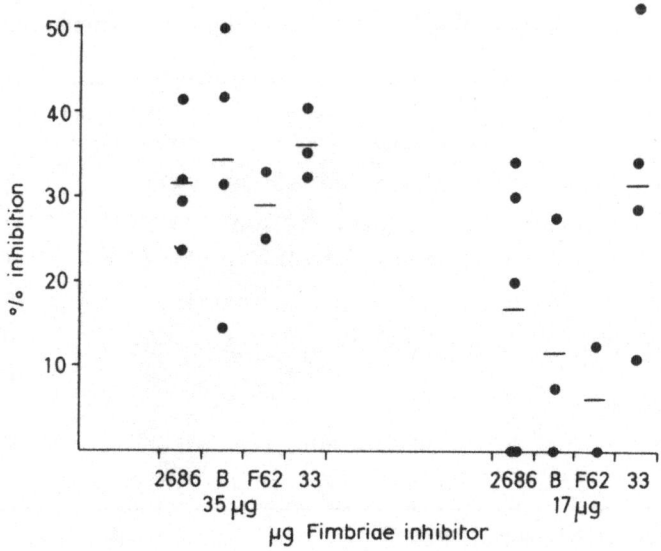

Fig. 10.4 Inhibition of attachment of ^{125}I-labeled fimbriae from *N. gonorrhoeae* strain 2686 to human buccal mucosal cells at 37°C by unlabeled fimbriae. Conditions were as in Fig. 10.3 except that unlabeled fimbriae from the four antigenically distinct strains (2686, B, F62 and 33) were used in 100- and 50-fold excess to inhibit the attachment of ^{125}I-fimbriae from strain 2686. ● = individual experiment at either 20 or 40 minutes incubation at 37°C after addition of inhibitor. — = arithmetic mean.

(0.1 fimbria μm^{-2}) or polymorphonuclear leukocytes (0.3 fimbria μm^{-2}) as summarized in Table 10.3.

These data suggest that more binding sites for fimbriae exist on the surface of cervical-vaginal cells, sperm and buccal cells than on the other cells studied. This observation of greatest 'receptor' density on cells that are histologically the most similar to the actual sites of human gonococcal infection is consistent with the hypothesis that fimbriae-mediated attachment of gonococci to human cells may determine, in part, the sites of eventual gonococcal infection (Pearce and Buchanan, 1978). The use of such binding studies to examine the nature of the receptor(s) for gonococcal fimbriae will be detailed below and in Section 10.4.

Isolated fimbriae attachment to epithelial cells was pH (3.5 < 4.5 > 5.5 > 7.5, Table 10.4) and temperature-dependent (37°C > 20°C > 4°C) (Pearce and Buchanan, 1978). This temperature dependence is thought to reflect membrane fluidity changes which inhibit receptor mobility, rather than an active cell endocytosis of fimbriae. Metabolic inhibitors such as 2, 4-dinitrophenol had no effect on attachment to viable fetal tonsil fibroblasts. Heating gonococci fimbriae for 1 h at 85°C or exposing them to UV irradiation for 2 h could greatly inhibit their attachment abilities (Pearce and Buchanan, 1978).

Table 10.3 Attachment of purified *N. gonorrhoeae* fimbriae to human cells at 37°C, pH 7.4*

Cell tested	No. of fimbriae attaching per cell †	Cell surface area (μm^2)	Fimbriae μm^{-2} †
Buccal mucosal epithelium	25 100 (8700—54 100)	5116	4.91 (10.6)
Cervical-vaginal epithelium	10 100 (6200—13 300)	4056	2.48 (3.27)
Sperm	86 (15—300)	84	1.03 (3.63)
Fallopian tube mucosa	—	—	(3.06—6.11)
Fetal tonsil fibroblasts	824 (500—1400)	1400	0.59 (1.03)
HeLa 'M' cells	189 (70—370)	600	0.32 (0.61)
Polymorphonuclear leukocytes	68 (50—90)	237	0.29 (0.37)
Erythrocytes	16 (6—70)	138	0.11 (0.49)

* After Pearce and Buchanan, 1978.
† Geometric mean values with range or maximum in parentheses.

Buchanan *et al.* (1978b) further investigated the characteristics of gonococcal fimbriae binding to human cells to probe the chemical nature of the reaction. As mentioned, at pH 4.5, the average pH of vaginal secretions, attachment was 3—4 fold greater than at pH 7.4. This optimum was sharp in either direction. Also, 0.1 mM ferric ion, pH 7.4, enhanced fimbriae attachment approximately twofold. Enhanced attachment in the presence of cations occurred at pH 7.4 with 0.1 mM Fe^{3+} > 0.1 mM Fe^{2+} > 1 mM Ca^{2+} or 1 mM Mg^{2+} (Pearce and Buchanan, 1978; Buchanan *et al.,* 1978b) as detailed in Table 10.4 below. James *et al.* (1976) had previously reported that 0.01—0.1 mM Fe^{3+} also enhanced the attachment of fimbriated gonococci to human sperm. However, this enhancement by ferric ion did not occur at pH 4.5 (Buchanan *et al.,* 1978b). Further, photooxidation of fimbriae in the presence of methylene blue to modify histidine residues prior to attachment blocked ferric ion enhancement at pH 7.4 (Table 10.4). This suggests that the enhanced attachment of fimbriae is mediated by ferric ions bound to fimbrial histidines when they are unprotonated at pH 7.4, in an analagous fashion to ferric ion binding in hemoglobin. This potential ion—histidine ligand formed between fimbriae and the cell surface, or between fimbriae themselves, appears specific since ferrous ions or 10-fold higher concentrations of Ca^{2+} or Mg^{2+} ions produced little enhancement of fimbriae attachment (Table 10.4). A similar charge bridge can be postulated to explain the 2.5-fold higher level of attachment in the presence of protamine sulfate, a highly positively charged molecule, at pH 7.4 (Table 10.4). Such a bridge could exist between negatively charged carboxyl groups of sialic acid and proteins as well as OSO_3^- groups on the cell surface (pKa of ~ 2.0, ~ 4.5 and < 1 respectively) to ionized carboxyl groups on fimbrial amino acid residues such as aspartic and glutamic acid.

Table 10.4 Enhancement of attachment of *N. gonorrhoeae* fimbriae to human buccal cells by ions or pH*

Condition	Enhancement †
Fe^{3+}	
0.01 mM	1.2
0.1 mM	2.5
0.1 mM‡	N.E. §
Ca^{2+}, 1.0 mM	1.2
Mg^{2+}, 1.0 mM	N.E.
Protamine sulphate, 1 mg ml^{-1}	2.5
pH 5.5	2
pH 4.5	3

* After Buchanan *et al.*, 1978b.
† Factor of enhancement as compared to attachment at pH 7.4 without added ions.
‡ Fimbriae were treated with 0.15 M methylene blue and 10 min of photo-oxidation prior to attachment experiment.
§ N.E. = no enhancement.

Various modifications of whole fimbriae or specific fimbria-reactive groups were performed to study the effects on subsequent fimbriae attachment (Buchanan *et al.*, 1978b; Table 10.5). Fimbria dissociated into subunits and denatured by boiling in 1% sodium dodecyl sulfate (SDS) were not able to adhere significantly to human buccal cells, suggesting that a specific fimbria conformation, possible requiring more than one subunit, is necessary for fimbriae to combine with receptors. However, fimbriae denatured with 6 M guanidine-hydrochloride are dissociated into subunits which retain all adherence capabilities. If this denaturation is less extensive than that produced with SDS, each subunit contains an entire attachment moiety. Alternatively, guanidine-HCl may expose normally internal hydrophobic regions of the fimbria subunit, which can then participate in non-specific, hydrophobic binding to the cell surface. Other inhibitor treatments were heating fimbriae at 85°C for 1−3 hours (denaturation) and modification of tryptophan residues (Table 10.5), suggesting a tryptophan is located near or in the fimbrial attachment moiety or is necessary for maintenance of the proper conformation of this moiety. Modification of histidine, lysine and tyrosine residues or alkylation of disulfide bonds had no effect on attachment (Buchanan *et al.*, 1978b). Buchanan and co-workers also found significant inhibition with antibody to fimbriae or buccal cells, by treating the epithelial cells with glycosidic enzymes, pronase, gonococcal enzymes or by hapten blockades with various carbohydrate-containing compounds as detailed in Section 10.3.4.

The gonococcus is not the only bacterium whose fimbriae have been isolated, purified and investigated for attachment capabilities. Brinton (1959) demonstrated

Table 10.5 Inhibition of attachment of *N. gonorrhoeae* fimbriae to human buccal cells by chemical or physical modification of the fimbriae*

Conditions of modification	% Inhibition †
Boiled in 1% sodium dodecyl sulfate, subunits isolated by preparative SDS-polyacrylamide gel electrophoresis, dialysis and chromatography on Sephadex G-50	90
Dissolved in 70% formic acid for 16 h at room temperature, dialyzed, treated with 6 M guanidine-HCl, 1 mM EdTA at pH 8.6 for 3 h at 37°C and then dialyzed	–
Heated at 85°C for 1 to 3 hours	56
Ultraviolet irradiation for 2 hours, 2 cm from light source (modifies tryptophan residues)	60
Exposed to 0.01 M Koshland's reagent (2-hydroxy-5-nitrobenzyl bromide) for 18 h at 25°C, unreacted Koshland's reagent removed on Sephadex G-50	80

* After Buchanan *et al.*, 1978b.

† Inhibition as compared to attachment of unmodified fimbriae at pH 7.4

that cell-free supernatants of fimbriated *E. coli* or partially purified fimbriae could agglutinate chicken erythrocytes. This hemagglutination activity of *E. coli* type 1 fimbriae has been conclusively demonstrated by Salit and Gotschlich (1977a) since purified type 1 *E. coli* K12 fimbriae alone were able to produce hemagglutination of guinea pig cells. This could be inhibited by antisera to isolated fimbriae. This was also the case for the attachment of these fimbriae to monkey kidney cells (Salit and Gotschlich, 1977b) or the attachment of intact fimbriated organisms. Adherence of isolated fimbriae to these cells occurred at a slightly slower rate than in the gonococcal experiments (Pearce and Buchanan, 1978). Also, Salit and Gotschlich (1977b) observed a pH optimum for isolated fimbriae attachment of pH 4–5, which they attribute to fimbriae having the lowest net charge at this pH allowing the most facile approach to a charged cell surface. This may also reflect fimbriae–fimbriae aggregation which they observed at this pH. Like gonococcal fimbriae binding, metabolic inhibitors had no effect on isolated *E. coli* type 1 fimbriae binding in this study, but attachment was not temperature-sensitive. Cell-free extracts or isolated 'colonization factor' fimbriae of enterotoxigenic *E. coli* have also been shown to agglutinate guinea pig (K88: Stirm *et al.*, 1967b; Jones and Rutter, 1974) and sheep (K99: Burrows *et al.*, 1976) erythrocytes and adhere to pig small intestine (Jones and Rutter, 1972).

A galactose-specific hemagglutinin of a pyocyanin-producing strain of *Pseudomonas aeruginosa* has been purified and shown to possess an agglutinating capacity for human erythrocytes equivalent to intact *Pseudomonas* bacteria

(Gilboa-Garber, 1972a; Gilboa-Garber *et al.*, 1972; Gilboa-Garber *et al.*, 1977). This hemagglutinin was also found to agglutinate human leukocytes and thrombocytes (Gilboa-Garber, 1972b). A second, mannose-specific hemagglutinin has been recently purified from the same *P. aeruginosa* strain grown in the presence of acetylcholine or choline (Gilboa-Garber *et al.*, 1977). This isolated hemagglutinin also displays activity for human erythrocytes. Both hemagglutinins are glycoproteins with estimated subunit molecular weights of 13 000–13 700 (Gilboa-Garber, 1972a) and 11 000 (Gilboa–Garber *et al.*, 1977) respectively. These sizes are conceivably in the correct range for fimbrial subunits (Section 10.4.2) and the agglutinins seem to be located on the cell surface (N. Gilboa-Garber, personal communication). However, the morphological characteristics of these molecules have not been determined, nor has an attachment role for them in the pathogenesis of infection been confirmed. Although cultural conditions that are maximal for hemagglutinin production (shaking for 72 h) do not coincide with optimal fimbriae expression for this strain, the possibility exists that these lectins could be fimbrial precursors (N. Gilboa-Garber, personal communication).

10.3.4 Chemical nature of receptors (binding sites) for bacterial fimbriae

Since the first observation that the hemagglutination produced by *E. coli* strains, which was later shown to be fimbria-mediated, could be inhibited by D-mannose (Collier and De Miranda, 1955) there has been an ever increasing interest in chemically specific inhibitors of adherence reactions mediated by fimbriae. The reasons for this are, first, as a method for distinguishing and classifying types of fimbriae in hemag-glutination or cell adherence assays. Second, such hapten blockades can be used to investigate the chemical nature of the binding reaction. The implicit hypothesis exists here that fimbriae mediate adherence by binding chemically specific and distinct receptors exposed on cell surface membranes. Methods for elucidating and distinguishing the lectin-like nature of fimbriae binding have primarily involved inhibition of the adherence of fimbriated bacteria or the isolated fimbriae themselves with various chemical compounds in a hapten blockade. These types of experiments will be considered in addition to other physical characteristics of fimbrial binding which have been derived from adherence tests.

Monosaccharides and oligosaccharides containing α-linked D-mannose inhibit from some to most fimbriae-mediated hemagglutination of *E. coli* (Collier and DeMiranda, 1955; Brinton, 1965; Duguid *et al.*, 1966; Salit and Gotschlich, 1977a), *Shigella* (Duguid and Gillies, 1957; Old, 1972), *Salmonella* (Duguid *et al.*, 1966; Old, 1972), *Klebsiella* and *Serratia* (Duguid, 1959; Cowan *et al.*, 1960) suggesting that mannose-containing oligosaccharides are found in the cell surface receptors for these fimbriae (Table 10.2). These substances do not inhibit one hemagglutinin produced by *Pseudomonas aeruginosa* (Gilboa-Garber, 1972a) described above as the galactose-specific hemagglutinin. Specific inhibition of this adhesive reaction is obtained with α-D-galactosides > D-galactose > lactose > *N*-acetyl-D-galactosamine

(Gilboa-Garber, 1972a). This is distinct from the mannose-specific hemagglutinin of the same *P. aeruginosa* strain. Inhibition by D-mannose far exceeded that produced by α-methyl-D-mannoside (Gilboa-Garber *et al.*, 1977). The size and functional similarities of these two agglutinins to fimbriae warrants further examination, but it can be speculated they are a type of fimbriae. The hemagglutination produced by *Vibrio* species is inhibited by L-fucose (Jones and Freter, 1976) more than by D-mannose (Tweedy *et al.*, 1968). The attachment of *V. cholerae* to brush border epithelial cells is best inhibited by L-fucosides followed by L-fucose, which are both at least 1000-fold more inhibitory than D-mannose (Jones and Freter, 1976), but, again, the specific attachment organelles, their receptors, and their identity or lack of it with fimbriae require further clarification (Nelson *et al.*, 1976; Chapter 8). The colonization factors produced by many human-associated enteropathogenic *E. coli*, which are morphologically indistinguishable from common fimbriae (Evans *et al.*, 1977; 1978; Evans and Evans, 1978), can mediate the hemagglutination of human and/or bovine and chicken cells, which is not blocked by D-mannose (Evans *et al.*, 1977, 1979; Orskov and Orskov, 1977; Evans and Evans, 1978). Similarly, K88 and K99 antigen-mediated hemagglutinations are mannose-resistant (Jones and Rutter, 1974; Orskov *et al.*, 1975; Burrows *et al.*, 1976). This suggests that the receptor(s) for these virulence-conferring 'fimbriae' that facilitate host-specific intestinal attachment and colonization is different from the receptors for common fimbriae. *E. coli* isolates from human extra-intestinal infections also possess a mannose-resistant hemagglutinin for human cells (Minshew *et al.*, 1978a,b). However, Svanborg–Edén and Hansson (1978) found evidence that the fimbria-mediated agglutination of guinea pig erythrocytes by *E. coli* isolates from infected urines was mannose-sensitive.

The hemagglutination mediated by *C. renale* fimbriae is limited to trypsinized sheep erythrocytes (Honda and Yanagawa, 1974). It is not inhibited by D-mannose, galactose, glucose or other hexoses or hexosamines, but periodate treatment of trypsinized red blood cells removes their agglutinability (Honda and Yanagawa, 1974). Thus, a carbohydrate-containing receptor is implicated.

The inhibition of fimbria-mediated hemagglutination by D-mannose and related compounds was investigated in some depth by Old (1972). He demonstrated the importance of the α configuration at the C-1 position, and of the hydroxyl positions at C-2, C-3, C-4 and C-6 for maximal inhibition of hemagglutination by type 1-fimbriated *Salmonella* or *Shigella*. The strongest inhibitors in this study were D-mannose, α-methyl-D-mannoside, 1,5-anhydromannitol, D-mannoheptulose and α-D-mannose-1-phosphate. Two other compounds, D-fructose and yeast mannan were moderately inhibitory at slightly higher concentrations (0.05–0.2% w/v). These results were comparable to those found by Salit and Gotschlich (1977a) for isolated *E. coli* common fimbriae, where inhibition of hemagglutination by α-methyl-D-mannoside > D-mannose > yeast mannan > D-fructose.

These same mannose-sensitive fimbriae can also be inhibited in attachment experiments using host tissues (Table 10.2). For example, mannose-sensitive fimbriae facilitated adherence of *Shigella flexneri* to human colonic epithelium (Duguid and Gillies, 1957) and of *E. coli* K12 organisms to monkey kidney cells in tissue culture (Salit and Gotschlich, 1977b). In the latter study, maximal inhibition was obtained with 0.5 μM α-methyl-D-mannoside which was 20 times and 100 times more potent than yeast mannan and D-mannose, respectively. Inhibitory saccharides were able to elute fimbriate bacteria from cells. That these *E. coli* fimbriae were binding specific mannose-containing receptors on the Vero cell surface was suggested by the inhibition of attachment by specific lectins. PHA had little effect, but Con A (20 μg ml^{-1}) or *Lens culinaris* (60 μg ml^{-1}) could block 90% and 80% of whole bacteria attachment respectively (Salit and Gotschlich, 1977b). Binding of the latter two lectins to Vero cells could be reversed by α-methyl-D-mannoside. Ofek *et al.* (1977) and Aronson *et al.* (1979) have performed similar experiments demonstrating the involvement of receptors containing D-mannose on human buccal mucosal cells for fimbriated *E. coli* adherence which was inhibited or reversed by α-methyl-D-mannoside. Other examples are outlined in Table 10.2.

In contrast, attachment to intestinal epithelium by enteropathogenic *E. coli* by colonization factor fimbriae was unaffected by 0.5% D-mannose (McNeish *et al.*, 1975). *E. coli* isolates not implicated in gastroenteritis did not adhere in the presence of D-mannose, possibly suggesting that cell surface receptors not containing D-mannose and the corresponding 'colonization factor' fimbrial attachment moiety determine tissue tropism for the site of infection with these enteropathogenic strains. Fimbrial adherence by *E. coli* strains causing urinary tract infections may also be specific for cell surface receptors not comprised of D-mannose. Svanborg-Edén and Hansson (1978) were unable to inhibit the adherence cells to human uroepithelial of *E. coli* isolates from urinary tract infections, which correlated with the presence of fimbriae, with D-mannose, α-methyl-D-mannoside, D-galactose or 6-deoxy-L-galactose. Nevertheless, D-mannose could inhibit the attachment of these and similar strains to buccal epithelium (Aronson *et al.*, 1979) and the agglutination of guinea pig erythrocytes (Svanborg-Edén and Hansson, 1978). This was not the case for human erythrocytes, however (Minshew *et al.*, 1978a, b). More recently, Aronson *et al.* (1979) were able to significantly block experimental ascending urinary tract infections in the mouse by including α-methyl-D-mannoside with *E. coli* isolates from human pyelonephritis injected into the bladder. α-Methyl-D-glucopyranoside was ineffective. Further specificity was demonstrated by the inactivity of either sugar against *Proteus mirabilis* experimental infections and adherence to buccal epithelium *in vitro*. Although the nature of the mediator or mediators participating in these adherence and colonization assays of urinary tract pathogens remains to be elucidated, several hypotheses can be made. First, attachment moieties carried by fimbriae most likely participate in the adherence of these *E. coli* isolates from urinary tract infections to human tissues and appear to be directed at receptors at the site of infection not containing D-mannose. Second, distinct mediators of adherence and receptors involving D-mannose-containing heterosaccharides may function in *E. coli* adherence to human tissues such as buccal epithelium, or, mouse

urinary tract epithelium and guinea pig erythrocytes. Good evidence exists to suggest
that these adherence moieties are carried by *E. coli* fimbriae (Salit and Gotschlich,
1977a,b). Recently Silverblatt *et al.* (1979) have shown that mannose-sensitive
fimbriae carried by *E. coli* isolates from blood and urine can interact with human
polymorphonuclear leukocytes to promote phagocytosis. They expect that this
aspects of fimbrial 'function' would decrease virulence, especially after bacteria have
penetrated, for example, renal parenchyma. Third, the potential exists for both organ-
and species-specific determinants of infection to be carried by fimbrial structures. An
understanding of these molecular mechanisms of micro-organism adherence to human
cells may suggest ways to interrupt attachment and pathogenesis.

Fimbria-mediated attachment of *N. gonorrhoeae* is mannose-resistant. Punsalang
and Sawyer (1973) found that hemagglutination due to gonococcal fimbriae was not
blocked by the simple sugars D-mannose, D-glucose, maltose, D-mannitol, dulcitol,
D-sorbitol, raffinose, saccharose, lactose, D-fructose, D-galactose and inulin. Using
isolated gonococcal fimbriae of sufficient purity to be sequenceable, Buchanan
et al. (1978b) have further studied the chemical nature of the binding reaction to
buccal mucosal epithelium to elucidate the identity of the receptor. Many compounds
capable of inhibiting fimbriae attachment were found. These include gangliosides,
heparin, fetuin, normal human serum and synovial fluid. It was anticipated that a
receptor for gonococcal fimbriae would be closely related or identical to compounds
frequently found on human cell surfaces. These include glycoproteins, glycosphingolipids and carbohydrate moieties within these complex molecules such as sialic acid,
galactose, fucose, mannose, glucose, *N*-acetyl-glucosamine and *N*-acetyl-glactosamine.
When tested, none of these monosaccharides alone could inhibit fimbriae attachment
(Buchanan *et al.*, 1978b). This suggests that if carbohydrate moieties compose part
of the fimbria receptor, oligosaccharide units might be involved, and that distinct
linkages within such an oligosaccharide might be necessary to serve as a fimbria receptor.

Treatment of buccal mucosal cells with a mixture of protease-free exoglycosidases
has been one approach to identify potential receptor structure. Marked inhibition of
attachment (30–60%) of fimbriae at pH 7.0 and 5.5 was obtained supporting the
concept of an oligosaccharide component on the receptor (Buchanan *et al.*, 1978b).
Inhibition of attachment at pH 4.5 was absent if sialidase was not present in the
mixture. Sialic acid may therefore be a component of the receptors at pH 4.5 or mask it, or,
alternatively, simply may be involved in non-specific, electrostatic binding (see
below) which supplements the specific receptor-directed adherence present at pH 7.4.

Since patients with disseminated gonococcal infection frequently have localization of infection within joints, we examined synovial fluid for the presence of a
fimbrial receptor. All synovial fluids examined were inhibitory in hapten blockade
experiments (45–60%, Table 10.6), but, in contrast, had no effect on the attachment
of purified *E. coli* type 1 fimbriae to human cells (Buchanan *et al.*, 1978b). Components of synovial fluid were then tested for inhibition. Hyaluronic acid, a mucopolysac-
charide, was not inhibitory. Of other mucopolysaccharides tested, only heparin
was inhibitory (36–43%, Table 10.6). Further, a heat stable, periodic acid- and
protease-sensitive component of human serum could also inhibit (26%) suggesting

Table 10.6 Inhibition of attachment of *N. gonorrhoeae* and *E. coli* fimbriae to human buccal cells at pH 7.4*

Compound	Concentration	% Inhibition†	
		N. gonorrhoeae	*E. coli*
Synovial fluid	14%	45−60	0
Heparin	1 mg ml⁻¹	36−43	NT ‡
Human serum pool			
Unmodified	14%	55	NT
Boiled 5 min	14%	26	NT
Treated with periodic acid, 0.1 M, 18 h, 4°C	14%	0	NT
Treated with protease, 1 mg ml⁻¹ x 30 min, then boiled 5 min	12%	0	NT
GM_1	1.4 μM	33	0−4
α-Methyl-D-mannoside	10 μM	0−1	16
Yeast mannan	860 μg ml⁻¹	19	27
Chitin oligosaccharides	1 mg ml⁻¹	21	7

*   After Buchanan *et al.*, 1978b.
†   Inhibition as compared to attachment without the addition of inhibitors.
‡   NT = not tested.

an oligosaccharide portion of a glycoprotein containing the components competing with the fimbrial receptor may be involved (Buchanan *et al.*, 1978b; Table 10.6). This confirmed the previous report by Punsalang and Sawyer (1973) that they could not inhibit attachment of fimbriated gonococci at pH 7.4 to erythrocytes after treating the cells with 1 mg ml⁻¹ trypsin or neuraminidase.

Gangliosides are sialic acid-containing glycosphingolipids (Table 10.7) known to be present on cell surfaces that can serve, for example, as the receptor for cholera toxin (GM_1, Lönnroth and Holmgren, 1975), tetanus toxin (GD_{1b} or GT_1, van Heyningen, 1974) or thyrotropin (GD_{1b}, Mullin *et al.*, 1976). Both pooled and purified gangliosides produced marked inhibition of gonococcal fimbrial attachment. At less than 1 μM concentration, GD_{1b} was most inhibitory, followed by GD_{1a} > GM_1 > GT (Tables 10.7 and 10.8). The fact that such low concentrations of gangliosides were required for inhibition of attachment suggests that the human cell surface receptor for gonococcal fimbriae is likely to resemble a ganglioside in structure. In contrast (Table 10.6), these same gangliosides had essentially no inhibitory effects on the attachment of *E. coli* type 1 fimbriae to buccal cells. However, α-methyl-D-mannoside, as reported by others (Ofek *et al.*, 1977; Salit and Gotschlich, 1977a, b), and yeast mannan were more inhibitory for *E. coli* fimbriae. Chitin oligosaccharides and synovial fluid had more effect on gonococcal fimbriae

Table 10.7 Ganglioside structures

```
          2                    3                      2
  Gal β1  → 3 GalNAc     β1 → Gal          β1 → 4 Glc → Ceramide
       3                    3
  1  ↑ a                   ↑ b
       2                    2             1
     NANA                 NANA 8 ← 2 NANA
                                  c
```

GM_1 contains only sialic acid (NANA) b.
GD_{1a} contains sialic acids a and b.
GD_{1b} contains sialic acids b and c.
GT contains sialic acids, a, b and c.

1 = Site of cleavage for sialidase.
2 = Site of cleavage for β-galactosidase.
3 = Site of cleavage for β-NAc-hexosaminidase.

attachment (Table 10.6). These differences in receptor specificity will be considered
further in Section 10.4 where the structural differences of these fimbria molecules
is discussed.

We have compared the structures of gangliosides (Table 10.7) to the inhibitory
effects on gonococcal fimbriae attachment produced by different gangliosides
(Table 10.8) and then used selective glycosidic enzymes to cleave cell surface
heterosaccharides to block fimbriae attachment as summarized in Table 10.9. The
enzymes used were highly selective for the sugar moiety and linkage indicated, and
contained less than 1% contamination with other glycosidic enzymes and no
proteolytic activity. The results indicate that sialic acid may constitute part of the
fimbria receptor at pH 4.5, but not at pH 7.4. This may be due to charge attraction,
since at pH 4.5 the net charge on fimbriae becomes positive (pI = 5.3, Robertson
et al., 1977) due to protonated histidines and a reduction in the negative charge
contributed by carboxyl groups of amino acid residues. However, at pH 4.5 the cell
surface charge would be expected to remain negative due to the low pKa of the
carboxyl groups of sialic acid and of the OSO_3^- groups of glycolipids. The effect of
β-galactosidase indicates that terminal galactose residues contribute to the fimbria
receptor at both pH 4.5 and 7.4 (Table 10.9). The ganglioside GD_{1b} produced more
inhibition of fimbriae attachment than the other gangliosides. GD_{1b} contains no
sialic acid at position 'a' in the structure above (Table 10.7), leaving a terminal,
β1 → 3-linked galactose, as it is on GM_1. The enzyme β-hexosaminidase produced no
effect by itself, which is not surprising since its substrate is not terminally located
on the molecule we have proposed, but, in combination with β-galactosidase, or with
β-galactosidase plus sialidase, it produced an additive effect. This suggests that, at
pH 4.5, the fimbria receptor may involve NANA 2 → 3 Gal/NAc β1 → 4 Gal, whereas, at
pH 7.4, the receptor may contain Gal β1 → 3 GalNAc β1 → 4 Gal (Tables 10.7 and 10.9).

Table 10.8 Inhibition of attachment of *N. gonorrhoeae* fimbriae to human buccal cells by gangliosides*

Ganglioside	Concentration (μM)	% Inhibition †
GM_1	160	38
	16	32
	1.4	30
	0.6	23
	0.3	22
GD_{1a}	74	63
	15	40
	1.4	31
	0.6	32
	0.3	20
GD_{1b}	74	55
	15	38
	1.4	30
	0.6	35
	0.3	33
GT	230	55
	23	36
	1.3	29
	0.6	18
	0.3	14

* After Buchanan *et al.*, 1978b.
† Inhibition as compared to attachment at pH 7.4 without added ganglioside.

Endo-β-NAc-glucosaminidase D or H had no inhibitory effect. Treatment of cells with glucosidase, fucosidase, xylosidase and mannosidase in addition to sialidase, β-galactosidase and β-NAc-hexosaminidase did not produce additional inhibition of fimbriae attachment. This is consistent with their respective substrates not comprising the fimbria receptor site (Table 10.9).

Studies of the receptors for non-gonococcal fimbriae have been initiated along similar lines. Salit and Gotschlich (1977a) found that the normally weak hemagglutination of human cells by isolated type 1 fimbriae of *E. coli* could be enhanced by proteolytic treatment of human erythrocytes beforehand. The effects of protease or trypsin pretreatment of guinea pig erythrocytes, which are normally readily agglutinated by these isolated fimbriae, were negligible. Individual glycosidases, or mixtures, did not alter hemagglutination except for α-mannosidase. This caused a twofold reduction in agglutination titers for isolated fimbriae, Con A or *Lens culinaris*. Neuraminidase caused an increase in red cell agglutinability and, interestingly, up to 50 mM concentrations of EDTA were ineffective on

Bacterial Adherence

Table 10.9 Effect of glycosidases to inhibit *N. gonorrhoeae* fimbriae attachment to human buccal cells

Enzymes*	% Inhibition †	
	pH 4.5	pH 7.4
Sialidase(s)	34	$\leqslant 5$
β-galactosidase	11	23
β-NAc-hexosaminidase	$\leqslant 5$	$\leqslant 5$
Sialidase + β-galactosidase	36	25
β-galactosidase + β-NAc-hexosaminidase	15	31
Sialidase + β-galactosidase + β-NAc-hexosaminidase	64	38
Endo-β-NAc-glucosaminidase D or H ‡	0	0
Sialidase + glycosidase mixture § (1 mg ml^{-1})	65	38

* Enzymes (0.01 units from Seikagaku Kogyo Co. via Miles Laboratories) reacted with buccal cells for 20 min at 37°C at the pH optimum of the enzymes used. Cells then were washed by centrifugation and resuspended in buffer for the attachment experiment conducted at pH 7.4 or 4.5. For method see Pearce and Buchanan (1978). Control cells treated identically except for omission of enzymes.

† Relative to the amount of fimbria attachment to cells treated identically except for the omission of glycosidic enzymes.

‡ Each of these enzymes cleaves glycopeptides containing high amounts of mannose. These enzymes require mannose residues for enzyme action and cleave glyco-proteins with the general structure: $(GlcNAc)_n (Man)_m$ GlcNAc - - ↑ - - GlcNAc Asn-Peptide at the site of the arrow. Terminal GlcNAc residues are not cleaved.

§ The glycosidase mixture contained significant activity for each of the following enzymes: α- and β-NAc mannosidase, α- and β-glucosidase, α- and β-galactosidase, α-L-fucosidase, β-xylosidase, α- and β-NAc-glucosaminidase, α- and β-NAc-galactosaminidase.

hemagglutination, even if EDTA was not removed (Salit and Gotschlich, 1977a). These investigators also found that neuraminidase increased the binding of isolated fimbriae to monkey kidney cells (25% enhancement), but pretreatment of cells with other isolated glycosidases, including α-mannosidase, had no effect (Salit and Gotschlich, 1977b). Trypsin and protease could slightly enhance binding. They attributed the increase of neuraminidase to a 'stripping' effect, the uncovering of cryptic binding sites, or to a reduction of net negative charge of the cell surface or a change of membrane protein mobility. Perhaps the α-mannosidase had no effect because mannose residues functioning as receptors may be subterminal or shielded from cleavage in these cells.

10.3.5 Non-specific fimbrial binding

Brinton (1965) observed that *E. coli* type 1 fimbriae could attach by their ends to polystyrene beads or any hydrophobic surface. Buchanan (1976) noted the same exclusively endwise adherence of isolated gonococcal fimbriae to these beads. Watt and Ward (1977) and Robertson *et al.* (1977) have proposed that gonococcal fimbriae may function in an analogously non-specific manner to effect adherence to cell surfaces. Fimbriae would bridge the barrier of electrostatic repulsion between bacteria and the cell surface, which normally both have a negatively charged character. The p*I* of gonococcal fimbriae or a fimbriated organism is 5.3, whereas non-fimbriate gonococci have a p*I* of 5.6 (Robertson *et al.*, 1977). Therefore the fimbriae dominate surface charge and are more negatively charged, but in themselves are no more electrostatically attractive. The fimbrial tip is proposed to be the point of initial contact and to penetrate an electrostatic barrier by means of its small surface area and radius of curvature, perhaps aided by a specialized structure rich in uncharged hydrophobic residues and therefore lower in charge density. These combine to reduce the electrostatic repulsion of fimbriae compared to the whole bacterial surface facilitating attachment (Brinton, 1965; Robertson *et al.*, 1977; Ward and Watt, 1977). A hydrophobic region on the fimbria tip could also become embedded in the lipid interior of host cell membranes for fimbria attachment (Watt and Ward, 1977). Such a hydrophobic portion on *Neisseria, Moraxella* and *Pseudomonas* fimbriae has been found (see Section 10.4.2). This proposed model does not exclude the existence of a chemically specific receptor for fimbriae, however. Alignment and interaction of the fimbrial attachment moiety with its receptor would be facilitated by these non-specific interactions, which might be thought of as early events. One interesting possibility is that the endwise attachment of fimbriae is the only possibility for strictly hydrophobic, non-specific binding reactions to polystyrene beads, for example. But attachment moieties existing along the length of a fimbria molecule may, by their specific receptor binding, allow fimbria adherence by more than just the tip. Salit and Gotschlich (1977b) have published electron micrographs which supports this possibility for type 1 fimbriae attachment to monkey kidney cells.

10.4 STRUCTURE OF BACTERIAL FIMBRIAE IN RELATION TO ADHERENCE

10.4.1 Antigenic structure

The earliest demonstrations of the antigenicity of fimbriae were the experiments designed to verify the role of fimbriae in adherence by inhibiting this function with

fimbrial antiserum (Section 10.3). These and subsequent studies have shown fimbriae
to possess immunogenicity, even during natural infections. For example, humans
with gonorrhea develop serum IgG (Buchanan *et al.*, 1973), as well as IgG and IgA
antibody in mucosal secretions (Tramont, 1977) to gonococcal fimbriae. Fimbriae
possess greater than one antigenic determinant per molecule, some of which are
shared between different serogroups, or even species and genera, while others are
strictly serotype-specific.

Gillies and Duguid (1958) utilized the previously demonstrated antigenicity of
type 1 fimbriae of enterobacteria (e.g. fimbrial antisera inhibit hemagglutination or
mediate bacterial agglutination) to study fimbrial antigens of *Shigella flexneri* and
their relationship to *E. coli* and other gram-negative fimbriae. The antigenic composition of all *S. flexneri* strains and serotypes were identical, containing a major
flexneri-specific antigen. *E. coli* fimbriae were found to also contain a major antigen,
but it was type-specific, being shared only within groups of related *E. coli* strains. In
addition, one or more minor *flexneri–coli* shared antigens were present in both
genera. *S. flexneri* fimbriae possessed no shared antigenicity with fimbriate cultures
of *Salmonella* species, *Enterobacter cloacae* or *Proteus* species (Gillies and Duguid,
1958). In a similar study Duguid and Campbell (1969) examined *Salmonella*
fimbriae, all of which carried a common antigen. This antigen was also present in
Arizona and *Citrobacter* fimbriae but not shared with other genera of enterobacteria
(i.e. *S. flexneri*, *E. coli*, *K. aerogenes*, *E. cloacae*). *Klebsiella* fimbriae shared antigens
with other *Klebsiella* strains tested, *S. flexneri* and *E. coli*, but not with *Salmonella*,
Arizona or *Citrobacter* (Duguid and Campbell, 1969).

Enterotoxigenic *E. coli* 'colonization factor', K88 and K99 fimbriae are antigenically distinct and distinct from common fimbriae (Stirm *et al.*, 1976b; Jones and
Rutter, 1974; Evans *et al.*, 1975, 1977, 1978; Burrows *et al.*, 1976; Ørskov and
Ørskov, 1977; Isaacson, 1977). Antigenic heterogeneity correlates with the species
specificity possessed by the strains' attachment and virulence potential mediated by
these antigens (Jones and Rutter, 1972; Wilson and Hohmann, 1974; McNeish *et al.*,
1975). Further, more than one 'colonization factor' antigen (CFA) and K88 antigen
have been found. CFA/I and CFA/II are not cross-reactive and do not occur on the
same strain (Evans and Evans, 1978). One or the other is found on nearly all human
enterotoxigenic *E. coli* isolates and these CFA$^+$ strains fall into one of two groups
possessing a limited number of 0-serotypes (Ørskov *et al.*, 1976; Evans *et al.*, 1977,
1978; Evans and Evans, 1978). The three known antigenic types of K88 antigen
(K88ab, K88ac, K88ad) are only partially cross-reactive (Wilson and Hohmann, 1974;
Guinée and Jansen, 1979). Antibodies to K88ab or K88ac can only neutralize
adherence mediated by the homologous antigen (Wilson and Hohmann, 1974).

As described in Section 10.3, antibody specific for gonococcal fimbriae can
inhibit hemagglutination by fimbriated bacteria or isolated fimbriae (Punsalang and
Sawyer, 1973; Buchanan and Pearce, 1976), inhibit the attachment of fimbriated
bacteria to sperm (James-Holmquest *et al.*, 1974) and buccal cells (Tramont, 1976),
and inhibit the attachment of isolated fimbriae to human epithelial cells (Pearce and

Buchanan, 1978). Antigenic relationships for fimbriae from different gonococcal strains have been studied using antiserum to purified and partially purified fimbriae, or whole gonococci. These investigations are summarized in Table 10.10. Each of the assays utilized indicated that gonococcal fimbriae are extremely antigenically heterogeneous. Probably more than 50 antigenic types exist (Brinton *et al.*, 1978), but some common antigens are shared by different fimbriae. Depending upon the assay system and the strains tested, estimates of the extent of shared antigenicity among fimbriae range from 1–10%. This extensive antigenic diversity among fimbriae is important with respect to vaccines utilizing fimbrial antigens. Protection induced by fimbrial antibodies blocking attachment would most likely result from shared fimbrial antigenic determinants in or near the attachment moiety. It is now clear that gonococcal fimbriae also possess antigenic determinants which stimulate and bind opsonic antibodies (Buchanan *et al.*, 1878a; Jones, R.B., Newland, J.C., Olsen, D.A. and Buchanan, T.M., unpublished data). The amount of cross-reactivity existing among these antigenic determinants, and therefore the effective range of immune-enhanced phagocytosis, may equally govern the level of protection afforded by fimbrial antibodies.

In one study (Buchanan and Pearce, 1976), fimbriae from each of four antigenically distinct gonococcal strains were examined using antisera specific for fimbriae to agglutinate bacteria, in Ouchterlony immunodiffusion and radioimmunoassay and to inhibit hemagglutination. Although some shared antigenicity was found by these tests (Table 10.10), the best inhibition of hemagglutination, for example, was by antisera to the homologous fimbrial type. In one case, a heterologous fimbrial antisera actually enhanced hemagglutination, suggesting a different location for the shared antigenic determinants interacting with antibodies inhibiting attachment on the two different fimbriae. In the homologous fimbriae, this would be at or near the attachment moiety, while in the heterologous, it may be far enough removed from the attachment moiety so that fimbrial aggregation by antibody could enhance adherence and hemagglutination. Inhibition of isolated fimbria attachment to epithelial cells confirmed the homologous > heterologous advantage but this advantage was less than that predicted by the amount of shared antigenicity on the basis of weight alone (Pearce and Buchanan, 1978). The inability to completely inhibit attachment with fimbrial antisera suggested the antigenic determinants of gonococcal fimbriae being detected may be removed from the attachment portion of the molecule. This was supported by the observation that antigenically different fimbriae appear to attach to similar or identical binding sites on human cells (Pearce and Buchanan, 1978). There may be more cross-reactivity between fimbrial antigens interacting with opsonic antibodies. Fimbriae from strains F62 and B have less than 2% shared antigenicity by weight. However, human antisera to F62 fimbriae produced 6–7-fold enhancement of phagocytosis of fimbriated F62 gonococci, but also 4-fold enhancement of phagocytosis of strain B-fimbriated organisms (Seigel, M.S. and Buchanan, T.M., unpublished observations). This suggests that the portion of fimbriae antigenicity that interacts with opsonins may show less antigenic heterogeneity between F62 and B than the whole F62 and B fimbriae molecules.

Table 10.10 Antigenic analysis of *N. gonorrhoeae* fimbriae from different strains using antiserum to fimbriae*

Method	Results	Reference
Direct agglutination of fimbriated gonococci	Mostly strain variations in fimbrial antigens	Buchanan and Pearce, 1976
Agglutination of fimbriae bundles: dark field microscopy	Mostly strain variations of fimbrial antigens with some antigenicity shared	Robertson *et al.*, 1977; C.C. Brinton, unpublished data
Ouchterlony immunodiffusion	Common and different fimbrial antigens among different strains	Buchanan and Pearce, 1976
Two-dimensional immuno-electrophoresis	Detectable fimbrial antigens vary among strains, fimbriae antiserum in high concentrations recognizes heterologous fimbriae	Buchanan, 1977
Immune electron microscopy	Mostly different, but some common antigens among different strains	Novotny and Turner, 1975
Radioactive antigen binding assay	Heterogeneity minimized; single strain of ^{125}I-labeled fimbriae detects antibody to fimbriae in 60–80% of women with gonorrhea	Buchanan *et al.*, 1973
Radio-immunoassay	Strain variations in fimbrial antigens; common antigens quantitated to be ≤ 2.5% by weight	Buchanan, 1975; Buchanan and Pearce, 1976
Hemagglutination inhibition	Mostly different antigens among fimbriae from different strains	Buchanan and Pearce, 1976
Inhibition of attachment of isolated ^{125}I-labeled fimbriae	Fimbrial antigens near attachment moiety vary among strains; antigenic variation less pronounced in domains near than in domains removed from attachment moiety	Pearce and Buchanan, 1978
Enzyme-linked immunosorbant assay (ELISA) using fimbriae as bound antigens	Up to 10% shared antigenicity in fimbriae: can serotype fimbriae on whole organisms	Buchanan, 1978

* After Buchanan, 1977.

10.4.2 Chemical structure

A full understanding of the structural characteristics and differences among bacterial fimbriae may explain their affinity for different cell surface receptors, allowing evaluation of the significance of non-specific forces involved in fimbrial binding (e.g. hydrophobic, electrostatic) and explain differences in antigenicity that occur even among fimbriae with similar adhesive activities. Table 10.11 summarizes the chemical and physical data for fimbriae described below.

The type 1 fimbriae of *E. coli* ($B_{am}P^+$) are entirely protein, containing no carbohydrate, lipid or nucleic acid, and have a minimum subunit molecular weight of 16 600 (Brinton, 1965). Salit and Gotschlich (1977a), working with type 1 fimbriae, isolated from another *E. coli* strain (K12), found a minimum subunit molecular weight of 17 099, an isoelectric point ranging between 4.5–5.1 and a density of 1.299 in cesium chloride. The amino acid composition of these fimbriae subunits, which contain at least 160 residues, reveals a high proportion of nonpolar residues (Brinton, 1965; Salit and Gotschlich, 1977a; Table 10.12). Brinton (1965) attributes the lateral aggregation seen for type 1 fimbriae, their adherence to latex spheres, and, at least, a component of their hemagglutinating ability to this hydrophobic character. The composition of *E. coli* type 1 fimbriae completely lacks methionine and tryptophan residues (Brinton, 1965; Salit and Gotschlich, 1977a), whereas all other fimbria described below have been found to contain these amino acids when they were determined. These fimbriae can be depolymerized by treatments which disrupt hydrogen or hydrophobic bonds (Brinton, 1965) suggesting that these forces are involved in subunit–subunit interactions. Brinton (1965, 1967) has further contributed electron microscopic, X-ray diffraction and crystallographic evidence, which he interpreted to demonstrate that type 1 fimbriae are rigid, right-handed α-helices containing an axial hole 2.0–2.5 nm in diameter, a 2.4 nm pitch and a helical repeat distance of 2.32 nm giving 3 1/8 subunit per turn.

The fimbriae of enterotoxigenic *E. coli* have larger subunit molecular weights than common fimbriae. K99 fimbriae are apparently composed of two subunits, a major one of 22 500 and a minor one of 29 500 molecular weight (Isaacson, 1977). These fimbriae have an isoelectric point greater than 10, contain 6.6% lipid by weight and less than 0.6% carbohydrate (Isaacson, 1977). K88 fimbriae are similar in having a 25 000 subunit molecular weight (Klemm, 1977) and they contain 3–4% lipid and 1.0% carbohydrate (Stirm *et al.*, 1976a) although the authors could not rule out contaminants as the source of lipid and neutral hexoses. Both K99 and K88 fimbriae contain a high proportion of hydrophobic amino acids (Table 10.12) and K99 is unique in having 3 residues of the unusual amino acid hydroxylysine per subunit (Isaacson, 1977).

Kumazawa and Yanagawa (1972) have investigated the chemical properties of *Corynebacterium renale* fimbriae (strain type II). They found them to contain only protein, and have a subunit molecular weight of 19 000–19 400. Antigenicity was thermostable in these fimbriae. The amino acid analysis of purified fimbriae showed

Table 10.11 Chemical and physical parameters of fimbriae

Fimbriae source	Subunit molecular weight	Isoelectric point	Density in CsCl	% Carbohydrate	Phosphate content	Reference
E. coli (type 1)	16 600–17 500	4.5–5.1 (3 bands)	1.299	None	–	Brinton et al., 1964; Brinton, 1965, 1967; Salit and Gotschlich, 1977a
E. coli (K99)	2 Subunits Major – 22 500 Minor – 29 500	10.1	–	<0.6	–	Isaacson, 1977
E. coli (K88)	25 000	–	–	1.0	0.04	Stirm et al., 1967a; Klemm, 1977
N. gonorrhoeae	19 000 ± 2500 (varies slightly for different strains)	5.3 (minor band at 4.9)	1.30–1.31	1.3–2.0 (1–2 hexoses/ subunit)	1–2 Phosphate groups/ subunit	Robertson et al., 1977; Buchanan, 1977; Pearce and Buchanan, 1978; T.M. Buchanan, unpublished data
P. aeruginosa K (PAK)	17 800 ± 300	3.9	1.295 (1.221 in sucrose)	None	None	Frost and Paranchych, 1977
M. nonliquefaciens	17 000	–	–	–	–	Frøholm and Sletten, 1977
C. renale (Strain type II)	19 000–19 400	4.35	–	None	–	Kumazawa and Yanagawa, 1972, 1973

a significant content of hydrophobic residents (Table 10.12). Structural differences were related to antigenic heterogeneity in *C. renale* fimbriae by Kumazawa and Yanagawa (1973), who determined the isoelectric points of fimbriae from each of three types of *C. renale* which bear antigenically different fimbriae. Isoelectric points of 3.75, 4.35 and 4.46 were found for fimbriae from types I, II and III respectively.

Frost and Paranchych (1977) studied the polar (PSA) fimbriae of *Pseudomonas aeruginosa* strain K (PAK) which serve as receptors for at least six bacteriophages. No phosphate, carbohydrate or lipid was found in these 17 800 ± 300 subunit molecular weight fimbriae. In contrast, the serologically distinct polar fimbriae of *P. aeruginosa* strain O (PAO) have a subunit molecular weight of only 15 500 (W. Paranchych, personal communication). The buoyant density of PAK fimbriae ranged from 1.221 on sucrose to 1.295 on cesium chloride and they had an iso-electric point of 3.9 (Frost and Paranchych, 1977). Upon amino acid analysis, 53% of the 172 residues were hydrophobic (Table 10.12) and a low level of α-helix was predicted by a large number of prolines (Frost and Paranchych, 1977). Although functionally it had first appeared that these fimbriae may be more similar to conjugative fimbriae, Frost and Paranchych (1977) concluded that the structure of PAK fimbriae was more closely related to that of type 1 rather than F fimbriae of *E. coli*. This was on the basis of molecular weight, phosphate and carbohydrate content and the proportion of hydrophobic residues. Further support for this similarity to fimbriae that function in adherence will be presented in the sequence data for PAK fimbriae below. Structurally similar fimbriae were characterized on *Moraxella nonliquefaciens* (Frøholm and Stetten, 1977). They have a 17 000 subunit molecular weight and a high content of non-polar residues, especially alanine, leucine and isoleucine, in addition to a high number of aspartic acid and threonine residues (Table 10.12).

Studies on isolated gonococcal fimbriae binding to human cells (Section 10.3) revealed that the attachment moiety is heat-labile and tryptophan residues may be present in or near this moiety as their modification reduces adherence (Buchanan *et al.*, 1978b). Also, histidine may be accessible on the surface of fimbriae to the formation of ferric ion ligands with the cell surface. Structural and chemical studies of gonococcal fimbriae are summarized in Table 10.11. The apparent subunit molecular weight, as measured by SDS-polyacrylamide gel electrophoresis, varies for antigenically distinct gonococcal fimbriae from as low as 17 000 to be maximum of 21 500 (Robertson *et al.*, 1977; Pearce and Buchanan, 1978; Buchanan, 1978a). An isoelectric point of 5.3, density on cesium chloride of 1.30–1.31, 1–2 hexose groups (1.3% carbohydrate content) and 1–2 phosphate groups per subunit for gonococcal fimbriae have been reported by Robertson *et al.* (1977). The carbohydrate group present is galactose with traces of glucose, but no heptose or amino sugars. The amino acid composition of one fimbriae type tested (strain 33) reveals approximately 55% hydrophilic and 45% non-polar residues (Hermodson *et al.*, 1978; Table 10.12). This composition generally agrees with that published for strain P9 and

Bacterial Adherence

Table 10.12 Hydrophobic content of bacterial fimbriae amino acid compositions

Fimbriae source (strain)	% Non-polar residues (Pro, Gly, Ala, Val, Met, Ile, Leu, Phe, Trp)	Reference
E. coli ($B_{am}P^+$, type 1)	54.0	Brinton, 1965
E. coli (K12, type 1)	56.3	Salit and Gotschlich, 1977a
E. coli (1474, K99)	54.7	Isaacson, 1977
E. coli (D520, K88)	47.8	Stirm *et al.*, 1967a
N. gonorrhoeae (33)	45.8	Hermodson *et al.*, 1978
N. gonorrhoeae (P9 and 201)	43.8–45.8	Robertson *et al.*, 1977
P. aeruginosa (K/2PfS)	53.7	Frost and Paranchych, 1977
M. nonliquefaciens (NCTC 7784 SC-c)	50.2*	Frøholm and Sletten, 1977
C. renale (46)	44.7	Kumazawa and Yanagawa, 1972

* Tryptophan not determined.

201 fimbriae by Robertson *et al.* (1977). Both groups reported the presence of at least two histidine and three tryptophan residues per subunit. The above table summarizes the known amino acid compositions of bacterial fimbriae that may function in adherence by their percentage of non-polar residues. Obviously, a common feature of these structures is a significant hydrophobic character.

At present, no fimbriae have been completely sequenced and only a few reports of the amino-terminal amino acid sequence are available. These include fimbriae from four different species and/or genera whose adherence is not inhibited by D-mannose and which, surprisingly, have similar amino-terminal amino acid sequences (Frøholm and Sletten, 1977; Hermodson *et al.*, 1978; Paranchych *et al.*, 1978). In contrast, the type 1 fimbriae of *E. coli* strain $B_{am}P^+$, whose attachment is inhibited by D-mannose containing carbohydrates, possess a totally unrelated amino-terminal amino acid sequence (Hermodson *et al.*, 1978). The partial sequence of the mannose-resistant, K88 fimbriae of *E. coli* strains D520 and D172 (Klemm, 1977) is also quite different from both the homologous group and the mannose-sensitive *E. coli* fimbriae. These findings are summarized in Table 10.13.

First, an unsual amino-terminal amino acid, *N*-methylphenylalanine, was found for all gonococcal and the meningococcal fimbriae (Hermodson *et al.*,1978), fimbriae from *P. aeruginosa* strain K (Frost *et al.*, 1978) and *P. aeruginosa* strain O (W. Paranchych, personal communication). This amino-terminus was also suggested for the fimbriae of *Moraxella nonliquefaciens* (Frøholm and Sletten, 1977) and recently confirmed (Frøholm, 1978). All *Neisseria* fimbria proteins had amino-terminal heterogeneity, half being *N*-methylphenylalanine and half threonine (the second amino acid in the longer protein). This heterogeneity was not found for PAK and *Moraxella* fimbriae, where 100% of fimbrial subunits contained *N*-methyl-phenylalanine. Also, the first 24 residues of *Neisseria* fimbriae are totally hydrophobic except for threonine and glutamic acid at positions 2 and 5 (Table 10.13). Paranchych *et al.* (1978) have determined the amino-terminal amino acid sequence for PAK fimbriae for the first 22 residues and Frøholm and Sletten (1977) have sequenced the first 49 residues for *M. nonliquefaciens* fimbriae. The latter group found that the hydrophobic character extended 40 residues into the amino-terminal sequence (Table 10.13).

The most striking finding shown by comparing the amino-terminal amino acid sequences in Table 10.13 is the extreme conservation of primary structure for fimbriae from bacteria quite taxonomically distinct from gonococci. Hermodson *et al.* (1978) found that each of four antigenically different gonococcal fimbriae and the meningococcal fimbria had identical sequence, a preservation of the structure of the first 29 residues out of a total of approximately 170. Recent results have shown that a fifth gonococcal fimbriae type from strain P9, obtained and purified at the laboratory of Michael Ward in Southampton, England using a different method than our own laboratory (Robertson *et al.*, 1977), also has the same sequence (M. A. Hermodson, unpublished results). Thus, the sequence shown in Table 10.13 may prove constant for all gonococcal fimbriae. The *Pseudomonas* (PAK) fimbriae differed by only four substitutions at positions 10 (Val for Ile), 13 (Ile for Val), 19 (Ile for Val) and 21 (Leu for Ile). Fimbriae purified from *M. nonliquefaciens* had only two substitutions compared to *Neisseria* fimbriae, and these were the same as the substitutions at residues 13 and 19 of the PAK fimbriae (Table 10.13). Further, each of these substitutions can result from a single base change in DNA sequence, and none affect the extremely hydrophobic nature of this portion of the fimbriae. Hermodson *et al.* (1978) interpreted the conservation of an amino-terminal sequence over strain, species and genera differences, as opposed to the significant antigenic heterogeneity (Section 10.4.1) to imply this region of the molecule is involved in an important way in the function of the protein. This common hydrophobic sequence may be necessary for fimbrial subunit synthesis, assembly, or maintenance of the proper secondary, tertiary or quaternary conformation for structural and functional integrity. Paranchych *et al.* (1978) suggest that this hydrophobic stretch may have a 'pilot' function which facilitates transport of subunits from the cytoplasm to their site of assembly. This conceivably involves passage through the cytoplasmic membrane from which fimbriae are known to originate (Hoeniger, 1965; Ottow, 1975). Brinton (1965) has proposed that fimbrial synthesis is likely to occur by the assembly of preformed subunits and the

Table 10.13 Fimbriae amino-terminal amino acid analysis

	1				5					10					15
Neisseria*	MePhe	Thr	Leu	Ile	Glu	Leu	Met	Ile	Val	Ile	Ala	Ile	Val	Gly	Ile
PAK†	MePhe	Thr	Leu	Ile	Glu	Leu	Met	Ile	Val	Val	Ala	Ile	Ile	Gly	Ile
Moraxella‡	MePhe	Thr	Leu	Ile	Glu	Leu	Met	Ile	Val	Ile	Ala	Ile	Ile	Gly	Ile
E. coli §	Ala	Ala	Thr	Thr	Val	Asn	Gly	Gly	Thr	Val	His	Phe	Lys	Gly	Glu
K88 ¶	Trp	Met	Thr	Gly	Asp	Phe	Asn	Gly	Ser	Val	Asp	Ile	Gly	Gly	Ser

					20					25				29
Neisseria:	Leu	Ala	Ala	Val	Ala	Leu	Pro	Ala	Tyr	Gln	Asp	Tyr	Thr	Ala
PAK:	Leu	Ala	Ala	Ile	Ala	Ile	Pro	–	–	–	–	–	–	–
Moraxella:	Leu	Ala	Ala	Ile	Ala	Leu	Pro	Ala	Tyr	Gln	Asp	Tyr	Ile	Ala
E. coli:	Val	Val	Asn	Ala	Ala	?	Ala	Val	Asp	–	–	–	–	–
K88:	Ile	Thr	Ala	Asx	Gly	–	–	–	–	–	–	–	–	–

* Amino-terminal amino acid sequence of antigenically distinct fimbriae from 4 different strains of *N. gonorrhoeae* (F62, B, 33 and 7122) and from strain 13090 of *N. meningitidis* (Hermodson *et al.*, 1978).

† Amino-terminal amino acid sequence for fimbriae from *P. aeruginosa* strain K (Paranchych *et al.*, 1978).

‡ Amino-terminal amino acid sequence of the fimbriae of *M. nonliquefaciens* strain 7784 SC-c (Frøholm and Sletten, 1977; Frøholm, 1978).

§ Amino-terminal amino acid sequence of the fimbriae of *E. coli* strain B$_{am}$P⁺ (Hermodson *et al.*, 1978).

¶ Amino-terminal amino acid sequence of the fimbriae of *E. coli* strains D520 and D1721 (K88 fimbriae; Klemm, 1977).

repolymerization of fimbrial subunits was observed to be spontaneous. After assembly, these non-polar regions would be likely to be buried within the subunit or involved in subunit—subunit interactions (Hermodson *et al.*, 1978; Paranchych *et al.*, 1978). In either event, the maintenance of this non-polar character is necessary for structural integrity.

Fimbriae from *N. gonorrhoeae* strain F62 have a subunit molecular weight of 16 700 and an estimated 150—160 residues per subunit. The remainder of the molecule must be quite hydrophilic, given the amino acid composition (Hermodson *et al.*, 1978; Table 10.12). It is possible to anticipate that both the fimbrial attachment moiety and the antigenic determinants will be within the remaining sequence, since these sites are likely to be exposed on the molecule's surface and therefore hydrophilic in nature. It is unlikely that the attachment moiety is present in the first one-third of the subunit primary sequence. We have predicted that tryptophan will be involved in the attachment portion of the molecule, and perhaps histidine as well. None of the three tryptophans or two histidines predicted by the amino acid composition of gonococcal fimbriae (Hermodson *et al.*, 1978) are present in the first 29 residues (Table 10.13). Also, these residues are not present in the sequence of *M. nonliquefaciens* fimbriae (Frøholm and Sletten, 1977) as far as residue 49 (Table 10.13), and the already demonstrated homology of *Moraxella* and gonococcal fimbrial sequences suggests this might also be the case for the gonococcus.

10.5 CONCLUSIONS

This chapter has reviewed the evidence supporting a function of bacterial fimbriae as mediators of adherence. More difficult to test, but also likely in many cases, is that the adherence advantage conferred by the presence of fimbriae enhances the virulence of potential pathogens. The occurrence, morphology, structure and binding properties of fimbriae have been shown to be distinctive for some genera and species, and yet the similarities among different fimbriae types suggest that general models of fimbriae structure—function relationships may eventually be developed. Further comparisons contrast the non-specific and specific aspects of fimbrial attachment and attempt chemical characterization of the receptor for specific fimbriae binding to human cells. For *N. gonorrhoeae* fimbriae, we have presented evidence of a carbohydrate-containing receptor somewhat resembling the carbohydrate moiety found on gangliosides. The lectin-like gonococcal fimbriae possess cell-surface specificity, with receptor densities on cells from the sites of gonorrheal infection being similar to densities found for many polypeptide hormones on their target cells. Additionally, electrostatic charge and hydrophobic interactions appear to significantly influence the non-specific binding of fimbriae to human cells.

The precise function of attachment moieties carried by fimbriae in influencing infection and disease remains to be determined. Hopefully, such studies will coincide with and complement further elucidations of eukaryotic cell membrane structure.

The potential exists for these mediators of adherence and their corresponding
receptors to profoundly influence, for example, the organ and species specificity
of infection. An understanding of these molecular mechanisms in bacterial adherence
to host tissues may then be used to suggest ways to interrupt attachment and
pathogenesis.

REFERENCES

Arai, H. and Sato, Y. (1976), *Biochim. biophys. Acta,* **444**, 765–782.

Aronson, M., Medalia, O., Schori, L., Mirelman, D., Sharon, N. and Ofek, I. (1979),
    J. Infect. Dis., **139**, 329–332.

Bøvre, K., Bergen, T. and Frøholm, L.O. (1970), *Acta path. microbiol. scand.,*
    Sect. B, **78**, 765–779.

Bøvre, K. and Frøholm, L.O. (1972), *Acta path. microbiol. scand.,* Sect. B, **80**,
    629–640.

Bradley, D.E. (1972), *Genet. Res.,* **19**, 30–51.

Bradley, D.E. and Pitt, T.L. (1974), *J. gen. Virol.,* **24**, 1–15.

Brinton, C.C. (1959), *Nature,* **183**, 782–786.

Brinton, C.C. (1965), *Trans. N.Y. Acad. Sci.,* Ser. II, **27**, 1003–1054.

Brinton, C.C. (1967), *The Specificity of Cell Surfaces,* (Davis, B.D. and Warren, L.,
    eds.), Prentice-Hall, Inc., Englewood Cliffs, N.J., 37–70.

Brinton, C.C., Bryan, J., Dillon, J.-A., Guerina, N., Jacobsen, L.J., Labik, A., Lee, S.,
    Levine, A., Lim, S., McMichael, J., Polen, S., Rogers, K., To, A.C.-C and
    To, S., C.-M. (1978), *Immunobiology of Neisseria gonorrhoeae,* (Brooks, G.F.,
    Gotschlich, E.C., Holmes, K.K., Sawyer, W.D. and Young, F.E., eds.),
    American Society for Microbiology, Washington, D.C., pp. 155–178.

Brinton, C.C., Gemski, P. and Carnahan, J. (1964), *Proc. natn. Acad. Sci., U.S.A.,*
    52, 776–783.

Brown, W.J., Lucas, C.T. and Kuhn, U.S.G. (1972), *Br. J. Vener. Dis.,* **48**, 177–178.

Buchanan, T.M. (1975), *J. exp. Med.,* **141**, 1470–1475.

Buchanan, T.M. (1976), *Microbiology,* (Schlessinger, D., ed.), American Society
    for Microbiology, Washington, D.C., 491–493.

Buchanan, T.M. (1977), *The Gonococcus,* (Roberts, R.B., ed.), John Wiley and Sons,
    New York, 255–272.

Buchanan, T.M. (1978), *J. Infect. Dis.,* **138**, 319–325.

Buchanan, T.M., Chen, K.C.S., Jones, R.B., Hildebrandt, J.H., Pearce, W.A.,
    Hermodson, M.A., Newland, J.C. and Luchtel, D.L. (1978a), *Immunobiology
    of Neisseria gonorrhoeae,* (Brooks, G.F., Gotschlich, E.C., Holmes, K.K.,
    Sawyer, W.D. and Young, F.E., eds.), American Society for Microbiology,
    Washington, D.C., pp. 145–154.

Buchanan, T.M. and Gotschlich, E.C. (1973), *J. exp. Med.,* **137**, 196–200.

Buchanan, T.M. and Pearce, W.A. (1976), *Infect. Immun.,* **13**, 1483–1489.

Buchanan, T.M., Pearce, W.A. and Chen, K.C.S. (1978b), *Immunobiology of
    Neisseria gonorrhoeae,* (Brooks, G.F., Gotschlich, E.C., Holmes, K.K.,
    Sawyer, W.D. and Young, F.E., eds.), American Society for Microbiology,
    Washington, D.C., pp. 242–249.

Buchanan, T.M., Swanson, J., Holmes, K.K., Kraus, S.J. and Gotschlich, E.C. (1973), *J. clin. Invest.*, **52**, 2896–2909.

Bumgarner, L.R. and Finkelstein, R.A. (1973), *Infect. Immun.*, **8**, 919–924.

Burrows, M.R., Sellwood, R. and Gibbons, R.A. (1976), *J. gen. Microbiol.*, **96**, 269–275.

Collier, W.A. and De Miranda, J.C. (1955), *Anton. Van. Leeuwen, J. Microbiol. Serol.*, **21**, 133–140.

Constable, F.L. (1956), *J. path. Bact.*, **72**, 133–136.

Cowan, S.T., Steel, K.J., Shaw, C. and Duguid, J.P. (1960),.*J. gen Microbiol.*, **23**, 601–612.

Devoe, I.W. and Gilchrist, J.E. (1974), *Infect. Immun.*, **10**, 872–876.

Devoe, I. W. and Gilchrist, J.E. (1975), *J. exp. Med.*, **141**, 297–305.

Duguid, J.P. (1959), *J. gen. Microbiol.*, **21**, 271–286.

Duguid, J.P. (1964), *Rev. Lat.-Amer. Microbiol.*, **7**, Suppl. 13–14, 1–16.

Duguid, J.P. (1968), *Arch. Immun. Ther. Exp.*, **16**, 173–188.

Duguid, J.P., Anderson, E.S. and Campbell, I. (1966),*J. Path. Bact.*, **92**, 107–138.

Duguid, J.P. and Campbell, I. (1969),*J. med. Microbiol.*, **2**, 535–553.

Duguid, J.P., Darekar, M.R. and Wheater, D.W.F. (1976),*J. med. Microbiol.*, **9**, 459–473.

Duguid, J.P. and Gillies, R.R. (1956), *J. gen. Microbiol.*, **15**, vi.

Duguid, J.P. and Gillies, R.R. (1957),*J. path. Bact.*, **74**, 397–411.

Duguid, J.P. and Gillies, R.R. (1958), *J. Path. Bact.*, **75**, 519–520.

Duguid, J.P., Smith, I.W., Dempster, G. and Edmunds, P.N. (1955),*J. Path. Bact.*, **70**, 335–348.

Evans, D.G. and Evans, D.J., Jr. (1978), *Infect. Immun.*, **21**, 638–647.

Evans, D.G., Evans, D.J., Jr. and Tjoa, W. (1977), *Infect. Immun.*, **18**, 330–337.

Evans, D.G., Evans, D.J., Jr., Tjoa, W.S. and DuPont, H.L. (1978), *Infect. Immun.*, **19**, 727–736.

Evans, D.G., Silver, R.P., Evans, D.J., Jr., Chase, D.G. and Gorbach, S.L. (1975), *Infect. Immun.*, **12**, 656–667.

Evans, D.J., Jr., Evans, D.G. and DuPont, H.L. (1979), *Infect. Immun.*, **23**, 336–346.

Frøholm, L.O. (1978), *Natn. Inst. Pub. Health Annals*, **1**, 35–47.

Frøholm, L.O. and Bøvre, K. (1972), *Acta path. Microbiol. Scand.*, Sect. B, **80**, 641–648.

Frøholm, L.O. and Sletten, K. (1977), *FEBS Letters*, **73**, 29–32.

Frost, L.S., Carpenter, M. and Paranchych, W. (1978), *Nature*, **271**, 87–89.

Frost, L.S. and Paranchych, W. (1977),*J. Bact.*, **131**, 259–269.

Fuerst, J.A. and Hayward, A.C. (1969),*J. gen. Microbiol.*, **58**, 227–237.

Gilboa-Garber, N. (1972a), *FEBS Letters*, **20**, 242–244.

Gilboa-Garber, N. (1972b), *Biochim. biophys. Acta*, **273**, 165–173.

Gilboa-Garber, N., Mizrahi, L. and Garber, N. (1972), *FEBS Letters*, **28**, 93–95.

Gilboa-Garber, N., Mizrahi, L. and Garber, N. (1977), *Can. J. Biochem.*, **55**, 975–981.

Gillies, R.R. and Duguid, J.P. (1958),*J. Hyg., Camb.*, 303–318.

Guinée, P.A.M. and Jansen, W.H. (1979), *Infect. Immun.*, **23**, 700–705.

Hermodson, M.A., Chen, K.C.S. and Buchanan, T.M. (1978), *Biochemistry*, **17**, 442–445.

Heumann, W. and Marx, R. (1964), *Arch. Mikrobiol.*, **47**, 325–337.

Hoeniger, J.F.M. (1965), *J. gen. Microbiol.*, **40**, 29–42.

Hohmann, A. and Wilson, M.R. (1975), *Infect. Immun.*, **12**, 866–880.

Honda, E. and Yanagawa. R. (1974), *Infect. Immun.*, **10**, 1426–1432.

Honda, E. and Yanagawa, R. (1975), *Am. J. Vet. Res.*, **36**, 1663–1666.

Honda, E. and Yanagawa, R. (1978), *Am. J. Vet. Res.*, **39**, 155–158.

Houwink, A.L. and van Iterson, W. (1950), *Biochim. biophys. Acta*, **5**, 10–44.

Isaacson, R.E. (1977), *Infect. Immun.*, **15**, 272–279.

James, A.N., Know, J.M. and Williams, R.P. (1976), *Br. J. Vener. Dis.*, **52**, 128–155.

James-Holmquest, A.N., Swanson, J., Buchanan, T.M., Wende, R.D. and Williams, R.P. (1974), *Infect. Immun.*, **9**, 897–902.

Jephcott, A.E., Reyn, A. and Birch-Andersen, A. (1971), *Acta path. microbiol. scand.*, Sect. B, **79**, 437–439.

Johnson, A.P., Taylor-Robinson, D. and McGee, Z.A. (1977), *Infect. Immun.*, **18**, 833–839.

Jones, G.W. (1977), *Microbial Interactions,* Receptors and Recognition, Series B, Vol. 3, (Reissing, J.L., ed.), Chapman and Hall, London, pp. 139–176.

Jones, G.W. and Freter, R. (1976), *Infect. Immun.*, **14**, 240–245.

Jones, G.W. and Rutter, J.M. (1972), *Infect. Immun.*, **6**, 918–927.

Jones, G.W. and Rutter, J.M. (1974), *J. gen. Microbiol.*, **84**, 135–144.

Kellogg, D.S., Jr., Cohen, I.R., Norins, L.C., Schroeter, A.L. and Reising, G. (1968), *J. Bact.*, **96**, 596–605.

Kellogg, D.S., Jr., Peacock, W.L., Jr., Deacon, W.E., Brown, L. and Pirkle, C.I. (1963), *J. Bact.*, **85**, 1274–1279.

Keogh, E.V. and North, E.A. (1948), *Aust. J. exp. Biol. Med. Sci.*, **26**, 315–322.

Klemm, P. (1977), Eleventh Meet. Fed. Europ. Biochem. Soc., Copenhagen, August 14–25, *Abstract* C-3, 223, 8/9.

Koransky, J.R., Scales, R.W. and Kraus, S.J. (1975), *Infect. Immun.*, **12**, 495–498.

Kumazawa, N. and Yanagawa, R. (1972), *Infect. Immun.*, **5**, 27–30.

Kumazawa, N. and Yanagawa, R. (1973), *Jap. J. Microbiol.*, **17**, 13–19.

Lönnroth, I. and Holmgren, J. (1975), *J. gen. Microbiol.*, **91**, 263–277.

Mårdh, P.-A. and Weström, L. (1976), *Infect. Immun.*, **13**, 661–666.

Marx, R. and Heumann, W. (1962), *Arch. Mikrobiol.*, **43**, 245–254.

Masry, F.L.G. (1952), *J. gen. Microbiol.*, **7**, 201–210.

McGee, Z.A., Dourmashkin, R.R., Gross, J.G., Clark, J.B. and Taylor-Robinson, D. (1977), *Infect. Immun.*, **15**, 594–600.

McNeish, A.S., Fleming, J., Turner, P. and Evans, N. (1975), *Lancet*, **II**, 946–948.

Minshew, B.H., Jorgensen, J., Counts, G.W. and Falkow, S. (1978a), *Infect. Immun.*, **20**, 50–54.

Minshew, B.H., Jorgensen, J., Swanstrum, M., Grootes-Reuvecamp, G.A. and Falkow, S. (1978b), *J. Infect. Dis.*, **137**, 648–654.

Morse, J.H. and Morse, S.I. (1970), *J. exp. Med.*, **131**, 1342–1357.

Mullin, B.R., Fishman, P.H., Lee, G., Aloj, S.M., Ledley, F.D., Winand, R.J., Kohn, L.D. and Brady, R.O. (1976), *Proc. natn. Acad. Sci. U.S.A.*, **73**, 842–846.

Nagy, B., Moon, H.W. and Isaacson, R.E. (1977), *Infect. Immun.*, **16**, 344–352.

Nelson, E.T., Clements, J.D. and Finkelstein, R.A. (1976), *Infect. Immun.*, **14**, 527–547.

Novotny, P. and Turner, W.H. (1975), *J. gen. Microbiol.*, **89**, 87–92.

Ofek, I., Mirelman, D. and Sharon, N. (1977), *Nature*, **265**, 623–625.

Old, D.C. (1972), *J. gen. Microbiol.*, **71**, 149–157.

Old, D.C., Corneil, I., Gibson, L.F., Thomson, A.D. and Duguid, J.P. (1968), *J. gen. Microbiol.*, **51**, 1–16.

Old, D.C. and Payne, S.B. (1971), *J. med. Microbiol.*, **4**, 215–225.

Ørskov, F., Ørskov, I., Evans, D.J., Jr., Sack, R.B., Sack, D.A. and Wadström, T. (1976), *Med. Microbiol. Immunol. (Berl.)*, **162**, 73–80.

Ørskov, I. and Ørskov, F. (1977), *Med. Microbiol. Immunol.*, **163**, 99–110.

Ørskov, I., Ørskov, F., Smith, H.W. and Sojka, W.J. (1975), *Acta path. microbiol scand.*, Sect. B, **83**, 31–36.

Ottow, J.C.G. (1975), *A. Rev. Microbiol.*, **29**, 79–108.

Paranchych, W., Frost, L.S. and Carpenter, M. (1978), *J. Bact.*, **134**, 1179–1180.

Pearce, W.A. and Buchanan, T.M. (1978), *J. clin. Invest.*, **61**, 931–943.

Pedersen, K.B., Frøholm, L.O. and Bøvre, K. (1972), *Acta path. microbiol. scand.*, Sect. B, **80**, 911–918.

Punsalang, A.P., Jr. and Sawyer, W.D. (1973), *Infect. Immun.*, **8**, 255–263.

Robertson, J.N., Vincent, P. and Ward, M.E. (1977), *J. gen. Microbiol.*, **102**, 169–177.

Salit, I.E. and Gotschlich, E.C. (1977a), *J. exp. Med.*, **146**, 1169–1181.

Salit, I.E. and Gotschlich, E.C. (1977b), *J. exp. Med.*, **146**, 1182–1194.

Sato, Y., Arai, H. and Suzuki, K. (1973), *Infect. Immun.*, **7**, 992–999.

Sato, Y., Arai, H. and Suzuki, K. (1974), *Infect. Immun.*, **9**, 801–810.

Sheddon, W.I.H. (1962), *J. gen. Microbiol.*, **28**, 1–7.

Shimono, E. and Yanagawa, R. (1977), *Infect. Immun.*, **16**, 263–267.

Silverblatt, F.J. (1974), *J. exp. Med.*, **140**, 1696–1711.

Silverblatt, F.J., Dreyer, J.S. and Schauer, S. (1979), *Infect. Immun.*, **24**, 218–223.

Silverblatt, F.J. and Ofek, I. (1976), *Clin. Res.*, **24**, 454A.

Stirm, S., Ørskov, I. and Mansa, B. (1967a), *J. Bact.*, **93**, 731–739.

Stirm, S., Ørskov, F., Ørskov, I. and Birch-Andersen, A. (1967b), *J. Bact.*, **93**, 740–748.

Svanborg-Edén, C., Eriksson, B. and Hanson, L.Å. (1977), *Infect. Immun.*, **18**, 767–774.

Svanborg-Edén, C., Eriksson, B., Hanson, L.Å., Jodal, U., Kaijser, B., Lidin-Janson, G., Lindberg, U. and Olling, S. (1978), *J. Ped.*, **93**, 398–403.

Svanborg-Edén, C., Hanson, L.Å., Jodal, U., Lindberg, U. and Åkerland, A.S. (1976), *Lancet*, **II**, 490–492.

Svanborg-Edén, C. and Hansson, H.A. (1978), *Infect. Immun.*, **21**, 229–237.

Swanson, J. (1973), *J. exp. Med.*, **137**, 571–589.

Swanson, J., King, G. and Zeligs, B. (1975), *Infect. Immun.*, **11**, 453–459.

Swanson, J., Kraus, S.J. and Gotschlich, E.C. (1971), *J. exp. Med.*, **134**, 886–906.

Tebbutt, G.M., Veale, D.R., Hutchinson, J.G.P. and Smith, H. (1976), *J. med. Microbiol.*, **9**, 263–273.

Thornley, M.J. and Horne, R.W. (1962), *J. gen. Microbiol.*, **28**, 51–56.

Tramont, E.C. (1976), *Infect. Immun.*, **14**, 593–595.

Tramont, E.C. (1977), *J. clin. Invest.*, **59**, 117–124.

Tweedy, J.M., Park, R.W.A. and Hodgkiss, W. (1968), *J. gen. Microbiol.*, **51**, 235–244.

van Heyningen, W.E. (1974), *Nature*, **249**, 415–417.

Waitkins, S.A. (1974), *Br. J. Vener. Dis.*, **50**, 272–278.

Ward, M.E. and Watt, P.J. (1977), *Gonorrhoea, Epidemiology and Pathogenesis*, (Skinner, F.A., Walker, P.D. and Smith, H., eds.), Academic Press, New York, pp. 83–95.

Ward, M.E., Watt, D.J. and Robertson, J.N. (1974), *J. Infect. Dis.*, **129**, 650–659.

Watt, P.J. and Ward, M.E. (1977), *The Gonococcus*, (Roberts, R.B., ed.), John Wiley and Sons, New York, pp. 355–368.

Weibull, C. and Hedvall, J. (1953), *Biochim. biophys. Acta*, **10**, 35–41.

Weiss, R.L. (1971), *J. gen. Microbiol.*, **67**, 135–143.

Wilson, M.R. and Hohmann, A.W. (1974), *Infect. Immun.*, **10**, 776–782.

Wistreich, G.A. and Baker, R.F. (1971), *J. gen. Microbiol.*, **65**, 167–173.

Yanagawa, R. and Honda, E. (1976), *Infect. Immun.*, **13**, 1293–1295.

Yanagawa, R. and Otsuki, K. (1970), *J. Bact.*, **101**, 1063–1069.

Yanagawa, R., Otsuki, K. and Tokui, T. (1968), *Jap. J. Vet. Res.*, **16**, 31–38.

11 Adherence of Marine Micro-organisms to Smooth Surfaces

MADILYN FLETCHER

11.1	Introduction	*page*	347
	11.1.1 Observations in natural environments		347
	11.1.2 Extracellular polymeric adhesives		348
	11.1.3 Stages of attachment		348
	11.1.4 Theoretical considerations of attachment and polymeric adhesives		349
11.2	The bacterial surface adhesive		350
	11.2.1 Electron microscopic evidence		350
	11.2.2 Experimental evidence		355
11.3	The substratum		358
	11.3.1 The physico-chemical properties of surfaces		359
	11.3.2 Adsorption of dissolved substances on surfaces		360
	11.3.3 Attachment of the marine pseudomonad to different substrata		360
	11.3.4 The influence of adsorbed polymers on bacterial attachment		363
11.4	Other factors influencing bacterial attachment		364
	11.4.1 The state of the organisms		
	11.4.2 Physicochemical factors		367
11.5	Summary and overview		370
	References		371

Acknowledgements
I should like to thank all those who have provided reprints and other information to
help with the preparation of this chapter. Particular thanks are due to G.I. Loeb,
N.G. Maroudas and K.C. Marshall for the inspirational part they have played in the
development of this research. I am also grateful to I.W. Sutherland for analysis of the
bacterial polymer and to T.L. Pitt for preparation of the antiserum. Some of the
unpublished studies cited here were supported by the Natural Environment Research
Council.

Bacterial Adherence
(*Receptors and Recognition,* Series B, Volume 6)
Edited by E.H. Beachey
Published in 1980 by Chapman and Hall, 11 New Fetter Lane, London EC4P 4EE
© 1980 Chapman and Hall

11.1 INTRODUCTION

Marine microbiologists have known for many decades that when an object is submerged in the sea, it is quickly colonized by marine bacteria which attach to its surface (ZoBell and Allen, 1935). Yet, we still know very little about the attachment mechanism(s), what conditions favor adhesion and how the attached, or sessile, micro-organisms are influenced by the proximity of the substratum. Moreover, we are further limited by not knowing the general applicability of the data which we do have. Are the few organisms which have been studied truly representative of most marine bacteria? Also are we justified in relating data from marine and freshwater environments, as is frequently done? These are important reservations to be kept in mind during the following chapter.

11.1.1 Observations in natural environments

A number of studies have been carried out looking at the attachment of marine bacteria *in situ* by immersing a variety of substrata in the sea and examining the attached population after a given time. Consistently, there is a successional development of attached organisms, and bacteria, the first colonizers, begin to attach within a few hours. Corpe (1972) reported that common marine chemo-organotrophs, primarily pseudomonads, were the first to become attached to glass substrata, but after 48 to 72 hours predominent colonizers were species of *Caulobacter, Hyphomicrobium* and *Saprospira.* Examples of the numbers of attached cells observed are more than 2×10^6 bacteria cm^{-2} on glass after a week of submersion (Corpe, 1972) and $10^7 \ cm^{-2}$ on dialysis membrane after 3 days (Vargo *et al.,* 1975). This formation of an initial bacterial film is then closely followed by the attachment of larger organisms, such as diatoms and ciliated and stalked protozoa (Corpe, 1972; Mitchell and Cundell, 1977). Such microbial films are often called 'slime layers' (Jones *et al.,* 1969) because of the copious amounts of 'slimey' extracellular polymers which may be produced. Slime layers are formed in both marine (Corpe, 1970) and freshwater (Jones *et al.,* 1969; Geesey *et al.,* 1977) environments, and the bacterial extracellular polymeric material may be largely acidic polysaccharide (Corpe, 1970a, 1972; Gessey *et al.,* 1977).

Although the build-up of bacterial films on inorganic surfaces is usually inevitable, bacterial attachment to biological surfaces, particularly macroalgae, can be prevented through production by the algae of inhibitors, such as acidic substances and polyphenols (Sieburth, 1968; Cundell *et al.,* 1977). Invertebrates, as well, often show bacterial colonization on one area of the body, while another area remains completely clean (cf. Sieburth, 1975). The mechanisms for most examples of biological

347

inhibition of attachment await explanation, and so far there is no evidence of specific adhesive interactions between aquatic micro- and macro-organisms.

11.1.2 Extracellular polymeric adhesives

A number of different modes of aquatic bacterial attachment have been described (Hirsch and Pankratz, 1970, Weise and Rheinheimer, 1978). Cell surface structures, described as 'blebs', 'droplets' and fibrils, have been observed and thought associated with attachment (Corpe *et al.*, 1976). Fimbriae have also been shown to mediate attachment, and since these are visualized only by electron microscopy, they may be a more common attachment mechanism in aquatic environments than is presently realized. Various types of holdfast materials have been described (Hirsch and Pankratz, 1970; Merkel, 1975); caulobacters, for example, attach by means of a substance produced at the distal end of the holdfast stalk (Corpe, 1970b). Also inorganic cements, such as iron deposits, have been observed (Hirsch and Pankratz, 1970). However, probably the most common adhesives are extracellular polymers which form a continuous 'coat' around the cell and are not concentrated in a specific holdfast region. These polymers may include long-chain polysaccharides generated by cell-surface polymerases (Costerton *et al.*, 1978) or possibly adhesives produced by organelles, such as the globlet-shaped cell-wall structures found in a marine *Flexibacter* species (Ridgway and Lewin, 1973). Since most of our information on marine bacterial adherence concerns extracellular polymeric adhesives, it is this mode of attachment which will be dealt with in this chapter. With further investigation we may find that additional attachment mechanisms are also common.

11.1.3 Stages of attachment

In general, attachment by extracellular polymeric adhesives involves three stages: reversible adhesion, irreversible adhesion and microcolony formation. In 1943, ZoBell pointed out that attachment was often reversible – that bacteria which appeared to have settled on a substratum could be washed off by a stream of water, and that firm attachment appeared to occur only after a bacterium had remained settled for several hours (ZoBell, 1943). The investigations of Marshall *et al.* (1971a) supported ZoBell's theory of two-stage attachment, and they called these stages 'reversible' and 'irreversible sorption'. Reversible sorption is the initial phase of bacterial attachment when the bacterium is only weakly held at the substratum; it exhibits Brownian motion and can be removed by washing with 2.5% NaCl. This is followed by 'irreversible sorption' in which the bacterium does not exhibit Brownian motion and is not removed by washing. Irreversible sorption was described as a time-dependent stage, which is due to the synthesis of extracellular polymers which bridge the bacterial and substratum surfaces. Since this two-staged attachment process has been frequently observed, it is probably quite common in aquatic environments. However, in some situations, reversible sorption is not a prerequisite for irreverisble sorption, and firm attachment can occur quite spontaneously, such as with the adhesion of certain marine bacteria to polystyrene and similar plastics (Fletcher, unpublished results).

After firm attachment has occurred there is usually a third stage, during which attached cells multiply and are joined by additional attaching cells, and this leads to the formation of microcolonies. Large amounts of extracellular polymers may also be produced, and electron microscopy has demonstrated that the bacteria in a micro-colony tend to be embedded in a polymeric matrix, which may be largely poly-saccharide (Jones *et al.*, 1969; Fletcher and Floodgate, 1973; Geesey *et al.*, 1977). Once a thick slime layer has built up, there is often sloughing off of the surface cells, which may then act as an inoculum for the aqueous phase. The characteristics of the microcolony will depend very much on the types of organisms involved, the polymers produced and environmental conditions, such as nutrient availability and strength of the water current.

11.1.4 Theoretical considerations of attachment and polymeric adhesives

Crisp (1972) has described two types of adhesives used to attach organisms to solid surfaces:

(i)   permanent cements which undergo a setting process, usually through the cross-linking of macromolecules, and
(ii)   temporary adhesives, which hold two surfaces together by the work done against viscosity to separate the surfaces. The latter type of adhesion, called Stefan adhesion, allows translational motion of the adhering organism across a surface, and has been used to explain the gliding motion of *Flexibacter* bacteria (Humphrey, Dickson and Marshall, personal communication).

It is not known whether bacterial adhesive polymers undergo a setting process and become permanent adhesives, although electron micrographs of intercellular matrices in slime layers suggest that the polymer can become quite rigid. Often where cells have died and undergone autolysis, cell-shaped holes are left in the matrix (Fletcher, unpublished observations), suggesting that the polymer has lost much of its original elasticity and that some cross-linkage has occurred, possibly involving divalent cations (Fletcher and Floodgate, 1976) (see Section 11.4.2). However, most adhesive polymers are probably sticky by virtue of their viscosity and their ability to be adsorbed onto the substratum.

Two principal theories have been used by cell biologists to explain the attachment of tissue cells to solid substrata. The first of these, the DLVO theory of lyophobic colloid stability (Shaw, 1970) accounts for the stability of hydrophobic colloidal particles of like charge in aqueous media, and has been extended to explain the adhesion of negatively charged cells to negatively charged substrata (Curtis, 1967). According to the theory, adhesion depends on a balance of short-range attractive forces (resulting from dispersion and dipole type interactions) and longer range

electrostatic repulsion forces. It predicts a strong attraction between surfaces at a
very close range, an electrostatic barrier at a longer range and then a weaker attraction
region (the 'secondary potential minimum') at a slightly greater separation of surfaces.
The magnitude of the repulsion barrier to very close approach is determined by the
ionic composition of the medium as well as the charge and radii of curvature of the
two surfaces. The weak attraction region at the secondary minimum allows close
association, but not actual contact, and was suggested by Marshall *et al.* (1971a) to
be the basis for the reversible sorption of bacteria.

A second theory used by cell biologists (Maroudas, 1975a) recognizes that the
DLVO theory alone cannot account for all types of cell adhesion, and that other
types of interactions are involved (see also Chapter 2). These include hydrogen or
hydrophobic bonding (Marshall and Cruickshank, 1973), co-ordination with metals or
other cations (Maroudas, 1975a), polar group interactions, steric interference
(Maroudas, 1973, 1975b) and specific reactions between surface functional groups
(Dazzo *et al.*, 1976). Electron microscopy has provided clear evidence that polymer
bridging is a principal mechanism in bacterial attachment (Marshall *et al.*,1971a, b;
Fletcher and Floodgate, 1973; Costerton *et al.*, 1978; Young, 1978).

Thus far we have seen that bacterial attachment occurs on surfaces in the marine
environment, that extracellular polymers are often found to act as adhesives and that
there is already a theoretical framework for dealing with observations on bacterial
attachment. However, as was pointed out at the beginning of this section, it is
impossible to make broad generalisations about marine bacterial adherence on the
basis of the small amount of information available. Thus the following discussion
will deal primarily with one organism, a marine *Pseudomonas* sp. (NCMB 2021)
isolated from the Menai Strait in North Wales. This will ensure that all the pieces
of data belong to the same puzzle, and that we are not trying to fit together fragments
from unrelated bacterial jigsaws.

The mechanism of marine bacterial adherence involves four main factors:

(i)   the bacterial surface,
(ii)   the substratum,
(iii) the medium separating the surfaces, and
(iv)   environmental factors which may affect any of the above components of the
system.

11.2 THE BACTERIAL SURFACE ADHESIVE

11.2.1 Electron microscopic evidence

The most striking evidence for bacterial surface polymeric adhesives is provided by
electron microscopy. A number of studies have clearly shown bacterial polymers
bridging the gap between the bacterial and substratum surfaces (Fletcher and Floodgate,

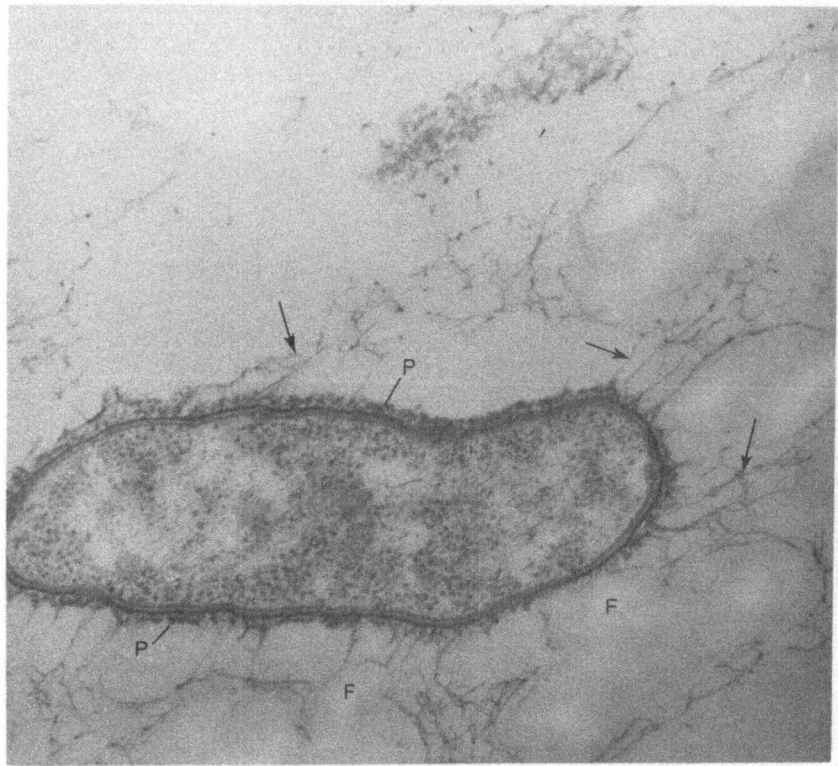

Fig. 11.1 An electron micrograph of the marine pseudomonad attached to a Millipore filter (F) which has been stained with ruthenium red. Note the polymeric coat (P), which appears to have stretched to bridge the gap between the bacterium and filter surfaces (arrows).

1973; Costerton *et al.*, 1978; Latham *et al.*, 1978). Often bridging polymers which are not immediately apparent can be demonstrated through the use of electron dense markers, and polycationic stains such as ruthenium red (Pate and Ordal, 1967; Fletcher and Floodgate, 1973) and alcian blue (Behnke, 1968; Fletcher and Floodgate, 1973) have been particularly useful. With ruthenium red, staining is combined with osmium fixation so that an osmium-ruthenium red marker visualizes the acidic groups present on a cell surface (Blanquet, 1976). The stain has been thought to be specific for acidic polysaccharides (Luft, 1966), but more recent evidence shows that there can be also be some staining of phospholipids (excluding lecithin) and the more soluble fatty acids (Luft, 1971). Alcian blue, a copper phthalocyanine dye, also combines with acidic groups. It can be used to specifically stain acidic poly-saccharides, but only when used at a low pH and short staining time, conditions generally not adhered to for ultrastructural histochemistry.

Bacterial Adherence

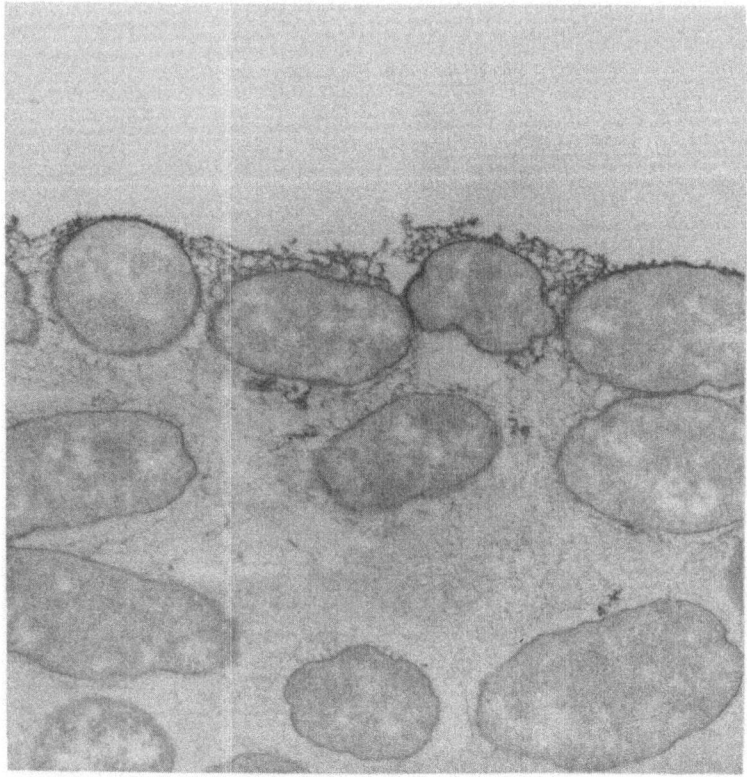

Fig. 11.2 Ruthenium red preparation of a microcolony which has developed on a Millipore filter in a peptone/yeast extract/seawater medium. Note the reticular polymer forming an intercellular matrix between the cells. (Reproduced with permission, Cambridge University Press).

Another stain used in electron microscopy to stain polysaccharides is silver methenamine (Dhir and Boatman, 1972). The staining reaction depends upon the presence of 1, 2 glycol groups (Pearce, 1968), and is analogous to the periodic acid-Schiff (PAS) reaction used in light histochemistry. Theoretically, acidic poly-saccharides, such as hyaluronic acid, should be PAS-positive, but, in practice, staining does not occur, possibly because periodate ions are excluded electrostatically by the charged acidic polysaccharides (Rambourg, 1971).

Figure 11.1 shows the marine pseudomonad, which is attached to a Millipore filter and stained with ruthenium red. The staining clearly shows a polymeric coat, approximately 15 to 25 nm thick, and in regions where the bacterium is attached to the filter, the polymer has stretched to bridge the gap between the two surfaces. Alcian blue staining also demonstrated an acidic surface coat (Fletcher and Floodgate, 1973), but there was no staining with silver methenamine (Fletcher, unpublished observations), suggesting that either the surface coat is not polysaccharide or that it contains a

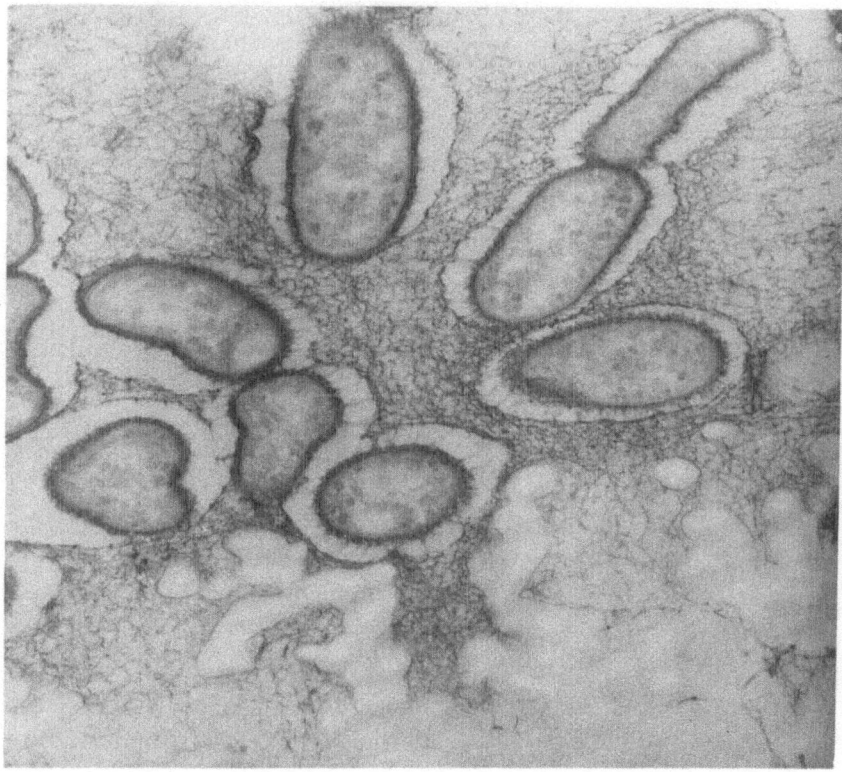

Fig. 11.3 Ruthenium red preparation of a microcolony which has developed on a Millipore filter in a glucose/seawater medium. Compare with Fig. 11.2, and note that both types of polymer, cell surface coat and reticular matrix, are present. (Reproduced with permission, Academic Press).

significant concentration of acidic groups. Thus the histochemical evidence strongly suggests that the bacterial adhesive is an acidic polymer, but there is no positive indication that it is a polysaccharide. An important point is that this adhesive is already present on the free-swimming cell, enabling it to attach if it comes into contact with a suitable surface (Fletcher and Floodgate, 1973).

After attachment has occurred, a second type of polymer is produced and this is also stained with ruthenium red (Fig. 11.2) and alcian blue (Fletcher and Floodgate, 1973). Although this second type of polymer appears to be derived from the initial polymeric coat (Fig. 11.1), the two polymers are quite distinct. The second polymer type is more diffuse and reticular than the polymeric coat, having the appearance of a hydrated gel framework. It is generally found with groups of cells as an intercellular matrix, and in a peptone/yeast extract/seawater medium it usually completely replaces the surface coat (Fig. 11.2). However in a glucose/

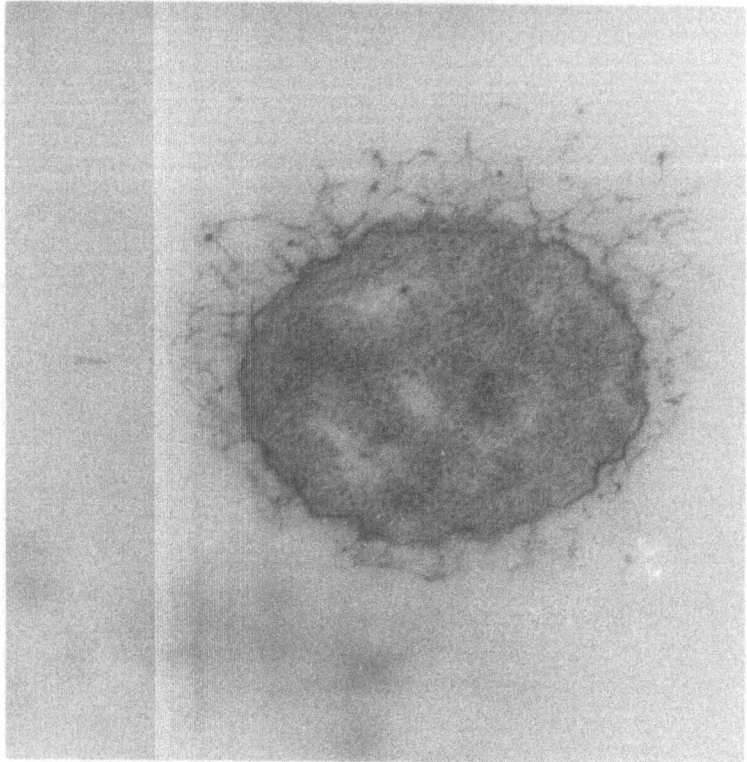

Fig. 11.4 Ruthenium red preparation of the marine pseudomonad attached to a Millipore filter in a medium of pH 8.6. Note the reticular appearance of the polymeric coat and compare with Fig. 11.1.

seawater medium, the situation is different (Fig. 11.3). The diffuse polymer is produced, but the polymeric surface coat is retained, and the bacteria rest in 'pockets' in the gel matrix. Clearly, attachment polymers depend not only upon the type of organism, but also upon its physiological state and on the environmental conditions. It is not known to what extent the two polymers — surface coat and intercellular matrix—differ, nor what initiates the change from the first form to the other, but physicochemical conditions of the microenvironment, such as pH, could be involved. For example, when the pseudomonad was grown at pH 8.6 (compared to pH 8.0) the polymeric coat on attached cells had a reticular appearance, somewhat like the gel matrix (Fig. 11.4).

11.2.2 Experimental evidence

(a) *Polymer analysis*

Having some electron microscopic evidence for an acidic, possibly polysaccharide, adhesive polymer, further experimental efforts were made to characterize the adhesive. An extracellular polymer was isolated from the pseudomonad by ethanol precipitation for analysis, and it was found to contain largely carbohydrate and protein, the protein component ranging from 50–80% of the dry weight (Fletcher, unpublished results). The carbohydrate fraction (kindly analysed by I.W. Sutherland) contained mannose, glucose, glucosamine, rhamnose, galactose and ribose, but no uronic acids were detected. However, infra-red spectrophotometric analysis of the polymer supported the presence of carboxyl groups (Fig. 11.5), which suggests

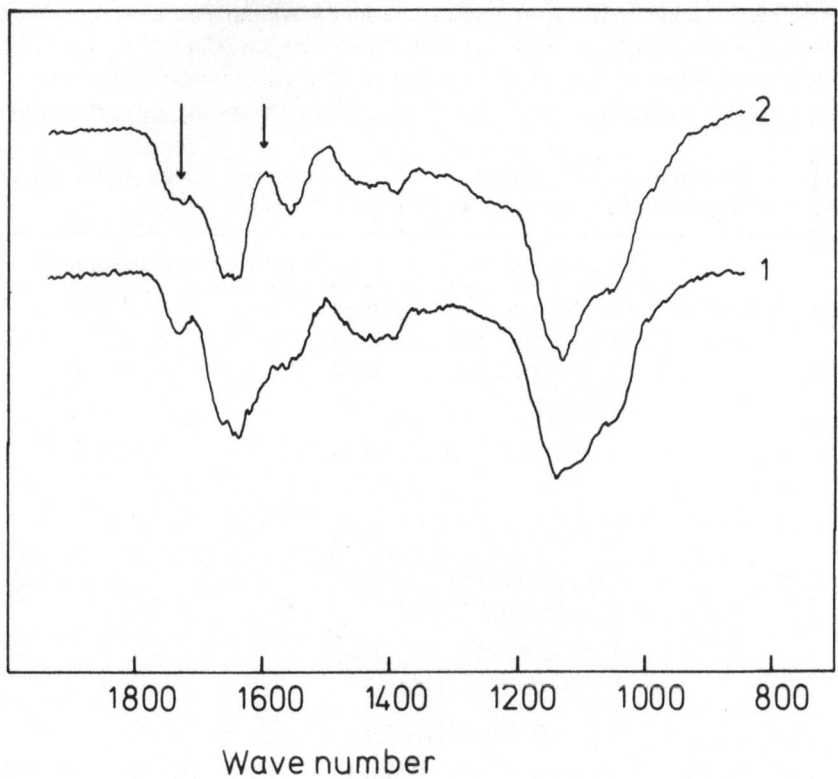

Wave number

Fig. 11.5 Two infra-red spectra of the isolated extracellular polymer dried from water onto silver chloride. One spectrum (1) is of untreated polymer whereas the second (2) is of polymer treated with 0.1N HC1 to depress ionization of any acidic groups. In (2), note the corresponding increase in the peak at 1730 cm^{-1} and the decrease in the peak at 1600 cm^{-1} indicating the presence of carboxyl groups.

that the acidic groups may be located in the protein fraction. Thus, although we have some experimental evidence for an extracellular polymer containing acidic groups, there is no suggestion of an acidic polysaccharide, and it is not certain that the isolated polymer is actually involved in attachment.

(b) *Chemical treatments and their effects on attachment*
Something can be learned about the attachment adhesive by chemically treating either

(i)   free-living cells and observing any influence on their subsequent attachment to surfaces, or
(ii)   attached bacteria to see if the attachment bond is consequently broken down.

Two chemicals known to denature polysaccharides — sodium periodate and disodium tetraborate — were found to inhibit or disrupt cell adhesion. Similarly, attachment was affected by a bacterial antiserum and bovine serum albumin (BSA). These results are given in Table 11.1, where the index of attachment, I_a, gives the ratio of

Table 11.1 The influence of periodate, borate, antiserum and bovine serum albumin (BSA) on attachment of the marine pseudomonad to glass.

	Treatment	I_a*	Influence on attachment[+]
Free-living bacteria	$NaI\,O_4$ (0.02 M)	0.05	—
	$NaI\,O_4$ (0.1 M)	0.03	—
	$NaI\,O_4$ (0.2 M)	0.02	—
	Antiserum		
	(0.5—0.025% v/v)	0.07	—
	BSA		
	(0.5—0.025% w/v)	0.3	—
Attached bacteria	NaCl (0.02 M)$\rightarrow$	0.4	—
	$NaI\,O_4$ (0.02 M)‡		
	NaCl (0.02 M)$\rightarrow$	0.02	—
	$Na_2\,B_4\,O_7$ (0.1 M)‡		
	$Na_2\,B_4\,O_7$ (0.1 M)	0.3	—
	Antiserum	60.1	+
	BSA	17.6	+

*   The index of attachment is the number of attached bacteria after treatment divided by the number of attached bacteria in an untreated control.
+   (−) reduction in attachment; (+) increase in attachment
‡   Bacteria were treated first in dilute NaCl to remove extraneous cations which might reduce effect of treatment.

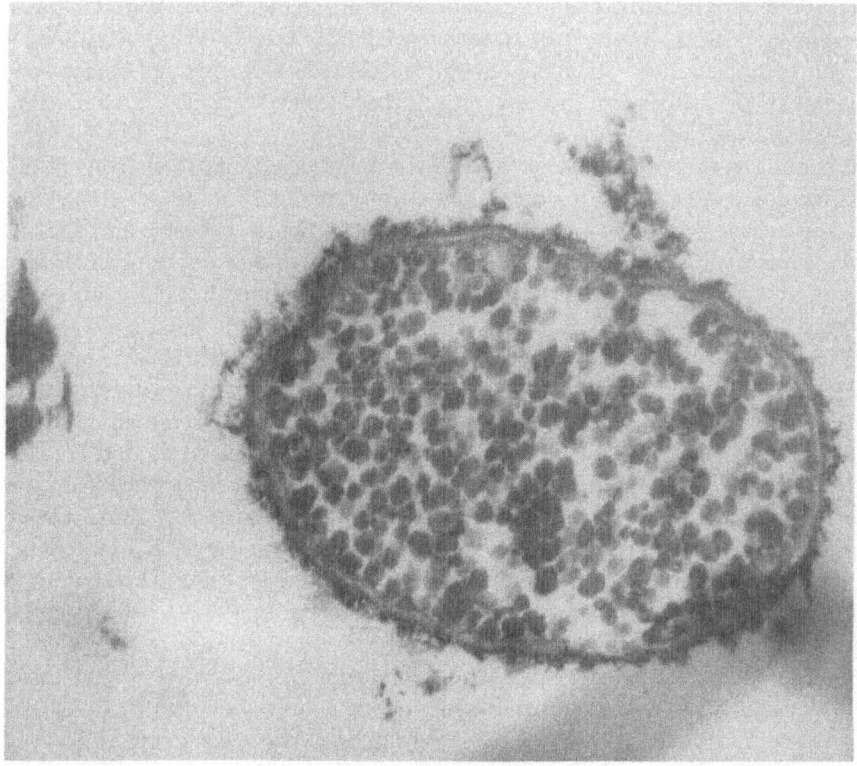

Fig. 11.6 Ruthenium red preparation of the pseudomonad treated with borate. Although the cell wall remains intact, both the polymer coat and the cell interior have been severely affected.

the number of attached bacteria after treatment to the number of attached bacteria in the corresponding control. Periodate treatment markedly inhibited subsequent bacterial attachment, and to a lesser extent caused the removal of attached bacteria. Marshall (1972) also found that, with some freshwater strains, attached bacteria could be removed by periodate treatment. The disruption of attachment by periodate suggests that a carbohydrate adhesive may be involved, since the characteristic action of periodate is the oxidation and cleavage of adjacent hydroxyl groups, found in sugars.

Treatment with borate also caused the removal of attached bacteria. Borate reacts with the adjacent hydroxyl groups of sugars (Zittle, 1951) to give negatively charged groups, which could enter into cross-linkage reactions or whose mutual electrostatic repulsion could disrupt the parent carbohydrate. Agglutination of certain bacterial strains can be induced through the addition of boric acid or disodium tetraborate (Zittle, 1951). Electron microscopy showed that borate had a marked effect on both extracellular and intracellular integrity (Fig. 11.6). Although the cell wall remained largely intact, the extracellular adhesive coat had been denatured and cell contents aggregated. Thus, the disruptive effects of periodate and borate on attachment add further tentative suggestions that a cell surface polysaccharide is involved in attachment.

It is not surprising that antiserum inhibited the attachment of free-living cells, and yet strengthened the adhesion of attached cells (Table 11.1). Antibodies reacting with the surface polymeric coat might be expected to block adhesive sites and impair subsequent attachment. On the other hand, when the surface polymer has already adhered to a substratum, attachment would be strengthened through cross-linking of the adhesive by antibodies.

The influence of BSA on attachment (Table 11.1) is more intriguing, since analogy with the antiserum results suggests that BSA is combining with sites on the cell surface. Other proteins, such as gelatin, fibrinogen, pepsin or protamine sulphate, do not inhibit the attachment of free-living cells (Fletcher, 1976), so that the clue to the effect of BSA might lie in its ability to bind non-polar substances (for example chloresterol and other steroids (Fruton and Simmonds, 1958)), suggesting that non-polar binding sites may occur on the bacterial surface adhesive. Such interactions are attributed to BSA's non-polar side chains, containing leucine, phenylalanine and serine. Although this suggestion of hydrophobic binding sites on the bacterial adhesive is only slight, it assumes greater significance in Section 11.3.3.

It was not possible to learn more about the chemical structure of this adhesive through enzymatic digestion, since none of a wide range of enzymes removed attached cells (pronase, trypsin, α- and β-amylases, cellulase, β-glucosidase, hyaluronidase, neuraminidase, pectinase, phospholipase C). However, pronase has been found to efficiently remove some strains of attached marine pseudomonads, whereas trypsin caused their removal to a lesser extent (Danielsson *et al.*, 1977). Additional evidence for a proteinaceous adhesive is the fact that firm adhesion of certain marine bacteria has been prevented by chloramphenicol, an inhibitor of protein synthesis (Marshall, 1972).

In summary, then, in many cases adhesion of bacteria is due to the presence of extracellular polymers, which may comprise polysaccharide and/or protein. These polymers have often been found to bear negative ionic groups, but it is not known whether these groups are actually active in adhesion. It is possible that non-polar groups may be involved.

11.3 THE SUBSTRATUM

The second main component of the mechanism of bacterial adhesion to surfaces is the substratum itself. The physico-chemical properties of various substrata can differ enormously and bacterial attachment to these surfaces could be expected to vary accordingly. A number of *in situ* studies have been carried out by submerging different materials in the sea, by retrieving these samples at time intervals and evaluating the organisms which have attached (O'Neill and Wilcox, 1971; Sechler and Gundersen, 1972; Dexter *et al.*, 1975; Loeb, 1977). In many cases, disparate materials showed different types and numbers of attached micro-organisms, especially during the first few days of submersion (Sechler and Gundersen, 1972).

11.3.1 The physico-chemical properties of surfaces

At the surface of a bulk solid, there are unsatisfied bonds, and this potential bonding energy of the surface is represented by its surface free energy. Inorganic solids with high melting points, e.g. metals, glass, tend to have high surface free energies ($5000-500$ erg cm^{-2}), whereas soft organic solids, with low melting points, usually have free energies of less than 100 erg cm^{-2} (Zisman, 1964). Several types of forces contribute to the free energy of a surface, and these include dispersion, dipole, electrostatic and metallic forces (Andrade, 1973). Electrostatic charge may be a particularly important component, and the surface charge of bacteria has been shown to affect their attachment (Heckels *et al.*, 1976). There is the tendency for the bonding potential of the surface to become partially satisfied (thereby reducing surface free energy) through the adsorption of substances onto the surface. The adsorption of material onto a surface will largely be determined by

(i) the surface free energy of the solid phase and the surface tension (i.e. surface free energy) of the liquid phase, and

(ii) the types of forces common to both phases, so they can thereby interact.

Since adsorption of a liquid onto a solid can only to some extent satisfy the bonding potential of both phases, the remaining unsatisfied bonding potential is known as the interfacial tension or interfacial free energy. This interfacial tension affects not only the adsorption of dissolved components at the interface, but has been shown to influence the accumulation or movement of bacteria at the phase boundary. There may be a more rapid and extensive accumulation of bacteria at a water/organic liquid boundary of high interfacial tension than at one of low tension (Marshall, 1976). Also when the interfacial tension is high, movement of bacteria along the boundary is jerky and spasmodic, whereas when low, there is vigorous and smooth streaming of bacteria along the boundary (Marshall, 1976). This behavior of bacteria is dependent not only upon the interfacial tension between the two liquid phases, but also on the surface characteristics of the bacteria. This is illustrated by the disposition of bacteria at the monolayer lipid films which occur at the surface of most natural bodies of water. Some bacteria (e.g. *Serratia marinorubra*, *Acholeplasma laidlawii*) may actually penetrate the film, whereas others (e.g. *Pseudomonas fluorescens*) form a layer underneath (Kjelleberg *et al.*, 1976). Similarly, the accumulation and movement of bacteria at solid/liquid boundaries should be influenced by interfacial tension and bacterial surface properties.

The free energy of a solid can be evaluated by measuring the contact angles of a series of liquids with a range of surface tensions. When the surface tensions of the liquids are plotted against the cosines of their respectives advancing contact angles on the solid, the intercept of the resultant straight line with a cosine of 1 gives the critical surface tension, an indication of surface free energy (Zisman, 1964). The measurement of water contact angles on solids is a less reliable indication of surface energy, since water is able to enter into certain specific interactions, such as hydrogen bonding and polar interactions. However, if it is actually the relationship between the solid and water which is of prime interest (as may be the case in aquatic studies), then it may be the contact angle of water, that is the hydrophobicity of the surface, which is the most pertinent measurement.

11.3.2 Adsorption of dissolved substances on surfaces

The intrinsic character of a submerged solid is invariably masked through the adsorption of dissolved inorganic and organic substances, as well as through the adsorption of water itself. With smaller molecules, an adsorption equilibrium is generally established, with both adsorption and desorption taking place. Polymers, however, tend not to desorb because of the large number of anchoring sites (Kipling, 1965). When solids are submerged in the sea, an adsorbed film rapidly forms, and the adsorbed material is largely organic with a moderate negative charge (Neihof and Loeb, 1972; Loeb and Neihof, 1975).

The attachment of bacteria to a solid surface may be thought of as the adsorption of the bacterial surface adhesive on the substratum. Thus, in a given system, there will be simultaneous, and possibly competitive, adsorption of the bacterial adhesive, other dissolved polymers and water. It is extremely difficult to predict which substances will be adsorbed, much less the sequence of adsorption, but generally, polar substances will be more strongly adsorbed than non-polar ones by polar surfaces, and the reverse is usually the case for non-polar substances. The capacity of a substratum to enter into more specific reactions, such as hydrogen bonding, may also be responsible for the preferential adsorption of a particular substance (Kipling, 1965). Is it possible then to find a relationship between bacterial attachment to surfaces and these theoretical bases for adsorption mechanisms, that is surface and interfacial free energies, hydrophobicity and polymer adsorption?

11.3.3 Attachment of the marine pseudomonad to different substrata

There is a clear relationship between the hydrophobicity of organic substrata and the attachment of this bacterium (Fletcher and Loeb, 1979). This is shown in Fig. 11.7, where the contact angles of water on different substrata have been plotted against the corresponding numbers of attached bacteria. Hydrophobic substrata such as polystyrene (non-wettable bacteriological dishes) show maximum numbers

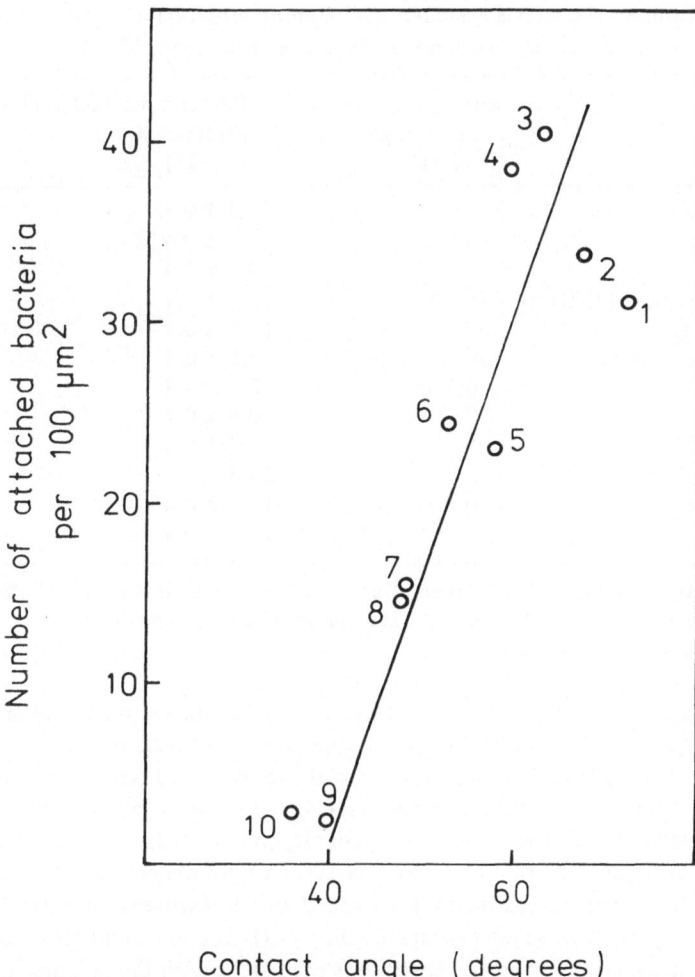

Fig. 11.7 The relationship between numbers of attached pseudomonads and advancing contact angles of water on different surfaces. The substrata are: 1, polytetrafluorethylene; 2, polyethylene; 3, polystyrene; 4, poly(ethylene terephthalate); 5, polyethylene recrystallized against gold foil to increase surface energy; 6, nylon 6.6; 7, epoxy resin; 8, nylon 6.6 recrystallized against gold foil to increase surface energy; 9, poly (ethylene terephthalate) treated in a radio-frequency plasma cleaning device to increase water wettability; 10, polystyrene, also treated in the radio frequency plasma cleaning device (after Fletcher and Loeb, 1979).

of attached bacteria; in a bacterial suspension containing approximately 10^9 bacteria ml^{-1}, the substratum is completely covered with attached bacteria within

Table 11.2 Number of bacteria attached to different substrata after two hours exposure to 10^9 bacteria ml^{-1} suspension (Fletcher and Loeb, 1979).

Substratum	Advancing water contact angle (degrees)	$\bar{x}$ Bacteria per 100 μm^2 ± confidence limits (p = 0.05)	I_a *
Polytetrafluoroethylene	105	31.2 ± 0.3	0.8
Polyethylene	95	33.8 ± 0.4	0.8
Polystyrene	86	40.5 ± 0.4	1.0
Poly(ethylene terephthalate)	79	38.6 ± 0.4	0.95
Epoxy resin	57	15.5 ± 0.3	0.5
Nylon 6.6	66	24.0 ± 0.4	0.9
Mica	0	2.6 ± 0.4	0.1
Glass	0	0.8 ± 0.2	0.02
Germanium†	0	28.2 ± 0.7	0.7
		23.3 ± 0.6	0.6
Platinum	0	13.4 ± 0.6	0.3
		15.5 ± 0.4	0.6

*   Index of attachment is the number of attached bacteria on the test substratum divided by the number of attached bacteria on a polystyrene control.
†   Cleaned in reducing conditions.

several hours, although the bacteria do not tend to touch one another and are separated by a finite space (Fletcher, 1977; Fletcher and Loeb, 1976.

Attachment to hydrophilic substrata is not so frequent and is more complex (Table 11.2). Numbers of attached bacteria are very low on negatively charged high energy materials, such as glass and mica. However, on essentially neutral (germanium) or positively charged (platimum) high energy surfaces, numbers of attached bacteria are significantly higher, suggesting that surface charge influences attachment to these high energy surfaces (Fletcher and Loeb, 1979). The low numbers of attached bacteria on the negatively charged substrata is consistent with the presence of a negatively charged surface polymer, since electrostatic repulsion could then account for inhibition of attachment.

The possibility of hydrophobic interactions being responsible for a wide range of attachment mechanisms in aqueous environments must be seriously considered. The literature contains a number of examples of attaching organisms showing a preference for hydrophobic substrata. These include bacteria attaching to siliconized germanium (Marshall, 1976), to glass coated with trimethylchlorosiline (Zvyagintsev, 1967), and to polystyrene (Loeb, 1977; Rutter and Abbott, 1978; Fletcher and Loeb, 1979). Other aquatic organisms, including bryozoa (Eiben, 1976; Loeb, 1977) and the alga *Clorella* (Loeb, 1977) have also been found to attach preferentially to hydrophobic substrata. Moreover essentially hydrophilic polymers, such as proteins, have been shown to adsorb to a greater extent on hydrophobic surfaces, as compared with hydrophilic substrata, corresponding to a greater reduction in interfacial energy (MacRitchie, 1972).

However, one investigation (Dexter *et al.*, 1975) reported higher numbers of attached marine bacteria on glass and other hydrophilic surfaces, a decrease in rate of attachment with a decrease in wettability, and a rise in attachment on very hydrophobic surfaces, such as polytetrafluoroethylene. Maximum numbers of attached bacteria occurred on surfaces with critical surface tensions ranging from 20–25 dyne cm^{-1}. This is in agreement with the figure of 20–30 dyne cm^{-1} which has been given as the range of critical surface tension giving minimum biological adhesion (Baier, 1972). There is not full agreement on this so-called 'biocompatibility range' of critical surface tension (Andrade, 1973), and it is quite possible that these discrepancies between reported data are determined by either dissolved components in individual systems or the conditioning and alteration of substrata through the adsorption of polymers, particularly protein (Baier and Weiss, 1975).

11.3.4 The influence of adsorbed polymers on bacterial attachment

Polymers adsorbed onto attachment substrata have been shown to inhibit subsequent attachment of bacteria. The proteins bovine serum albumin, gelatin, fibrinogen and pepsin inhibited the attachment of this pseudomonad to polystyrene (Fletcher, 1976), and albumin, concanavalin A, whole saliva and serum interferred with the attachment of streptococci to glass (Ørstavik, 1977). The presence of an adsorbed protein film may inhibit the attachment of the pseudomonad to polystyrene simply by converting the favourable hydrophobic surface to an unfavourable hydrophilic one. This would also explain the levelling off of numbers of attached bacteria once a monolayer of attached bacteria has formed (Fig. 11.8). However, adsorbed polymers may inhibit the subsequent adsorption of bacterial adhesive polymers by other mechanisms, such as steric or entropic effects (Hesselink, Vrij and Overbeek, 1971; Maroudas, 1973, 1975b; Napper, 1977). It is well established that non-ionic polymers can stabilize colloidal suspensions, preventing mutual adsorption of particles and flocculation (Napper, 1977). The influence of adsorbed polymers on bacterial attachment is an extremely complex aspect, which requires much more detailed study.

There are two further, and somewhat speculative points, which should be mentioned. First, the tendency for a bacterium to become attached to a particular substratum may be quite distinct from the subsequent strength of adhesion in the long term. As the bacteria attached to a surface metabolize, cell surface polymers may alter or diffuse away from the cells, so that once efficient adhesives become less efficient, and the bacteria may be desorbed (Zvyagintsev *et al.*, 1971, 1977). The time-dependent desorption of *Bacillus mycoides* and *Serratia marcescens* from various substrata has been demonstrated and attributed to exoenzymes which broke down the adhesive (Zvyagintsev *et al.*, 1977).

Secondly, a particular organism may be capable of attaching to surfaces by more than one type of mechanism. The marine pseudomonad attaches spontaneously to

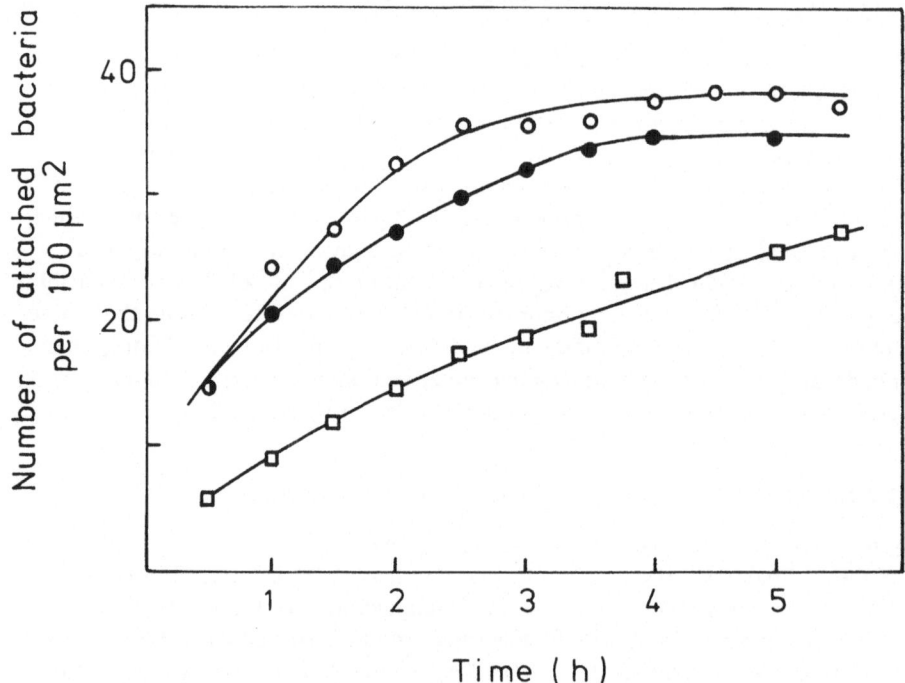

Fig. 11.8 The relationship between culture concentration and the number of pseudomonads which become attached to polystyrene after two hours. Note the decrease in attachment with increase in culture age and the levelling off of the curves as the substratum becomes covered with bacteria. (O) Log phase, (●) stationary phase, (□) death phase (after Fletcher, 1977).

hydrophobic surfaces, but with time it can also attach to hydrophilic materials. Although numbers of attached bacteria on hydrophilic surfaces is initially small, there may be a build-up of cells so that a bacterial film is formed within a few days. This slower attachment to hydrophilic surfaces is probably due to the time-dependent production of extracellular polymers. Thus attachment may not always be dependent upon just one mechanism.

11.4 OTHER FACTORS INFLUENCING BACTERIAL ATTACHMENT

11.4.1 The state of the organisms

The tendency for bacteria to become attached and the strength of attachment depends to some extent on their physiological activity. This is particularly noticeable with batch

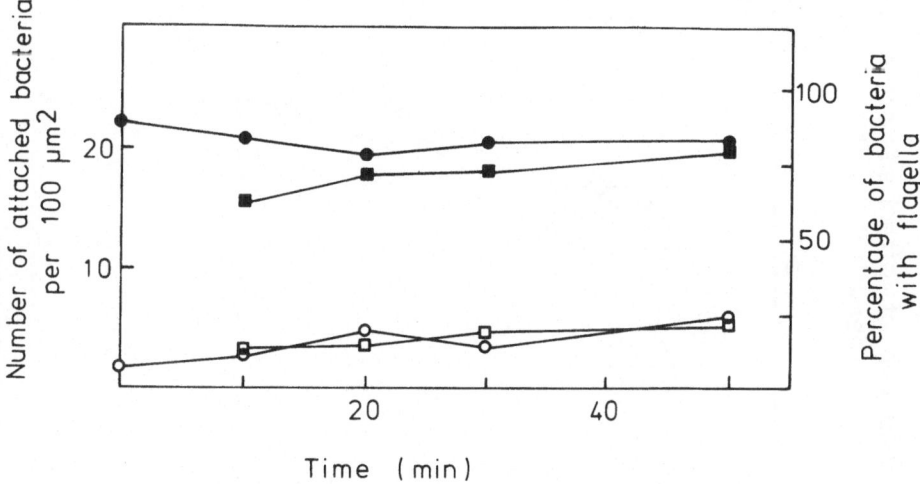

Fig. 11.9 The influence of removal of flagella by homogenization on attachment. Percentage of bacteria with flagella: (●) control; (○) homogenized suspension; Number of attached bacteria: (■) control; (□) homogenized suspension (after Fletcher, 1979).

cultures, where attachment tendencies can vary with the growth phase. With the marine pseudomonad, bacteria from log phase cultures have been found to have the greatest tendency to attach to surfaces, and there was a progressive decline in numbers of bacteria which become attached after the onset of stationary phase and during death phase (Fig. 11.8) (Fletcher, 1977). Similar studies have shown a decrease in attachment strength with increase in culture age (Zvyagintsev *et al.*, 1971, 1977). The most likely explanation for such a decline is a progressive change in quality or quantity of cell surface polymer. However, with the marine pseudomonad, cell motility also may have been a factor, because the decrease in attachment which occurred at the onset of the stationary phase coincided with a decrease in motility (Fletcher, 1977). Moreover, if the flagella were removed from log phase organisms, there was a marked decrease in their subsequent attachment (Fig. 11.9). Cell motility could facilitate attachment either by increasing the number of bacterial collisions with the substratum, thereby increasing the statistical probability of attachment, or by increasing the force with which the bacterium encounters the surface, helping to overcome electrostatic repulsive forces.

Bacterial attachment is also affected by the nutrients available to the organisms. Marshall *et al.* (1971a) found that the attachment of a marine pseudomonad was dependent upon the concentration of the carbon source glucose. At concentrations of 7 mg l^{-1}, attachment was favored, whereas attachment was reduced at higher levels of 14 and 21 mg l^{-1}, and it was completely inhibited at 30 and 70 mg l^{-1}. At such low glucose levels, it is likely that these differences in attachment are due to

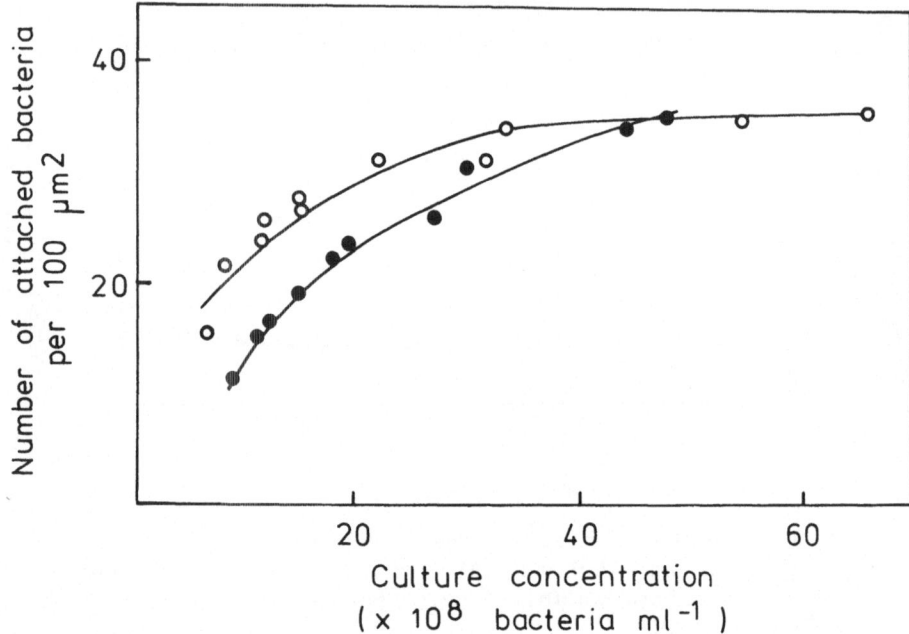

Fig. 11.10 Relationship between temperature and the number of pseudomonads attached to polystyrene. (O) 20°C ± 1; (●) 3°C ± 1.

consequent variations in the organisms rather than to an influence on the substratum through adsorption. An attempt was made to correlate the higher levels of attachment at low glucose levels with polymer production; at 7 mg l^{-1}, electron microscopy demonstrated many polymeric fibrils attaching the bacteria to the substratum, whereas at higher glucose levels, polymer production was variable and not obviously correlated with attachment numbers (Marshall *et al.,* 1971a). Adhesive polymer production certainly can be influenced by the nature of the carbon source, as was shown in the micrographs of microcolonies of the marine pseudomonad (Figs. 11.2 and 11.3).

Bacterial attachment properties are also dependent on whether carbon or nitrogen is the limiting nutrient. When an enrichment culture from river water was grown under nitrogen limitation in a chemostat, few bacteria attached to aluminium, but these had produced abundant quantities of extracellular polymer. By contrast, with carbon limitation there were many attached bacteria and no apparent accumulation of polymer on the aluminium substratum (Brown *et al.,* 1977). Indeed, it is probably not the quantity of polymer which is important in determining attachment, but its actual adsorption and solubility characteristics in the system.

Temperature may also influence the tendency for bacteria to attach (Fig. 11.10). Increased attachment of the pseudomonad at 20°C, as compared to 3°C, was probably due to a physiological response, or possibly to an alteration in polymer viscosity, hence adhesive efficiency (Fletcher, 1977). Temperature has been shown to influence both the quantity of extracellular polymer produced (Rose, 1967) and its apparent stickiness (Stanley and Rose, 1967).

11.4.2 Physicochemical factors

(a) *Cation concentration*

The presence of dissolved substances in the medium is a critical factor in attachment, through their influence on interactions between the bacterial and substratum surfaces. The importance of dissolved polymers, particularly proteins, was stressed in Section 11.3.2; however, cation concentration may be equally important. Cations can influence the attachment of marine bacteria

(i)   indirectly by influencing cell physiology or membrane permeability (Drapeau and MacLeod, 1965),

(ii)   directly through their accumulation at surfaces and formation of the electric double layer (Shaw, 1970), thereby influencing repulsion forces between bacterial and substratum surfaces, or

(iii)   by helping to maintain the structural integrity of charged adhesive polymers through cross-linking of acidic groups.

Ca^{2+} and Mg^{2+} concentrations have been shown to influence the attachment of marine bacteria. For example, a marine pseudomonad would not become firmly attached in the absence of both cations, but would adhere when either Ca^{2+} or Mg^{2+} was present (Marshall *et al.*, 1971a). Also the initial reversible adsorption of an *Achromobacter* species has been shown to correlate with cation concentration (Marshall *et al.*, 1971a), and this was accounted for by a decrease in thickness of the electric double layer. As double layer thickness at a surface decreases, repulsion forces between that surface and an opposing surface are accordingly reduced (Shaw, 1970).

Sequestering agents have also been used to inhibit or disrupt attachment (Table 11.3). The attachment of the marine pseudomonad to glass was inhibited by ethylenediamine tetraacetic acid (EDTA), and the degree of inhibition was related to EDTA concentration. EDTA treatment would not remove attached pseudomonads, but treatment with pyrophosphate was effective. Treatment with pyrophosphate has also been necessary for the subsequent removal of attached marine bacteria by periodate (Marshall, 1972).

The ability of the sequestering agent to remove attached cells suggests that cations are necessary for the efficiency of the adhesive polymer, possibly by cross-linking or screening acidic groups on the adhesive. This is supported by the ability of Al^{3+} and La^{3+} to inhibit attachment of the marine pseudomonad (Fig. 11.11) (Fletcher, 1979). These trivalent cations might be expected to facilitate attachment by decreasing the thickness of the electric double layer and reducing repulsive forces. However, their inhibitory effect suggests that they can combine with negatively

Table 11.3 The effects of sequestering agents on the attachment of the marine pseudomonad to glass

	Treatment	I_a*	Influence on attachment†
Free-living bacteria	EDTA $(1 \times 10^{-5} M)$	0.7	−
	EDTA $(5 \times 10^{-5} M)$	0.4	−
	EDTA $(1 \times 10^{-4} M)$	0.4	−
	EDTA $(1 \times 10^{-3} M)$	0.01	−
Attached bacteria	NaCl (0.02 M) → Na$_4$P$_2$O$_7$ (0.1 M)‡	0.003	−
	EDTA (0.01 M)	1.2	0
	NaCl (0.02 M) → EDTA (0.1 M)	1.1	0

* The index of attachment is the number of attached bacteria after treatment divided by the number of attached bacteria in an untreated control.
† (−) reduction in attachment; (0) no effect.
‡ Bacteria were treated first in dilute NaCl to remove extraneous cations which might reduce effect of treatment.

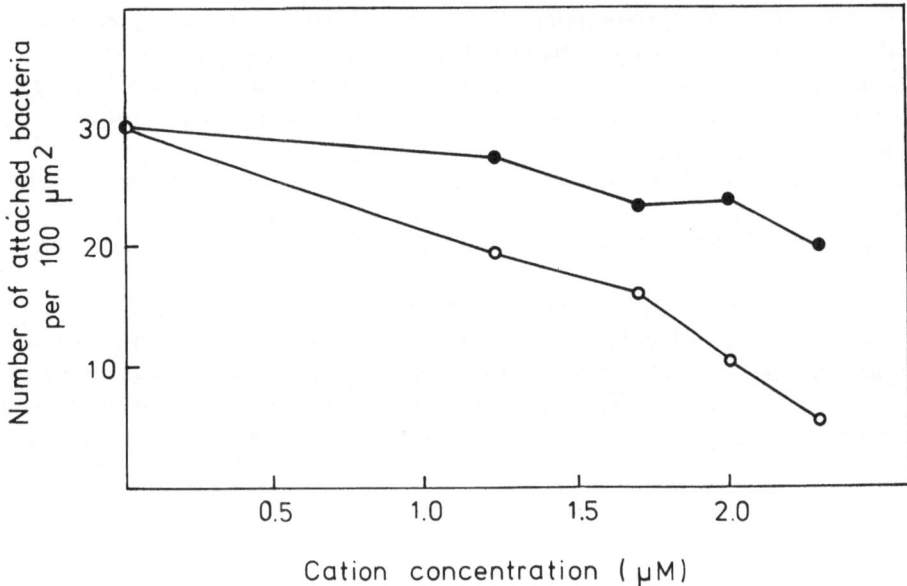

Fig. 11.11 The influence of La^{3+} (○) and Al^{3+} (●) on attachment of the pseudomonad to polystyrene. Note the decrease in attachment with the increase in concentration of trivalent cation, particularly lanthanum.

charged sites on the adhesive (possibly displacing less strongly bound divalent and monovalent cations), denaturing it and reducing its efficiency.

There is also electron microscopic evidence that Ca^{2+} and Mg^{2+} may be important in maintaining structural integrity of the intercellular polymeric matrix within attached microcolonies of the marine pseudomonad (Fletcher and Floodgate, 1976). When these microcolonies are transferred to media deficient in Ca^{2+} and Mg^{2+} there is almost immediate denaturation of the polymer (Fig. 11.12). It loses its fine reticular

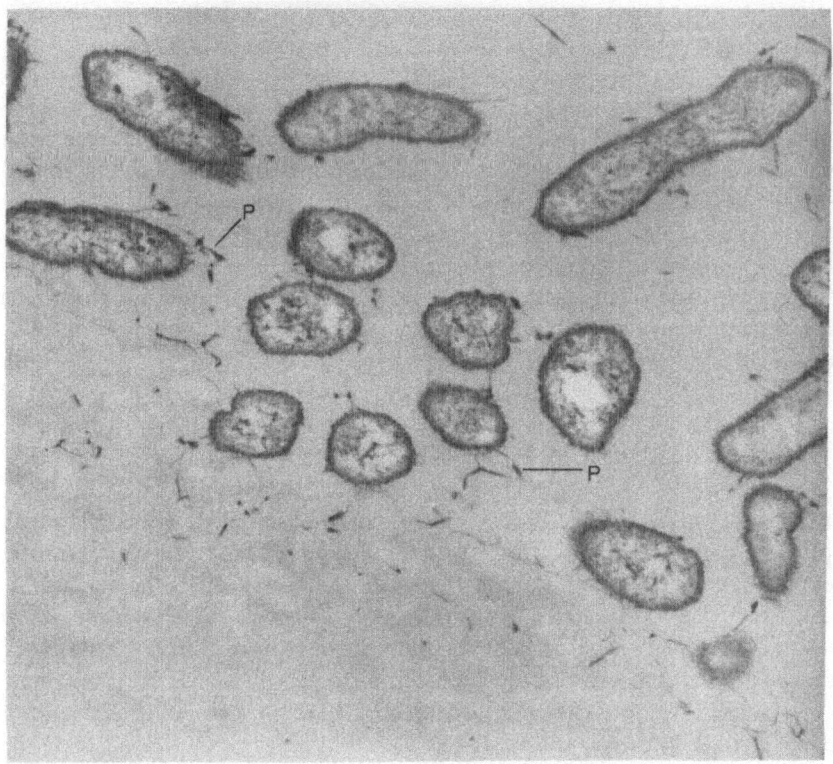

Fig. 11.12 Ruthenium red preparation of a microcolony transferred to a medium deficient in Ca^{2+} and Mg^{2+} and left for 5 minutes before fixation. Note that the intercellular polymer (P) has condensed, whereas the polymeric coat remains unaffected.

appearance and forms condensed globules in the intercellular spaces. This could be explained by a depletion of the Ca^{2+} and Mg^{2+} within the matrix, which had been screening acidic groups on the polymer. When the polymer acidic groups were exposed, they were then precipitated by the ruthenium red stain. Note, however, in Fig. 11.12, that the polymeric cell surface coat has not been affected by the cation depletion, further indicating a fundamental difference between the two types of polymer.

(b) *Temperature*

Temperature has been mentioned as a factor which can influence attachment indirectly by influencing the physiology of the organism (Section 11.4.1). However, temperature can also have a direct influence by affecting chemical or physical adsorption of adhesive polymers. The effects of temperature on adsorption processes are complex, depending not only on the adsorbate and adsorbent but also on the complexity of the liquid phase (Kipling, 1965). Generally, however, physical adsorption is favoured by lower temperatures, whereas chemisorption is favoured by higher temperatures.

(c) *Hydrographical factors*

In natural environments especially, attachment of bacteria may be influenced by water movements and strength of current (Skerman, 1956). This is well illustrated by the large numbers of attached bacteria which tend to accumulate in depressions and fissures on surfaces, such as sand grains (Weise and Rheinheimer, 1978), whereas surfaces exposed to scouring by water currents and suspended solids tend to be clean. Such observations emphasize the extreme complexity of studying attachment in natural environments, where conditions may be variable and unpredictable and their significance may be extremely hard to assess.

11.5 SUMMARY AND OVERVIEW

One of the most critical questions concerning marine bacterial attachment is whether it is a passive process requiring no energy expenditure on the part of the bacterium, or whether it is an active process, suggesting that attachment offers the organism some advantage which allowed its evolutionary development. In fact, both types of attachment seem to occur. An example of spontaneous attachment is the marine pseudomonad which encounters polystyrene and frequently is adsorbed immediately. Active attachment also occurs, and is described by the two stages of reversible and irreversible attachment.

It is not yet certain what advantages the state of attachment could offer bacteria. Attachment could encourage growth by providing an environment which is rich in adsorbed macromolecular nutrients, as compared to the surrounding aqueous environment, or which provides more favourable physicochemical conditions, such as pH or oxygen tension.

The ability to attach to solid surfaces also could provide a survival advantage, in that the development of a thick slime layer provides a community environment which is protected from fluctuating, and often stressful, conditions in the surrounding medium. Growth in the matrix may be limited, due to slow diffusion of nutrients and metabolic products, but some bacteria survive until they are sloughed off or exposed to the medium by a grazing animal. Thus the slime layer acts as a bacterial reservoir capable of repopulating the aquatic environment when conditions favour growth.

In summary, there are two basic attachment mechanisms which utilise extracellular polymeric adhesives:

(1)   Spontaneous attachment occurs when a bacterium encounters a substratum and comes close enough for its extracellular polymeric coat to be adsorbed. Contact between the two surfaces can be prevented by electrostatic repulsive forces, but if these are not prohibitively large, attachment then depends on the adsorption interactions between the substratum and the bacterial adhesive. Hydrophobic interactions may be favored, otherwise the tendency for a bacterium to remain in suspension could be equal to its tendency to adhere to a highly hydrated hydrophilic substratum.

(2)   Time-dependent attachment involves the physiological production of extracellular adhesives. Such polymers tend to be highly hydrated gels which may eventually form an intercellular matrix within a group of cells. These polymers often bear a large number of acidic groups, which may require cations for screening or crosslinking interactions to maintain structural integrity of the adhesive. Spontaneous attachment can also be followed by polymer production, which may either strengthen or weaken the attachment bond.

A great many questions remain to be answered, many of these being relevant to bacterial attachment in other environments. Attachment by extracellular adhesives appears to be widespread, but we do not know which of the two modes — spontaneous or time-dependent attachment — occurs more frequently. Moreover, much more information is needed on the composition of cell surface polymers and their adsorption and solubility properties. We should look more closely at the importance of substratum characteristics to attachment in natural environments, particularly with respect to dissolved medium components and their conditioning of substrata through adsorption. Finally, data on attachment to marine plants and animals may help shed light on similar interactions in other environments by giving information on possible specific adhesive interactions or on the prevention of attachment by the production of inhibitors.

REFERENCES

Andrade, J.D. (1973), *Med. Instr.*, **7**, 110–119.

Baier, R.E. (1972), In: *Proc. 3rd Int. Congress on Marine Corrosion and Fouling.* (Acker, R.F., Brown, B.F., De Palma, J.R. and Iverson, W.P., eds), Northwestern University Press, Evanston, Ill., pp. 633–639.

Baier, R.E. and Weiss, L. (1975), *Adv. Chem. Ser.*, **145**, 300–307.

Behnke, O. (1968), *J. Ultrastruct. Res.*, **24**, 51–69.

Blanquet, P.R. (1976), *Histochemistry*, **47**, 175–189.

Brown, C.M., Ellwood, D.C. and Hunter, J.R. (1977), *FEMS Microbiol. Letters*, **1**, 163–166.

Corpe, W.A. (1970a), *Dev. Indust. Microbiol.*, **11**, 402–412.

Corpe, W.A. (1970b). In: *Adhesion in Biological Systems.* (Manly, R.S., ed.), Academic Press, New York, pp. 73–87.

Corpe, W.A. (1972), In: *Proc. 3rd Int. Congress on Marine Corrosion and Fouling.* (Acker, R.F., Brown, B.F., De Palma, J.R. and Iverson, W.P., eds), Northwestern University Press, Evanston, I 11. pp. 598–609.

Corpe, W.A., Matsuuchi, L. and Armbruster, B. (1976), In: *Proc. 3rd Int. Biodegradation Symposium.* (Sharpley, J.M. and Kaplan, A.M., eds), Applied Science Publishers, London, pp. 433–442.

Costerton, J.W., Geesey, G.G. and Cheng, K.J. (1978), *Sci. Am.,* **238**, 86–95.

Crisp, D.J. (1972), In: *Proc. 3rd Int. Congress on Marine Corrosion and Fouling.* (Acker, R.F., Brown, B.F., De Palma, J.R. and Iverson, W.P., eds), Northwestern University Press, Evanston, I 11., pp. 633–639.

Cundell, A.M., Sleeter, T.D. and Mitchell, R. (1977), *Microb. Ecol.,* **4**, 81–91.

Curtis, A.S.G. (1967), *The Cell Surface.* Academic Press, London.

Dazzo, F.B., Napoli, C.A. and Hubbell, D.H. (1976), *App. env. Microbiol.,* **32**, 166–171.

Danielsson, A., Norkrans, B. and Björnsson, A. (1977), *Bot. Mar.,* **20**, 13–17.

Dexter, S.C., Sullivan, J.D., Williams, J. III and Watson, S.W. (1975), *App. Microbiol.,* **30**, 298–308.

Dhir, S.P. and Boatman, E.S. (1972), *J. Bact.,* **111**, 267–271.

Drapeau, G.R. and MacLeod, R.A. (1965), *Nature,* 206–531.

Eiben, R. (1976), *Mar. Biol.,* **37**, 249–254.

Fletcher, M. (1976), *J. gen. Microbiol.,* **94**, 400–404.

Fletcher, M. (1977), *Can. J. Microbiol.,* **23**, 1–6.

Fletcher, M. (1979), In: *Adhesion of Microorganisms to Surfaces.* (Ellwood, D.C., Melling, J. and Rutter, P.R. eds), Academic Press, London.

Fletcher, M. and Floodgate, G.D. (1973), *J. gen. Microbiol.,* **74**, 325–334.

Fletcher, M. and Floodgate, G.D. (1976), In: *Microbial Ultrastructure.* (Fuller, R. and Lovelock, D.W., eds), Academic Press, London, pp. 101–107.

Fletcher, M. and Loeb, G.I. (1976), In: *Colloid and Interface Science.* (Kerker, M., ed.), Vol. III, Academic Press, New York, pp. 459–469.

Fletcher, M. and Loeb, G.I. (1979), *App. env. Microbiol.,* **37**, 67–72.

Fruton, J.S. and Simmonds, S. (1958), *General Biochemistry,* 2nd edn., Chapman and Hall, London.

Geesey, G.G., Richardson, W.T., Yeomans, H.G., Irvin, R.T. and Costerton, J.W. (1977), *Can. J. Microbiol.,* **23**, 1733–1736.

Heckels, J.E., Blackett, B., Everson, J.S. and Ward, M.E. (1976), *J. gen. Microbiol.,* **96**, 359–364.

Hesselink, F.T., Vrij, A. and Overbeek, J.T.G. (1971), *J. Phys. Chem.,* **75**, 2094–2103.

Hirsch, P. and Pankratz, S.H. (1970), *Z. Allg. Mikrobiol.,* **10**, 589–605.

Jones, H.C., Roth, I.L. and Sanders, W.M. III (1969), *J. Bact.,* **99**, 316–325.

Kipling, J.J. (1965), *Adsorption from Solutions of Non-Electrolytes.* Academic Press, London.

Kjelleberg, S., Norkrans, B., Löfgren, H. and Larsson, K. (1976), *App. env. Microbiol.,* **31**, 609–611.

Latham, M.J., Brooker, B.E., Pettipher, G.L. and Harris, P.J. (1978), *App. env. Microbiol.,* **35**, 156–165.

Loeb, G.I. (1977), Naval Research Laboratory Memorandum Report 3665.

Loeb, G.I. and Neihof, R.A. (1975), *Adv. Chem. Ser.*, **145**, 319–335.

Luft, J.H. (1966), *Ruthenium Red and Violet. I. Chemistry, Purification, Methods of Use and Mechanism of Action.* University of Washington Press, Seattle.

Luft, J.H. (1971), *Anat. Rec.*, **171**, 347–368.

MacRitchie, F. (1972), *J. Coll. Interf. Sci.*, **38**, 484–488.

Marshall, K.C. (1972), In: *Proc. 3rd Int. Congress on Marine Corrosion and Fouling.* (Acker, R.F., Brown, B.F., De Palma, J.R. and Iverson, W.P. Northwestern University Press, Evanston, Ill., pp. 625–632.

Marshall, K.C. (1976), *Interfaces in Microbial Ecology.* Harvard University Press, Cambridge, Mass.

Marshall, K.C. and Cruickshank, R.H. (1973), *Arch. Mikrobiol.*, **91**, 29–40.

Marshall, K.C., Stout, R. and Mitchell, R. (1971a), *J. gen. Microbiol.*, **68**, 337–348.

Marshall, K.C., Stout, R. and Mitchell, R. (1971b), *Can. J. Microbiol.*, **17**, 1413–1416.

Maroudas, N.G. (1973), *Nature*, **244**, 353–354.

Maroudas, N.G. (1975a), *J. theor. Biol.*, **49**, 417–424.

Maroudas, N.G. (1975b), *Nature*, **254**, 695–696.

Merkel, G.J. (1975), *Water Res.*, **9**, 881–885.

Mitchell, R. and Cundell, R.M. (1977), Office of Naval Research Contract N00014-76-C-0042 NR-1 04-967 Technical Report No. 3.

Napper, D.H. (1977), *J. Coll. Interf. Sci.*, **58**, 390–407.

Neihof, R.A. and Loeb, G.I. (1972), *Limnol. Oceanog.*, **17**, 7–16.

O'Neill, T.B. and Wilcox, G.L. (1971), *Pacific Sci.*, **25**, 1–12.

Ørstavik, D. (1977), *Acta path. microbiol. scand.* Sect. B., **85**, 47–53.

Pate, J.L. and Ordal, E.J. (1967), *J. Cell Biol.*, **35**, 37–51.

Pearse, A.G.E. (1968), *Histochemistry: Theoretical and Applied.* Vol. 1. J.A. Churchill, London.

Rambourg, A. (1971), In: *International Review of Cytology.* (Bourne, G.H. and Danielli, J.F., eds), Academic Press, New York, pp. 57–114.

Ridgway, H.F. and Lewin, R.A. (1973), *J. gen. Microbiol.*, **79**, 119–128.

Rose, A.H., ed. (1967), *Thermobiology.* Academic Press, London.

Rutter, P.R. and Abbott, A. (1978), *J. gen. Microbiol.*, **105**, 219–226.

Sechler, G.E. and Gundersen, K. (1972), In: *Proc. 3rd Int. Congress on Marine Corrosion and Fouling.* (Acker, R.F., Brown, B.F., De Palma, J.R. and Iverson, W.P. eds), Northwestern University Press, Evanston. Ill., pp. 610–616.

Shaw, D.J. (1970), *Introduction to Colloid and Surface Chemistry*, 2nd edn., Butterworths, London.

Sieburth, J.M. (1968), *Adv. Microbiol. Sea*, **1**, 63–94.

Sieburth, J.M. (1975), *Microbial Seascapes.* University Park Press, Baltimore, Md.

Skerman, T.M. (1956), *N. Z. J. Sci. Tech.*, Sec. B, **38**, 44–57.

Stanley, S.O. and Rose, A.H. (1967), *J. gen. Microbiol.*, **48**, 9–23.

Vargo, G.A., Hargraves, P.E. and Johnson, P. (1975), *Mar. Biol.*, **31**, 113–120.

Weise, W. and Rheinheimer, G. (1978), *Microb. Ecol.*, **4**, 175–188.

Young, L.Y. (1978), *Microb. Ecol.*, **4**, 267–277.

Zisman, W.A. (1964), *Adv. Chem. Ser.*, **43**, 1–51.

Zittle, C.A. (1951), *Adv. Enzymol.*, **12**, 493–527.

ZoBell, C.E. (1943), *J. Bact.*, **46**, 39–56.

ZoBell, C.E. and Allen, E.C. (1935), *J. Bact.*, **29**, 239−251.
Zvyagintsev, D.G. (1967), *Lab. Delo*, **6**, 345−348.
Zvyagintsev, D.G., Guzev, V.S. and Guzeva, L.S. (1977), *Microbiology*, **46**, 245−249.
Zvyagintsev, D.Z., Pertsovskaya, A.F., Yakhnin, E.D. and Averbakh, E.I. (1971), *Microbiology*, **40**, 889−893.

12 Microbial Adherence in Plants

JAMES A. LIPPINCOTT
and
BARBARA B. LIPPINCOTT

12.1	Introduction	*page*	377
	12.1.1 General background		378
	12.1.2 Plant surfaces		378
12.2	*Agrobacterium*–host adherence		378
	12.2.1 Adherence required in tumor initiation		378
	12.2.2 Bacterial lipopolysaccharide active in adherence		380
	12.2.3 Host component of adherence		382
12.3	*Rhizobium*–host attachment		385
	12.3.1 Evidence for adherence		385
	12.3.2 Bacterial antigens and surface polysaccharides		386
	12.3.3 Lectins and host–pathogen cross-reactive antigens		387
12.4	Microbial adherence in the induction of plant hypersensitive necrosis		389
	12.4.1 Evidence for adherence		389
	12.4.2 Role of bacterial lipopolysaccharides and host lectins in adherence		390
12.5	Additional examples of adherence in micro-organisms–plant interactions		391
	12.5.1 Bacteria		391
	12.5.2 Fungi		392
12.6	Summary and prospectus		393
	References		395

Acknowledgement
Work from our laboratory was aided by Public Health Service research grant AI12149 from the National Institute for Allergy and Infectious Deseases.

Bacterial Adherence
(*Receptors and Recognition,* Series B, Volume 6)
Edited by E.H. Beachey
Published in 1980 by Chapman and Hall, 11 New Fetter Lane, London EC4P 4EE
© 1980 Chapman and Hall

12.1.1 General background

The importance of adherence in the establishment of plant–microbe interactions and the alteration of plant metabolism has only recently begun to be appreciated and to be a direct subject of investigation. A search for factors responsible for specificity in host–pathogen, host–symbiont interactions and for the mechanisms involved in the induction of disease symptoms and host defense reactions provided the initial incentive for most of the work which comprises this developing field. A few reviews have covered certain aspects of this problem, e.g., Albersheim and Anderson-Prouty (1975), Marshall (1976), Raa *et al.* (1977), Bauer (1977), Sequeira (1978), Beringer *et al.* (1979) and Dazzo (1979). The present review will focus primarily on interactions of micro-organisms with seed plants; adherence interactions involving micro-organisms with algae, yeasts and fungi or between different portions of the plant life cycle are outside the scope of this review.

The remarkable specificity shown in many plant–microbe interactions is easily rationalized by assuming that complementary molecular structures each with a high affinity for the other occur on the surfaces of the interacting organisms. Because these outer surfaces constitute the first point of direct physical interaction between the two organisms, they would thus function as a molecular recognition mechanism and consequently might also determine specificity. As long as one considers only the adherence of organisms to cells such a direct correlation between adherence, recognition and specificity poses no problem. When more complex developments are used to measure the interaction such as disease symptom formation or the induction of host defense systems, many factors in addition to adherence may function to determine if the specific response sought will occur. In such cases, adherence may constitute only one small part of the specificity-recognition system, or, in the extreme and even though essential, an adherence mechanism could prove so non-specific that it would ordinarily have no role in specificity-recognition phenomena.

By far the most progress in elucidating microbial–plant adherence mechanisms has been made in studies of the two genera of the *Rhizobiaceae, Agrobacterium* and *Rhizobium,* which respectively induce tumors or dinitrogen-fixing nodules in plant hosts, and of the hypersensitive response (HR) that occurs between pathogens and incompatible host tissues, resulting in a rapid collapse, dessication and browning of cells surrounding the point of microbial invasion. In each case these are complex host responses which undoubtedly develop only after a series of sequential events and may depend on much more from the micro-organism than simple contact adherence. These three systems will be considered in detail below, followed by several specific micro-organism—plant interactions where important, but less complete, results on adherence have appeared.

12.1.2 Plant surfaces

Plant surfaces to which micro-organisms potentially might adhere are rather poorly characterized. Surfaces of the aerial portion of plants are covered by a cuticle layer composed largely of fatty materials and waxes secreted by the underlying epidermal cells and often covered by a layer of crystalline wax (Holloway, 1971; Cutter, 1976; Jeffree *et al.*, 1976). A thinner version of this cuticle seems to cover the intercellular spaces in leaves which are in direct contact via stomatal openings with the external atmosphere (Goodman *et al.*, 1976). In addition, many plants have special secretory cells and hair cells or trichomes variously distributed over areas of the leaf and stem surfaces and organized groups of secretory cells are often associated with specific portions of the plant, e.g., flowers. The materials secreted by these cells and glands is varied and, in general, not well characterized. Plant roots are exposed to a very different environment and are covered by a mucilagenous layer, the 'mucigel', composed in part of acidic polysaccharides (Dazzo, 1979). A chemically character-ized polysaccharide slime is also secreted by cells of the root cap covering the extreme tips of the growing root system (Paull and Jones, 1975). Beneath this mucigel and slime the normal components of plant cell walls appear to constitute the first 'solid' surface.

Wounds expose additional surfaces that are essential for the establishment of many plant—microbe associations. The particular nature of the wound surfaces responsible for recognition undoubtedly varies depending on the particular micro-organism and will be considered with individual microbe—plant interactions. It is important to recognize that, because of their sessile mode of life, plants are readily subject to mechanical wounding by wind and large animals and by feeding nematodes, worms, insects and vertebrates. Also, self-inflicted growth wounding occurs rather commonly in plants so that wound-exposed surfaces are a normal feature of plant life.

12.2 *AGROBACTERIUM*—HOST ADHERENCE

12.2.1 Adherence required in tumor initiation

Crown-gall induced by *Agrobacterium tumefaciens* is a commercially important plant disease characterized by the appearance of autonomous host outgrowths which, after initiation, are independent of the bacterium and, analogous to cancer cells, independent of the host (Lippincott and Lippincott, 1975). This gram-negative bacterium typically induces symptoms only when introduced into wound areas and does not appear to have any invasive powers. Using a bioassay analogous to a phage plating assay in which the number of tumors formed on bean leaves is proportional to the number of bacteria in the inoculum, Lippincott and Lippincott (1969) showed that certain avirulent strains of *Agrobacterium* as well as killed cells of virulent strains could inhibit tumor initiation. The inhibition was proportional to the number of avirulent or killed cells and occurred only when they were applied before or in conjunction with virulent tumor-inducing bacteria. When added only 15 minutes later, there was no inhibition, showing that a very early step in the tumor initiation process was involved and that inhibition was not dependent upon cell viability. Cells of representative species of *Corynebacterium, Bacillus, Escherichia, Pseudomonas* and *Rhizobium,* however, were non-inhibitory showing the phenomenon was dependent on specific features of the agrobacteria and, correspondingly, of the plant wound, which were not met by the other bacteria tested. Based on these data it was proposed that an initial step in *Agrobacterium* infections was the attachment of the bacterium to a wound-exposed host site.

These results were confirmed by Schilperoort (1969), Kerr (1969), Manigault (1970), Beiderbeck (1976), Glogowski and Galsky (1978) and Matthysse *et al.* (1978) and electron photomicrographs showing *Agrobacterium* adhering to plant cell walls

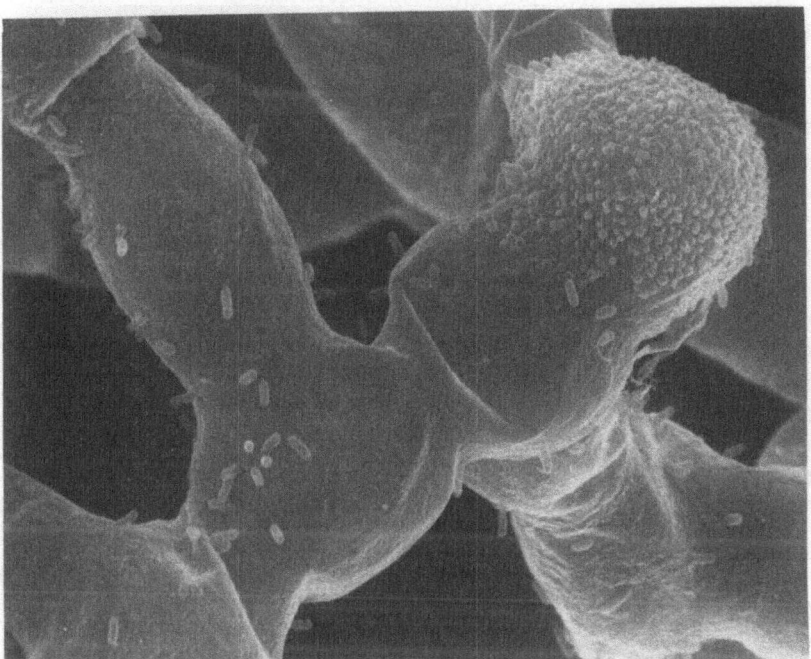

Fig. 12.1 Adherence of *Agrobacterium tumefaciens* strain B6 to the moss *Pylaisiella selwynii.* Examples of both polar and prone attachment are seen. The knobby structure in the upper left is the moss spore coat. SEM x 2280.

or to cultured cells have appeared (Bogers, 1972; Spiess *et al.,* 1977; Matthysse *et al.,* 1978; Ohyama *et al.,* 1979). Fig. 12.1 shows an example of *Agrobacterium* adhering to moss plants as visualized by scanning electron photomicrographs.

12.2.2 Bacterial lipopolysaccharide active in adherence

In a follow-up of these results, Whatley *et al.* (1976) showed that *Agrobacterium* cell envelope preparations, Boivin antigen preparations and lipopolysaccharide (LPS) from *Agrobacterium* were effective inhibitors of tumor initiation when applied with or before the infecting bacteria but, as with whole cells, were not effective when added 15 minutes later. Titration of the LPS inhibition showed that concentrations as low as 1 ng ml^{-1} were sufficient to inhibit the initiation of tumors by 35%. Since LPS molecules could account for the activity of both the cell envelope fragments and the Boivin antigen preparations, this component of the outermost membrane of gram-negative cell envelopes appears to be the major *Agrobacterium* component involved in wound site adherence. The inhibition by LPS is dependent on molecular structure since LPS from non-site binding strains of *Agrobacterium* is not inhibitory in these tumorigenicity assays.

The polysaccharide or O-antigen portion of *Agrobacterium* LPS, when isolated by acetic acid hydrolysis, is essentially equally as inhibitory as intact LPS molecules (Whatley and Lippincott, in preparation). The lipid A moiety solubilized with bovine serum albumin, however, is non-inhibitory, as are peptidoglycan fractions of the *Agrobacterium* envelope. Thus, it appears that, although the lipid A portion anchors the LPS molecule in the outer membrane, only the externally projecting LPS-poly-saccharide chains are involved in adherence. Tumorigenic agrobacteria have both chromosomal and plasmid-borne genes, each capable of specifying site-binding ability (Whatley *et al.,* 1978). It has not been determined whether identical carbohydrate structures are specified or if chemically distinct isofunctional structures account for these results.

As long as six hours may be required for adherence of all agrobacteria when inoculated on decapitated pea seedlings held in a moist enclosure (Kurkdjian, 1971). By washing the inoculated pea seedlings at various times, Kurkdjian and Manigault (1969) showed that the optimal number of tumors formed were only in those seedlings not washed until a full six hours or longer after inoculation. Glogowski and Galsky (1978), however, found that adherence of tumorigenic or killed *Agrobacterium* to discs of tissue cut from potato tubers is essentially complete at 10–15 minutes after addition, as judged by tumor initiation results (Fig. 12.2).

Agrobacterium adherence assays using cultured plant cells have also been developed. Matthysse *et al.* (1978) followed adherence of *Agrobacterium* to cells of tobacco and carrot in liquid medium by determining through direct plate counts the number of bacteria not adhering to plant cells collected on a Miracloth filter. Maximum adherence occurred after about two hours, and involved from 20–50% of the applied bacteria. The adherence was specific and was inhibited by LPS isolated

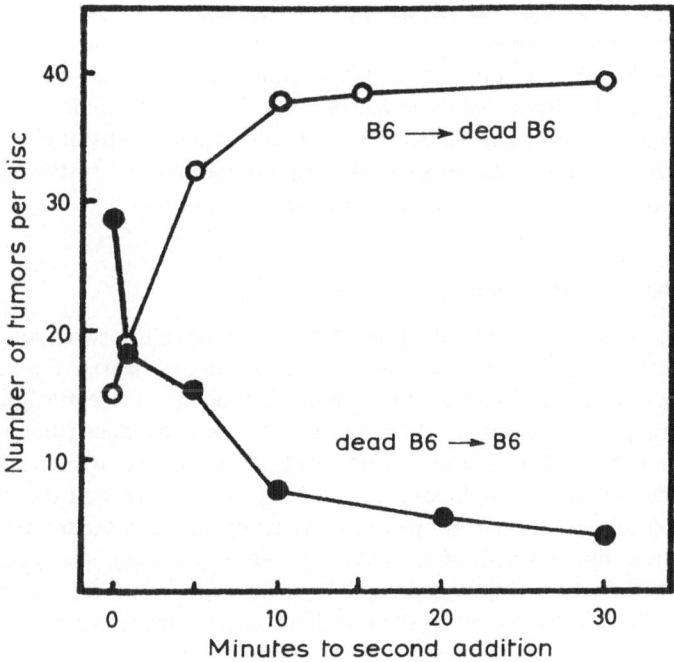

Fig. 12.2 Effect of adding heat-killed *Agrobacterium tumefaciens* strain B6 at various times to potato tuber discs inoculated with viable B6 (○) and of the reciprocal order of addition (●) on the mean number of tumors formed per disc. Data plotted from Tables 3 and 4 of Glogowski and Galsky (1978). No significance change in tumor number was observed in either type experiment when the second application of bacteria was made 15 to 20 minutes after the first.

from *Agrobacterium*, but not by LPS from non-binding strains. Results obtained with a particular plasmid-cured strain of *Agrobacterium* suggest that this assay may differentiate plasmid-determined and chromosome-determined attachment.

Ohyama *et al.* (1979) followed the adherence of ^{32}P-labeled *Agrobacterium* in liquid medium to suspension cultures of *Datura* by a filtration technique rather similar to that of Matthysse *et al.* (1978). Adherence was maximal after two hours co-incubation and was proportional to the bacterial concentration. Only about 10% of the bacteria introduced into the cultures over the range where this proportional relationship held were bound by the plant cells. Binding was optimal at pH 6 and neither Mg^{2+} nor EDTA had significant effects on adherence. Ca^{2+} at a 10 mM concentration, however, did inhibit binding. Binding showed no apparent temperature dependence between 24°C and 35°C; at 0°C adherence was reduced by about 50% and to less than 10% at 47°C. Attempts to compete for binding sites using killed or avirulent strains gave very minimal effects, possibly because the concentrations of bacteria employed were too low.

Agrobacterium has also been found to induce gametophore formation in certain mosses grown in liquid culture (Spiess *et al.*, 1971) and this can be inhibited by *Agrobacterium* LPS (Whatley and Spiess, 1977). Using parabiotic chambers, Spiess *et al.* (1976) found that gametophores were initiated only when the bacterium and the moss were in direct physical contact in the same chamber.

12.2.3 Host component of adherence

The composition of the wound site to which *Agrobacterium* adheres was sought by again employing the quantitative tumor initiation assay on bean leaves, the basic concept being that plant fractions containing binding sites would effectively compete with the intact sites on the leaves for the bacteria, and thereby reduce tumor initiation. Of the various plant fractions tested, only those containing plant cell wall fragments proved inhibitory (Lippincott *et al.*, 1977). Purified cell wall fragments (CWF) were inhibitory at concentrations as low as 50 μg ml^{-1} and the inhibitory activity could be neutralized by treatment of the CWF with either site-binding avirulent agrobacteria or LPS from site-binding strains. They were non-inhibitory when applied 15 minutes after the bacteria, showing their action to be limited to the site-binding step.

The ability of whole bacteria or LPS from these bacteria to neutralize the inhibitory action of the plant CWF proved identical to their ability to compete for tumor sites (Table 12.1). Thus, the *in vivo* adherence sites and the CWF sites are similar by these criteria. CWF from many plants have been tested for *Agrobacterium* adherence sites with some unusual results (Lippincott and Lippincott, 1978). CWF from 10 species of dicotyledonous plants all showed bacterial binding ability, while those from 10 monocotyledonous plants showed none. Since the latter rarely, if ever, develop the crown-gall disease, whereas most dicotyledonous plants are susceptible, this correlation suggests that the presence of adherence sites may be a primary determinant of crown-gall disease. CWF from both crown-gall tumors and embryonic tissues of dicotyledonous plants, however, were also unable to bind *Agrobacterium*, indicating that tumorous conversion alters plant cell wall metabolism in the direction of that found in plant embryos. The net result of crown-gall formation, therefore, is that any agrobacteria multiplying in the wound soon become surrounded by dividing tumor cells to which they can no longer adhere. This could be of advantage both for the survival and multiplication of the bacteria and also to the host since it effectively walls off the bacteria and prevents any further expression of their tumorigenic activity.

Polygalacturonic acid, a component of the plant cell wall, was found to be an excellent inhibitor of *Agrobacterium* initiated tumors, giving total inhibition at 10 mg ml^{-1} and 50% inhibition at 1 μg ml^{-1} (Lippincott and Lippincott, 1977). It is inhibitory only when applied prior to or simultaneously with the bacterium and

Table 12.1 Ability of whole bacteria and LPS of certain agrobacteria to neutralize the inhibitory effect of bean leaf cell walls on tumor initiation

Bacterial strain	Virulence	Compete with B6 for tumor sites *in vivo*	Cell wall neutralization	
			Bacteria	LPS
A. radiobacter S1005	−	−	−	−
A. radiobacter 6467	−	−	−	−
A. radiobacter TR1	−	−	−	−
A. tumefaciens Ag19	−	+	+	+
A. tumefaciens NT1	−	+	+	+
A. tumefaciens IIBNV6	−	+	+	+
A. tumefaciens C58−3	+	+	+	+
A. tumefaciens B6	+	+	+	+

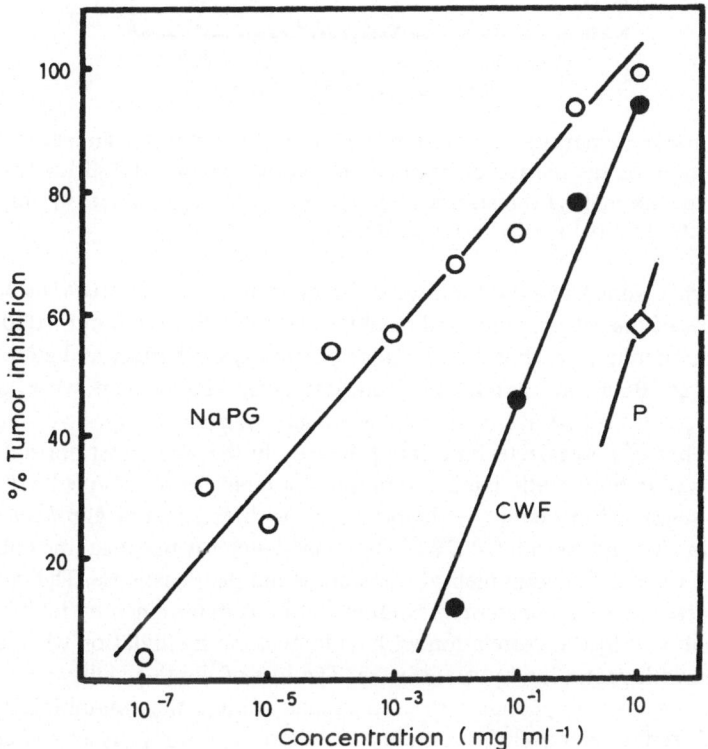

Fig. 12.3 Inhibition of tumor initiation by *Agrobacterium tumefaciens* strain B6 on bean leaves as a function of the concentration of (○) sodium polygalacturonate (NaPG); (●) bean leaf cell wall fragments (CWF); and (◇) pectin (P).

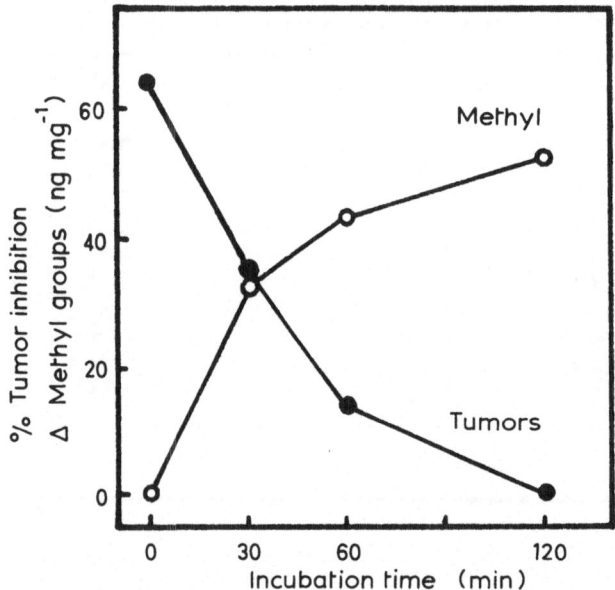

Fig. 12.4 Change in methyl ester content and ability to inhibit bean leaf tumor initiation on incubation of bean leaf cell wall fragments (CWF) with mung bean pectin methyl transferase and S-adenosyl-L-methionine: (●) inhibition of tumor initiation; (○) methyl ester content.

is ineffective when added 15 minutes after the bacterium. A pectin preparation with about 30% of the galacturonic acid residues methylated was about a 1000-fold less effective as an inhibitor, (Fig. 12.3). Other plant polysaccharides and gums were non-inhibitory at 10 mg ml^{-1} and lectins from jack bean, kidney bean, wheat germ, gorse seed and potato tubers at 0.1 mg ml^{-1} were also non-inhibitory.

The occurrence of polygalacturonic acid primarily in the outermost portion of the plant cell wall in both methylated and non-methylated forms, coupled with the above results, suggested this could be the basis for the differences in *Agrobacterium* binding both *in vivo* and to isolated CWF. Direct evidence for this idea was obtained by treating the CWF with pectin methyl transferase and pectinesterase. The data in Fig. 12.4 illustrate that *Agrobacterium* binding ability as determined by inhibition of tumor initiation is lost in correlation with an increase in methylation when the CWF is treated with pectin methyl transferase. The loss of binding ability was reversed by treatment with pectinesterase. In similar fashion, the non-inhibitory CWF obtained from monocotyledons, embryonic dicotyledonous tissues and crown-gall tumor tissues became inhibitory after treatment with pectinesterase, consistent with the appearance of *Agrobacterium* attachment sites. The latter is also reversible by subsequent treatment with pectin methyl transferase.

The adherence step essential for *Agrobacterium* infections thus depends upon the O-antigen portion of the LPS in the outer membrane of the bacterial cell envelope and this polysaccharide is apparently involved in a molecular conformation-type binding mechanism with polygalacturonic acid molecules in the outer portion of the plant cell wall, possibly exposed by the wounding process required for infection. No evidence for lectin participation in this adherence has been found. Thus, a direct polysaccharide—polysaccharide conformation mechanism for *Agrobacterium* adherence is indicated by the available evidence.

12.3 *RHIZOBIUM*–HOST ATTACHMENT

12.3.1 Evidence for adherence

The ability of rhizobia to induce root nodules on leguminous plants, where the bacterium—host symbiotic dinitrogen fixing association develops, depends initially on the successful adherence and invagination-penetration of root cells by the bacteria. The stringent host specificity shown in particular by different species of fast-growing rhizobia has suggested that adherence may involve specific molecular recognition on the part of both the host and the bacterium. The typical pathway of nodule formation involves attachment of the bacterium near the tip of a root hair, a long narrow tubular extension of single epidermal cells that develop near the root apex. There-after, the hair curls around the bacterium, the hair cell wall is breached and the bacterium forms an infection thread which progresses by a growing invagination of the host cytoplasmic membrane as the bacterium enlarges and divides (Fahraeus and Ljunggren, 1968).

The species classification of the rhizobia is based primarily on host-pathogen specificity and while certain strains, especially of the slow-growing rhizobia, may exhibit a somewhat broader host range, the general degree of specificity observed is very high. The bacteria show chemotactic movement toward legume roots and a specific glycoprotein produced by the roots has been implicated in this process (Currier and Strobel, 1977). The specificity of this chemotaxis is insufficient, how-ever, to account for the degree of host—pathogen specificity. Yao and Vincent (1969) and Li and Hubbell (1969) found that the root hair curling response (tip curvature to form a tight crozier hook) exhibits the degree of specificity required and that bacterial culture filtrates were active in inducing this response. There is evidence to indicate that the active substance is an extracellular polysaccharide (Hubbell, 1970; Solheim and Raa, 1973). Host specificity, thus, has been suggested to be determined prior to the invagination and infection thread stages.

Early electron microscope studies of Sahlman and Fahraeus (1963) indicated that adhering rhizobia are surrounded by apparent capsular zones and secondarily by a 'mucilaginous sheath' presumed to be of plant origin. The bacteria were polarly attached, suggesting a similarly localized area of molecular affinity for host surface structures. This polar attachment was also shown by Marshall *et al.* (1975) for rhizobia adhering to clover, pea and soybean. A distinct fibrous material develops between the tip of the bacterium and the host cell wall or between the bacterium and a solid surface, suggesting the extracellular, apparently adherent, material to be of bacterial origin. Bohlool and Schmidt (1976), using fluorescent antibodies and lectins, demonstrated that various rhizobia preferentially bind lectin at one pole of the bacterium. The adherence of *R. trifolii* to clover roots is typically polar and shows a fibrillar capsule in contact with globular electron dense granules on the periphery of the cell wall (Dazzo and Hubbell, 1975a; Dazzo *et al.,* 1976).

Purchase and Nutman (1957) pioneered the use of mixtures of avirulent and virulent rhizobia to dissect the nodule initiation process and demonstrated initiation site competition. They concluded that initiation occurs at discrete foci which could be blocked by the avirulent strain, when present in great excess, and that relatively few bacteria were sufficient to initiate a single nodule. Lim (1963) extended these studies by showing that for initial infections at low bacterial densities, the number of nodules was directly dependent on the number of bacteria and was thus consistent with a one bacterium per nodule initiation process. The three species of clover studied showed differences in number of sites, site distribution and the apparent susceptibility of sites as the inoculum concentration was increased. Thus, a variety of discrete sites perhaps with somewhat different molecular adherence properties may exist.

12.3.2 Bacterial antigens and surface polysaccharides

Early studies of the antigenic properties of the rhizobia on the whole failed to demonstrate a clear relation between bacterial strain and host specificity. Charudattan and Hubbell (1973) found that three species of *Rhizobium* each showed cross-reactive antigens with each of their three respective hosts and in most cases with five non-host species of legumes. No such antigenic relations, however, were found with eight species of non-leguminous plants. They suggest these common antigenic relations may be involved in a post-infection—process such as host—pathogen tolerance. However, in comparing four infective and four non-infective mutants of *R. trifolii*, Dazzo and Hubbell (1975a) discovered an antigen common to the infectious strains which was absent in the mutants. Antisera to the *R. trifolii* clover host reacted with both virulent and, to a somewhat lesser degree, with the avirulent strains. After pre-treatment with avirulent cells, however, the remaining antibody was specific for virulent *R. trifolii* (Dazzo and Hubbell, 1975b). When this 'purified' antiserum was tested against five non-compatible species of infectious *Rhizobium*, no significant interaction was found. Using immunofluorescent techniques they showed that antisera to capsular polysaccharides of infectious *R. trifollii* bind to the surface of clover root cells and most intensely to root hair tips. This capsular antigen was reported to be a polysaccharide of ca. 5 megadaltons containing β-linked 2-deoxy-glucose, galactose, glucose and glucuronic acid and to be active in inducing root hair curling.

The results of Hamblin and Kent (1973) and Bohlool and Schmidt (1974) dramatically focused attention on the ability of plant lectins to bind bacterial polysaccharides and the relation of this binding to *Rhizobium* specificity. Dazzo and Hubbell (1975b) subsequently isolated a lectin from clover roots that specifically agglutinated infectious, but not non-infectious, *R. trifolii* mutants or other species of rhizobia. Agglutination by this lectin was inhibited by 2-deoxyglucose and *N*-acetyl-galactosamine suggesting that the former, due to its occurrence in the capsular poly-saccharide, might be involved in the lectin—bacterium interaction. From these results they proposed a lectin cross-bridging model to account for the adherence specificity of *R. trifolii*. Adherence of *R. trifolii* to clover roots was inhibited by 2-deoxyglucose (Dazzo *et al.*, 1976) but not by related sugars. Adherence of *R. meliloti* to alfalfa root hairs was not affected by 2-deoxyglucose, suggesting different sugars and/or polysaccharide linkages may be utilized in the adherence of different *Rhizobium* species. Bhuvaneswari *et al.* (1977) isolated a soybean lectin that specifically bound *R. japonicum* and could be eluted with D-galactose or *N*-acetyl-D-galactosamine.

12.3.3 Lectins and host—pathogen cross-reactive antigens

Trifoliin, the clover lectin that agglutinates *R. trifolii*, has been purified and shown to be a glycoprotein with a subunit molecular weight of ca. 50 000 daltons (Dazzo *et al.*, 1978). It is eluted from clover roots by 2-deoxyglucose and occurs in highest concentration near the ends of root hairs (Dazzo and Brill, 1977). Soil levels of nitrate or ammonium ion inhibit *Rhizobium*-induced nodulation at levels much lower than that capable of inhibiting either bacterium or host growth. The amount of trifoliin produced by clover seedlings was shown to decrease as the level of nitrate or ammonium ion in the inoculation medium was increased. Correspondingly fewer *R. trifolii* were able to bind to the roots of the treated plants (Dazzo and Brill, 1978). Thus, soil levels of fixed nitrogen may determine the extent of nodulation through regulatory effects on plant synthesis of lectins involved in the *Rhizobium*—legume adherence process.

Wolpert and Albersheim (1976) chromatographed lipopolysaccharides of four species of *Rhizobium* on lectin-agarose columns prepared with lectins isolated from each of the four plant host species. The different LPS preparations adhered only to the lectin column corresponding to the *in vivo* symbiotic relationship. The poly-saccharide portion of the LPS was responsible for the binding. However, Sanders *et al.* (1978) showed that mutational loss and restoration of infectivity in *R. leguminosarum* was correlated with a decrease and restoration in the level of extracellular polysaccharide production without any obvious change in the composi-tion of the LPS. Planque and Kijne (1977; Planque *et al.*, 1979) fractionated *R. leguminosarum* LPS on a molecular sieve column and found a polysaccharide that differed in composition from that of either the LPS or the extracellular polysaccharide. Pea lectins agglutinated the LPS polysacharide and the larger LPS-associated poly-saccharide, but did not interact with the extracellular polysaccharide.

Dazzo and Brill (1979) have recently described two acidic polysaccharides that occur in the capsular material of *R. trifolii* and which both bind to trifoliin and react with anti-clover root antibody. The least abundant polysaccharide contained 2-keto-3-deoxyoctulosonic acid, heptose and lipid A, indicating a similarity to typical LPS of gram-negative organisms, while the major trifoliin-binding polysaccharide lacked these components. Thus, an antigenically similar and perhaps structurally identical polysaccharide occurring in both the free and LPS-associated form, may account for some of the different results thus far observed. The major trifoliin-reactive polysaccharide has an apparent molecular weight of 10^6 daltons and does not appear to be a precursor of LPS polysaccharide (Dazzo, personal communication). Maier and Brill (1978) found that non-nodulating mutants of *R. japonicum* differ from the nodulating parental strains in the polysaccharide portion of the LPS. Carlson *et al.* (1978) nevertheless could find no obvious correlation between the composition of purified LPS obtained from three strains each of *R. leguminosarum, R. phaseoli,* and *R. trifolii.* This suggests that *Rhizobium* host adherence specificity may reside in relatively minor differences in the sequence of sugar residues, linkage or side groups. DNA from *R. japonicum* transforms mutant *Azotobacter vinelandii* to produce the nodulating-type of *R. japonicum* LPS polysaccharide in association with about 6% of the isolates where transformation restored nitrogen fixation (Maier *et al.,* 1978). *Rhizobium* genes for host binding specificity, therefore, may be closely linked with those for enzymes directly involved in dinitrogen fixation.

Ferritin-labeled soybean lectin binds specifically to capsular material of *R. japonicum* (Calvert *et al.,* 1978). Bal *et al.* (1978) found that only encapsulated forms of *R. japonicum* containing storage granules bound soybean lectin and this binding was localized in the capsular material. Less than 1% of the bacterial population occurred in this form which was also the form that bound to soybean hairs, though no apparent polarity in either type of binding was observed. Differences in lectin binding can depend on culture conditions. Culture age is critical as the number of *R. japonicum* showing soybean lectin varies from 0–70% with maximum binding occurring in early to mid-log phase growth (Bhuvaneswari *et al.,* 1977). Bhuvaneswari and Bauer (1978) grew 11 strains of *R. japonicum* in soybean root exudates, in the presence of hydroponically grown soybean roots and on synthetic medium. The eleven strains all bound soybean lectin when grown on the first two media while only five of the strains grown on synthetic medium bound the lectin. Strains of *R. phaseoli, R. meliloti, R. lupini,* and *R. trifolii* grown in the presence of soybean roots or root exudates, however, did not produce soybean lectin binding materials. Dazzo *et al.* (1979) found that the number of *R. trifolii* showing lectin binding receptors changes during the growth cycle in culture, varying from only a few percent of the cells in the lag and late log phases to 90% or more of the cells in early to mid-log and increasing to a sharp maximum again at the beginning of the stationary phase. Metabolic alteration or turnover of the surface capsular structure important in adherence is suggested by these results. The amount of soybean lectin on primary roots is highest during the first week after planting and declines to an undetectable level by day 15 after planting (Pueppke *et al.,* 1978).

The overall picture to emerge from these studies is a dynamic one where bacterial and host surface molecules involved in *Rhizobium*–host adherence vary with age and culture conditions of each organism. Bacterial polysaccharides occurring in the capsule and/or the LPS portion of the envelope which are antigenically similar to polysaccharides on the host hair cell surfaces are involved in adherence. Specific multivalent host lectins that adhere to each of these common antigens to provide a molecular linkage mechanism, as originally proposed by Dazzo and Hubbell (1975b) account for the initial recognition and probably much of the specificity in the host-symbiotic association. The localized interaction of these bacterial polysaccharides with the root hairs induces curling and may activate production or activity of a host polygalacturonase and the synthesis of callose (Kumarasinghe and Nutman, 1977).

12.4 MICROBIAL ADHERENCE IN THE INDUCTION OF PLANT HYPERSENSITIVE NECROSIS

12.4.1 Evidence for adherence

The plant hypersensitive reaction (HR) is a rapid (6–12 h) localized dessication and necrotic response to microbial invasion of intercellular spaces which appears to be the result rather than the cause of an incompatible pathogen–host interaction (Kiraly *et al.*, 1977). Successful pathogens fail to elicit this response and multiply in the host, whereas the incompatible HR-inducing strains do not multiply and lose viability as the response develops. Heat-killed *Pseudomonas solanacearum*, *P. lachrymans, Xanthomonas axonopodis,* but not *Escherichia coli*, were found by Lozano and Sequeira (1970) to prevent the induction of HR by viable cells, demonstrating bacterial viability to be essential for HR and raising the possibility that site attachment might be involved in the response (Sequeira, 1976). Comparable results were obtained with heat-killed *P. tabaci* and *P. glycinea* on the ability of *P. glycinea* to induce HR in tobacco (Sleesman *et al.*,1970).

Goodman *et al.* (1976) showed by electron microscopy that an incompatible strain of *Pseudomonas pisi* was immobilized in tobacco leaf tissue by adherence and envelopment by internal plant cuticle-like material at the surface of the cells. Neither a compatible pathogen (*Pseudomonas tabaci*) nor a saprophytic strain (*Pseudomonas fluorescens*) produced this response. Sequeira *et al.* (1977) obtained similar results in the interaction of avirulent and incompatible strains of *Pseudomonas solanacearum* with tobacco leaves; also, challenge by live HR-inducing bacteria 24 h after treatment with heat-killed bacteria resulted in the lack of attachment of the live bacteria. Saprophytic *E. coli* and *B. subtilis* also showed attachment and envelopment, although these bacteria failed to induce a HR. Sing and Schroth (1977) described the adherence and envelopment of a saprophytic *P. putida* on bean leaf cells while a compatible pathogen, *P. phaseolicola*, and a pathogen for tomato but not bean, *P. tomato*, showed neither response. Klement (1977) concluded that adherence-recognition is involved in the induction of HR by *P. pisi* as opposed to production of a HR-inducing toxin. Bacterial contact with host cell walls is essential for development of the HR in pepper plants induced by *X. vesicatoria* (Stall and Cook, 1979).

12.4.2 Role of bacterial lipopolysaccharides and host lectins in adherence

The virulence of *P. solanacearum* for tobacco is positively correlated with the presence of an extracellular polysaccharide slime (Hussain and Kelman, 1958) and loss of slime production results in loss of virulence. Sequeira and Graham (1977) demonstrated that 55 virulent and 24 avirulent isolates of *P. solanacearum* could be distinguished by their reaction to a lectin isolated from potato tubers. The virulent cells failed to agglutinate and this was correlated with their formation of an extracellular polysaccharide which was not produced by the avirulent, HR-inducing, lectin-agglutinated strains. Because potato plants react similarly to tobacco when inoculated with *P. solanacearum*, this lectin could be directly involved in the adherence of HR-inducing strains (Sequeira, 1978). A lectin which reacts similarly with compatible and incompatible pseudomonads was extracted from tobacco leaves by infiltrating saline into intercellular spaces (Sequeira and Graham, 1977).

Mazzacchi and Pupillo (1976) extracted a protein-LPS complex from *Erwinia chrysanthemi* with EDTA that prevented the HR response of tobacco leaves to *P. syringae* injected 48 hours later. Partial inhibition was obtained in the 3 to 50 μg ml^{-1} range and complete inhibition at higher concentrations. Similar results were obtained with purified protein-LPS preparations from virulent *Pseudomonas tabaci* and with incompatible strains of *P. aptata* and *P. lachrymans* in preventing the HR response of tobacco leaves to seven incompatible pseudomonads. These preparations also delayed the development of a susceptible reaction induced by five compatible pseudomonads (Mazzucchi *et al.*, 1979). The portion of the protein-LPS complexes primarily responsible for this effect was not determined.

Sleesman *et al.* (1970) reported that a phenol—water extract of two strains of *Pseudomonas* inhibited the HR response of tobacco infiltrated with *P. glycinea*. Graham *et al.*, (1977) using purified LPS from *P. solanacearum* showed it prevented HR in tobacco leaves challenged 24 h later with an incompatible strain of the same species. LPS from rough as well as smooth strains gave complete protection at concentrations of 50 μg ml^{-1} indicating that the 0-antigen portion may not be required for activity. Acetic acid hydrolysis resulted in loss of activity as did alkaline deacylation which suggests that both the fatty acids of the lipid A and the core polysaccharide lipid A linkages are required for activity. The specificity of the response seems rather low as LPS from both *E. coli* and *S. marcescens* also induced resistance to HR. This result is compatible with the possibility that the HR response may be part of a generalized defense mechanism to invading micro-organisms.

LPS from an HR-inducing strain of *P. solanacearum* was agglutinated by both potato tuber lectin and a tobacco leaf lectin which differentiates pseudomonads giving compatible and non-compatible reactions on tobacco (Sequeira and Graham,

1977). However, these lectins also agglutinated the isolated lipid A solubilized with bovine serum albumin, whereas this type of preparation does not protect against the HR response.

Protection against compatible 'non-binding' pathogens can also be induced by pre-inoculation of plants with dead bacteria (Lovrekovich and Farkas, 1965; Lozano and Sequeira, 1970; Sleesman *et al.,* 1970; Carroll and Lukezic, 1972; Sequeira and Hill, 1974; Verma *et al.,* 1978) in much the same way that dead bacteria and LPS prevent HR induction by incompatible strains. Both this response and the HR may stimulate production of lectin-like substances responsible for protection. Huang *et al.* (1975) have shown that apple petioles inoculated with an avirulent strain of *Erwinia amylovora* produce an agglutinin which causes clumping of the HR-inducing strain. Virulence in *E. amylovora* as in *P. solanacearum* is highly correlated with extracellular polysaccharide production (Bennett, 1978). Results similar to those of Huang *et al.* (1975) have been reported by Horino (1976) with *Xanthomonas oryzae* in rice and by Main (1968) working with *P. solanacearum* and tobacco. Goodman *et al.* (1977) obtained an agglutinin from tobacco petioles infiltrated with *Pseudomonas pisi* that specifically agglutinated this organism. Sing and Schroth (1977) infiltrated bean leaves with LPS isolated from HR-inducing *P. putida* and one hour later the leaves were challenged with either virulent or HR-inducing strains. Within six hours the virulent as well as the avirulent strains were immobilized by encapsulation. The protein-LPS complexes of pseudomonads described above (Mazzucchi *et al.,* 1979) also protected against both compatible and incompatible pathogens. However, clear evidence that the induced production of a lectin is responsible for subsequent resistance of plants to virulent compatible micro-organisms or to HR-inducing organisms is lacking.

Many plant diseases are characterized by localized necrosis, chlorosis or water soaking, responses not unlike HR. Klement *et al.* (1978) have compared the requirement shown for HR with those for the 'normosensitive' response induced by the pathogen *P. tabaci* on tobacco. Both responses have a similar induction time, time of appearance and a requirement for live inducing bacteria, are inhibited by heat-killed cells and at a temperature of 37°C, and are associated with increased host cell permeability. The only difference detected was the 5–100-fold higher concentration of the virulent pathogen necessary to obtain the normosensitive response. The events detected in the HR response system, therefore, may be identical with those occurring in diseases characterized by this group of symptoms.

12.5 ADDITIONAL EXAMPLES OF ADHERENCE IN MICRO-ORGANISM–PLANT INTERACTIONS

12.5.1 Bacteria

The ability of roots and leaves to accumulate choline sulfate from solution is markedly stimulated by an inductive mechanism on exposure to many gram-negative

bacteria (Nissen, 1971, 1973) including *Agrobacterium, Rhizobium* and *Pseudomonas* sp.
Viable bacteria and direct cell—cell contact are required for this response, as the response was not elicited when the bacteria and plant tissue were separated by membrane
filters in a parabiotic chamber. Evidence for a calcium-mediated adherence mechanism
was obtained. Rumen bacteria active in digesting plant components in the bovine
intestine have been shown to adhere to plant cell walls and to cellulose fibrils, which
presumably facilitates their subsequent maceration of these materials (Akin *et al.*,
1974; Patterson *et al.*, 1975). An isolated glycoprotein aggregate closely associated
with LPS receptors for a phage that specifically infects virulent strains of *Pseudomonas
marsprunorum* was found to induce several symptoms characteristic of disease when
injected into cherry leaves (Hignett and Quirk, 1979). This protein also inhibited the
hypersensitive response of tobacco leaves inoculated with HR-inducing pseudomonads
and thus may interact with potential *Pseudomonas* adherence sites.

Extracellular polysaccharides of *Xanthomonas* exist in ordered conformations and
can react in a highly specific way with certain unlike plant cell wall polysaccharides
to form a stable gel (Morris *et al.*, 1977; Dea *et al.*, 1977). The authors speculate
that this interaction could serve as a molecular holdfast in the case of plant pathogenic
species to anchor the bacteria to its host. Exopolysaccharides of *Arthrobacter stabilis*
and *Erwinia carotovora* do not show this gel formation with these polysaccharides.
These results demonstrate that recognition systems based on carbohydrate—carbohydrate interactions can be a specific and effective mechanism for micro-organism
adherence.

The frequent occurrence of common or cross-reactive antigens in compatible host—
pathogen combinations has been suggested as potential evidence for a recognition
factor important to the specificity of the host—pathogen interaction (De Vay and
Adler, 1976). While studies on *Rhizobium* adherence have brought attention to these
antigens as potential lectin-binding sites that might be involved in a similar adherence
mechanism, with the exception of the results from Sequeira's laboratory on
P. solanacearum no convincing evidence for this type of function has been found.
Common antigens are found between *Agrobacterium* and some of its hosts but all
evidence indicates lectins are not involved in the adherence of this pathogen. The
alternative possibility (De Vay and Adler, 1976) that these antigens may be basic
compatibility factors bypassing a 'non-self' recognition-defense mechanism seems
a more probable function for such antigens in most host-pathogen relationships.

12.5.2 Fungi

Several reports indicate that adherence may be important in fungal colonization of
plant hosts. Gupta and Heale (1971) demonstrated that *Verticillium albo-atrum*
produced a disease associated cellulase only when the hyphae were in direct contact
with cellulose and Cooper (1977) has obtained electron microscopic evidence for
adherence of this pathogen to cell walls of tomato. Raa *et al.* (1977) found that the
germ tube of *Cladosporium cucumerinum* adheres firmly to the surface of cucumber

hypocotyls, whereas saprophytes and non-host pathogens do not. Recently, Mendgen (1978) showed that a cell wall preparation from the bean rust pathogen *Uromyces phaseoli* labeled with fluorescein isothiocyanate adheres to the cell walls of host tissue and poorly or not at all to cell walls of non-host tissues. The author suggests that the specificity in haustoria formation may depend on an induction mechanism dependent on this specific adherence. Huemme *et al.* (1978) used a similar cell wall preparation from this pathogen to induce resistance to subsequent challenge with the pathogen.

Melon seedlings infected with the fungus *Colletotrichum lagenarium* causing anthracnose show a 10-fold increase in the amount of extensin, a hydroxyproline-rich wall glycoprotein (Esquerré-Tugayé and Toppan, 1977). The possibility was considered that the extensin might be involved in direct binding of fungal wall components. Kleinschuster and Baker (1974) found differences in the lectin-binding specificity of spores from two host-distinct species of pathogenic fusaria. They speculate this difference may relate to host specificity through an adherence—recognition interaction. Kojima and Uritani (1974) compared the ability of extracts of different hosts to agglutinate germinated spores of 7 host-specific strains of *Ceratocystis fimbriata,* the black rot fungus. With three exceptions, these extracts agglutinated spores of the non-compatible pathogens and failed to agglutinate spores of the compatible strains. The agglutinating factor from sweet potato had an apparent molecular weight greater than 15 000 and was inactivated by treatment with protease or polysaccharide-hydrolyzing enzymes, consistent with the possibility that a lectin-type plant glycoprotein was responsible for this response. On purification, this lectin was shown to lose pathogen specificity and to have a molecular weight of 1.6 megadaltons and require divalent cations for activity (Kojima and Uritani, 1978a). A low molecular weight compound present in heated extracts of crude lectin preparations restored the specificity of the purified lectin for the strain pathogenic for sweet potato. Conconavalin A binding to these seven strains of *C. fimbriata* was shown to vary considerably and binding was reversible by α-methyl-D-mannoside and α-methyl-D-glucoside (Kojima and Uritani, 1978b).

12.6 SUMMARY AND PROSPECTUS

Smith (1977) describes the following non-exclusive categories of ways in which microbial surfaces may influence host pathogen relations:

(1)   host adherence and entry;
(2)   multiplication *in vivo*;
(3)   interference with host defenses;
(4)   host damage; and
(5)   host and tissue specificity.

The host—pathogen systems examined above, though primarily considered relative to adherence, illustrate aspects of each of these categories. In the case of *Agrobacterium*, adherence is essential but involves a relatively non-specific host cell wall component, in keeping with its broad host range. The specificity of *Rhizobium* adherence, however, is strikingly specific and involves bacterium and host-cross-reactive antigens linked by a glycoprotein that specifically binds these antigens to each other. A stable cross linkage between the bacteria and host cells leads to bacterial penetration and multiplication within host cells. The hypersensitive reaction (HR), however, differentiates adhering, non-disease-producing organisms from the non-adhering pathogens which are thus able to multiply and cause disease. In several cases, the extracellular polysaccharides of these successful pathogens serve not only to protect the pathogen by interfering with this defense mechanism but are also directly involved in damaging host tissues.

While the major details of adherence in *A. tumefaciens* and *R. trifolii* infections and in *P. solanacearum*-induced HR seem clear, the specific chemistry of adherence in each of these model systems remains unresolved. The initial adherence seems to involve a polysaccharide–polysaccharide interaction in *Agrobacterium*, a polysaccharide–lectin–polysaccharide interaction in *Rhizobium* and a lipid A plus core polysaccharide–lectin–unknown interaction in the case of HR induced by *P. solanacearum*. On the part of the bacteria, either capsular polysaccharides or LPS components may be necessary for adherence depending on the organism. On the part of the host, cell wall polygalacturonic acid, lectins and cell surface polysaccharides have thus far been implicated in these adherence mechanisms. Additional components involved in plant–micro-organism adherence will undoubtedly be found as more systems are examined and cases where adherence depends on a active, enzymatically catalyzed, chemical bonding of molecules at the surface of the micro-organism with those on the surface of the plant seem a distinct, but undemonstrated, possibility.

The specific mutual recognition that occurs between host and pathogen has long been an intriguing phenomenon, which in the case of *Agrobacterium* and *Rhizobium* apparently involves, in the first instance, surface molecules of both host and pathogen acting in a lock and key type molecular recognition-adherence manner. For a major group of plant pathogens that must multiply in host intercellular spaces to produce disease, recognition apparently depends on the absence of adherence, at least in the early stages of etiology. Adherence in these cases results in a plant 'non-self' recognition response as shown by closely related incompatible pathogens which trigger potential defense reactions. As these results and those with animal pathogens amply indicate, pathogen–host strategies seem to have developed along all possible lines of interactions. Further research will undoubtedly bring new examples of each type of mechanism. New mechanisms of adherence may be found as well. The nature of the events elicited by the specific molecules active in adherence in both the pathogen and host have only been touched upon in this review, but exciting developments occurring in this area greatly add to the attraction and importance of continued investigation of microbial–plant adherence.

REFERENCES

Akin, D.E., Burdick, D. and Michaels, G.E. (1974), *Appl. Microbiol.,* **27,** 1149–1156.

Albersheim, P. and Anderson-Prouty, A.J. (1975), *A. Rev. Plant Physiol.,* **26,** 31–52.

Bal, A.K., Shantharam, S. and Ratnam, S. (1978), *J. Bact.,* **133,** 1393–1400.

Bauer, W.D. (1977), *Basic Life Sci.,* **9,** 283–297.

Beiderbeck, R. (1976), *Z. Naturforsch.,* **31c,** 317–318.

Bennett, R.A. (1978), *Proc. 4th Int. Conf. Plant Path. Bact.,* **2,** 479–481.

Beringer, J.E., Brewin, N., Johnston, A.W.B., Schulman, H.M. and Hopwood, D.A. (1979), *Proc. R. Soc. Lond.,* Ser. B **204,** 219–233.

Bhuvaneswari, T.V. and Bauer, W.D. (1978), *Plant Physiol.,* **62,** 71–74.

Bhuvaneswari, T.V., Pueppke, S.G. and Bauer, W.D. (1977), *Plant Physiol.,* **60,** 486–491.

Bogers, R.J. (1972), *Proc. 3rd Int. Conf. Phytopath. Bact.,* pp. 239–250.

Bohlool, B.B. and Schmidt, E.L. (1974), *Science,* **185,** 269–271.

Bohlool, B.B. and Schmidt, E.L. (1976), *J. Bact.,* **125,** 1188–1194.

Calvert, H.E., Lalonde, M., Bhuvaneswari, T.V. and Bauer, W.D. (1978), *Can. J. Microbiol.,* **24,** 785–793.

Carlson, R.W., Sanders, R.E., Napoli, C. and Albersheim, P. (1978), *Plant Physiol.,* **62,** 912–917.

Carroll, R.B. and Lukezic, F.L. (1972), *Phytopathology,* **62,** 555–564.

Charudattan, R. and Hubbell, D.H. (1973), *Antonie van Leeuwenhoek J. Microbiol. Serol.,* **39,** 619–627.

Cooper, R.M. (1977), In: *Cell Wall Biochemistry related to Specificity in Host– Plant Pathogen Interactions.* (Solheim, B. and Raa, J. eds.), Universitatsforlaget, Oslo. pp. 163–206.

Currier, W.W. and Strobel, G.A. (1977), *Science,* **196,** 434–435.

Cutter, E.G. (1976), In: *Microbiology of Aerial Plant Surfaces.* (Dickinson, C.H. and Preece, T.F., eds.), Academic Press, New York, pp. 1–40.

Dazzo, F.B. (1979), In: *Adsorption of Micro-organisms to Surfaces.* (Britton, G. and Marshall, K.C., eds.), J. Wiley and Sons, Inc., New York, (In press).

Dazzo, F.B. and Brill, W.J. (1977), *Appl. Environ Microbiol.,* **33,** 132–136.

Dazzo, F.B. and Brill, W.J. (1978), *Plant Physiol.,* **62,** 18–21.

Dazzo, F.B. and Brill, W.J. (1979), *J. Bact.,* **137,** 1362–1373.

Dazzo, F.B. and Hubbell, D.H. (1975a), *Appl. Microbiol.,* **30,** 172–177.

Dazzo, F.B. and Hubbell, D.H. (1975b), *Appl. Microbiol.,* **30,** 1017–1033.

Dazzo, F.B., Napoli, C.A. and Hubbell, D.H. (1976), *Appl. Environ. Microbiol.,* **32,** 166–171.

Dazzo, F.B., Urbano, M.R. and Brill, W.J. (1979), *Curr. Microbiol.,* **2,** 15–20.

Dazzo, F.B., Yanke, W.E. and Brill, W.J. (1978), *Biochim. biophys. Acta,* **539,** 276–286.

Dea, I.C.M., Morris, E.R., Rees, D.A., Welsh, E.J., Barnes, H.A. and Price, J. (1977), *Carbohyd. Res.,* **57,** 249–272.

De Vay, J.E. and Adler, H.E. (1976), *A. Rev. Microbiol.,* **30,** 147–168.

Esquerré-Tugayé, M.T. and Mazau, D. (1974), *J. exp. Bot.,* **25,** 509–513.

Esquerré-Tugayé, M.T. and Toppan, A. (1977), In: *Cell Wall Biochemistry related to Specificity in Host–Plant Pathogen Interactions.* (Solheim, B. and Raa, J., eds.), Universitetsforlaget, Oslo. pp. 299–303.

Fahreus, G. and Ljunggren, H. (1968), In: *The Ecology of Soil Bacteria.* (Gray, T.R.G. and Parkinson, D., eds.), Liverpool University Press, Liverpool. pp. 396–421.

Glogowski, W. and Galsky, A.G. (1978), *Plant Physiol.,* **61**, 1031–1033.

Goodman, R.N., Huang, P. Y., Huang, J.S. and Thaipanich, V. (1977), In: *Biochemistry and Cytology of Plant–Parasite Interaction,* (Tomiyama, K., Daly, J.M., Uritani, I., Oku, H. and Ouchi, S., eds.), Elsevier Publishing Co., New York. pp. 35–42.

Goodman, R.N., Huang, P.Y. and White, J.A. (1976), *Phytopathology,* **66**, 754–764.

Graham, T.L., Sequeira, L. and Huang, T.-S. R. (1977), *Appl. Environ. Microbiol.,* **34**, 424–432.

Gupta, D.P. and Heale, J.B. (1971), *J. gen. Microbiol.,* **63**, 163–173.

Hamblin, J. and Kent, S.P. (1973), *Nature New Biology,* **245**, 28–30.

Hignett, R.C. and Quirk, A.V. (1979), *J. gen. Microbiol.,* **110**, 77–81.

Holloway, P.J. (1971), In: *Ecology of Leaf Surface Micro-organisms.* (Preece, T.F. and Dickinson, C.H., eds.), Academic Press, London. pp. 39–53.

Horino, O. (1976), In: *Biochemistry and Cytology of Plant–Parasite Interaction.* (Tomiyama, K., Daly, J.M., Uritani, I., Oku, H. and Ouchi, S., eds.), Elsevier Publishing Co., New York. pp. 43–55.

Huang, P.Y., Huang, J.S. and Goodman, R.N. (1975), *Physiol. Plant Path.,* **6**, 283–287.

Hubbell, D.H. (1970), *Bot. Gaz.,* **131**, 337–342.

Huemme, B., Hoppe, H.H. and Heitfuss, R. (1978), *Phytopathol. Z.,* **92**, 281–284.

Hussain, A. and Kelman, A. (1958), *Phytopathology,* **48**, 155–165.

Jeffree, C.E., Baker, E.A. and Holloway, P.J. (1976), In: *Microbiology of Aerial Plant Surfaces.* (Dickinson, C.H. and Preece, T.F., eds.), Academic Press, New York. pp. 119–158.

Kerr, A. (1969), *Aust. J. Biol. Sci.,* **22**, 111–116.

Kiraly, Z., Hevesi, M. and Klement, Z. (1977), *Acta Phytopath. Acad. Sci. Hung.,* **12**, 247–256.

Kleinschuster, S.J. and Baker, R. (1974), *Phytopathology,* **64**, 394–399.

Klement, Z. (1977), *Acta Phytopath. Acad. Sci. Hung.,* **12**, 257–261.

Klement, Z., Hevesi, M. and Sasser, M. (1978), *Proc. 4th Int. Conf. Plant Path. Bact.,* **2**, 679–685.

Kojima, M. and Uritani, I. (1974), *Plant Cell Physiol.,* **15**, 733–737.

Kojima, M. and Uritani, I. (1978a), *Plant Physiol.,* **62**, 751–753.

Kojima, M. and Uritani, I. (1978b), *Plant Cell Physiol.,* **19**, 1099–1101.

Kumarasinghe, R.M.K. and Nutman, P.S. (1977), *J. exp. Bot.,* **28**, 961–976.

Kurkdjian, A. (1971), *Recherche sur la Participation de l'Agrobacterium tumefaciens à l'induction tumorale chez les vegetaux.* Thesis, University of Paris. 110 pp.

Kurkdjian, A. and Manigault, P. (1969), *C.R. Acad. Sci. Paris,* **268**, 2756–2757.

Li, D. and Hubbell, D.H. (1969), *Can. J. Microbiol.,* **15**, 1133–1136.

Lim, G. (1963), *Ann. Bot. N.S.,* **27**, 55–67.

Lippincott, B.B. and Lippincott, J.A. (1969), *J. Bact.,* **97**, 620–628.

Lippincott, B.B., Whatley, M.H. and Lippincott, J.A. (1977), *Plant Physiol.,* **59**, 388–390.

Lippincott, J.A. and Lippincott, B.B. (1975), *A. Rev. Microbiol.,* **29**, 377–405.

Lippincott, J.A. and Lippincott, B.B. (1977), In: *Cell Wall Biochemistry related to Specificity in Host–Plant Pathogen Interactions.* (Solheim, B. and Raa, J., eds.), Universitetsforlaget, Oslo. pp. 439–451.

Lippincott, J.A. and Lippincott, B.B. (1978), *Science,* **199,** 1075–1078.

Lovrekovich, L. and Farkas, G.L. (1965), *Nature,* **205,** 823–824.

Lozano, J.C. and Sequeira, L. (1970), *Phytopathology,* **60,** 875–879.

Maier, R.J., Bishop, P.E. and Brill, W.J. (1978), *J. Bact.,* **134,** 1199–1201.

Maier, R.J. and Brill, W.J. (1978), *J. Bact.,* **133,** 1295–1299.

Main, C.E. (1968), *Phytopathology,* **58,** 1058–1059.

Manigault, P. (1970), *Ann. Inst. Pasteur,* **119,** 347–359.

Marshall, K.C. (1976), *Interfaces in Microbial Ecology.* Harvard University Press, Cambridge, Mass.

Marshall, K.C., Cruickshank, R.H. and Bushby, H.V.A. (1975), *J. gen. Microbiol.,* **91,** 198–200.

Matthysse, A.G., Wyman, P.M. and Holmes, K.V. (1978), *Infect. Immun.,* **22,** 516–522.

Mazzucchi, U., Bazzi, C. and Pupillo, P. (1979), *Physiol. Plant Path.,* **14,** 19–30.

Mazzucchi, U. and Pupillo, P. (1976), *Physiol. Plant Path.,* **9,** 101–112.

Mendgen, K. (1978), *Arch. Microbiol.,* **119,** 113–117.

Morris, E.R., Rees, D.A., Young, G., Walkinshaw, M.D. and Darke, A. (1977), *J. mol. Biol.,* **110,** 1–16.

Nissen, P. (1971), In: *Symposium on Informative Molecules in Biological Systems.* (Ledoux, L., ed.), North-Holland Publishing Co., Amsterdam, pp. 201–212.

Nissen, P. (1973), *Scientific Reports of the Agricultural University of Norway,* **52,** 1–52.

Ohyama, K., Pelcher, L.E., Schaefer, A. and Fowke, L.C. (1979), *Plant Physiol.,* **63,** 382–387.

Patterson, H., Irvin, R., Costerton, J.W. and Cheng, K.J. (1975), *J. Bact.,* **122,** 278–287.

Paull, R.E. and Jones, R.L. (1975), *Plant Physiol.,* **56,** 302–312.

Planque, K. and Kijne, J.W. (1977), *FEBS Letters,* **73,** 64–66.

Planque, K., Van Nierop, J.J., Burgers, A. and Wilkinson, S.G. (1979), *J. gen. Microbiol.,* **110,** 151–159.

Pueppke, S.G., Bauer, W.D., Keegstra, K. and Ferguson, A.L. (1978), *Plant Physiol.,* **61,** 779–784.

Purchase, H.F. and Nutman, R.S. (1957), *Ann. Bot. N.S.,* **21,** 439–454.

Raa, J., Robertsen, B., Solheim, B. and Tronsmo, A. (1977), In: *Cell Wall Biochemistry related to Specificity in Host–Plant Pathogen Interactions.* (Solheim, B. and Raa, J., eds.,) Universitetsforlaget, Oslo. pp. 11–28.

Sahlman, K. and Fahraeus, G. (1963), *J. gen. Microbiol.,* **33,** 425–427.

Sanders, R.E., Carlson, R.W. and Albersheim, P. (1978), *Nature,* **271,** 240–242.

Schilperoort, R.A. (1969), *Investigations on Plant Tumors. Crown Gall. On the Biochemistry of Tumor Induction by Agrobacterium tumefaciens.* Thesis. University of Leiden.

Sequeira, L. (1976), In: *Specificity in Plant Diseases,* (Wood, R.K.S. and Graniti, A. eds.), Plenum Press, New York, pp. 289–306.

Sequeira, L. (1978), *A. Rev. Phytopath.,* **16,** 453–481.

Sequeira, L., Gaard, G. and De Zoeten, G.A. (1977), *Physiol. Plant Path.*, **10**, 43–50.
Sequeira, L. and Graham, T.L. (1977), *Physiol. Plant Path.*, **11**, 43–54.
Sequeira, L. and Hill, L.M. (1974), *Physiol. Plant Path.*, **4**, 447–455.
Sing, V.O. and Schroth, M.N. (1977), *Science*, **197**, 759–761.
Sleesman, H.C., Perley, J.E. and Hoitink, H.A.J. (1970), *Phytopathology*, **60**, 1314.
Smith, H. (1977), *Bact. Rev.*, **41**, 475–500.
Solheim, B. and Raa, J. (1973), *J. gen. Microbiol.*, **77**, 241–247.
Spiess, L.D., Lippincott, B.B. and Lippincott, J.A. (1971), *Am. J. Bot.*, **58**, 726–731.
Spiess, L.D., Lippincott, B.B. and Lippincott, J.A. (1976), *Am. J. Bot.*, **63**, 324–328.
Spiess, L.D., Turner, J.C., Mahlberg, P.G., Lippincott, B.B. and Lippincott, J.A. (1977), *Am. J. Bot.*, **64**, 1200–1208.
Stall, R.E. and Cook, A.A. (1979), *Physiol. Plant Path.*, **14**, 77–84.
Verma, J.P., Chowdhury, H.D. and Singh, R.P. (1978), *Proc. 4th Int. Conf. Plant Path. Bact.*, **2**, 795–802.
Whatley, M.H., Bodwin, J.S., Lippincott, B.B. and Lippincott, J.A. (1976), *Infect. Immun.*, **13**, 1080–1083.
Whatley, M.H., Margot, J.B., Schell, J., Lippincott, B.B. and Lippincott, J.A. (1978), *J. gen. Microbiol.*, **107**, 395–398.
Whatley, M.H. and Spiess, L.D. (1977), *Plant Physiol.*, **60**, 765–766.
Wolpert, J.S. and Albersheim, P. (1976), *Biochem. biophys. Res. Commun.*, **70**, 729–737.
Yao, P.Y. and Vincent, J.M. (1969), *Aust. J. Biol. Sci.*, **22**, 413–423.

13 Cell Recognition Systems in Eukaryotic Cells

DAVID R. PHILLIPS
and
T. KENT GARTNER

13.1	Introduction	*page*	401
13.2	Non-developmentally regulated cell recognition systems		403
	13.2.1 Yeast mating factors		403
	13.2.2 Egg–sperm interaction		404
	13.2.3 Agglutination of marine sponge cells		407
	13.2.4 Opsonic activity of soluble fibronectin		410
13.3	Developmentally regulated cell recornition systems		411
	13.3.1 Slime molds		411
	13.3.2 Galactolectins		413
	13.3.3 Embryonic chick retina		416
	13.3.4 The role of non-tissue-specific cell adhesion molecules in development		419
13.4	Platelet interactions		420
	13.4.1 Platelet membrane composition and structure		421
	13.4.2 Platelet aggregation		422
	13.4.3 Platelet lectin activity		422
	13.4.4 Platelet lectin receptor		426
	13.4.5 Speculations concerning the mechanism of platelet aggregation		429
13.5	Concluding remarks		431
	References		433

Acknowledgements

Research support for unpublished data included in this review was provided through project grants HL 15616, HL 21487, and HL 23010 from the U.S. Public Health Service, and Career Development Award HL 00080. The authors thank Mr D.C. Williams for drawing the figures.

Bacterial Adherence

(*Receptors and Recognition,* Series B, Volume 6)

Edited by E.H. Beachey

Published in 1980 by Chapman and Hall, 11 New Fetter Lane, London EC4P 4EE

© 1980 Chapman and Hall

13.1 INTRODUCTION

Direct interactions between cells are essential to the proper functioning of virtually all multicellular organisms. Study of these interactions extends to most disciplines in biology. In addition to bacterial adherence, the subject of this volume, excellent examples are to be found in embryology (interactions in developing tissues), hematology (hemostasis reactions), immunology (lymphocyte and macrophage interactions) and virology (viral infectivity). Although apparently unrelated, each of these processes depends on direct physical interactions between two or more membrane surfaces, and the ability of cells to discriminate between different membranes.

Understanding the selectivity of cell—cell interaction requires clarification of its molecular basis. That such interactions are mediated by specific molecules has become apparent from many findings, particularly those demonstrating tissue and species specificity. Classical studies have shown that similar cells can 'sort out' or re-associate, even when placed in a complex mixture of cells (e.g., species-specific aggregation in a mixture of dissociated red and yellow sponge cells (Wilson, 1907), re-assortment of cells from dissociated amphibian embryos (Townes and Holfreter, 1955); and specific aggregation of platelets to the exclusion of other blood cells (Eberth and Schimmel-busch, 1886)).

A convenient illustration of cellular interaction is provided by the 'lock and key' models, such as those used to depict antigen—antibody interactions (as suggested by Tyler, 1947, and Weiss, 1947) and enzyme—substrate complex formation (Roseman, 1970). This view assumes that complementary molecules (termed cell ligands (Moscona, 1968), lectins and lectin receptors (Rosen *et al.*, 1974, Simpson *et al.*, 1977)), or agglutinins are present on membrane surfaces and that these molecules mediate interactions between membrane surfaces. The specific features of these interactions may be quite different. In Fig. 13.1a, for instance, the complementary molecules of heterologous cells are components of the cell surfaces. The egg—sperm interaction in the fertilization of sea urchin eggs (Vacquier and Moy, 1977), and the sexual mating of yeast cells (Crandall and Brock, 1968) are examples of cellular interactions directly mediated by asymmetrically distributed membrane bound components. A slight modification of this system is illustrated by the random distribution of lectin and lectin receptors on the two membrane surfaces shown in Fig. 13.1b, as found in cohesive amoebae of slime molds (Rosen *et al.*, 1974; Simpson *et al.*, 1977). Alternatively, both the lectin and its receptor could be on one membrane-bound molecule (Fig. 13.1c; Müller and Giersch, 1978). Not all direct cell—cell interactions are mediated exclusively by membrane-bound components. Extramembrane molecules can bridge interacting cells as shown in Fig. 13.1d. An example of this type of crosslinking molecule is the aggregation factor of marine

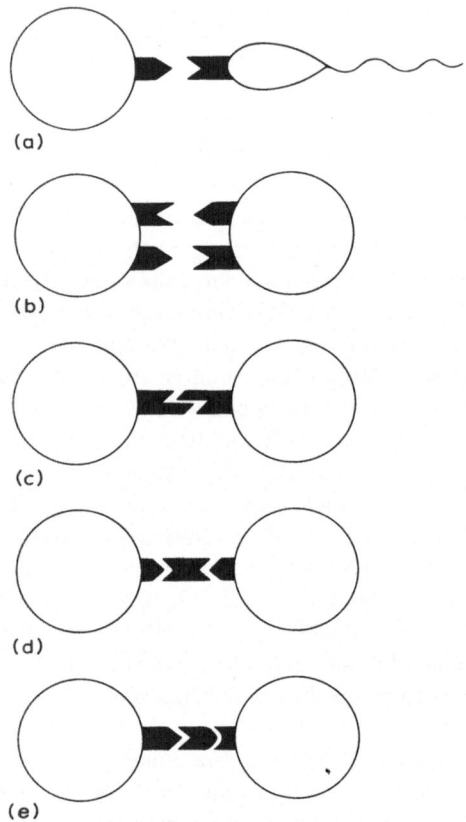

Fig. 13.1 This figure depicts alternative distributions of components which mediate cellular interactions. Explanations and examples of each are provided in the text.

sponges, which is polyvalent with respect to binding sites, and mediates the inter-action of identical cells (Humphreys, 1963; Moscona, 1963). Such bridging molecules can also be polyvalent with non-identical sites and cause the interaction of two different cell types (Fig. 13.1e). One example of this model appears to be the soluble fibronectin-mediated phagocytosis of damaged cells by macrophages (Saba *et al.,* 1978).

Selected eukaryotic systems of animal cell—cell interaction are reviewed in this chapter. The review is not intended to be comprehensive, but rather to include those eukaryotic systems that best fit the mechanisms of cell—cell interaction illustrated in Fig. 13.1. Special emphasis is given to platelet—platelet interactions, because the membrane surface of platelets can mediate their specialized function only following platelet activation.

Non-developmentally regulated systems are considered first. Included under this heading are the sexual mating of yeast, the interaction of sperm with egg, agglutination of marine sponge cells, and the opsonic activity of fibronectin. This is followed by a consideration of developmentally regulated cell recognition systems including slime mold, skeletal muscle cells and retina. The balance of the review presents details concerning the interaction of human blood platelets.

13.2 NON-DEVELOPMENTALLY REGULATED CELL RECOGNITION SYSTEMS

13.2.1 Yeast mating factors

The mating factors present in the cell wall of the yeast *Hansenula wingei* are in example of asymmetrically distributed complementary molecules that mediate cell–cell interaction. This is the first cellular system in which such complementary molecules were identified and they have been isolated and extensively characterized.

H. wingei lives in a symbiotic relationship with bark beetles. Although the wild-type strain is diploid (Wickerham, 1958), two haploid strains (types 5 and 21), which conjugate with each other to form the diploid strain, have been isolated. Before conjugation, several type 21 cells surround and aggregate with one type 5 cell. In more than 90% of the matings, a diploid cell which loses its affinity for both type 21 and type 5 cells is produced. (It is thought that the genes for both complementary aggregating factors are not expressed in the diploid cell.) Polyploidy occurs when more than one of the type 21 cells fuse with a type 5 cell. (For a review of the life cycle and genetics of *H. wingei* see Wickerham and Burton, 1962 and Crandall and Brock, 1968).

(a) *Aggregation factor on type 5 cells*

The aggregation factor on type 5 cells was identified after its release from the cell surface by subtilisin digestion and its specific adsorption to type 21 cells (Taylor and Orton, 1967). This aggregation factor, termed the 5-factor, was well-characterized physically and chemically (Crandall and Brock, 1968; Taylor and Orton, 1970; Yen and Ballou, 1974b). It had a high molecular weight (mol. wt. $\cong 10^6$) and was 85% carbohydrate (mostly mannose), 10% protein and 5% phosphate. The mannose chains were highly branched, containing 1–15 mannose residues per chain, and where linked to the protein by alkali-labile linkages. Approximately 60% of the amino acid residues were linked to carbohydrate (Ballou, 1976). This distribution was different from the other mannan-containing molecules in the cell surface, which had longer manno-oligosaccharide chains and were coupled to protein in an alkali-stable linkage (Yen and Ballou, 1974a). The 5-factor was found to be polymeric, consisting of a large central core linked to six glycopeptide subunits by disulfide bonds. Each of the six subunits had a molecular weight of about 12 500 and contained 28 amino acids and 60 mannose molecules.

The six mannan-peptide subunits appeared to contain the binding sites for type 21 cells (i.e., the 'active sites'), since the mannan-peptides (Taylor and Orton, 1971), but not the large central core of the 5-agglutinin, adsorbed onto the type 21 cells. The polyvalency of the intact 5-factor, which permitted it to agglutinate type 21 cells, was attributed to the proper alignment of the six mannan-peptides in the molecules. Both pronase and disulfide-cleaving reagents inactivated the 5-factor, indicating that intact protein structures were necessary (Taylor, 1964). Exo-α-mannanase and periodate oxidation also inactivated the 5-factor, demonstrating that an intact carbohydrate moiety was also required for the binding activity (Yen and Ballou, 1974b).

(b) *Aggregation factor on type 21 cells*

The mating factor released from the surface of type 21 cells was identified after treatment of these cells with trypsin. Trypsin treatment of type 21 cells rendered them incapable of interacting with type 5 cells (Brock, 1958). The material released by proteolysis inhibited the activity of the 5-factor, but did not agglutinate type 5 cells, suggesting that it bound the 5-factor and was monovalent (Crandall and Brock, 1968). The purified 21-factor was an acidic glycoprotein with a molecular weight of 40 000 (Crandall and Brock, 1968; Crandall *et al.*, 1974). It should be noted, however, that this material was isolated after trypsin hydrolysis, so its native molecular weight was probably larger. Unlike the 5-factor, it was heat-labile and stable to reduction (Crandall and Brock, 1968). Further characterization of the 21-factor awaits increased purification of the molecule.

21-factor and 5-factor are complementary molecules, as judged from their ability to form complexes detectable by gel filtration (Crandall *et al.*, 1974). The standard free energy of the interaction of intact cells was -14.5 kcal mol^{-1}, comparable to the free energy of hapten—antibody interactions (Taylor and Orton, 1970). Assuming that the interaction between cells is mediated exclusively by associations between 5-factor and 21-factor, this value represents the additive effects of the molecular interactions, each of which has a binding energy in the range of -5 to -9 kcal mol^{-1} (Taylor and Orton, 1971).

The mating of yeast cells has proven a fruitful system for investigating the molecular mechanism of cell—cell interactions. It was the first cellular system that permitted complementary molecules to be identified and served as a prototype for studies of many eukaryotic and prokaryotic cellular interactions.

13.2.2 Egg—sperm interaction

The interaction of sperm with egg is another biological system in which cell surface molecules that mediate intercellular recognition have been identified. The system is particularly useful because egg—sperm interactions show species specificity and the cells can be readily isolated without contamination.

A favored organism for study has been the sea urchin: it produces large, readily available quantities of egg and sperm and its production of gametes can be controlled *in vitro*. Although a complete discussion of morphological and biochemical features related to fertilization is beyond the scope of this review, a brief statement of the events occurring in the fertilization of sea urchin eggs, as they relate to cell—cell interaction, is appropriate (see Epel, 1977 and Summers *et al.*, 1975 for a more complete discussion). Homologous sperm first penetrate the jelly coat surrounding the egg, followed by the interaction of the acrosome region (apex) of the sperm with a thin, glycoprotein-rich coat (vitelline layer) which covers the egg. One sperm penetrates the vitelline layer and its plasma membrane fuses with the egg's plasma membrane. It is the acrosome—vitelline layer interaction that imparts the species specificity to fertilization (Aketa *et al.*, 1972).

(a) *Sperm receptor on egg*

Treatment of dejellied, unfertilized eggs with *N*-bromosuccinimide 'stabilized' the cells and allowed the vitelline layers to be isolated intact. These layers had the shape of empty sacks with the dimensions of an egg. The isolated vitelline layer was about 200–300 Å thick but only 100 Å thick when not removed from the egg (Glabe and Vacquier, 1977). This material was 90% protein and 3.5% carbohydrate and contained a variety of polypeptides. Sperm bound only on one side (external surface) of this reticulum, indicating that it retained its molecular asymmetry. A soluble extract prepared from these isolated 'layers' appeared to contain sperm receptors, since molecules in the extract bound to sperm and prevented them from fertilizing eggs (Schmell *et al.*, 1977). Interestingly, species specificity was retained in that the soluble receptor inhibited only homologous sperm. The receptor was judged to be a glycoprotein because

(i)   the activity was released from the egg by trypsin, and
(ii)   the receptor adhered to concanavalin A—Sepharose, a carbohydrate-binding affinity column.

The glycoprotein receptor in the vitelline layer did not appear to have a catalytic function since glutaraldehyde-fixed eggs still bound sperm (Vacquier and Payne, 1973). Oxidation of fixed eggs with metaperiodate, however, decreased sperm binding. In addition, pretreatment of eggs with concanavalin A rendered them incapable of being fertilized (Schmell *et al.*, 1977). Thus the sperm receptor function appeared to reside on the carbohydrate of the glycoprotein. The inability of trypsin to inactivate soluble receptor (Aketa, 1978) is consistent with this view. Simple mono- and disaccharides had no effect on sperm—egg interaction, suggesting that the functional structure on sperm-binding protein is a complex carbohydrate.

Two approaches were used to isolate sperm receptors from sea urchin eggs. Either the proteolytic fragment released into suspension by trypsin (Aketa, 1978) was isolated or the receptor was solubilized from the vitelline layer with urea and separated by protein fractionation procedures (Tsuzuki *et al.*, 1977). The latter

procedure produced a reasonably pure, high-molecular-weight glycoprotein
(mol. wt $> 220\ 000$) which inhibited fertilization by homologous sperm without
altering their mobility. Interestingly, antiserum against this material blocked the binding of only homologous sperm indicating that interactions due to the glycoprotein
are species-specific. This material was difficult to characterize, however, because it
did not penetrate sodium dodecyl sulfate gels and appeared in the void volume of
Sepharose 4B, even in the presence of 6 м guanidine HC1. Because of these characteristics, it may be that the proteolytic fragment will prove simpler to characterize.

Despite criticisms of this work (Schmell *et al.*, 1977), a recent report supported
the notion that the receptor is the high-molecular-weight glycoprotein. Sea urchin
eggs have a potent protease that is released after fertilization (Epel, 1977). The
protease functions to remove sperm receptors so that the egg cannot bind additional
sperm. The presence of this enzyme was evident, since vitelline layers isolated by
Glabe and Vacquier (1977) in the absence of protease inhibitors did not bind sperm.
These authors also investigated which vitelline layer proteins were susceptible
to proteolysis by the egg protease. Using lactoperoxidase-catalyzed iodination, they
found that the major component on the exposed surface of the vitelline layer was
a very high-molecular-weight glycoprotein. Although vitelline layers isolated in the
presence of protease inhibitors contained this glycoprotein, a preparation devoid
of this component was obtained when the inhibitors were omitted. Thus, the available data indicate that the sperm receptor on sea urchin eggs is a high-molecular-
weight glycoprotein in the vitelline layer of the egg and that the receptor function
is, in part, dictated by the carbohydrate in this molecule.

(b) *Bindin – the egg receptor on sperm*
While the sperm receptor on egg has proved difficult to isolate and characterize, the
egg-binding protein on sperm was remarkably simple to obtain. A membrane-bound
acrosome vesicle is located at the apex of the head of sea urchin sperm. Vacquier
and co-workers (Vacquier and Moy, 1977; Bellet *et al.*, 1977; Glabe and Vacquier
1977), removed these vesicles from sperm by solubilizing them in 2.5% triton X-100.
The acrosome material, which consisted of only one protein, was subsequently
isolated by differential centrifugation. Since the authors believed that this relatively
insoluble protein was the substance that binds sperm to egg, they named it 'bindin'.
Isolated bindin is polyvalent and exists as a macromolecular complex of 30 500
dalton proteins in aqueous solution. For example, the complex was found to agglutinate unfertilized, dejellied eggs as well as isolated vitelline layers. It appeared that
bindin, interacting with specific receptors on the egg, caused this agglutination,
since agglutination was inhibited by material released by trypsin from isolated vitelline
layers.

The available evidence indicates that the molecules responsible for egg–sperm
adhesion are capable of a high degree of species discrimination. Not only are sperm
receptors on eggs unique to each species tested thus far but bindins differ according
to species (Bellet *et al.*, 1977). It is tempting to conclude that the complementary

specificities of these molecules prevent interspecies mating.

Vacquier and co-workers (Glabe and Vacquier, 1977; Bellet *et al.*, 1977) have postulated that the egg—sperm interaction in animal groups other than sea urchin may also occur because of specific protein—carbohydrate interactions. For example, mammals also have species-specific attachment of sperm to the zona pellucida of the egg. They suggest that if this interaction is mediated by a mammalian bindin, this protein could be used as an antigen 'to induce immunosterility without risking the development of auto-immune complications'. A more conservative, but nonetheless potentially useful, approach would be to use synthetic glycopeptide sperm receptors as contraceptives.

13.2.3 Agglutination of marine sponge cells

Marine sponges have long been a favorite tool with which to study cellular interactions. Early observations of Wilson (1907) demonstrated their potential by showing that single cells from sponges will aggregate to form complete functional sponges. This aggregation is caused by specific recognition sites on the cell surface, since aggregation occurs only between cells of the same species. Striking examples of this species specificity are the formation of single color cell aggregates from mixtures of dissociated cells from different color sponges (Wilson, 1907; Humphreys, 1970).

In an attempt to understand the molecular basis for this species-specific aggregation, efforts have been made to isolate the molecules responsible. An early observation by Humphreys (1963) established the prototype for these isolations. In these experiments, Humphreys mixed sponges in Ca^{2+} and Mg^{2+}-free sea water and observed that when the dissociated cells were removed from solution, a soluble component remained. Since this component promoted the aggregation of dissociated cells it was termed an 'aggregation factor'. Such aggregation factors have been characterized from several sponge species and each appears to preferentially aggregate the same species from which it was isolated (Humphreys, 1963, 1970; Moscona, 1963; Gasic and Galanti, 1966; MacLennon and Dodd, 1967; McClay, 1974).

Isolation and characterization of specific sponge aggregation factors in soluble fractions was first attempted from *Microcronia prolifera* by Margoliash *et al.* (1965), who showed that the factor was about 50% protein and 50% carbohydrate. More complete biochemical characterization of the factor from *M. parthena* (Henkart *et al.*, 1973; Cauldwell *et al.*, 1973) showed it to be a large acidic proteoglycan complex with a molecular weight of about 20×10^6. The complex was 47% by weight amino acids (containing a large amount of aspartic and glutamic acid residues) and 49% carbohydrate (containing primarily galactose, mannose, uronic acid, glucosamine, and galactosamine). Electron microscopy disclosed that aggregation factor from *M. parthena* had a unique organization, consisting of a central ring about 800 Å in diameter with 15 to 16 arms radiating outward about 1100 Å. The term 'sunburst' was applied to this molecule in reference to its shape.

The large size of the molecule indicates that only one aggregation factor can be accommodated between any two interacting cells. Studies on subunit organization within the aggregation factor were complicated by the low solubility of the factor. Most detergents, including sodium dodecyl sulfate, did not dissociate the subunits of this factor; however, chelation of the divalent cations with ethylenediamine tetraacetate for 4 weeks at $0°C$ caused the factor to dissociate (Humphreys *et al.*, 1977). When the factor from *M. prolifera* was treated in this manner, the central circle and the 'rays' of the 'sunburst' were dissociated and subsequently isolated on a micropore galss column. Further characterization of each component was hindered by their low solubilities.

The molecular basis of the interaction between the aggregation factor and the membrane surface remains unsolved, even though it has been the subject of numerous investigations. Because the aggregation factor from *M. prolifera* contained a high amount of carbohydrate and was sensitive to periodate, Turner and Burger (1973) investigated the effects of various carbohydrates on the aggregation-enhancing activity of this factor. Glucuronic acid was the only sugar found to be an effective inhibitor. Inhibition was observed only when dissociated cells were incubated with glucuronic acid before the addition of the aggregation factor suggesting that glucuronic acid residues on the aggregation factor are required for the interaction of the factor with *M. prolifera* cells. This conclusion is not valid for all species, since glucuronic acid did not affect the aggregation of cells from *Cliona celata* (Turner and Burger, 1973).

Müller and co-workers (Müller and Zahn, 1973; Müller *et al.*, 1974; Müller *et al.*, 1977) isolated the aggregation factor from cells of *Geodia cydonium* species and began studies on the molecular basis of its activity. They found that the factor was similar to the one from *M. parthena*. It had a molecular weight of approximately 20×10^6 and consisted of a closed circle with the 'sunburst' organization. Interestingly, sialyl transferase activity was purified with the *Geodia* aggregation factor. This enzyme catalyzes the transfer of sialic acid (*N*-acetylneuraminic acid) from cytidine monophosphate-sialic acid to the terminal position of oligosaccharides on proteins and lipids. Müller and his collaborators became aware of this activity when they observed incorporation of $[^{14}C]$-sialic acid caused by reaggregation of desialated sponge cells in the presence of $[^{14}C]$-cytidine monophosphate-sialic acid. Only slight incorporation was observed in reaggregated control cells. Sialyltransferase was readily separated from the sunburst structure with nonidet P-40, a non-ionic detergent that caused a 680-fold purification of the factor from the intact cells. Since the sialyl transferase was associated with aggregation factor, Müller and co-workers suggested that the function of sponge aggregation factor is related to its sialyl transferase activity.

The contention that a glycosyl transferase mediates cell—cell interactions is similar to the view originally presented by Roseman (1970), which proposed that the adhesive force between cells can be attributed to the enzyme—substrate interactions between glycosyl transferases on one cell with saccharide-containing substrates for these enzymes on the surface of another cell. Although having little experimental support, this idea remains attractive because

(i)   it allows for cell interaction specificity by utilizing cell-specific glycoproteins, and
(ii)   both the enzymes and their substrates have been detected on the surfaces of many types of cells.

Müller *et al.* (1977) applied this idea, suggesting that the sponge aggregation factor functions because of its sialyltransferase activity. In this model, binding of the factor to cells is mediated by an enzyme substrate complex between the sialyltransferase of the aggregation factor and the oligosaccharide chain of the glycoprotein acceptor on the sponge cells.

Sialoglycoproteins from the surface of sponge cells were isolated to identify the substrate for the sialyltransferase associated with the aggregation factor (Müller *et al.*, 1976). One sialoglycoprotein, which was removed from the cell either by a Ca^{2+}-free medium or by incubation of sponge cells with trypsin, was identified. The isolated material was 81% neutral carbohydrate and 7.5% protein, but contained 73% of the total sialic acid in the sponge. Removal of this glycoprotein from the cell surface diminished the aggregation potency of the cells (Müller *et al.*, 1976). Interestingly, this isolated glycoprotein was found to interact with purified aggregation factor in solution. Desialylation reduced the adhesiveness of *Geodia* cells. These observations supported the suggestion that this sialoglycoprotein on the surface of sponge cells was the aggregation receptor and that the sialyltransferase was the functional component in the aggregation factor from *Geodia* cells.

The membrane receptor for the aggregation factor in *M. prolifera* was also studied by a different method. Weinbaum and Burger (1973) observed that dissociated cells exposed to hypotonic conditions lost protein and were no longer agglutinated by aggregation factor. The conclusion that the material released by hypotonic shock was the aggregation receptor (termed 'baseplate' by Weinbaum and Burger, 1973) is based on:

(i)   addition of baseplate to shocked cells restored their ability to be aggregated by aggregation factor;
(ii)   baseplates neutralized aggregation factor;
(iii) baseplates covalently attached to sepharose beads were agglutinated by aggregation factor.

Whether the baseplate present in *M. prolifera* is similar to the aggregation factor receptor in *G. cyclonium* is not known.

Clearly, studies on the interaction of sponge cells have done much to clarify the specific interactions of eukaryotic cells: methods for studying cellular interactions have been developed; molecular entities responsible for cellular interactions have been identified; the mechanisms of these interactions have been partially characterized. These studies indicate that the system which mediates the interaction of sponge cell contains at least two components. One is an aggregation factor that dissociates in a solution low in divalent cations. The aggregation factor appears to be polyvalent and thereby binds two cells. The selectivity of aggregation between sponge cells indicates that the aggregation factor is species-specific. Further work is required to determine the identity of the cell surface receptor.

13.2.4 Opsonic activity of soluble fibronectin

Fibronectin is another protein known to be involved in cell adhesion which has been extensively studied. Two types of fibronectins have been identified: soluble fibronectin (also termed cold-insoluble globulin, CIg) and cell-surface fibronectin (also termed cell surface protein, CSP, large external transformation-sensitive protein, LETS, and other labels; Mosesson, 1977). Purified preparations of soluble fibronectin contain several different polypeptides all of which are immunologically identical (Mosesson, 1977). The basis of this heterogeneity is not understood, although proteolysis of circulating soluble fibronectin *in vivo* has been suggested (Mosesson, 1978). By contrast, purified cell-surface fibronectin is homogenous (Yamada *et al.*, 1977). The fibronectins are dimeric glycoproteins with disulfide-linked subunits with molecular weights ranging from 200 000–250 000. The soluble and cell surface fibronectins are immunologically similar although subtle differences have been observed among them (Yamada *et al.*, 1978; Yamada and Olden, 1978). The fibronectins are thought to mediate cell interactions, both cell–cell and cell-substratum. The techniques used to purify the fibronectins, their physical characteristics, distribution *in vivo*, functions in cell adhesion and relationship to cancer have been the subjects of excellent and extensive reviews (Grinnell, 1978; Hynes, 1976; Mosesson, 1977; Vaheri and Mosher, 1978; Yamada and Olden, 1978; Yamada and Pastan, 1976). Hence, we will only consider the recently recognized opsonic function of soluble fibronectin; the facilitation of non-immune phagocytosis by macrophages of the reticuloendothelial system.

Serum contains a high molecular weight glycoprotein thought to mediate non-immune phagocytosis of damaged cells and cell debris. Although early reports termed this glycoprotein α_2-surface binding glycoprotein, three recent reports (Saba *et al.*, 1978; Blumenstock *et al.*, 1978; Dessau *et al.*, 1978) pointed out that soluble fibronectin and α_2-surface binding glycoprotein were physically and immunologically similar. The glycoprotein functions as a non-immune opsonin by recognizing certain forms of collagen in broken cells and assisting in their phagocytosis by macrophages of the reticuloendothelial system (Blumenstock *et al.*, 1976; Filkens and Smith, 1965; Hopper *et al.*, 1976; Molnar *et al.*, 1977; Saba *et al.*, 1977; Saba and Di Luzio, 1969; Wagner and Iio, 1964).

Depletion of circulating serum fibronectin resulted in a depression of the reticuloendothelial system function (Di Luzio and Lindsey, 1973; Saba, 1975; Saba *et al.*, 1978). Fibronectin depletion was observed in patients after surgery, traumatic shock, or other injury (Saba *et al.*, 1978). Such changes have also been observed in rats and other experimental animals (Saba, 1975). The decreased concentration of fibronectin correlated with a decrease in phagocytotic capacity of liver Kupffer cells, which was corrected by addition of exogenous soluble fibronectin.

Surgery often causes pulmonary insufficiency, possibly because of cellular debris lodging in the pulmonary blood vessels (Fulton and Jones, 1975). This restriction of pulmonary blood vessels is thought to stem, at least in part, from a deficiency of the reticuloendothelial system. The implication was that a decrease in circulating fibronectin following surgery resulted in decreased efficiency of phagocytosis, thereby allowing cell debris to accumulate in the pulmonary vessels. Support for this view was obtained from the report that post-surgical intravenous infusion of soluble fibronectin (cryoprecipitate) caused an improvement of pulmonary function in postoperative patients (Saba *et al.*, 1977).

The therapeutic potential of infusion of soluble fibronectin is not limited to overcoming the consequences of post-surgical trauma. Injection of leukemic but not normal lymphocytes caused a severe depression of reticuloendothelial function in rats (Di Luzio *et al.*, 1972). Likewise, human patients with advanced malignant disease had decreased levels of circulating fibronectin (Di Luzio, 1975). These findings indicate that fibronectin may have some usefulness in the treatment of patients with malignant disease.

The molecular basis of the cell recognition function of soluble fibronectin in phagocytosis appears to be complex. Apparently, the protein possesses two distinct cell recognition activities, one that recognizes the abnormal cellular material to be cleared from the blood; and one that recognizes macrophages (Saba *et al.*, 1977). To function efficiently, the opsonic protein must discriminate between abnormal and normal cells. The basis of this selectivity is not understood, and much work will be needed to learn how the non-immune opsonic protein accomplishes the specific recognition of both abnormal cells and macrophages.

Cell surface fibronectin has been shown to possess hemagglutinating activity (Yamada *et al.*, 1975). Although the opsonic activity of soluble fibronectin may depend on a similar function, it has a very low level of hemagglutination activity. (Yamada and Kennedy, 1978). Thus, there appears to be a functional difference between the soluble and cell-surface forms of fibronectin. This difference is intrinsic to the two forms of fibronectin and does not result from different methods of purification.

13.3 DEVELOPMENTALLY REGULATED CELL RECOGNITION SYSTEMS

13.3.1 Slime molds

The cellular slime mold, *Dictyostelium discoideum*, is an organism that has proven useful for studying the basis of direct cell–cell interaction in development. Although the subject of recent reviews (Simpson *et al.*, 1978; Frazier, 1976), the topic will be discussed briefly here in order to outline important achievements.

Slime mold amoebae exist in various forms depending on their nutrient supply. When the supply of bacteria (the food source of slime mold cells) is plentiful, the vegetative cells are noncohesive. When the food supply is depleted, a chemotactic stimulus, cyclic adenosine monophosphate, is emitted and the cells migrate towards each other. During this time, the membrane surface of the cell is altered so that the cells become cohesive and are therefore able to aggregate. The multicellular structure differentiates morphologically and functionally leading to the formation of fruiting bodies (Bonner, 1967).

(a) *Discoidin*

Examination of the developmental changes of the lectin activity of amoebae before and after starvation has resulted in identification of molecules responsible for the interaction of cohesive slime mold cells. For this purpose, the soluble extract obtained by homogenizing either vegetative or cohesive cells in 0.15 M saline was used to agglutinate trypsinized, formalinized sheep erythrocytes (Rosen *et al.*, 1973). Between 3 and 9 hours after vegetative cells were deprived of food, a lectin was produced. Its activity was blocked by *N*-acetyl-galactosamine, 6-deoxy-L-galactose (L-fucose) and to a lesser extent by other saccharides containing galactose, but not by other sugars. The carbohydrate-binding property of the lectin permitted its isolation by affinity chromatography. Since this protein was present on the membrane surface of cohesive, but not on vegetative cells, it was suggested that this protein, later termed discoidin (Simpson *et al*, 1974), may mediate cohesion between the developing amoebae. Discoidin has a molecular weight of 100 000 under nondissociating conditions and appears to be composed of four subunits (Simpson *et al.*, 1974). Subsequent studies showed that there are two developmentally regulated lectins in discoidin, one termed discoidin I (mol. wt. = 26 000 under denaturing conditions) and the other discoidin II (mol. wt = 23 000) (Frazier *et al.*, 1975). Both appeared responsible for the lectin activity of discoidin. The aggregation factor from the cellular slime mold *Polysphondylium pallidum*, termed pallidin, and discoidin were species-specific agglutinins (Rosen *et al.*, 1974); univalent antibodies to pallidin inhibited agglutination (Rosen *et al.*, 1976). However, the significance of the anti-pallidin inhibition has been questioned (Müller and Gerisch, 1978).

Cohesiveness of *D. discoideum* slime mold cells is not controlled solely by the appearance of discoidin, but is also regulated by expression of lectin receptors on the cell surface. This has been demonstrated in experiments in which purified discoidin did not agglutinate glutaraldehyde-fixed, noncohesive cells, but did agglutinate fixed, cohesive cells (Reitherman *et al.*, 1975).

These studies have led to the lectin–lectin receptor model for cell–cell interaction (Rosen *et al.*, 1974; Simpson *et al.*, 1977). According to this model, the aggregation of cohesive amoebae results from the interactions between lectins and lectin receptors on adjacent cells, much as outlined in Fig. 13.1b.

(b) *Contact site A*

A second approach to isolating aggregation factors in *D. discoideum* is based on immunologic principles. This approach was based on the observation that monovalent antibody derivatives (Fab fragments) against membrane antigens of cohesive cells completely inhibit end-to-end cell contact without affecting chemotactic response to cyclic adenosine monophosphate or cell motility (Beug *et al.*, 1970). Fab fragments of antibodies specific for other membrane sites failed to affect adhesion (Beug *et al.*, 1973). The Fab fragments were apparently directed against a species-specific aggregation site, since they failed to affect the interaction of cohesive amoebae of another slime mold, *Polysphondylium pallidum* (Bozzaro and Gerisch, 1978).

The membrane antigen responsible for generation of the inhibitory antibody was purified from isolated membranes by using reversal of immunospecific Fab inhibition of cell aggregation as an assay. With this procedure, it was possible to isolate a membrane glycoprotein (termed cs-A) that differed from discoidin and had a molecular weight of 80 000, as determined after reduction by sodium dodecylsulfate gel electrophoresis. The non-reduced molecular weight was 120 000–130 000 as determined in deoxycholate on Sephadex G-200 (Huesgen and Gerisch, 1975; Müller and Gerisch, 1978c). Cs-A was free of discoidin and did not bind to this protein, indicating that cs-A was distinct from discoidin and its receptor.

The identification of an additional adhesion molecule on cohesive *Dictyostelium* cells indicates that either multiple systems are involved in the interaction of these cells or that these aggregation factors are functionally related in some as yet undefined way. However, the available data do not indicate any functional relationship between these aggregation factors, leading Gerisch and co-workers (see, e.g., Müller and Gerisch, 1978) to conclude that end-to-end aggregation of amoebae is mediated by the interaction of identical molecules. The discoidin research, on the other hand, indicates that non-identical complementary molecules interact to mediate aggregation (Simpson *et al.*, 1978). Further work is required to clarify the relative physiological importance of these two aggregation systems in the development of slime mold.

13.3.2 Galactolectins

In 1975, Teichberg *et al.*, reported that extracts from the electric organs of the electric eel agglutinated trypsinized rabbit erythrocytes. Since this activity was similar to other lectins, it was termed 'electrolectin'. Lactose and other saccharides containing a terminal non-reducing D-galactosyl residue in the β-linkage inhibited hemagglutination by electrolectin. Interestingly, the most potent inhibitor of all the compounds tested was thiodigalactoside (D-galactosyl-β $(1 \rightarrow 1)$ – thiogalacto-pyranoside), an analogue of lactose. Electrolectin had a molecular weight of approximately 33 000 and was activated by reducing agents. Other tissues containing lectin with similar sugar specificity include mouse N-18 neuroblastoma cells, embryonic chick pectoral muscle, adult chick rectus muscle, and adult rat heart, diaphragm, and soleus muscles. The tissue with the highest measured specific activity was embryonic chick pectoral muscle.

Subsequent studies disclosed that, although the lectins in tissues from different species have similar sugar specificities, they have different molecular weights. In contrast, lectins from various tissues of the same animal have identical molecular weights. For example, the lectin from calf heart and lung (Briles *et al.*, 1979; De Waard *et al.*, 1976) have a lower molecular weight than the lectins from embryonic chick brain, liver, striated muscle, and the liver from 7-day-old chicks (Nowak *et al.*, 1977; Den and Malinzak, 1977; MacBride and Pzybylski, 1978; Kobiler and Barondes, 1977). Lectins from all these tissues agglutinate rabbit erythrocytes and are inhibited by thiodigalactoside and other galactose-containing saccharides, indicating that they comprise a family of proteins with similar specificity. Since the proteins responsible for the lectin activity show some species differences in molecular size, their nomenclature should not refer to the original tissue of discovery, but rather to their sugar specificity. Accordingly, these molecules will be referred to as galactolectins.

The role of galactolectins in cell—cell interactions has been extensively studied in developing striated muscle. Muscle cell interaction was particularly interesting because it is essential to the development of mature skeletal muscle and because this interaction can be observed in tissue cultures. Skeletal muscle is composed of long, cylindrical muscle fibers called myotubes. Each skeletal muscle is a composite structure of many bundles of muscle fibers (Myotubes) held together by connective tissue. Each myotube is a multinucleated cell formed by the fusion of many mononucleated myoblasts. It is this fusion during differentiation that requires specific cell—cell interactions by the developing myoblasts (Capers, 1960; Firket, 1958; Konigsberg *et al.*, 1960; Lash *et al.*, 1957; Stockdale and Holtzer, 1961 and Wilde, 1959). An important feature of muscle development is that morphological and functional differentiation of myoblasts into myotubes depends on direct inter-actions of specific cell types. A demonstration of this specificity was the observation that myoblasts will fuse only with myoblasts of the same or different species; they will not fuse with other cell types, even those from the same species (Yaffe and Feldman, 1965).

The lectin activity of developing myoblasts is conveniently studied with L_6, a permanent cell line established by Yaffe (1968) from embryonic rat thigh muscle. L_6 is a useful cell line because it offers a continuous source of myoblasts which proliferate and differentiate *in vitro*. Upon differentiation, the outer surface of the plasma membrane changes so that the cells become cohesive. The myoblasts stick together and the membranes fuse to form differentiated, multinucleated, electrically active and chemically sensitive myotubes (Harris *et al.*, 1971; Yaffe, 1968).

The presence of lectin activity in L_6 was demonstrated in homogenates of myoblasts (Gartner and Podleski, 1975). Both the soluble and particulate fractions of these homogenates agglutinated formaldehyde-fixed, trypsinized rabbit erythrocytes. Hemagglutination was blocked by both lactose and thiodigalactoside, but not by α-methyl glucose, α-methyl mannose or any other compounds tested. Intact myoblasts that were removed from the tissue culture dishes either chemically with EDTA or mechanically with a rubber policeman were found to agglutinate rabbit erythrocytes. This membrane surface lectin appeared to be identical with the one present in cell homogenates, since it was also inhibited by thiodigalactoside.

Thiodigalactoside and other sugars were tested as inhibitors of myoblast fusion and myotube formation to determine if the myoblast galactolectin was involved in cell–cell recognition (Gartner and Podleski, 1975). Myoblasts cultured in the presence of thiodigalactoside failed to fuse and form myotubes, while cultures incubated in the absence of thiodigalactoside or with other sugars developed normally. Thiodigalactoside did not decrease the growth rate of the myoblasts or cause any obvious deleterious physiological effects. These observations lead to the hypothesis that the cell–cell interactions of developing muscle are mediated by an interaction between myoblast galactolectins and their receptors on differentiating myoblasts.

Subsequent work supported the observation that galactolectin mediates myoblast interaction. Den and Malinzak (1977) using the L_8 embryonic rat myoblast cell line (Yaffe, 1968), also found that thiodigalactoside blocks fusion of myoblasts. Merlie and Teichberg (1978) additionally reported that L_5, a non-fusing clone derived from embryonic rat striated muscle (Yaffe, 1968), possessed very little galactolectin activity.

If the galactolectin is responsible for myoblast fusion, its expression should occur before myotube formation. In studies of the agglutination of rabbit erythrocytes, Gartner and Podleski (1976) and Nowak *et al.* (1976) found that the specific activity of galactolectin increased during culture, reaching a maximum about $1-1.5$ days before early fusion. By contrast, another lectin, which is not affected by thiodigalactoside, reached maximum specific activity at the time of early fusion (Gartner and Podleski, 1975). The activities of both lectins decreased with further development of the cultures. These data show that the temporal expression of galactolectin is consistent with its postulated role in myoblast fusion.

The developmental regulation of galactolectin in embryonic chick pectoral muscle was also evaluated *in ova* by (Nowak *et al.,* 1976). These workers showed that the muscle from 8-day-old embryos had a much lower lectin activity than muscle tissue derived from older embryos. The specific activity of the lectin activity apparently reached a maximum in 16-day-old embryos. This increase of galactolectin activity correlated with formation of muscle *in ova,* with myotube formation being most pronounced by day 16. Thereafter, the galactolectin activity decreased with increasing age.

In a similar study, Den *et al.* (1976) showed that galactolectin also increases with time in the thigh muscle of embryonic chick. The lectin activity of extracts from this tissue increased from day 7, attaining a maximum at day 13 and decreasing thereafter. The maximum activity coincided with the time of myoblast fusion *in ova.* In contrast to earlier reports (Gartner and Podleski, 1975), however, Den *et al.* (1976) found that thiodigalactoside and lactose failed to inhibit the fusion of embryonic chick thigh myoblasts. Den *et al.* (1976) concluded that galactolectin is not responsible for the interaction of myoblasts. This is a critical observation concerning the function of galactolectin in myoblast fusion, so it is important to examine the conditions used in the study. Interestingly, primary cultures were obtained from 12-day-old embryos. Since embryos of this age have nearly maximum expression of galactolectin, their development may be too advanced for galactose derivatives to affect fusion. It seems more likely that myoblast recognition, mediated by lectin—lectin receptor interactions, occurred by day 12 and that fusion is a subsequent event.

The importance of galactolectin in muscle development was also demonstrated by the studies of MacBride and Przybylski (1978), who showed that galactolectin, purified from either pectoral muscle of 14 to 16-day-old chick embryos or day-5 muscle culture inhibited myotube formation *in vitro*. This inhibition was reversed by lactose. These results support the notion that galactolectin is involved not only in the fusion of the tissue culture cell lines L_5, L_6, and L_8, but also in myoblast fusion of primary cultures of embryonic chick pectoral muscle. Thus, it appears that under the appropriate conditions, the galactolectins can alter the course of myotube formation in chick primary cultures and may play a role in the differentiation of muscle *in vivo*.

Mir-Lechaire and Barondes (1978) showed that, in addition to galactolectin, another lectin was present in embryonic chick pectoral muscle, one that was not affected by galactose-containing sugars, but was inhibited by N-acetyl-D-galactosamine. This lectin appeared to be developmentally regulated as it reached a maximum activity at the time of extensive myoblasts fusion. Because of its lack of stability, this lectin has proven difficult to isolate. In this regard, it was similar to a lectin that was present in L_6 myoblasts, but not in myotubes and was not affected by thiodigalactoside (Gartner and Podleski, 1976). The relationship of the unstable lectin in embryonic chick pectoral muscle to differentiation has not been evaluated.

13.3.3 Embryonic chick retina

In 1967, Lilien and Moscona observed that embryonic chick retina cells in monolayer culture released proteins into the culture medium which supported cell adhesion. Since this culture medium appeared to be a rich source of molecules involved in cell adhesion, it has been used as a source of retinal cell adhesion molecules.

(a) *CAM protein*

Rutishauser *et al.* (1976) isolated several proteins present either in the culture medium of embryonic chick retina or monolayers of these cells. Fluorescent derivatives of antibodies to these proteins were prepared and two were found to bind to the surfaces of retina cells. The surface location of the antigens was identified by iodinating the surface proteins on intact cells with lactoperoxidase and precipitating the extracted membrane proteins with the antisera. These procedures resulted in the identification of a surface membrane protein that was termed CAM (cell adhesion molecule). Two observations suggested that CAM is either directly or indirectly involved in adhesion between retina cells. First, the binding of retina cells and chick embryonic brain cells to immobilized retina cells was inhibited by anti-CAM antibody. Second, the binding of cultured retina cells to nylon fibers coated with anti-CAM antibody reached a maximum with cells from 8-day old embryos, the embryonic age at which cultured retina cells show the maximum adhesiveness for each other. The function of CAM in cell adhesion was not determined, but it was suggested that CAM molecules on adjacent cells might be able to bind each other and thereby mediate cell interaction. Other mechanisms were not excluded.

In the study by Rutishauser *et al.* (1976), antibodies were prepared against proteins isolated from 'conditioned' cell culture medium. In a subsequent study (Brackenbury *et al.*, 1977), antibodies were prepared (called anti-R_{10}) against intact retina cells from 10-day old embryonic chicks. Anti-R_{10} was found to inhibit the adhesion of embryonic chick retina cells. Interestingly, monovalent Fab' fragments of anti-R_{10} blocked aggregation of 10-day old retina cells, whereas succinyl concanavalin A and Fab' fragments of antibodies against cell surface carbohydrates failed to block aggregation of the same cells. This differential activity occurred even though all three substances bound to the cell surface in comparable amounts. Thus, the ability of the anti-R_{10} preparation to inhibit cell adhesiveness was due to binding of antibodies to specific cell surface components, rather than non-specific inhibition caused simply by the presence of antibodies on the surfaces of the cells.

This specific binding of anti-R_{10} antibodies to cell surface components was used as the basis for identifying molecules involved in cell adhesiveness. The assay used was to measure the specific neutralization of the anti-aggregation activity of the Fab' fragments of anti-R_{10} preparations. Using this procedure, Thiery *et al.* (1977) identified three peptides with neutralizing activity in the culture medium 'conditioned' by retina tissue. These peptides were purified and antibodies against them inhibited cell adhesion as measured by the aggregation test (Brackenbury *et al.*, 1977). Unlike the anti-R_{10} antibody preparation, however, antibodies against the purified proteins specifically precipitated only three peptides from the 'conditioned' tissue culture medium. When proteins extracted from the plasma membranes of retina cells were treated with this antibody, only two proteins were precipitated, a minor component thought to be actin and a major component with a molecular weight of 140 000, which was identical to the CAM protein. Thus, these studies and the technique as originally developed by Gerisch and associates (Beng, *et al.*, 1970; Beng *et al.*, 1973; Huesgen and Gerisch, 1975) have resulted in the development of a procedure that may be of general significance for identifying cell surface components active in cell adhesion. In addition, use of the procedure confirmed the involvement of CAM in the adhesiveness of embryonic chick retina cells. These studies did not, however, clarify the function of CAM in cell adhesion.

A subsequent study from the same laboratory (Rutishauser *et al.*, 1978a), confirmed the external plasma membrane location of the CAM protein and demonstrated that it is present in a variety of neural tissues. Aggregated retina cells from 10-day old chick embryos were examined for cell surface CAM by use of specific anti-CAM antibody. CAM was found in nearly all regions of the cell surface, and there was no evidence that cell adhesion had resulted in changes of the surface distribution of this protein. Furthermore, no striking difference in the amount of CAM in intact retina tissue from 8-day-old and 14-day-old embryos was demonstrable by the techniques used in this study. Thus, regulation of the activity of CAM receptors may play a critical role in CAM function. Staining of frozen sections of a variety of tissues with anti-CAM antibody disclosed that 6- to 14-day old embryos contained CAM in retina, brain, optic nerve, spinal cord and both sympathetic root ganglia. By contrast, liver and muscle cells contained little, if any, CAM. CAM was also identified on isolated neurite outgrowths of cultured retina cells. In this regard, anti-CAM antibodies inhibited the formation of membrane—membrane contacts between neurites formed by cultured aggregates of retina cells from 8-day-old embryos. This wide distribution of CAM in neural tissues coupled with the inhibitory effects of anti-CAM antibodies on cell aggregation and mebrane-membrane interaction between neurites led the authors to postulate that CAM-mediated cell adhesiveness may be a general feature of neural tissue.

Observations on the effect of anti-CAM antibodies on the formation of nerve bundles has led to the conjecture that CAM may mediate the morphogenesis of most nerve tissues during embryogenesis of chick embryos (Rutishauser *et al.*, 1978b). Thus, anti-CAM antibodies inhibited the formation of nerve bundles (fascicles) by neurites from different classes of spinal ganglia from 10-day old chick embryos. The important feature of this inhibition was that the antibodies appeared to inhibit membrane—membrane interactions between adjacent neurites rather than inhibiting either neurite growth cone or nerve growth factor functions. Interestingly, the inhibitory effect of anti-CAM antibodies on fascicle formation was reversible. Thus, removal of the antibody resulted in side-by-side association of the neurites and formation of fascicles. On the basis of these and other observations (Rutishauser *et al.*, 1978a), it was suggested that CAM-mediated membrane—membrane interactions between adjacent neurites may be essential for nerve trunk formation (Rutishhauser *et al.*, 1978b).

(b) *Retina-specific cell-aggregating factor*

Hausman and Moscona (1973, 1975, 1976) also isolated a soluble factor that aggregated retina cells. This aggregation activity was isolated either from 'conditioned' culture medium or from retina tissue membranes of 10-day old chick embryos. These activities are believed to be caused by identical molecules. The aggregating factor has an estimated molecular weight of 50 000 and is glycosylated. Aggregation of retina cells, but not optic tectum cells or cerebrum cells, was enhanced by this aggregation factor. The factor could only be isolated from retina tissue obtained from embryos no older than 13 days. Likewise, it only mediated aggregation of retina cells from embryos younger than 13 days. Unfortunately, the specificity of this aggregation factor was not tested on heterologous tissue of different ages. This may be an important point since age appears to be a critical determinant in the expression of cell recognition components. Further work is required to elucidate the role of this cell aggregation factor in the development of neural tissue.

13.3.4 The role of non-tissue-specific cell adhesion molecules in development

It has been suggested that direct cell—cell interactions occurring during embryogenesis are mediated by tissue-specific cell adhesion molecules (see Hausman and Moscona, 1976, for discussion of this point). While this idea has a certain attraction for explaining cell sorting during development, experimental evidence substantiating this view is not available. For example, embryonic retina, an extensively studied developmental tissue, appears to produce both tissue-specific and non-tissue-specific cell adhesion molecules (Gottlieb and Glaser, 1975; Hausman and Moscana, 1976; Merrell and Glaser, 1973, Rutishauser *et al.*, 1976).

An alternative to the concept of tissue-specific aggregation factors is that the same recognition system mediates the interaction of cells of more than one tissue. This hypothesis was supported by Rutishauser *et al.*, (1976) who demonstrated that a common recognition system facilitated the development of two tissue types from the same embryo. These authors showed that anti-CAM antibody, prepared against the surface protein on embryonic chick retina cells, not only inhibited adhesion of embryonic chick retina cells, but also of embryonic chick brain cells. This and other observations lead to the conclusion that the same cell adhesion factor (i.e., the CAM protein) was involved in the development of both types of tissue (Rutishauser *et al.*, 1976).

Since dissociated embryonic cells have the ability to 'sort out' or associate with cells from the same tissue, one must explain how one recognition system can be responsible for homologous interactions when two different cell types are present. A mechanism that satisfactorily resolves this question is a tissue-specific temporal regulation of the cell recognition system. When more than one cell type uses identical complementary molecules for cellular recognition, each tissue has a unique time for expression of these molecules. That this situation holds for both retina and brain cells was demonstrated in two ways. First, the binding of cultured cells from embryos of different ages to nylon fibers coated with anti-CAM antibodies varied with age. For example, the binding of embryonic retina cells increased from day 4 to day 8, but decreased thereafter. Although the binding of embryonic brain cells also increased from day 4, their binding capacity reached a maximum at day 6. Another method used to demonstrate differences in temporal development of cell adhesiveness was to measure the direct binding of cells to other immobilized cells. The adherence of cultured retina cells for each other reached a maximum in cultured cells from 8-day embryos. Similarly, adherence of cultured brain cells to each other was most pronounced in cells from 6-day embryos. Thus, the ages of the embryos producing cells with the maximum cell—cell binding capacity in the *in vitro* assay was coincident with the ages imparting maximum binding of cells to antibody-coated fibers. Such relationships may have significance for cell adhesiveness *in vivo*, since cultured 8-day retina cells bound to cultured 6-day brain cells to approximately the same extent as they bound to homologous cell types. By contrast, 8-day retina cells did not efficiently bind to either 13-day retina cells or 10-day brain cells, nor did 10-day brain cells bind efficiently to each other. Importantly, both homologous and heterologous cell interactions were inhibited by anti-CAM antibody.

The basis of the temporal regulation of the common cell recognition system was not determined. Further experimentation is required to reveal if development of cell adhesiveness results from increased amounts of CAM, the activation of pre-existing CAM or the appearance of CAM receptors. Any of these changes could account for the temporal control of cell adhesiveness.

The data indicate that two tissues from the same embryo use the same cell recognition system during development. The key to the successful utilization of this non-tissue-specific cell recognition system in development appears to be the staggered time of expression of the cell recognition system by different tissues. Such a system may be genetically frugal for developing organisms since the same genetic information could be used by different cell types as a consequence of a tissue-specific modulation of its time of expression. This point may also be relevant to the function of galactolectins which are present in a wide variety of tissues.

13.4 PLATELET INTERACTIONS

Platelets perform a variety of functions in the maintenance of vascular integrity. Many of these functions depend on specific receptors on the platelet for various cells and tissues. For example, platelets interact with endothelium, a process with the potential of causing thrombotic lesions; platelets interact with subendothelial tissue, a process critical for the initiation of a hemostatic plug; platelets aggregate with each other, a reaction essential for the arrest of blood flow (hemostasis) and the formation of arterial thrombi (thrombosis); platelets adhere to smooth muscle cells, a process implicated in wound healing; and platelets interact with neoplastic cells, which may have a bearing on tumor metastasis.

Platelets are useful for studying homologous cell–cell interactions not only because of the importance of platelet aggregation in maintaining vascular integrity, but also because platelets do not have to be disrupted before studies can be performed. Most investigations of homologous interactions of metazoan cells require the disruption of tissue followed by an examination of subsequent cellular associations. Disruption techniques are varied with the most widely used techniques including removal of metal ions by chelation, mechanical agitation, and, most commonly, enzyme hydrolysis. Evaluation of the significance of subsequent adhesion assays is dependent upon proper controls to ensure that the surface properties of the cells have recovered from the disruption trauma. Very often, cell adhesion studies overlook this obvious, yet critical, point. Blood platelets, by contrast, exist as non-cohesive free cells in their native state. Platelets aggregate rapidly after stimulation (within minutes) and the conversion of platelets to the cohesive form can be experimentally controlled *in vitro*. These features, coupled with their availability in sufficient quantities for biochemical studies, make blood platelets an ideal tissue for examining the mechanism of interaction of homologous cells.

13.4.1 Platelet membrane composition and structure

The platelet plasma membrane is composed of lipid, protein and carbohydrate. The membrane is 50% by weight lipid, containing both neutral and phospholipids (Barber and Jamieson, 1970). The arrangement of the phospholipids within the membrane is similar to that in other plasma membranes. The polar groups are asymmetrically distributed about the two faces of the membrane; phosphatidyl serine and phosphatidyl ethanolamine oriented to the cytoplasmic face; phosphatidyl choline and sphingomyelin oriented to the outside of the cell (Schick *et al.*, 1976; Zwaal *et al.*, 1977). Thus, phospholipids make a negligible contribution to the net charge of the outer membrane surface. The phospholipids also function as a source of arachadonic acid, the precursor of prostaglandins which are important to platelet stimulation.

Since the composition and distribution of lipids in platelets is similar to that in other cells, it seems that the specialized functions of the platelet membrane surface are due to the exposed membrane proteins. In 1968, Behnke reported that anion-specific stains reacted with the membrane surface of the platelet and correctly interpreted this finding as evidence that the outer surface of the platelet had a net negative charge, but incorrectly assumed that the coat consisted of sulfated mucopolysaccharides. Subsequent studies showed that the coat was composed primarily of glycoprotein (Pepper and Jamieson, 1969). Proteolysis of the platelet membrane demonstrated that several different glycoprotein species were accessible on the membrane surface. However, the large number of glycoproteins present became apparent only with the advent of surface-labeling techniques. These surface probes are designed to label protein and carbohydrate residues on the outer surface of membranes. Some of these probes are enzymatic and are restricted to altering the outer surface of the membrane if the experiments are performed before endocytosis or cell disruption. Other probes are slowly penetrating chemicals that tend to preferentially label surface components. Surface-labeling techniques which have been applied to platelet membrane studies include iodination of tyrosine residues by lactoperoxidase (Phillips, 1972; Phillips and Agin, 1977a), labeling of galactose residues using neuraminidase-galactose oxidase, and (^{3}H)NaBH$_4$ (Phillips and Agin, 1977b), labeling of sialic acid residues by periodate and (^{3}H)NaBH$_4$ (Andersson and Gahmberg, 1978), and labeling of amino residues using either diazotized diiodosulfanylic acid (George *et al.*, 1976) or amino transferase (Okumura and Jamieson, 1976).

Several generalizations can be made concerning the platelet membrane proteins. All of the major proteins exposed on the outer surface are sialoglycoproteins. Some of these glycoproteins are transmembranous, that is, they span the thickness of the membrane. The carbohydrate-containing portions of these molecules are exposed to the outside of the cell, with little or no carbohydrate on the membrane's inner face. The two major non-glycosylated proteins, actin and myosin, which are invariably isolated with the membranes, are not labeled on the intact platelet, and their location in the membrane remains to be determined. Glycoproteins found in platelet membranes are not found in other blood cells (Gates *et al.*, 1975). It can be anticipated, therefore, that the interactions of platelets with themselves and other cell types will be mediated by specific membrane glycoproteins.

13.4.2 Platelet aggregation

Perhaps the most important function of platelets is aggregation. Although aggregation can be initiated by a variety of stimuli, e.g., prostaglandins, collagen, adenosine diphosphate, and epinephrine, the most potent physiological agent is thrombin. Thrombin is a proteolytic enzyme (mol. wt. = 36 000) which is produced from its zymogen precursor at the time of vascular trauma. Binding studies showed that thrombin interacted with specific receptors on platelets (Tollefsen *et al.*, 1974) resulting in the cleavage of one surface glycoprotein (glycoprotein V, mol. wt. = 89 000; Phillips and Agin, 1977c). These events cause a change in the platelet shape, the release of the contents of the two storage organelles (dense bodies and α-granules), and platelet cohesiveness which results in aggregation.

Several lines of evidence indicated that the membrane surface of these 'activated platelets' had a different molecular composition than that of unstimulated platelets. Thrombin-stimulated platelets had more receptor sites for the lectin Concanavalin A (Phillips and Agin, 1973) and lentil phytohemagglutin (Majerus and Brodie, 1972) than unstimulated platelets. Thrombin treatment of platelets also increased the number of factor Xa binding sites on the membrane surface (Miletich *et al.*, 1977), an event responsible for the procoagulant activity of the platelets. Finally, lacto-peroxidase-catalyzed iodination disclosed that thrombin induced a change in the molecular organization of the platelet plasma membrane proteins since different proteins were labeled following thrombin stimulation (Phillips and Agin, 1973).

13.4.3 Platelet lectin activity

Since the membrane surfaces of activated platelets are cohesive while those of unstimulated platelets are not, it is apparent that an aggregation function accompanies the expression of membrane molecular changes following platelet activation. Stated in terms of the lectin—lectin receptor model of cell—cell interaction, activation of platelets should result in either the expression of a lectin or its receptor, or both.

To determine if the appearance of lectin activity was correlated with platelet activation, lectin activity was determined in unstimulated and thrombin-activated human platelets by measuring the ability of these platelets to agglutinate trypsinized, formalinized erythrocytes from several different species including cow, electric eel, rabbit, catfish, human and sheep (Gartner *et al.*, 1978). Erythrocytes from most species were weakly agglutinated by human platelets, with unstimulated and thrombin-activated platelets producing essentially the same titers. Thrombin-activation of platelets caused an enhanced platelet lectin activity when cow or sheep erythrocytes were used as a source of receptors. This observation was of fundamental importance to initial studies on the test of the lectin–lectin receptor hypothesis for platelet aggregation as it was the first indication that the membrane surface of activated platelets was capable of binding specific receptors on other cells.

The specificity of the lectin was characterized to determine if the thrombin-enhanced lectin mediated cell–cell contact during aggregation (Gartner *et al.,* 1978). A variety of sugars and amino acids were tested for inhibition of the thrombin-enhanced lectin activity. The enhanced lectin activity was inhibited by 30 mM of the amino sugars galactosamine, glucosamine and mannosamine and the basic amino acids, arginine and lysine. None of these compounds inhibited the lectin activity of unstimulated platelets, demonstrating again that the lectin activity on thrombin-activated platelets was different from that on unstimulated platelets. Neutral sugars, *N*-acetylated amino sugars and other amino acids had no effect on either activity. It appeared, therefore, that the lectin had a cation binding site, but may have some sugar selectivity since the order of amino sugar inhibition was mannosamine $>$ glucosamine $>$ galactosamine.

The lectin model for platelet aggregation would predict that the same substances that inhibited thrombin-induced lectin activity would also inhibit thrombin-induced platelet aggregation. Indeed, it was found that the amino sugars and basic amino acids were potent inhibitors of platelet aggregation; neutral sugars were without effect (Gartner *et al.,* 1978). Inhibition was due to binding of the inhibitors to specific aggregation sites on the membrane and not to an inhibition of thrombin activation since

(i)   the aggregation induced by adenosine diphosphate, a non-proteolytic stimulus, was also inhibited by the amino sugars,
(ii)   the amino sugars and basic amino acids had little effect on thrombin-induced serotonin secretion, an index of platelet activation, and
(iii)   inhibition of the thrombin-enhanced lectin activity was observed after platelet activation had occurred.

Thus, the data demonstrated that platelet activation caused the expression of a previously inactive lectin that appeared to mediate at least the initial phases of platelet aggregation by specifically binding to receptors on adjacent platelets.

(a) *Membrane-bound lectin*
The lectin activity of isolated human platelet membranes was evaluated using trypsinized, formalinized erythrocytes from several species (Gartner *et al.,* 1977). These membranes demonstrated lectin activity which was species-specific; cow erythrocytes were agglutinated to the greatest extent with human, sheep, electric eel and rabbit erythrocytes showing only slight agglutination. Human erythrocyte membranes had little lectin activity, indicating that the lectin activity was

tissue-specific. Since the lectin activity of the platelet membranes was inhibited by the same substances that inhibited the thrombin-enhanced lectin activity of washed platelets, it appeared that these two activities were similar.

(b) *Secreted lectin*

Although the thrombin-enhanced lectin was isolated with the plasma membranes, it appeared that this activity came from intracellular proteins. Thrombin activation causes the selective secretion of several macromolecules from α-granules. We have found (Williams, Phillips and Gartner, manuscript in preparation) that the supernatant from thrombin-stimulated platelets agglutinated formaldehyde-fixed bovine erythrocytes and platelets. The importance of this activity was demonstrated by using platelets from patients with the bleeding disorder (Gray Platelet Syndrome', which is characterized by defective thrombin-induced platelet aggregation and an absence of α-granules (White and Gerrard, 1976). These platelets did not have thrombin-enhanced lectin activity as measured by the agglutination of fixed erythrocytes and platelets. Since the glycoprotein composition of the membranes (as detected by SDS gels) and the expression of these glycoproteins (as determined by lacto-peroxidase-catalyzed iodination) was the same as normal platelets, it would seem that the lectin causing the enhanced agglutination was derived from within the platelet. These data indicate that the lectin was secreted from the α-granules and bound to the surface of stimulated platelets. If true, secretion from α-granules must preceed aggregation suggesting that the kinetics of α-granule release is a promising area of future investigation.

The lack of a demonstrable thrombin-enhanced lectin in platelets from Gray Platelet Syndrome has provided a clue to the identity of the platelet lectin. α-granules contain a number of platelet-specific proteins, including 'thrombin-sensitive protein' (also termed glycoprotein G and thrombospondin), platelet factor 4 (PF4), β-thromboglobulin, platelet growth factor and other proteins which are also found in plasma such as fibrinogen, coagulation factors V and VIII, collagenase, and heparinase. We have begun to assess the lectin activity of the platelet-specific proteins and have found that some of these proteins can agglutinate platelets.

Platelet factor 4 (PF4) is a protein released from the α-granules following platelet stimulation (Niewiarowski and Thomas, 1969; Lüscher and Käser-Glanzmann, 1975). This protein contains 70 amino acids and sequence analysis has shown that the charge distribution is highly asymmetric with 5 of the 8 negatively charged residues clustered near the NH_2 terminus and 10 of the 13 positively charged residues in clusters elsewhere in the protein (Hermodson *et al.*, 1977; Walz *et al.*, 1977; Deuel *et al.*, 1977). This feature initiated the idea* that PF4 might have lectin activity. Indeed, when PF4 was analyzed for agglutinating activity, it was found that PF4 agglutinated formaldehyde-fixed, trypsin-treated bovine erythrocytes (Gartner, Phillips and Walz, in preparation). It was also found that PF4 agglutinated washed platelets and formaldehyde-fixed washed platelets.

* The idea that PF4 can act as a platelet agglutinin originated with Dr John Fenton II.

The mechanism of interaction of PF4 with the platelets was examined by use of inhibitors. PF4 is also termed 'anti-heparin factor' because it forms a complex with heparin and neutralizes its activity (Niewiarowski and Thomas, 1969). We have found that heparin is a potent inhibitor of the PF4-mediated platelet agglutination. This agent not only inhibited PF4 agglutination of platelets, but also caused previously formed agglutinates to dissociate. Arginine (60 mM final concentration) also inhibited PF4-mediated platelet agglutination; however, a higher concentration was required for this inhibition than is required to block either platelet aggregation in plasma or the thrombin-enhanced lectin activity of washed platelets. Although lysine also inhibited, it was less effective than arginine. At this concentration, glutamine and other amino acids were without effect. The amino sugars galactosamine, glucosamine and mannosamine only slightly inhibited PF4-mediated platelet agglutination. The PF4 receptor on fixed platelets appeared to be a protein since PF4 failed to agglutinate trypsin-treated platelets.

If charged residues on PF4 are responsible for lectin activity, then selective modification of these residues should affect the lectin activity. When arginine residues of PF4 were modified with cyclohexanedione, less than 25% of the platelet-agglutinating activity remained (Gartner, Phillips, and Walz, manuscript in preparation). By contrast, when lysine residues of PF4 were modified with O-methylisourea, the platelet agglutination activity was unaffected. These results suggested that the arginine residues may play a critical role in the ability of PF4 to crosslink platelets.

Although PF4 crosslinked platelets, three observations indicated that PF4 was not the only lectin in thrombin-treated platelets. First, a higher concentration of arginine was required to inhibit 50% of the PF4-mediated platelet agglutination than was required to inhibit 50% of the thrombin-enhanced lectin activity of washed platelets. Second, relative to inhibition by arginine, PF4-mediated platelet agglutination was less sensitive to inhibition by amino sugars than thrombin-enhanced lectin activity. Third, thrombin-enhanced lectin activity was destroyed by trypsin, whereas the platelet agglutinating activity of PF4 was not. Thus, even through PF4 had the capability to agglutinate platelets, it was not clear to what extent PF4 performs this role *in vivo*.

Since the function of PF4 did not appear identical with the thrombin-enhanced activity, other α-granule proteins were tested for lectin activity. Neither β-thrombo-globulin, which had a similar, but non-identical amino acid sequence as PF4 (Begg *et al.*, 1978), nor platelet growth factor, agglutinated fixed platelets. Fibrinogen is another α-granule protein (Day and Solum, 1973). Although this protein has not yet been tested in microtiter assays, many authors have speculated that fibrinogen crosslinks platelets since (1) fibrinogen was required for ADP-induced aggregation (McLean *et al.*, 1964) and (2) Fab fragments of antifibrinogen antibodies inhibited thrombin-induced aggregation (Tollefson and Majerus, 1975). Consistent with this view, extracellular fibrinogen was incorporated into aggregates of platelets (Mustard *et al.*, 1978). This and other high molecular weight proteins secreted by α-granules will be subjects of future investigations.

13.4.4 Platelet lectin receptor

The postulate that interaction of the platelet lectin with its receptor mediates aggregation was based on the observation that inhibitors of endogenous platelet lectin inhibited platelet aggregation. Although the expression of the lectin was dependent upon thrombin stimulation, the effect of thrombin on the expression of lectin receptors had not been determined. Consequently, we developed an assay for the lectin receptor to determine if its expression was dependent on platelet activation.

(a) *Lectin receptor on platelet membranes*

Since receptor for lectins on erythrocytes (Lis and Sharon, 1972) and slime mold amoebae (Reitherman *et al.,* 1975) are not inactivated by formaldehyde fixation, formaldehyde-fixed platelets were examined as a source of lectin receptor (Phillips and Gartner, manuscript in preparation). The assay was possible because fixed platelets, when treated with thrombin, neither agglutinated themselves nor bovine erythrocytes. However, the fixed platelets were found to contain functional lectin receptors since the fixed platelets caused a four-fold enhancement of the agglutination titer of the thrombin-activated platelets. This increase in titer was attributed to agglutination of the fixed platelets by both released substances and the lectin molecules on the plasma membranes of thrombin-activated platelets because the fixed platelets lacked lectin activity. Platelets fixed with formaldehyde either before or after platelet stimulation by adenosine diphosphate or thrombin were active. Thus, the lectin receptor was distinct from the lectin in that the latter was dependent on platelet activation for its expression and was inactivated by formaldehyde.

In the determinations discussed above, it appeared that the fixed platelets were agglutinated directly with the thrombin-activated washed platelets. This was shown conclusively by fluorescently labeling the fixed platelets with ethidium bromide before mixing them with activated platelets, which allowed the visual distinction of fixed platelets from washed platelets in the microtiter assay. As with unlabeled platelets, the ethidium bromide-labeled platelets enhanced the agglutination of thrombin-activated washed platelets. Also, agglutination of the labeled platelets was inhibited by the same sugars and amino acids that inhibited the agglutination of fixed platelets and the aggregation of washed platelets. Therefore, the binding specificity of the receptors on fixed platelets appeared indistinguishable from that of either washed platelets or platelets in plasma.

The importance of lectin receptors in hemostasis was shown by the incorporation of formaldehyde-fixed platelets into aggregates formed by washed platelets. This was achieved by adding thrombin to a stirred suspension containing both washed and fixed platelets. Most of the formaldehyde-fixed platelets were present in the thrombin-induced aggregates, but aggregation did not occur unless washed platelets were present. When fixed platelets were pretreated with trypsin before being mixed with washed platelets, they were not incorporated into aggregates. These results indicated that the incorporation of fixed platelets into aggregates was mediated by specific protease-sensitive receptors on the membrane surface. Thus, in platelets without measurable lectin activity, the binding capacity of the receptor for the lectin on platelets appeared adequate to allow participation in aggregation in the presence of normal platelets.

(b) *Platelet defects in Glanzmann's thrombasthenia*
One approach to identifying the molecular entities involved in platelet specific reactions is to examine the platelets from patients with inherited bleeding disorders caused by defective platelets. An analogous approach has been helpful in identifying many of the factors involved in blood coagulation (Ratnoff, 1972) and in other biological systems. Many of the platelet bleeding disorders have a primary functional defect in the platelet, which makes them useful for assigning structure—function relationships. For example, platelets from patients with Bernard-Soulier syndrome lack membrane glycoprotein Ib (Jenkins *et al.*, 1976) and do not adhere to the subendothelium (Weiss *et al.*, 1974); other platelets lack functional dense bodies and are defective in secondary aggregation to some stimuli (White and Gerrard, 1976).

Glanzmann's thrombasthenia, another inherited bleeding disease is also character-ized by a primary functional defect caused by platelets. This disorder, transmitted by an autosomal recessive gene, is diagnosed by a prolonged bleeding time, abnormal clot retraction, normal platelet count, and the absence of platelet aggregation (Hardisty *et al.*, 1964; Caen *et al.*, 1966). Although the incidence of this disorder has not been established, it would no doubt be low since the family histories of thrombasthenic individuals often include relatives who have died at birth of an unknown bleeding disorder (Caen *et al.*, 1966). Analysis of thrombasthenic platelets has shown that thrombasthenic platelets adhere normally to exposed endothelium (Tschopp *et al.*, 1975), change shape in response to ADP and undergo a release reaction in response to thrombin and collagen (Caen *et al.*, 1966; Caen and Michel, 1972; Zucker *et al.*, 1966). The primary functional defect of thrombasthenic platelets is their inability to aggregate (the abnormal clot retraction is no doubt linked to the failure to aggregate). Since the functional defect of these platelets is so limited, it appears that molecular studies on thrombasthenic platelets might disclose the importance of the endogenous platelet lectin and its receptor in platelet aggregation and also aid in the assignment of structure—function relationships for the lectin or its receptors.

Many studies have shown molecular abnormalities associated with the throm-basthenic platelets. For example, decreased concentrations of ATP (Gross, 1960), fibrinogen (Weiss and Kochwa, 1968), glutathione peroxidase (Karpatkin and Weiss, 1972), glyceraldehyde-3-phosphate dehydrogenase, and pyruvate kinase (Gross *et al.*, 1960) have been reported. However, none of these deficiencies were found in all thrombasthenic samples examined, suggesting that these changes may be secondary to the primary lesion.

Several lines of evidence indicate that the primary lesion causing thrombasthenia may result in an alteration of the outer surface of the platelet plasma membrane.

(i)   Thrombasthenic platelets lack a membrane surface function, that of aggregation (Hardisty *et al.*, 1964; Caen *et al.*, 1966).

(ii)   Compared to normal platelets, thrombasthenic platelets have a decreased affinity for plasma fibrinogen (Zucker *et al.*, 1966).

(iii)   The abnormal platelets also have less than normal clot promoting activity on their membrane surfaces (Walsh, 1972).

(iv)   Adenosine diphosphate will modify the surface of thrombasthenic platelets so that neither ristocetin nor bovine factor VIII will induce aggregation (Cohen *et al.*, 1975).

(v)   Finally, freeze-fracture studies have demonstrated a reduced concentration of intramembraneous particles within thrombasthenic platelet membranes (Reddick and Mason, 1973).

(c) *Lectins and lectin receptors in thrombasthenic platelets*

The inability of thrombasthenic platelets to aggregate may be due to a defect of either the lectin or its receptor. In an attempt to distinguish between these alternatives, the interactions of thrombasthenic platelets were studied in microtiter plates. In a study including platelets from five thrombasthenic individuals, thrombasthenic platelets were found to have normal thrombin-enhanced lectin activity. The enhanced lectin activity was demonstrated using formaldehyde-fixed, trypsin-treated bovine erythrocytes and formaldehyde-fixed normal platelets as sources of receptors.

Although thrombasthenic platelets express normal lectin activity, they are markedly deficient in functional lectin receptors. This was shown by the inability of formaldehyde-fixed thrombasthenic platelets to be agglutinated by thrombin-activated control platelets (Gartner, Phillips, Gerrard and White, manuscript in preparation). Therefore, it appears that the inability of thrombasthenic platelets to aggregate may be caused by a platelet lectin receptor deficiency.

(d) *Membrane molecular defects in Glanzmann's thrombasthenia*

Since the results indicate that thrombasthenic platelets lack the receptor for the endogenous platelet lectin, any molecular defect of the plasma membranes of thrombasthenic platelets might disclose the identity of the lectin receptor. A fundamental observation was made by Nurden and Caen (1974) who observed that thrombasthenic platelets lacked one of the three major membrane glycoproteins. This observation was based on a decreased amount of material which reacted with the periodic acid-Schiff's stain on an acrylamide gel and was the first convincing demonstration of a molecular defect in the membrane surface.

Subsequent studies using lactoperoxidase-catalyzed iodination gave a more complete analysis of molecular deficiencies (Phillips *et al.*, 1975; Phillips and Agin, 1977b). These studies showed that there is a paucity not only of glycoprotein IIb, but also of glycoprotein III in thrombasthenic platelets. Thus far, the platelets from twenty different thrombasthenic individuals have been examined by the iodination procedure and all had a marked reduction of these glycoproteins. Carriers of thrombasthenia, i.e., parents of homozygous patients, have intermediate amounts of these two components (Phillips, 1976). Comparison of surface-labeling data of control and thrombasthenic platelets using lactoperoxidase-catalyzed iodination and neuraminidase/galactose oxidase (^{3}H)NaBH$_4$ labeling showed that all cell surface proteins, except IIb and III, were present in apparently normal amounts in thrombasthenic platelets. Thus, a characteristic feature of Glanzmann's thrombasthenia is a decreased concentration of two membrane glycoproteins.

The orientation of these glycoproteins in platelet membranes has been extensively studied. Both are integral membrane glycoproteins since they cannot be removed from the membrane unless its structure has been disrupted with detergent (Phillips, unpublished observations). The exposure of these glycoproteins to the outer membrane surface is amply documented since both are labeled by membrane surface probes such as lactoperoxidase (Phillips and Agin, 1977b), transaminase (Okumura and Jamieson, 1976, and diiodo-diazosulfanylic acid (George *et al.,* 1976). Both glycoproteins contain sialic acid (Phillips and Agin, 1977b). The carbohydrate-containing portion of these glycoproteins is expressed entirely on the outer surface of the platelet since it can be removed by chymotrypsin, a proteolytic enzyme which does not induce the platelet release reaction (Jenkins *et al.,* 1976). Interestingly, neither glycoprotein is affected by thrombin on the intact platelet, but both are hydrolyzed by this protease on isolated membranes indicating that these glycoproteins span the thickness of the plasma membrane (Phillips and Agin, 1974). Glycoprotein IIb is a dimeric glycoprotein with a nonreduced molecular weight of 142 000 and a reduced molecular weight of 132 000. Both subunits are glycosylated and the carbohydrate-containing regions of each are exposed to the outside of the cell. Glycoprotein III contains at least two intrachain disulfides and has a reduced molecular weight of 114 000 (Phillips and Agin, 1977a).

The specific absence of two membrane surface glycoproteins in thrombasthenic platelets indicates that one, or both, of these glycoproteins is associated with platelet aggregation. Since thrombasthenic platelets also lack the lectin receptor activity, it may further indicate that the lectin receptor is one of the thrombasthenic glycoproteins. Further studies will establish the validity of this speculation.

13.4.5 Speculations concerning the mechanism of platelet aggregation

The ideas resulting from the analysis of the endogenous platelet lectin and its receptor prompted the formulation of an hypothetical model describing certain features of platelet aggregation (Fig. 13.2). Many features of this model were derived not only from the lectin—lectin receptor study, but also from studies by numerous investigators working in this area. It is hoped that formulation of this model will stimulate future research on the mechanism of platelet aggregation.

The relevant features of this model are:

(1) A membrane-bound lectin is expressed following thrombin stimulation.

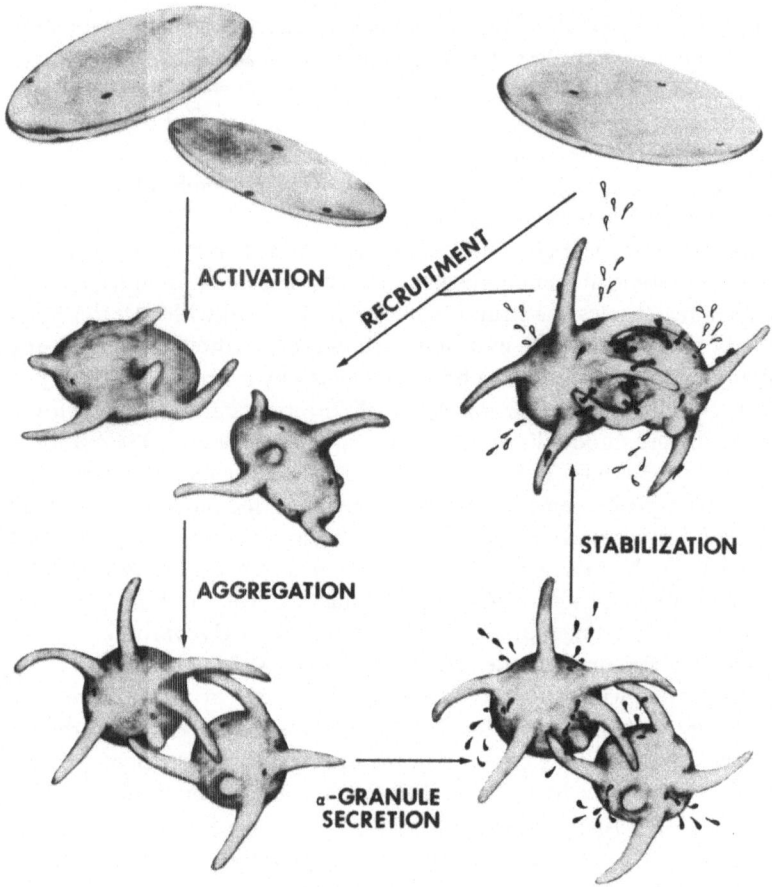

Fig. 13.2 Pictorial representation of the sequence of events occurring during platelet aggregation. The experimental evidence in support of this model is discussed in the text.

(2) The function of the lectin receptor is not dependent upon platelet stimulation. Although there may be an increase in receptor function following interaction with thrombin, platelets fixed with formaldehyde before stimulation have sufficient receptors to participate in aggregation.

(3) The lectin receptor appears to be glycoprotein IIb and/or III. These surface glycoproteins are missing in the inherited bleeding disorder Glanzmann's thrombasthenia; platelets from these patients lack the lectin receptor function. This observation stresses the importance of the lectin receptor in normal platelet aggregation.

(4) Initial aggregation appears to be mediated by lectin—lectin receptor interactions. Since inhibitors of the thrombin-stimulated platelet lectin, as measured by the agglutination of bovine erythrocytes, also inhibit platelet aggregation, it would

appear that the same activity is responsible for both functions. Therefore, the observation that unstimulated platelets have functional receptors indicates that these platelets are non-cohesive because they lack this membrane-bound lectin.

(5) Stabilization of platelet aggregates may be mediated by proteins released from the α-granules. Although one of the proteins released from the α-granules during thrombin aggregation, platelet factor 4 (PF4), readily agglutinates platelets, it has different properties from the lectin responsible for primary aggregation. Therefore, PF4 is a likely candidate for stabilization of preformed aggregates as its release most likely follows initial aggregation. The role of another α-granule constituents, fibrinogen, in the stabilization of platelet aggregates is already well-documented (Tollefsen and Marjerus, 1975).

(6) Release of dense bodies is involved in recruitment of additional platelets to aggregate. Elegant experiments by Charo *et al.* (1977) demonstrated that aggregation preceded the release of dense body constituents (adenosine diphosphate, adenosine triphosphate, Ca^{2+}, K^+ and serotonin), indicating that they have little, if any, role in thrombin aggregation. A likely role for these products, particularly adenosine diphosphate and serotonin, is to stimulate other platelets and thereby recruit them to participate in the formation of hemostatic plugs.

13.5 CONCLUDING REMARKS

Although precise information concerning the molecular mechanism for the direct interaction of any eukaryotic cell is not available, recent studies have identified molecules important to this process. This data, discussed in this review and summarized in Table 13.1, indicate that a variety of mechanisms are involved in these interactions. For example, some cells apparently interact exclusively by membrane-bound factors, while others also require an extra molecule which 'bridges' the membrane surfaces. The specificities of these agglutinins indicate that, in some but not all, cases different cells use different molecules to mediate cell interactions. Agglutinins have been demonstrated with specificity for galactose (Gartner and Podleski, 1975), amino sugars (Gartner *et al.*, 1978), glucuronic acid (Turner and Burger, 1973), fucose (Rosen *et al.*, 1973), and N-acetyl galactosamine (Asao and Oppenheimer, 1978), while others are unaffected by any simple sugar. These data suggest that the mechanism for the interaction of each cell type must be individually determined.

The adhesion molecules can have a variety of distributions between interacting cells. In some instances, symmetrical distributions have been observed with identical complementary molecules, others with identical, noncomplementary molecules which are linked by bridging molecules, and still others with non-identical, complementary molecules on both interacting membrane surface. Homotypic interactions i.e. the adherence of like cells, are characterized by such distributions. Alternatively, these molecules can be asymmetrically distributed, with one type of molecule existing on one membrane and a different, complementary molecule on another membrane. This

Table 13.1 Recognition molecules involved in the interaction of eukaryotic cells

Molecule	Properties	Source	Process mediated
5-Factor	Polyvalent mannopeptide, heat-stable, inactivated by reduction	*Hansenula wingei* yeast cells	Mating
21-Factor	Monovalent acidic glycoprotein, heat-labile, stable to reduction	*Hansenula wingei* yeast cells	Mating
Sperm receptor	High mol. wt. glycoprotein	Sea urchin eggs	Fertilization
Bindin	30 500 dalton protein	Sea urchin sperm	Fertilization
Sponge aggregation	Polyvalent, extramembranous bridging molecules	Marine sponges	Cell associations
Fibronectin	Dimeric glycoprotein	Serum	Phagocytosis
Discoidin (Pallidin)	Developmentally regulated lectin	Cohesive slime mold amoebae	Aggregation
CSA	Developmentally regulated antigen	Cohesive slime mold amoebae	Aggregation
Galactolectins	Developmentally regulated lectins inhibited by galactose-containing oligosaccharide	Muscle, electric organ, etc.	Cell–cell interaction
CAM protein	140 000 dalton glycoprotein	Embryonic neural retina tissue	Mediates fascicle formation
Retina-specific cell-aggregating factor	50 000 dalton glycoprotein	Embryonic chick retina	Cell–cell interaction
Glycoprotein IIb and III	Intrinsic membrane glycoproteins	Human platelet	Aggregation

asymmetric distribution of adhesive molecules is essential for heterotypic cell interactions as it allows only the interaction of dissimilar cells. Thus, the difference in the distribution of cell-adhering molecules is an important feature controlling the specificity of cell interactions.

Although these studies have just begun to unravel the molecular complexities of cell—cell interactions, the central role of membrane—membrane contacts in biology promises that this area of research will continue to be very active.

REFERENCES

Aketa, K. (1978), *Exp. Cell Res.*, **80**, 439–440.

Aketa, K., Onitake, K. and Tsuzuki, H. (1972), *Exp. Cell Res.*, **71**, 27–32.

Andersson, L.C. and Gahmberg, C.G. (1978), *Blood*, **52**, 57–67.

Asao, I. and Oppenheimer, S.B. (1978), *J. Cell Biol.*, **79**, #2 pt 2, MF 1527.

Ballou, C. (1976), *Adv. Microb. Physiol.*, **14**, 93–158.

Barber, A.J. and Jamieson, G.A. (1970), *J. biol. Chem.*, **245**, 6357–6365.

Begg, G.S., Pepper, D.S., Chesterman, C.N. and Morgan, F.J. (1978), *Biochemistry*, **17**, 1739–1744.

Behnke, O.J. (1968), *J. ultrastruct. Res.*, **24**, 51–59.

Bellet, N.F., Vacquier, J.P. and Vacquier, V.D. (1977), *Biochem. biophys. Res. Commun.*, **79**, 159–165.

Beug, H., Gerisch, G., Kempff, S., Ridel, V. and Cremer, G. (1970), *Exp. Cell Res.*, **63**, 147–158.

Beug, H., Katz, F.E., Stein, A. and Gerisch, G. (1973), *Proc. natn. Acad. Sci. U.S.A.*, **70**, 3150–3157.

Blumenstock, F., Saba, T.M., Webber, P. and Cho, E. (1976), *J. Reticuloendothel. Soc.*, **19**, 157–172.

Blumenstock, F.A., Saba, T.M., Weber, P. and Laffin, R. (1978), *J. biol. Chem.*, **253**, 4287–4291.

Bonner, J.T. (1967), *The Cellular Slime Molds*, Princeton University Press, Princeton, New Jersey.

Bozzaro, S. and Gerisch, G. (1978), *J. mol. Biol.*, **120**, 265–279.

Brackenbury, R., Thiery, J.P., Rutishauser, U. and Edelman, G.M. (1977), *J. biol. Chem.*, **252**, 6835–6840.

Briles, E.B., Gregory, W., Fletcher, P. and Kornfeld, S. (1979), *J. Cell Biol.*, **81**, 528–537.

Brock, T.D. (1958), *J. Bact.*, **76**, 334–335.

Caen, J.P., Castaldi, P.A., Leclerc, J.C., Inceman, S., Larrieu, M.-J., Probst, M. and Bernard, J. (1966), *Am. J. Med.*, **41**, 4–26.

Caen, J.P. and Michel, H. (1972), *Nature*, **240**, 148–149.

Capers, C.R. (1960), *J. Biophys. Biochem. Cytol.*, **7**, 559–566.

Cauldwell, C.B., Henkart, P. and Humphreys, T. (1973), *Biochemistry*, **12**, 3051–3055.

Charo, I.F., Feinman, R.D. and Detwiler, T.C. (1977), *J. clin. Invest.*, **60**, 866–873.

Cohen, I., Glaser, T. and Seligsoh, U. (1975), *Br. J. Haematol.*, **31**, 343–347.

Grandall, M.A. and Brock, T.D. (1968), *Bact. Rev.*, **32**, 139–163.

Grandall, M., Lawrence, L.M. and Saunders, E.R.M. (1974), *Proc. natn. Acad. Sci. U.S.A.*, **71**, 26–29.

Day, H.J. and Solum, N.O. (1973), *Scand. J. Haematol.*, **10**, 136–143.

Den, H. and Malinzak, D.A. (1977), *J. biol. Chem.*, **252**, 5444–5448.

Den, H., Malinzak, D.A. and Rosenberg, A. (1976), *Biochem. biophys. Res., Commun.*, **69**, 621–627.

Dessau, W., Jilek, F., Adelmann, B.C. and Hörmann, H. (1978), *Biochim. biophys. Acta*, **533**, 227–237.

Deuel, T.F., Keim, P.S., Farmer, M. and Heinrikson, R.L. (1977), *Proc. natn. Acad. Sci. U.S.A.*, **74**, 2256–2258.

De Waard, A., Hickman, S. and Kornfeld, S. (1976), *J. biol. Chem.*, **251**, 7581–7587.

Di Luzio, N.R. and Lindsey, E.S. (1973), *Proc. Soc. exp. Biol. Med.*, **143**, 715–718.

Di Luzio, N.R., McNamer, R., Miller, E.F. and Pisano, J.C. (1972), *J. Reticuloendothel. Soc.*, **12**, 314–323.

Di Luzio, N.R. (1975), *The Reticuloendothelial System*, The Williams and Wilkins Co., Baltimore, Maryland.

Epel, D. (1977), *Sci. Am.*, **237**, 128–138.

Eberth, C.J. and Schimmelbusch, C. (1886), *Virchows Arch. Path. Anat. Physiol.*, **105**, 331–350.

Filkens, J.P. and Smith, J.J. (1965), *Proc. Soc. exp. Biol. Med.*, **119**, 1181–1184.

Firket, H. (1958), *Arch. Biol.*, **69**, 1–166.

Frazier, W.A., Rosen, S.D., Reitherman, R.W. and Barondes, S.H. (1975), *J. biol. Chem.*, **250**, 7714–7721.

Frazier, W.A. (1976), *Trends in Biochem. Sci.*, **1**, 130–133.

Fulton, R.L. and Jones, C.E. (1975), *Rev. Surg.*, **32**, 84–85.

Gartner, T.K. and Podleski, T.R. (1975), *Biochem. biophys. Res. Comm.*, **67**, 972–978.

Gartner, T.K. and Podleski, T.R. (1976), *Biochem. biophys. Res. Commun.*, **70**, 1142–1149.

Gartner, T.K., Williams, D.C. and Phillips, D.R. (1977), *Biochem. biophys. Res. Commun.*, **79**, 592–599.

Gartner, T.K., Williams, D.C., Minion, F.C. and Phillips, D.R. (1978), *Science*, **200**, 1281–1283.

Gasic, G.J. and Galanti, N.L. (1966), *Science*, **151**, 203–205.

Gates, R.E., Phillips, D.R. and Morrison, M. (1975), *J. Biochem.*, **147**, 373–376.

George, J.N., Potterf, R.D., Lewis, P.C. and Sears, D.A. (1976), *J. Lab. Clin. Med.*, **88**, 232–246.

Glabe, C.G. and Vacquier, V.D. (1977), *Nature*, **267**, 836–837.

Glabe, C.G. and Vacquier, V.D. (1977), *J. Cell Biol.*, **75**, 410–421.

Gottlieb, D.I. and Glaser, L. (1975), *Biochem. biophys. Res. Comm.*, **63**, 815–821.

Grinnell, F. (1978), *Int. Rev. Cytol.*, **53**, 65–144.

Gross, R., Gerok, W., Loehr, G.W., Vogell, W., Waller, H.O. and Theopold, W.T. (1960), *Klin. Wochenschr.*, **38**, 194–206.

Hardisty, R.M., Dormandy, K.M. and Hutton, R.A. (1964), *Br. J. Haematol.*, **10**, 371–387.

Harris, A.J., Heinemann, S., Schubert, D. and Tarakis, H. (1971), *Nature*, **231**, 296–301.

Hausman, R.E. and Moscona, A.A. (1973), *Proc. natn. Acad. Sci. U.S.A.*, **70**, 3111–3114.

Hausman, R.E. and Moscona, A.A. (1975), *Proc. natn. Acad. Sci. U.S.A.*, **72**, 916–920.

Hausman, R.E. and Moscona, A.A. (1976), *Proc. natn. Acad. Sci. U.S.A.*, **73**, 3594–3598.

Henkart, P., Humphreys, S. and Humphreys, T. (1973), *Biochemistry*, **12**, 3045–3050.

Hermodson, M., Schmer, G. and Kurachi, K. (1977), *J. biol. Chem.*, **252**, 6276–6279.

Hopper, K.E., Adelmann, B.C., Gentner, G. and Gay, S. (1976), *Immunology*, **30**, 249–259.

Huesgen, A. and Gerisch, G. (1975), *FEBS Letters*, **56**, 46–49.

Humphreys, T. (1963), *Dev. Biol.*, **8**, 27–47.

Humphreys, T. (1970), *Nature*, **228**, 685–686.

Humphreys, S., Humphreys, T. and Sano, J. (1977), *J. Supramol. Struct.*, **7**, 339–351.

Hynes, R.O. (1976), *Biochim. biophys. Acta*, **458**, 73–107.

Jenkins, C.S.P., Phillips, D.R., Clemetson, K.J., Meyer, D., Larrieu, M.-J. and Lüscher, E.F. (1976), *J. clin. Invest.*, **57**, 112–124.

Karpatkin, S. and Weiss, J.J. (1972), *N. Engl. J. Med.*, **287**, 1062–1066.

Kobiler, D. and Barondes, S.H. (1977), *Dev. Biol.*, **60**, 326–330.

Konigsberg, I.R., McElvain, N., Tootle, M. and Herrmann, H. (1960), *J. Biophys. Biochem. Cytol.*, **8**, 333–343.

Lash, J.W., Holtzer, H. and Swift, H. (1957), *Anat. Rec.*, **128**, 679–697.

Lilien, J.E. and Moscona, A.A. (1967), *Science*, **157**, 70–72.

Lis, H. and Sharon, N. (1972), *Methods Enzym.*, **28**, 360–365.

Lüscher, E.F. and Käser-Glanzmann, R. (1975), *Thromb. Haemostas.*, **33**, 66–72.

MacBride, R.G. and Przybylski, R.J. (1978), *J. Cell Biol.*, **79**, #2 pt 2, MU 2031.

McClay, D.R. (1974), *J. exp. Zool.*, **188**, 89–101.

McLean, J.R., Maxwell, R.E. and Hertler, D. (1964), *Nature*, **202**, 605–606.

McLennan, A.P. and Dodd, R.Y. (1967), *J. Embryol. exp. Morph.*, **17**, 473–480.

Majerus, P.W. and Brodies, G.N. (1972), *J. biol. Chem.*, **247**, 4253–4257.

Margoliash, E., Schenck, J.R., Hargie, M.P., Burokas, S., Richter, W.R., Barlow, G.H. and Moscona, A.A. (1965), *Biochem. biophys. Res. Commun.*, **20**, 383–388.

Merlie, J.P. and Teichberg, V.I. (1978), *Methods Enzym.*, **50**, 301.

Merrell, R. and Glaser, L. (1973), *Proc. natn. Acad. Sci. U.S.A.*, **70**, 2794–2798.

Miletich, J.P., Jackson, C.M. and Majerus, P.W. (1977), *Proc. natn. Acad. Sci. U.S.A.*, **74**, 4033–4036.

Mir-Lechaire, F.J. and Barondes, S.H. (1978), *Nature*, **272**, 256–258.

Molnar, J., McLain, S., Allen, C., Laga, H., Gara, A. and Gelder, F. (1976), *Biochim. biophys. Acta*, **493**, 37–59.

Moscona, A.A. (1963), *Proc. natn. Acad. Sci. U.S.A.*, **49**, 742–747.

Moscona, A.A. (1968), *Dev. Biol.*, **18**, 250–277.

Mosesson, M. (1977), *Thromb. Haemostas.*, **38**, 742–750.

Mosesson, M.W. (1978), *Ann. N.Y. Acad. Sci.*, **312**, 11–30.

Müller, K. and Gerisch, G. (1978), *Nature*, **274**, 445–449.

Müller, W.E.G. and Zahn, R.K. (1973), *Exp. Cell Res.*, **80**, 95–104.

Müller, W.E.G. and Müller, I. and Zahn, R.K. (1974), *Experientia*, **30**, 899–902.

Müller, W.E.G., Müller, I., Zahn, R.K. and Kurelec, B. (1976), *J. Cell Sci.*, **21**, 227–241.

Müller, W.E.G., Arendes, J., Kurelec, B., Zahn, R.K. and Müller, I. (1977), *J. biol., Chem.*, **252**, 3836–3842.

Mustard, J.F., Packham, M.A., Kinlough-Rathbone, R.L., Perry, D.W. and Regoeczi, E. (1978), *Blood*, **52**, 453–466.

Niewiarowski, S. and Thomas, D.P. (1969), *Nature,* **222,** 1269–1270.

Nowak, T.P., Haywood, P.L. and Barondes, S.H. (1976), *Biochem. biophys. Res. Commun.,* **68,** 650–657.

Nowak, T.P., Kobiler, D., Roel, L.E. and Barondes, S.H. (1977), *J. biol. Chem.,* **252,** 6026–6030.

Nurden, A.T. and Caen, J.P. (1974), *Br. J. Haematol.,* **28,** 253–260.

Okumura, T. and Jamieson, G.A. (1976), *J. biol. Chem.,* **251,** 5944–5949.

Pepper, D.S. and Jamieson, G.A. (1969), *Biochemistry,* **8,** 3362–3369.

Phillips, D.R. (1972), *Biochemistry,* **11,** 4582–4588.

Phillips, D.R. and Agin, P.P. (1973), *Ser. Haematologica,* **6,** 292–310.

Phillips, D.R. and Agin, P.P. (1974), *Biochim. biophys. Acta,* **352,** 218–227.

Phillips, D.R., Jenkins, C.S.P., Lüschen, E.F. and Larrieu, M.J. (1975), *Nature,* **257,** 599–600.

Phillips, D.R. (1976), *Platelet Function Testing.* DHEW Publication No. (NIH), 204–215.

Phillips, D.R. and Agin, P.P. (1977a), *J. biol. Chem.,* **252,** 2121–2126.

Phillips, D.R. and Agin, P.P. (1977b), *J. clin. Invest.,* **60,** 535–545.

Phillips, D.R. and Agin, P.P. (1977c), *Biochem. biophys. Res. Commun.,* **75,** 940–942.

Ratnoff, O.D. (1972), *Progress in Hemostasis and Thrombosis,* Grune and Straton, New York.

Reddick, R.L. and Mason, R.G. (1973), *Am. J. Path.,* **70,** 473–488.

Reitherman, R.W., Rosen, S.D., Frazier, W.A. and Barondes, S.H. (1975), *Proc. natn. Acad. Sci. U.S.A.,* **72,** 3541–3545.

Roseman, S. (1970), *Chem. Phys. Lipids,* **5,** 270–297.

Rosen, S.D., Kafka, J.A., Simpson, D.L. and Barondes, S.H. (1973), *Proc. natn. Acad. Sci., U.S.A.,* **70,** 2554–2557.

Rosen, S.D., Simpson, D.L., Rose, J.E. and Barondes, S.H. (1974), *Nature,* **252,** 128, 149–150.

Rosen, S.D., Haywood, P.L. and Barondes, S.H. (1976), *Nature,* **263,** 425–427.

Rosen, S.D., Chang, C.M. and Barondes, S.H. (1977), *Dev. Biol.,* **61,** 202–213.

Rutishauser, U., Thiery, J.P., Brackenbury, R., Sela, B.A. and Edelman, G.M. (1976), *Proc. natn. Acad. Sci., U.S.A.,* **73,** 577–581.

Rutishauser, U., Thiery, J.P., Brackenbury, R. and Edelman, G.M. (1978a), *J. Cell Biol.,* **79,** 371–381.

Rutishauser, U., Gall, W.E. and Edelman, G.M. IV. (1978b), *J. Cell Biol.,* 382–393.

Saba, T.M. and Di Luzio, N.R. (1969), *Am. J. Physiol.,* **216,** 197–205.

Saba, T.M. (1975), *Circ. Shock,* **2,** 91–108.

Saba, T.M., Blumenstock, F.A., Bernard, H. and Kaplan, J.E. (1977), *J. Reticuloendothel. Soc.,* **22,** 16a.

Saba, T.M., Blumenstock, F.A., Weber, P. and Kaplan, J.E. (1978), *Ann. N.Y. Acad. Sci.,* **312,** 43–55.

Schick, P.K., Kurica, K.B. and Chacko, G.K. (1976), *J. clin. Invest.,* **57,** 1221–1226.

Schmell, E., Earles, B.J., Breaux, C.B. and Lennarz, W.J. (1977), *J. Cell Biol.,* **72.** 35–46.

Simpson, D.L., Rosen, S.D. and Barondes, S.H. (1974), *Biochemistry,* **13,** 3487–3493.

Simpson, D.L., Thorne, D.R. and Loh, H.H. (1977), *Nature,* **266,** 367–369.

Simpson, D.L., Thorne, D.R. and Loh, H.H. (1978), *Life Sci.,* **22,** 727–748.

Stockdale, F.E. and Holtzer, H. (1961), *Exp. Cell Res.*, **24**, 508–520.

Summers, R.G., Hylander, B.L., Colwin, L.H. and Colwin, A.L. (1975), *Am. Zool.*, **15**, 523–551.

Taylor, N.W. (1964), *J. Bact.*, **87**, 863–866.

Taylor, N.W. and Orton, W.L. (1967), *Arch. Biochem. Biophys.*, **120**, 602–608.

Taylor, N.W. and Orton, W.L. (1970), *Biochemistry*, **9**, 2931–2934.

Taylor, N.W. and Orton, W.L. (1971), *Biochemistry*, **10**, 2043–2049.

Teichberg, V.I., Silman, I., Beitsch, D.D. and Resheff, G. (1975), *Proc. natn. Acad. Sci. U.S.A.*, **72**, 1383–1387.

Thiery, J.P., Brackenbury, R., Rutishauser, U. and Edelman, G.M. (1977), *J. Biol. Chem.*, **252**, 6841–6845.

Tollefsen, D.M., Feagler, J.R. and Majerus, P.W. (1974), *J. biol. Chem.*, **249**, 2646–2651.

Tollefsen, D.M. and Majerus, P.W. (1975), *J. clin. Invest.*, **55**, 1259–1268.

Townes, P.L. and Holtfreter, J. (1955), *J. exp. Zool.*, **128**, 53–120.

Tschopp, T.G., Weiss, H.J. and Baumgartner, H.R. (1975), *Experientia*, **31**, 113–116.

Tsuzuki, H., Yoshida, M., Onitake, K. and Aketa, K. (1977), *Biochem. biophys. Res. Commun.*, **76**, 502–511.

Turner, R.S. and Burger, M.M. (1973), *Nature*, **244**, 509–510.

Tyler, A. (1947), *Growth*, **10**, 7–18.

Vacquier, V.D. and Payne, J.E. (1973), *Exp. Cell Res.*, **82**, 227–235.

Vacquier, V.D. and Moy, G.W. (1977), *Proc. natn. Acad. Sci. U.S.A.*, **74**, 2456–2460.

Vaheri, A. and Mosher, D.F. (1978), *Biochim. biophys. Acta*, **516**, 1–25.

Wagner, H.N. and Iio, M. (1964), *J. clin. Invest.*, **43**, 1525–1532.

Walsh, P.N. (1972), *Br. J. Haemat.*, **23**, 553–569.

Walz, D.A., Wu, V.Y., deLamo, R., Dene, H. and McCoy, L.E. (1977), *Thromb. Res.*, **11**, 893–898.

Weinbaum, G. and Burger, M.M. (1973), *Nature*, **244**, 510–512.

Weiss, H.J. and Kochwa, S. (1968), *J. Lab. Clin. Med.*, **71**, 153–165.

Weiss, H.G., Tschopp, T.B. and Baumgartner, H.R. (1974), *Am. J. Med.*, **57**, 920–925.

Weiss, P. (1947), *Yale J. Biol. Med.*, **19**, 235–278.

White, J.G. and Gerrard, J.M. (1976), *Am. J. Path.*, **83**, 590–632.

Wickerham, L.J. (1958), *Science*, **128**, 1504–1505.

Wickeham, L.J. and Burton, K.A. (1962), *Bact. Rev.*, **26**, 382–397.

Wilde, C.E. (1959), *Cell Organism and Milieu*, Ronald Press, New York.

Wilson, H.V. (1907), *J. exp. Zool.*, **5**, 245–258.

Yaffe, D. (1968), *Proc. natn. Acad. Sci. U.S.A.*, **61**, 477–483.

Yaffe, D. and Feldman, M. (1965), *Dev. Biol.*, **11**, 300–317.

Yamada, K.M., Yamada, S.S. and Pastan, I. (1975), *Proc. natn. Acad. Sci. U.S.A.*, **72**, 3158–3162.

Yamada, K.M. and Pastan, I. (1976), *Trends in Biochem. Sci.*, **1**, 222–224.

Yamada, K.M., Schlesinger, D.W., Kennedy, D.W. and Pastan, I. (1977), *Biochemistry*, **16**, 5552–5559.

Yamada, K.M., Olden, K. and Pastan, I. (1978), *Ann. N.Y. Acad. Sci.*, **312**, 256–277.

Yamada, K.M. and Olden, K. (1978), *Nature*, **275**, 179—188.
Yamada, K.M. and Kennedy, D.W. (1978), *J. Cell Biol.*, **80**, 492—498.
Yen, P.H. and Ballou, C.E. (1974a), *Biochemistry*, **13**, 2420—2437.
Yen, P.H. and Ballou, C.E. (1974b), *Biochemistry*, **13**, 2428—2437.
Zucker, M.B., Pert, J.H. and Hilgartner, M.W. (1966), *Blood*, **28**, 524—534.
Zwaal, R.F.A., Comfurius, P. and van Deenen, L.L.M. (1977), *Nature*, **268**, 358—360.

14 Prospects for Preventing the Association of Harmful Bacteria with Host Mucosal Surfaces

ROLF FRETER

14.1	Introduction	*page*	441
14.2	The role of mucus gel in mucosal association with bacteria		442
14.3	Sequential reactions leading to mucosal association with bacteria		445
14.4	Interference with bacterial adherence by non-immunological means		448
	14.4.1 Anti-adhesive substances in milk		448
	14.4.2 Anti-adhesive substances in meat		449
	14.4.3 Lectins as anti-adhesive substances in foodstuffs		450
	14.4.4 Taxins as anti-adhesive substances in foodstuffs		450
	14.4.5 Supplementation of natural foodstuffs		451
	14.4.6 Other mechanisms		452
	14.4.7 General comments		453
14.5	Interference with bacterial adherence by local immune-mechanisms		454
14.6	Summary and conclusions		455
	References		456

Bacterial Adherence
(*Receptors and Recognition,* Series B, Volume 6)
Edited by E.H. Beachey
Published in 1980 by Chapman and Hall, 11 New Fetter Lane, London EC4P 4EE
© 1980 Chapman and Hall

14.1 INTRODUCTION

The notion that bacteria which inhabit or infect the body surfaces of a metazoan host, must somehow 'stick' to these surfaces in order to avoid physical removal, is exceedingly simple and intuitively obvious. An analogous concept has been generally accepted for many decades by marine biologists. It is therefore difficult to explain why this idea – and especially its experimental exploration in medical bacteriology – had remained confined to isolated esoteric studies until only a few years ago. Indeed, the published literature contains few relevant references until the early 1960's when, among other contributions, the concept was formulated on theoretical grounds (Dixon, 1960; Lankford, 1960; Freter *et al.*, 1961; Smith and Halls, 1967), when colonization of body surfaces was demonstrated histologically (La Brec *et al.*, 1965; Dubos *et al.*, 1965; Hoffman and Frank, 1966) and when the protective effect of local immunity was correlated with inhibition of adhesion (Freter, 1969). In spite of the mounting evidence, the current 'explosion' of interest in the role of bacterial adherence in host–parasite interactions of body surfaces did not take place until well into the 1970's. It is fascinating to speculate what circumstance suddenly motivated so many workers to study bacterial adhesion to host tissues. Most likely, responsibility for this trend cannot be assigned solely to any single precipitating cause, but the prolific and provocative publications of Gibbons and associates must have contributed to it in a major way. These workers demonstrated a close association between adhesive capacity and *in vivo* distribution of a variety of oral bacteria (reviewed by Gibbons and van Houte, 1975). This work, as well as that of earlier investigators (Hoffman and Frank, 1966), indicates that bacterial adhesion in the oral cavity involves a direct interaction, presumably based on receptor–ligand binding, between the bacterial surface and that of mucosal cells or tooth enamel. The persuasive example of this first large-scale ecological study of bacterial adhesion on oral surfaces may well have influenced subsequent investigators who were interested in bacterial colonization of areas other than the mouth, to similarly concentrate their studies on the role of surface-ligand interactions. This preferential attention, amply documented in the present volume, is certainly not misplaced, in that surface–ligand interactions no doubt contribute to bacterial colonization of many body surfaces.

Whatever the merits may be of the above analysis, it remains a fact that the current emphasis on research of cell surface–ligand interactions has somewhat obscured the recognition and study of other factors which in many instances may be of equal or even major importance in the association of bacteria with body surfaces. Moreover, a number of such interactions are preliminary to the eventual ligand binding of bacteria to epithelial cell surfaces, and the possibility of deliberate

441

manipulation of these preliminary interactions offers a prime opportunity for future study by those interested in the prevention of adherence of harmful bacteria to host tissues. In view of the scant attention given the subject in the past, an introductory part of this article will be devoted to an account of the various host–bacteria interactions that may occur at body surfaces. In view of the fact that most instances of primary bacterial colonization of the body involve mucosal surfaces, this article will emphasize bacteria–host interactions at these sites. This will serve as a necessary preliminary to the subsequent discussion of possible strategies that may be employed for the prevention of bacterial adherence. In view of the fact that the primary ecological function of bacterial adherence to body surfaces, i.e. resistance to physical removal, may be accomplished by means other than surface–ligand interactions, the term 'association' with the mucosa will be employed where the nature of the relevant interactions is not known or not specified (Freter, 1979). In contrast, the terms 'adhesion' or 'adherence' will be used synonymously to specifically denote surface–ligand binding between bacterial and mammalian cells. As proposed by Jones (1977), 'adhesin' will describe a substance on the bacterial surface that mediates the specific attachment of the bacterium to a surface receptor. Conversely, 'receptors' are components of the animal cell surface that react specifically with bacterial adhesins. A taxin is a substance which will either attract or repel a chemotactic bacterium traveling through a concentration gradient of the taxin.

14.2 THE ROLE OF MUCUS GEL IN MUCOSAL ASSOCIATION WITH BACTERIA

By definition, mucosae are epithelia moistened by 'mucus' (Warwick and Williams, 1973). The substances to which this term is collectively applied must be grouped into two major categories. The first consists of glycoproteins and glycolipids which appear to be synthesized by the epithelial cell and which form an integral part of the cell membrane. The carbohydrate–rich surface layer formed in this manner is the 'glycocalyx' (Bennet, 1963). In addition, specialized cells secrete glycoproteins which differ in various characteristics from those of the glycocalyx. This material, hereafter referred to as 'mucus gel' often forms a layer covering the epithelium. It is unfortunate, in view of the different properties of these two types of 'mucus', that a distinction between them is not always made in the literature. As reviewed by Jones (1977), there is evidence to support the view that heterosaccharides of the glycocalyx act as receptors for bacterial adhesins. The possible role of adherent mucus gel in bacterial associations will be discussed later in this article.

On some mucosal surfaces, such as those of the gasto-intestinal tract, mucus gel is thought to form a major barrier against penetration of bacteria to the level of the epithelial cells. For example, Edwards (1978) considered mucus to be a 'particle- and macromolecule-proof coating for cell surfaces' through which bacteria such as *Vibrio cholerae* must 'bore a channel' by means of special virulence mechanisms. Actually, there is considerable published evidence showing that macromolecules can cross mucosal barriers with some regularity (reviewed by Warshaw *et al.*, 1977). In contrast, the degree to which mucus gel inhibits the access of colloidal particles to the epithelial cells is not well established and relatively little work has been published on the subject. Florey (1933) made direct stereomicroscopic observations on live intestinal mucosa and concluded that the mucus gel represented a barrier in the form of a discontinuous lacework of strands, which in certain areas was penetrable to colloidal particles. The active motion of the villi caused particles to be 'rolled up' in mucus gel and thus to be expelled into the intestinal lumen (*loc. cit.*). The work of Schrank and Vervey (1976) suggested that mucus gel prevents the penetration of carbon particles. In contrast, cholera vibrios were able to penetrate the mucus gel, but active penetration could not be detected until at least several hours after infection. The massive layers of bacteria which cover the large intestinal mucosa of rodents and, presumably, of man as well are thought by some to be embedded in mucus gel (Dubos *et al.*, 1965; Hartley *et al.*, 1979). The paucity of investigations concerned with the role of mucus gel as a temporary or permanent habitat of bacteria associated with mucosal surfaces stems, no doubt, from the difficulty of preserving the mucus gel in standard light- and electronmicroscopic sections. For example, some investigators have been puzzled by their findings that they could cultivate bacteria from mucosal specimens, but could not visualize them on scanning electron micrographs (Nelson and Finkelstein, 1977). The best method for demonstrating mucus gel and its associated bacteria appears to be in frozen sections (Davis, 1976) which are stained with concentrated dye without prior fixation (Freter *et al.*, 1979a).

Convincing evidence of the importance of the mucus gel in bacterial association with the mucosa was presented by Schrank and Vervey (1976) who showed that efficient penetration required actively motile cholera vibrios. Non-motile vibrios did not demonstrably penetrate the mucus gel. They suggested that local antibacterial immunity, which previously had been shown to inhibit mucosal association of vibrios (Freter, 1969), may do so by inhibiting bacterial motility. The relationship between bacterial motility and virulence has been studied extensively by Berry and co-workers (Yancey and Berry, 1978).

Recent work from this writer's laboratory has indicated that a small fraction of polystyrene spheres presented to the lumenal side of the mucosa can penetrate rapidly, i.e. within 15 minutes or less, into the intervillous spaces and crypts of the mouse or rabbit small intestine. This effect was observed with intestinal slices exposed to bacterial suspensions *in vitro*, as well as with intact animals (Allweiss *et al.*, 1977; Freter *et al.*, 1978a,b; 1979a,b). It is difficult to quantify the fraction of particles or bacteria that actually penetrated, but this was certainly small. In order to reliably observe penetration in histologic sections, about 10^9 particles had to be placed into 5–10 cm-long loops of mouse or rabbit intestine. Larger particles such as yeast cells penetrated less efficiently than 1.1 μm diameter polystyrene spheres (*loc. cit.*). Non-motile mutants of *V. cholerae* showed very slow penetration, at a rate resembling that of yeast cells. Penetration of normally motile but non-chemotactic vibrio mutants was only slightly more efficient than that of non-motile bacteria. In contrast, wild-type cholera vibrios of normal motility and normal chemotactic responsiveness showed rapid penetration into the intervillous spaces at a rate significantly faster than that of the 1.1 μm diameter polystyrene spheres. These differences in rate of penetration of the various vibrio mutants correlated with their ability to colonize gnotobiotic mice or intestinal loops of adult rabbits (*loc. cit.*). More recent data show that smaller polystyrene spheres (0.44 μm diameter) can penetrate into the intervillous mucus gel of mouse small intestine at a significantly higher rate than the larger (1.1 μm diameter) particles. In fact, the rate of penetration by 0.44 μm spheres approached that observed with wild-type (motile and chemotactic) cholera vibrios (Freter, 1979; unpublished).

The evidence reviewed is consistent with the conclusions of some earlier workers that the mucus gel represents a barrier to the penetration of particles in the size range of bacteria. This barrier appears to be imperfect, however, because even inert particles may penetrate to the base of the villi within a period of 15 minutes or less. The rate of such passive penetration is relatively low, but can be increased significantly when bacteria are motile and are guided by chemotactic stimuli towards the bases of the villi. In this respect, bacterial motility without chemotactic guidance appears to be ineffective. This evidence was obtained with intestinal mucosa, but it seems reasonable to assume that similar effects would occur in the mucus gel of other mucosae as well. For example, the work of Parsons *et al.* (1978) suggests that the mucin layer which coats the transitional epithelium of the rabbit bladder, represents a strong barrier to bacterial attachment. Treatment of bladders with 0.3 N HCl removed the mucus and resulted in more than ten-fold increase in adhesion of *E. coli* to the denuded epithelium. Other workers have shown that *E. coli* strains from acute urinary tract infections have a special ability to associate with bladder epithelial cells (Svanborg-Eden *et al.*, 1977), suggesting that adhesion to bladder epithelium is an important feature of such infections and suggesting further that certain special bacterial strains must be able to overcome the blocking effect of the mucin barrier overlaying the bladder wall.

It seems self-evident that active or passive penetration of mucus gel by bacteria will have a similar ecological function as binding to the epithelial cell surfaces: in either case the bacteria are protected from physical removal by virtue of their association with the mucosa. By necessity, penetration of the mucus gel must be the first stage of mucosal association. In some instances, trapping in the mucus gel may even represent the *only* manner in which bacteria associate with the mucosa. Even when adhesion to epithelial cells constitutes the final mechanism of colonization, the first stage of bacterial association with the mucus gel may be quite prolonged and for this reason may be of considerable importance in pathogenesis. For example, Nelson *et al.* (1976) have shown that bacterial adhesion to intestinal epithelial cells in experimental cholera is not demonstrable until 2 hours after infection suggesting that the cholera vibrios must have remained in the mucus gel until that time. It also seems obvious that adhesion to the epithelial surface would be impossible without penetration of the mucus gel, except where distribution of the mucus gel is inhomogeneous.

14.3 SEQUENTIAL REACTIONS LEADING TO MUCOSAL ASSOCIATION WITH BACTERIA

It is apparent from the above discussion, that the seemingly simple process of bacterial adherence actually represents a complex sequence of host—bacterium interactions, each of which tends to either promote or restrain progress towards an eventual surface-to-surface ligand binding between bacteria and the epithelial cells of the mucosa. The more obvious mechanisms of interaction will be listed briefly below. It should be emphasized that not all of these mechanisms need to function in infection with any one pathogenic bacterium. On the other hand, each of these interactions has either been demonstrated directly or can be assumed on theoretical grounds to have some importance in the mucosal association of at least some pathogens.

(1) In the case of motile bacteria, a suitable environment must exist *in vivo* (or outside the body before infection) for the synthesis of flagellae and of certain receptors for bacterial chemotaxis. *In vitro* synthesis of flagellae by *E. coli*, for example, depends critically on cultural conditions (Adler and Templeton, 1967). Also, certain taxin-receptors of *E. coli* are inducible, i.e. are not present on the surface of bacteria grown in the absence of the specific taxin (Adler *et al.*, 1973). For this reason, bacteria classified by *in vitro* criteria as motile and chemotactic may not exhibit either of these traits in certain *in vivo* environments and vice versa.

(2) Bacteria that are present in the vicinity of a mucosal surface may be attracted to it by means of a chemotactic gradient. Such attraction has been observed with *E. coli, Salmonella typhimurium* and cholera vibrios (Allweiss *et al.*, 1977). Repulsion from the mucosa by negative chemotaxis has also been observed (*loc. cit.*). This interaction requires, of course, the presence of positive or negative taxin gradients and, on the part of the bacteria, an ability to respond chemotactically to these taxins.

(3) Penetration of the mucus gel by bacteria located at the gel surface constitutes the next step of association. As discussed above, this may occur passively at a slow rate, but is significantly enhanced when bacterial motility is guided by chemotactic gradients. These interactions, like those discussed in the preceding paragraph, require the production of taxins by the host as well as the corollary responsive capabilities on the part of the bacteria. Recent data (Freter *et al.*, 1979a,b) indicate that the taxin gradient in mucus gel of small intestinal mucosa of mice and rabbits extends into the deep intervillous spaces and crypts and, for this reason, can indeed promote deep bacterial penetration. Conceivably, the taxins which attract bacteria to the mucosal surface may differ from those present in the deeper regions of the mucus gel.

(4)   Microscopic observations in this author's laboratory (unpublished) of the mucus gel of rabbit intestinal slices or of intestinal loops which had been exposed to cholera vibrios, often show phase-dense material, embedded in the gel, around which numerous vibrios are aggregated. These aggregated bacteria have become immobilized and appear to adhere to the phase-dense material. This phenomenon has not been studied further but the possibility of bacterial adhesins reacting with receptor substances in the mucus gel must certainly be considered as a potential additional mechanism which would promote mucosal association by retaining bacteria in the mucus gel.

(5)   Bacterial adhesion to epithelial cell surfaces represents the final step in the process of association with the mucosa. As discussed above, it is also the one mechanism which had been identified earliest and which, consequently, has been studied most widely. The mechanisms of those adhesive interactions which have been identified to date appear to involve specific receptor—ligand binding. A non-specific mechanism of adhesion i.e. one which is not based on the interaction of specific adhesin-receptor systems, but which depends instead on differences in surface charges, on hydrophobic interactions or other forces, which may be generated with equal efficiency by a large variety of different surface components present on a variety of cells, has not been demonstrated to date in the interaction of unmodified bacteria and host cells. Such a mechanism has been considered by several workers, however, and the suggestion has been made that non-specific surface interactions may be *preliminary* events which promote subsequent more specific mechanisms of adhesion (Brinton, 1965; Isaacson, 1977; Smyth *et al.*, 1978). A good case may actually be made for the idea that non-specific mechanisms will often be necessary to stabilize specific bacterial adhesion to cell surfaces *in vivo*: In the natural environment, epithelial cell surfaces are bathed in glycoproteins from such sources as mucus gel, saliva, etc. Many of these glycoproteins share receptor specificities with the glyco-calyx of epithelial cells (Williams and Gibbons, 1975; Etzler, 1979). Forstner (1971) showed that glycoproteins of the glycocalyx may be released by mild proteolytic digestion, as may occur *in vivo*. In the *in vivo* environment, substances from glycocalyx and mucus gel must therefore be expected to compete for bacterial adhesins with receptors on the epithelial cell surfaces. Moreover, data presented by Forstner *et al.* (1973) and Etzler (1979) indicate that mucus gel may adhere to the epithelial cell surface with considerable tenacity, such that it can be removed only by vigorous washing. Consequently, bacteria may adhere *in vivo* to epithelial cells not only via receptors of the glycocalyx but also by reacting with receptors of mucus gel that is coating the cell surface. For this reason, again, mucus gel overlaying the epithelium must be expected to compete *in vivo* with the epithelial cell surface for bacterial adhesins. It seems reasonable to assume, therefore, that specific binding to epithelial cells *in vivo* would often be relatively weak and highly reversible, unless the specific bond could be reinforced at the cell surface by a second adhesive mechanism such as hydrophobic interactions (which, according to this hypothesis would have to *follow* specific binding). Not infrequently, then, the *in vivo* interactions of bacteria

with epithelial cells may be quite different from those observed *in vitro* with isolated cells or cell membranes suspended in buffer.

Whether or not the adhesive mechanism is specific, it requires the synthesis of the appropriate receptors and adhesins and their deposition on the respective bacterial or mammalian cell surfaces. It has been demonstrated repeatedly, that the phenotypic expression of bacterial adhesive capacity depends on environmental conditions (Jones *et al.*, 1976) and that certain *in vivo* environments differ in that respect from *in vitro* cultures (Nagy *et al.*, 1977). The role of dietary sucrose in glucan synthesis and dental plaque formation by *Streptococcus mutans* is, of course, a classical example of environmental influence on bacterial adhesive capacities (reviewed by Gibbons and van Houte, 1975). Conversely, some individuals of a given host species may lack certain epithelial cell receptors and by virtue of this defect will be resistant to a pathogen which requires these receptors for colonization (Sellwood *et al.*, 1975).

(6)   It is important to realize that bacterial enzymes may modify several host factors of importance in mucosal association. For example, the neuraminidase of cholera vibrios, also called the 'receptor-destroying enzyme', may indeed destroy red blood cell receptors for influenza virus or for *Mycoplasma pneumoniae*, but it can also *enhance* the adhesion of some bacteria to the altered cell surface (Sobeslawsky *et al.*, 1968). Enhancement of adhesion may occur under these circumstances because repulsive negative charges are removed or because cleavage of terminal neuraminic acid uncovers new determinant groups which form receptors of different specificities. Proteolytic enzymes also release neuraminic acid (*loc. cit.*) and glycoproteins (Forstner, 1971) from cell surfaces. Peptic digestion of mucosal scrapings has been shown to release materials which inhibit cholera vibrios with respect to both chemotaxis and adhesion to isolated brush border membranes of epithelial cells (Freter and Jones, 1976; Allweiss *et al.*, 1977). Presumably, such inhibitors represent taxins and receptors which inhibit competitively by blocking the corollary determinants on the bacterial surface. It is extremely interesting in this regard to note Schneider and Parker's (1978) work, which demonstrates that some protease- and neuraminidase-deficient mutans of *V. cholerae* show a 'dramatic' loss of virulence in experimental cholera. Parsons *et al.* (1978) also noted that neuraminidase enhanced the adhesion of *E. coli* to rabbit urinary bladders. While the mechanisms involved in these phenomena obviously require further investigation, the above-mentioned examples illustrate that promotion as well as inhibition of bacterial association with body surfaces by microbial or other enzymes is not an unlikely possibility.

It is apparent from the above discussion that there must be a large number of different potentially effective strategies that can be employed to reduce or prevent bacterial association with mucosal surfaces. In fact, the deliberate interference with any one of the above mentioned mechanisms of association may be expected to accomplish this end. As emphasized above, no single pathogenic bacterium is likely to possess the complete array of all possible attributes that promote association with the mucosa. It seems intuitively obvious that the presence of one or a few traits which strongly promote mucosal association should be sufficient to counterbalance the weakness or absence of some of the others in a given pathogenic bacterium. For example, an *E. coli* strain which is strongly adhesive by virtue of the K88 adhesin may not have to be motile in order to colonize the mucosa. Conceivably, the few *E. coli* cells which passively traverse the mucus gel (with the same low efficiency as inert particles) would be sufficient to adhere, multiply and give rise to strongly adherent bacterial populations of increasing size. In contrast, a relatively weakly adherent bacterium, such as the cholera vibrio, may be crucially dependent on motility and chemotaxis for successful association, because its adherence alone would not be sufficiently effective in counteracting the forces (e.g. mucus flow) which remove bacteria from intervillous spaces. For this reason, one must expect that the deliberate interference with a given mechanism that promotes bacterial association with the mucosa will be more or less successful in preventing bacterial colonization, depending on the extent to which a given pathogen relies on that particular mechanism for achieving association with the mucosa. This reservation should be kept in mind when reading the remainder of this article, which is devoted to a discussion of various promising strategies that may be employed to deliberately inhibit the interaction of harmful bacteria with body surfaces. For obvious reasons, most of this discussion has to remain in the realm of speculation, and most of the points that will be made should be taken principally as suggestions for future basic studies which are required before it will be possible to formulate more realistic and practical approaches to a control of adhesive pathogenic bacteria in man and animals.

14.4 INTERFERENCE WITH BACTERIAL ADHERENCE BY NON-IMMUNOLOGICAL MEANS

Most bacterial adhesins which have been studied in this respect, appear to react with carbohydrate receptors on the mammalian cell surface (Jones, 1977). Complex or single carbohydrates are used experimentally in inhibition studies aimed at identification of the active moities on adhesive cell surfaces (*loc. cit.*). Since most natural foodstuffs contain sugar, polysaccharides, glycoproteins or glycolipids, one must assume that many of these will antagonize the adhesion of bacteria in the mouth and gastro-intestinal tract by competitive inhibition. Milk probably represents the most persuasive example of this. It will be argued below, however, that other foodstuffs are equally likely contenders and that competitive inhibition is by no means the only potential mechanism by which dietary substances may antagonize bacterial association with the mucosa.

14.4.1 Anti-adhesive substances in milk

Milk fat forms colloidal size globules which are surrounded by a unit membrane that has many similarities to cell membranes. For example, fat globules show serologic cross-reactions with red blood cells (Dowben *et al.*, 1967). Reiter and Brown (1976) reported that hemagglutination of red blood cells by *E. coli* K88 and K99 antigens can be inhibited by milk fat globules or fat globule membranes. They postulated, therefore, that the fat globule membranes of milk 'may, at least partially, interfere with the attachment of *E. coli* possessing capsular antigens to the brush border of the intestinal wall and thus have a protective activity'. To the extent that the milk fat globule membrane carries other cell receptor moieties in addition to those reacting with K88 and K99 antigens, Reiter and Brown's statements may be extended to include other adhesive pathogens as well. Anti-adhesive substances and toxins (see below) must therefore be added to the list of known antimicrobial factors of milk (Reiter, 1978; McClelland *et al.*, 1978). The likelihood that such factors will match the receptor-specificities of the host's epithelial cells is of course highest with milk from the same species – another argument in favor of mother's milk.

14.4.2 Anti-adhesive substances in meat

While bacterial adhesins exhibit a remarkable degree of species specificity, there are also a surprising number of cross-reactions. For example, *V. cholerae* and human enteropathogenic *E. coli* strains adhere to the rabbit small intestine, and agglutinate red blood cells from a variety of animal species. It is likely that a number of such reactions with cells of other species involve adhesin – receptor systems different in specificity from those operating in infection of the natural host (Freter, 1979; Jones, 1977 and this volume). A meat diet may therefore contain receptors which are different from those of the host, but which may still react with certain adhesins of bacteria present in the host. Obviously, such receptors cannot inhibit bacterial adhesion to host cells by competitive inhibition in the strict biochemical sense. Nevertheless, bacteria bound to cells or cell debris of dietary meat products are likely to be diverted from association with mucosal epithelium, regardless of the specificities of the adhesive reactions involved. Moreover, deposition on a bacterial surface of molecules which are able to react with one set of specific adhesins, will often inhibit the reactivity of other adhesins as well. This may be the case when the interfering receptor molecules have a strong electrostatic charge, when they are surface-active or when their large size inhibits reactions with other surfaces by steric hindrance, as discussed below for lectins and antibodies. Thus, receptor substances in the diet which do not match host cell receptors may still inhibit bacterial adhesion, as long as they are able to bind in some manner to the bacterial surface.

Basic proteins from dietary meat represent another group of potential inhibitors. These substances may complex with negatively charged molecules, such as lipoteichoic acids, as has been described by Ofek *et al.* (1979). If such a reaction, were to occur on a bacterial surface it might well inhibit bacterial adhesion, in a similar manner as basic protein can interfere with antibody-mediated passive hemagglutination of red blood cells coated with lipoteichoic acid (*loc. cit.*). Lipids from meat and other dietary sources are still another class of substances with potential anti-adhesive properties. Pertinent here is the observation by Ofek *et al.,* (1975) that pretreatment of epithelial cells with the lipid component of streptococcal lipoteichoic acid inhibited streptococcal adhesion. Other lipids, such as may be found in dietary substances, can have similar effects. For example, Ellen and Gibbons (1974) showed an inhibitory effect of phospholipids on streptococcal adhesion to buccal mucosal cells.

14.4.3 Lectins as anti-adhesive substances in foodstuffs

The term lectin was applied originally to substances derived from plants, but similarly active carbohydrate-binding proteins have also been isolated from animal sources (Simpson *et al.,* 1978). Lectins may agglutinate cells, as well as inhibit adhesion. For example, concanavalin A inhibited adhesion of lactobacilli to epithelial cells (Fuller, 1975) and a variety of lectins inhibited the adhesion of *E. coli* to epithelial cell brush border membranes (Jones, 1977). The latter author noted that the reaction was not specific, in that lectins with specificities for different carbohydrates inhibited adhesion of *E. coli* equally well, as long as the lectin was able to bind to the brush borders. For this reason, he attributed the inhibition of adhesion to steric hindrance (*loc. cit.*). This lack of specificity is unfortunate if one wishes to use lectins experimentally to identify the ligands in adhesive reactions. On the other hand, the non-specific nature of lectin-inhibition marks this reaction as one of potentially broad significance in the inhibition of bacterial adhesion by dietary substances. Brady *et al.* (1978) showed that wheat germ agglutinin, a lectin specific for *N*-acetylglucosamine, could be detected in the feces of human subjects consuming a diet containing wheat germ. They concluded, 'it is feasible that orally ingested wheat germ agglutinin and other plant lectins which interact with a wide variety of cell membranes may alter intestinal epithelial or bacterial cell function in the human bowel'. They review evidence showing that some dietary lectins may be toxic whereas others are not (*loc. cit.*). It should also be noted that many lectins are agglutinins for red blood cells. If dietary lectins were to bring about the clumping of bacteria in the gastro-intestinal tract, this would inhibit bacterial association with the mucosa because, as reviewed earlier, large particles penetrate the mucus gel poorly. Regarding a somewhat analogous situation, Marx (1977) reviews observations suggesting that tobacco plants may protect themselves against invading *Pseudomonas solanacearum* by clumping these bacteria with lectin and attaching them to plant cell walls.

14.4.4 Taxins as anti-adhesive substances in foodstuffs

As reviewed by Adler (1975) only small molecules, such as monosaccharides and amino acids have been shown to attract or repel bacteria chemotactically. It is not known whether larger molecules can also function as taxins. Be this as it may, it seems obvious that practically all foodstuffs which are being digested as they pass through the mouth and gastro-intestinal canal, must elaborate taxins which can affect the behavior of motile bacteria *in vivo*. Such taxins may function in several ways:

(a)   They may create a taxin gradient which has its highest concentration in the lumen, thereby drawing bacteria away from the mucus gel into the lumen, where they are subject to elimination by peristaltic movements.

(b)   Alternately, taxin present in relatively high concentrations may block the corresponding receptors on the bacterial surface, thereby abolishing the ability of the bacterium to respond chemotactically to that specific taxin.

If the blocking taxin or taxins are specific for the same bacterial receptors as are the taxins which attract the bacteria into the mucus gel, active penetration of the mucus gel by the bacteria will be abolished. It is likely that this latter mechanism is responsible for the blocking effect of mucosal extracts on the penetration of cholera vibrios into intestinal mucosa (Freter *et al.*, 1979b). In contrast, we have speculated that the mechanism which infant mice employ in protection against chemotactic cholera vibrios involves the situation described under (a) above (Freter *et al.*, 1978a).

14.4.5 Supplementation of natural foodstuffs

When more specific knowledge has accumulated concerning the *in vivo* mechanisms of bacterial association, it may become possible to enrich foodstuffs with anti-adhesive substances in such a manner as to achieve increased prophylactic effectiveness agains colonization by harmful adhesive bacteria. Basically, foodstuffs may be enriched with (a) bacterial adhesins, (b) mucosal receptors, (c) taxins or (d) substances which clump bacterial or render them non-motile. Most likely, dietary administration of bacterial adhesins may be a less practicable long-term measure, because these substances would react continually with epithelial cell receptors, a process which might produce undesirable side-reactions. Also, a local immune response would probably render such additives ineffective as anti-adhesive agents, but the local immunity may induce resistance against infection with the bacteria themselves. In this sense, adhesins would represent non-viable oral vaccines which have been shown to induce significant levels of local antibodies in human volunteers (Freter, 1962). Candidate bacterial adhesins would be, for example, the CF-antigens of *E. coli* (Evans *et al.*, 1978), bacterial pili which inhibit adhesion (Isaacson *et al.*, 1978), cholera vibrio hemagglutinin (Finkelstein *et al.*, 1977) and low molecular weight dextrans (Gibbons and Keyes, 1969). The second approach, that of adding an excess of epithelial cell receptors to the diet appears to be a more suitable non-immunological approach. Since these substances would not react with host cells and would not elicit and immune response, safe long-term administration appears to be feasible. It is not possible at the moment to predict likely sources of receptor material because most host cell receptors have not yet been identified. Mannans may be mentioned as one possibility (Aronson *et al.*, 1979), but these substances would be limited in effectiveness to a small group of bacteria, of course, namely those carrying type 1 fimbriae (Jones, 1977).

In view of the fact that the known taxins for bacteria are small molecules, one may wonder whether direct oral administration would be effective at all, because much of these materials would be absorbed before reaching the intended site of action.

Potentially more effective would be larger molecules, e.g. proteins or carbohydrates which could slowly release active products during transit through the gastro-intestinal tract. As mentioned earlier, the generation of large-size nonmotile particles by clumping of bacteria must be expected to reduce or inhibit mucosal association by these micro-organisms. Clumping may be brought about by substances such as lectins or basic proteins and by the administration of receptor substances in insoluble form. Insoluble receptor material may occur naturally, e.g. chitin, or may be prepared by chemical modification or by coating particulate substrates with active materials, that will then bind bacteria carrying the appropriate adhesin (Jones and Freter, 1976).

14.4.6 Other mechanisms

In view of the introductory discussion of the various steps involved in bacterial association with the mucosa, a large number of additional approaches suggest themselves as obvious possibilities for the prevention of such association. The practical exploitation, however, of any of these would require considerably more insight into the underlying physiological and biochemical mechanisms of association than is currently available, thus precluding for the present time fruitful speculations as to the realistic prospects for prophylaxis against infections with adhesive bacteria. For example, the possible addition to the diet of enzyme inhibitors would be a promising line of investigation to pursue if and when the enzymes involved in bacterial association are identified. The same may be said for agents which inhibit bacterial motility or which affect the *in vivo* synthesis of adhesins, taxin receptors or flagellae. Agents affecting mucus flow (Parke, 1978), peristalsis or intestinal pH can also be expected to affect bacterial association. If, as suggested earlier, hydrophobic interactions are indeed important for *in vivo* bacterial adhesion to epithelial cells, the administration of surface-active agents may be of protective value to the host.

A special case must be made for the effect of the indigenous microflora. Components of this flora have been shown to degrade glycoproteins, presumably as substrates for growth. In fact, Hoskins and Boulding (1976) have presented evidence to the effect that the human intestine selects for indigenous bacteria which are capable of utilizing blood group-specific glycoproteins of the host. Systemic infections with *Bacteroides* species are known to bring about phenotypic changes in a patient's blood group (T or Tk polyagglutination) by cleaving terminal sugar residues on the red blood cells (Inglis *et al.*, 1975). There can be no doubt, then, that the nature of host cell receptors in areas normally inhabited by an indigenous microflora must be determined to a considerable extent by the activities of that flora. Moreover, since the indigenous flora forms thick, multilayered populations on some mucosal surfaces (Savage, 1972), initial adhesion of other micro-organisms may be to the indigenous bacteria themselves, rather than to the epithelial surface. This may conceivably account for the failure to detect an adhesive capacity of pneumococci for human pharyngeal cells, a recent finding which surprised the investigators who reported it (Slinger and Reed, 1979). It may be appropriate in this context to mention the report

by Sanford *et al.,* (1978), who showed that infection with influenza A virus of a
stable line of canine kidney cells resulted in the adhesion of certain streptococci
which would not adhere to uninfected cells. The authors suggested that the presence
of viral hemagglutinin on the cell surface might be responsible for streptococcal
adhesion. The intestinal flora also affects the levels of mucosal enzymes and the rates
of mucosal cell turnover and intestinal peristalsis (reviewed in Kawai and Morotomi,
1978), each of which can, in turn, affect mucosal association of bacteria. Consequently,
modification of the indigenous flora by dietary, pharmacological or other means may
be expected to result in drastic changes in the mucosal cell receptors as well as in the
overlaying mucus gel. Unfortunately, deliberate, controlled modification of the
indigenous microflora is not yet a practical possibility (Freter, 1976).

14.4.7 General comments

In discussing the role of adherence in the ecology of oral microflora, Gibbons and
van Houte (1975) pointed out that there appears to be an equilibrium between the
mechanisms which promote adhesion and those which inhibit it. Among the latter
they noted local antibodies and certain salivary glycoproteins which counteract
adhesion by competitive inhibition. One must strongly suspect from the above
discussion that analogous equilibria exist throughout the gastro-intestinal tract and
that foodstuffs contribute to this system a number of mechanisms, most of which
appear to shift the balance in the direction of reduced adhesion. As pointed out
earlier (Allweiss *et al.,* 1977), such mechanisms may relate to certain aspects of the
long-recognized, but still enigmatic, relationship between diet and susceptibility to
infection. The present discussion also suggests that the most promising strategies for
deliberate non-immunological interference with the adherence of harmful bacteria
to body surfaces appear to be those aimed at augmenting inhibitory mechanisms
which are already operating in the normal host. For obvious reasons, epithelia whicn
are sheltered from contact with dietary substances present much fewer opportunities
for deliberate non-immunological intervention in adhesive processes. Here only a
pharmacologically oriented approach to the delivery of inhibitory substances appears
to be of promise. This would of course require a precise definition of the inhibitors
to be delivered, based on an understanding of bacterial association at a level which
has not been attained at present. In the case of livestock, the breeding of strains
which lack phenotypic expression of epithelial receptors (Sellwood *et al.,* 1975) is
an obvious possibility which may be applied on an empirical basis.

As a final thought it may be pointed out that there are probably instances where
association of harmful bacteria with mucosal surfaces would actually be desirable.
This would be the case when mucosa-associated bactericidal mechanisms are present,
such as have been found in immunized animals (see below) or in the stomach and
small intestine of infant mice infected with cholera vibrios (Freter, unpublished).
In such instances, deliberate attempts to *increase* bacterial association with the
mucosa may be in the best interest of the host.

14.5 INTERFERENCE WITH BACTERIAL ADHERENCE
BY LOCAL IMMUNE-MECHANISMS

Hoffman and Valdina (1968) considered it 'most probable that the attachment of
bacteria to oral epithelial cells is the result of antigen—antibody reactions'. In view
of the binding and agglutinating properties of antibodies, this was a most reasonable
assumption. It became soon apparent, however, that the usual function of local
antibody was exactly opposite, namely that of preventing bacterial association with
mucosa (Freter, 1969). This finding has subsequently been extended to a large
number of different adhesive bacteria as well as to macromolecules (Walker *et al.,*
1976). Inhibition of association may be brought about by a variety of mechanisms.
Among these are bacterial clumping (Steele *et al.,* 1975) and inhibition of bacterial
motility (Williams *et al.,* 1973; Bellamy *et al.,* 1975; Eubanks *et al.,* 1977), the latter
possibly being a consequence of the former in some instances. Antibody may protect
against experimental infection and inhibit bacterial adhesion also directly, i.e. in the
absence of clumping, as evidenced by direct observations (Freter and Jones, 1976)
and by the use of non-agglutinating Fab fragments (Steele *et al.,* 1975). Some authors
have implied that antibody directed against the adhesive determinant on the bacterial
surface would be most effective (e.g. Rutter and Jones, 1973; Nagy *et al.,* 1978).
This may well be the case in some instances, especially where bacterial adhesin
prevents access of antibody to other surface antigens. In instances where that situation does not obtain, antibody to surface antigens that are not involved in adhesion
may also be expected to affect bacterial association with the mucosa, either by the
mechanisms mentioned above, by steric hindrance or by alterations in surface charge
and hydrophobic characteristics of the bacterial surface (van Oss, 1978), in a similar
manner as discussed above for lectins. For example, antibodies specific for the heat
labile as well as the heat stable antigens of V. cholerae are protective (Bellamy *et al.,*
1975; Freter, 1964), even though the latter have no known function in pathogensis.

In view of the potentially important role of mucus gel in mucosal association
discussed above, it was interesting to note a report by Walker *et al.* (1977) showing
that soluble immune complexes can stimulate the release of mucus gel from intestinal
goblet cells. More recently, this group (Lake *et al.,* 1979) has extended this work to
show that soluble antigen alone will have this effect in orally immunized rats, but
not in normal or systemically immunized animals. If bacteria entering the mucus
gel of the immune host can also trigger this reaction, one would have to consider it a
potentially effective immune mechanism that counteracts mucosal association with
harmful micro-organisms. Intestinal mucus also appears to function as a reservoir
of local antibodies. Walker *et al.* (1974) reported that antibody activity could be
removed from the mucosa of rat intestine only by a method which removed the
mucus as well.

The question whether there can be bactericidal mechanisms operating on the
mucosal surface of immunized hosts has been debated for many years. Evidence for
such mechanisms has been reported (Freter, 1970; Chaicumpa and Rowley, 1972; Knop

and Rowley, 1975) and denied (Bloom and Rowley, 1977). Bellamy and Nielsen (1974) reported that the injection of bovine serum albumin into intestinal loops of guinea pigs, which had been immunized with this antigen, resulted in the emigration of large numbers of neutrophils into the lumen. Recent data on colibacillosis in calves indicate that such leukocytes were functional with respect to phagocytosis of bacteria (Bellamy, personal communication). In addition to immune reactions, bacterial chemotactic factors are thought to stimulate leukocyte emigration even in the absence of complement components (Bellamy, personal communication). If confirmed in other systems, this reaction would have to be regarded as a potentially effective bactericidal mechanism against mucosa-associated bacteria in immune as well as in non-immune animals.

It was formerly thought that the induction of local antibody production at a given mucosal site required the application of the immunizing antigen to that same site. Recent evidence (reviewed by Kleinman and Walker, 1979) suggests that there are exceptions to this rule. For example, antibodies may appear in the colostrum of females which had been immunized by the oral route. This mechanism offers important new possibilities for the protection of newborns by maternal antibodies in the colostrum and milk, because the production of such antibodies may be induced simply by oral vaccination of the mother.

14.6 SUMMARY AND CONCLUSIONS

This article emphasizes the implications of recent data which pertain to the functions of mucus gel on mucus membranes. Some of these functions may promote bacterial association. For example, bacteria are able to enter the mucus gel and thereby resist physical removal. Thus, ligand-binding to the epithelial cell surface is not the only mechanism by which bacteria can associate with a mucosal surface. Mucus gel also contains taxins which guide motile chemotactic bacteria deeply into the mucus. On the other hand, the mucus gel can also inhibit bacterial association with the mucosa. It serves, for example, to physically restrain the progress of bacteria towards the epithelial cell surface. In addition, mucus gel contains glycoproteins, some of which are competitive inhibitors of the reaction between bacterial adhesins and epithelial cell receptors. The presence of antigen on the mucosa of an immunized host can result in an increase in mucus flow, which would tend to remove bacteria from the epithelial surface. Mucus appears to be a reservoir of local antibodies which have anti-adhesive properties. Foodstuffs and other materials located on the surface of the mucus gel can furnish a variety of substances which have anti-adhesive effects by virtue of a variety of mechanisms. It is concluded, therefore, that a harmful bacterium must proceed sequentially through a large number of different reactions, some of which promote or retard its progress in approaching the epithelial cells, whereas others promote or antagonize its eventual adhesion to the epithelial cell surface. Bacterial association with the mucosa is therefore determined by the final equilibrium which is established as a consequence of mutually supportive and opposing reactions. Most of these reactions cannot be evaluated adequately *in vitro* by a study of bacterial adhesion to host cells suspended in buffer solutions. This article concludes that prospects for the prevention of association of harmful bacteria with host mucosal surfaces are optimal, when such attempts are directed towards reinforcing one or several of the anti-adhesive mechanisms which are already functioning to some degree in the normal or immunized host.

REFERENCES

Adler, J. (1975), *A. Rev. Biochem.*, **44**, 341–356.

Adler, J. and Templeton, B. (1967), *J. gen. Microbiol.*, **46**, 175–184.

Adler, J., Hazelbauer, G.L. and Dahl, M.M. (1973), *J. Bact.*, **115**, 824–847.

Allweiss, B., Dostal, J., Carey, K.E., Edwards, T.F. and Freter, R. (1977), *Nature*, **266**, 448–450.

Aronson, M., Medalia, O., Schori, L., Mirelman, D., Sharon, N. and Ofek, I. (1979), *J. Infect. Dis.*, **139**, 329–332.

Bellamy, J.E.C., Knop, J., Steele, E.J., Chaicumpa, W. and Rowley, D. (1975), *J. Infect. Dis.*, **132**, 181–188.

Bellamy, J.E.C. and Nielsen, N.O. (1974), *Infect. Immun.*, **9**, 615–619.

Bennett, H.S. (1963), *J. Histochem. Cytochem.*, **11**, 14–23.

Bloom, L. and Rowley, D. (1977), *Aust. J. exp. Biol.*, **55**, 385–391.

Brady, P.G., Vannier, A.M. and Banwell, J.G. (1978), *Gastroenterol.*, **75**, 236–239.

Brinton, Jr., C.C. (1965), *Trans. N.Y. Acad. Sci.*, **27**, 1003–1054.

Chaicumpa, W. and Rowley, D. (1972), *J. Infect. Dis.*, **125**, 480–485.

Davis, C.P. (1976), *Appl. Env. Microbiol.*, **31**, 304–312.

Dixon, J.M.S. (1960), *J. Path. Bact.*, **79**, 131–140.

Dowben, R.M., Brunner, J.R. and Philpott, D.E., (1967), *Biochim. Biophys. Acta*, **135**, 1–10.

Dubos, R.J., Schaedler, R.W., Costello, R. and Hoet, P. (1965), *J. exp. Med.*, **122**, 67–76.

Edwards, P.A.W. (1978), *Br. med. Bull.*, **34**, 55–56.

Ellen, R.P. and Gibbons, R.J. (1974), *Infect. Immun.*, **9**, 85–91.

Etzler, M.E. (1979), *J. clin. Nutr.* **32**, 133–138.

Eubanks, E.R., Gunetzel, M.N. and Berry, L.J. (1977), *Infect. Immun.*, **15**, 533–538.

Evans, D.G., Satterwhite, T.K., Evans, Jr., D.J. and DuPont, H.L. (1978), *Infect. Immun.*, **19**, 883–888.

Finkelstein, R.A., Arita, H., Clements, J.D. and Nelson, E.T. (1977), *Proc. 13th Joint Conf. on Cholera. U.S.A.*, Japan Cooperative Medical Science Progr. DHEW Publication No. (NIH) 78–1590, pp, 137–151.

Florey, H.W. (1933), *J. Path. Bact.*, **37**, 283–289.

Forstner, G.G. (1971), *Biochem. J.*, **121**, 781–789.

Forstner, J., Taichman, N., Kalnins, V. and Forstner, G. (1973), *J. Cell Sci.*, **12**, 585–602.

Freter, R. (1962), *J. Infect. Dis.*, **111**, 37–48.

Freter, R. (1964), *Bull. World Health Org.*, **31**, 825–834.

Freter, R. (1969), *Texas Rep. Biol. Med.*, **27** (Suppl. 1), 299–316.

Freter, R. (1970), *Infect. Immun.*, **2**, 556–562.

Freter, R. (1976), In: Proceedings *Microbial Aspects of Dental Caries*. (Stiles, H.M., Loesche, W.J. and O'Brien, T.C., eds), Information Retrieval, Inc., Washington, D.C., pp. 109–120.

Freter, R. (1979), *Proc. Nobel Symp.*, **43**, S. Karger, Basel (in press).

Freter, R., Allweiss, B., O'Brien, P.C.M. and Halstead, S.A. (1978a), *Proc. 13th Joint Conf. on Cholera*. U.S.A. Japan Cooperative Medical Science Program (U.S. Governement Printing Office 1978) pp. 152–181.

Freter, R. and Jones, G.W. (1976), *Infect. Immun.*, **14**, 246–256.

Freter, R., O'Brien, P.C.M. and Halstead, S.A. (1978b), In: Secretory Immunity and Infection, (McGhee, J.R., Mestecki, J. and Babb, J.L., eds), *Adv. exp. Med.*, **107**, 429–437.

Freter, R., O'Brien, P.C.M. and Macsai, M.S. (1979a), *J. clin. Nutr.*, **32**, 128–132.

Freter, R., O'Brien, P.C.M. and Macsai, M. (1979b), *Proc. 14th Joint Conf. on Cholera*. U.S.A., Japan Cooperative Medical Science Program (Japanese Cholera Panel 1979), (in press).

Freter, R., Smith, Jr., W.L. and Sweeney, Jr., F.J. (1961), *J. Infect. Dis.*, **109**, 35–42.

Fuller, R. (1975), *J. gen. Microbiol.*, **87**, 245–250.

Gibbons, R.J. and van Houte, J. (1975), *A. Rev. Microbiol.*, **29**, 19–44.

Gibbons, R.J. and Keyes, P.H. (1969), *Arch. Oral Biol.*, **14**, 721–724.

Hartley, C.L., Neuman, C.S. and Richmond, M.H. (1979), *Infect. Immun.*, **23**, 128–132.

Hoffman, H. and Frank, M.E. (1966), *Acta Cytol.*, **10**, 272–285.

Hoffman, H. and Valdina, J. (1968), *Acta Cytol.*, **12**, 37–41.

Hoskins, L.C. and Boulding, E.T. (1976), *J. clin. Invest.*, **57**, 74–82.

Inglis, G., Bird, G.W.G., Mitchell, A.A.B., Milne, G.R. and Wingham, J. (1975), *J. clin. Pathol.*, **28**, 964–968.

Isaacson, R.E. (1977), *Infect. Immun.*, **15**, 272–279.

Isaacson, R.E., Fusco, P.C., Brinton, C.C. and Moon, H.W. (1978), *Infect. Immun.*, **21**, 392–397.

Jones, G.W. (1977), In: *Microbial Interactions*. (Reissig, J.L., ed.), Chapman and Hall, London, pp. 139–176.

Jones, G.W., Abrams, G.D. and Freter, R. (1976), *Infect. Immun.*, **14**, 232–239.

Jones, G.W. and Freter, R. (1976), *Infect. Immun.*, **14**, 240–245.

Kawai, Y. and Morotomi, M. (1978), *Infect. Immun.*, **19**, 771–778.

Kleinman, R.E. and Walker, W.A. (1979), *J. Peds.*, **94**, (in press).

Knop, J. and Rowley, D. (1975), *Aust. J. exp. Biol.*, **53**, 155–165.

LaBrec, E.H., Sprinz, H., Schneider, H. and Formal, S.B. (1965), *Proc. Cholera Symp*, Honolulu, Hawaii. U.S. Govt. Printing Office, pp. 272–276.

Lake, A.M., Bloch, K.J., Neutra, M.R. and Walker, W.A. (1979), *J. Immunol.*, **122**, 834–837.

Lankford, C.E. (1960), *Ann. N.Y. Acad. Sci.*, **88**, 1203–1212.

McClelland, D.B.L., McGrath, J. and Samson, R.R. (1978), *Acta path. scand.*, Suppl. **271**, 1–20.

Marx, J.L. (1977), *Science,* **196**, 1429–1478.

Nagy, B., Moon, H.W. and Isaacson, R.E. (1977), *Infect. Immun.,* **16**, 344–352.

Nagy, B., Moon, H.W., Isaacson, R.E., To, C.C. and Brinton, C.C. (1978), *Infect. Immun.,* **21**, 269–274.

Nelson, E.T. and Finkelstein, R. (1977), Abstr. Ann. Meeting Am. Soc. Microbiol., p. 19 (B25).

Nelson, E.T., Clements, J.D. and Finkelstein, R.A. (1976), *Infect. Immun.,* **14**, 527–547.

Ofek, I., Beachey, E.H., Jefferson, W. and Campbell, G.L. (1975), *J. exp. Med.,* **141**, 990–1003.

Ofek, I., Whitaker, J.N., Campbell, G.L. and Beachey, E.H. (1979), *J. Infect. Dis.,* **139**, 93–96.

van Oss, C.J. (1978), *A. Rev. Microbiol.,* **32**, 19–39.

Parke, D.V. (1978), *Br. med. Bull.,* **34**, 89–94.

Parsons, C.L., Shrom, S.H., Hanno, P.M. and Mulholland, S.G. (1978), *Invest. Urol.,* **16**, 196–200.

Reiter, B. (1978), *J. Dairy Res.,* **45**, 131–147.

Reiter, B. and Brown, T. (1976), *Proc. Soc. gen. Microbiol.,* **3**, 109.

Rutter, J.M. and Jones, G.W. (1973), *Nature,* **242**, 531–532.

Sanford, B.A., Shelokov, A. and Ramsay, M.A. (1978), *J. Infect. Dis.,* **137**, 176–181.

Savage, D.C. (1972), *22nd Sympos. Soc. gen. Microbiol.* pp. 25–57.

Schneider, D.R. and Parker, C.D. (1978), *J. Infect. Dis.,* **138**, 143–151.

Schrank, G.D. and Vervey, W.F. (1976), *Infect. Immun.,* **13**, 195–203.

Selinger, D.S. and Reed, W.P. (1979), *Infect. Immun.,* **23**, 545–548.

Sellwood, R., Gibbons, R.A., Jones, G.W. and Rutter, J.M. (1975), *J. med. Microbiol.,* **8**, 405–411.

Simpson, D.L., Thorne, D.R. and Loh, H.H. (1978), *Life Sci.,* **22**, 727–748.

Smith, H.W. and Halls, S. (1967), *J. Path. Bact.,* **93**, 499–529.

Smyth, C.J., Jonsson, P., Olsson, E., Soderlind, O., Rosengren, J., Hjerten, S. and Wadstrom, T. (1978), *Infect. Immun.,* **22**, 462–472.

Sobeslavsky, O., Prescott, B. and Channock, R.M. (1968), *J. Bact.,* **96**, 695–705.

Steele, E.J., Chaicumpa, W. and Rowley, D. (1975), *J. Infect. Dis.,* **132**, 175–180.

Svanborg-Eden, C., Eriksson, B. and Hanson, L.A. (1977), *Infect. Immun.,* **18**, 767–774.

Walker, W.A., Abel, S.N., Wu, M. and Bloch, K.J. (1976), *J. Immunol.,* **117**, 1028–1032.

Walker, W.A., Isselbacher, K.J. and Bloch, K.J. (1974), *J. clin. Nutr.,* **27**, 1434–1440.

Walker, W.A., Wu, M. and Bloch, K.J. (1977), *Science,* **197**, 370–372.

Warshaw, A.L., Bellini, C.A. and Walker, W.A. (1977), *Am. J. Surg.,* **133**, 55–58.

Warwick, R. and Williams, P.L. (1973), *Gray's Anatomy,* W.B. Saunders Co., Philadelphia, p. 23.

Williams, H.R., Vervey, W.F., Schrank, G.D. and Hurry, E.K. (1973), *Proc. 9th Joint Conf. on Cholera.* U.S., Japan Coop. Med. Sci. Prog., Dept. of State Publ. 8762, pp. 161–173.

Williams, R.C. and Gibbons, R.J. (1975), *Infect. Immun.,* **11**, 711–718.

Yancey, R.J. and Berry, L.J. (1978), *Infect. Immun.,* **22**, 387–392.

Index

Acquired pellicle of teeth,
  and bacterial adherence, 65, 66, 73,
    77, 86
  properties and composition, 73, 74
Actinomyces naeslundii,
  glucan synthesis, 111
Actinomyces viscosus,
  accumulation on teeth, 95
  glucan synthesis, 111
  glucosyltransferase binding, 109
Adherence, *see* Bacterial adherence,
Aeromonas liquefaciens,
  fimbriae, 300, 310
Agrobacterium spp.,
  polygalacturonic acid and tumor
    inhibition, 382, 383, 384
Agrobacterium tumefaciens,
  adherence to plant cell walls,
    379–382, 385
  induction of crown gall, 378, 382
  lipopolysaccharide, 380, 385
Amphophiles,
  biological properties, 146
  disease process mediation, 146
Antibiotics,
  and bacterial adherence, 25

Bacterial adherence,
  anti-adhesive substances,
    lectins, 450
    in milk, 448, 449
    in meat, 449, 450
    taxins, 450, 451
  bacterial adhesins, 448, 451
  carbohydrate receptors, 448, 451
  and chemotaxis, 236, 445
  DLVO theory, 273–275, 282
  electrostatic attraction, 82, 83

genetic variables, 11, 19, 39
  and the infectious process, 18, 24
  infection site, 20
  inhibition of, 8, 17, 127, 128,
    448–453
    by immune-mechanisms, 454, 455
    by interference, 448, 451, 452
    by D-mannose, 8, 199
  mechanisms of interaction, 445–447
  mucus gel taxins, 445
  phenotypic variables, 11–13, 19, 39
  polymeric bridging, 81, 83, 91, 94
  polysaccharide binding, 81, 83, 89
  role of mucus gel, 442–446, 454,
    455
  sequence of interactions, 445–447
  versus infectivity, 18–23
  see also Fimbriae
Bacterial adherence to plants,
  hypersensitivity reaction, 389–392,
    394
  nature of plant surfaces, 378, 392
  role of bacterial lipopolysaccharides,
    390–392, 394
  role of plant lectins, 390–392, 394
Bacterial motility and chemotaxis,
  and mucosal association, 236, 445
Bacteroides melaninogenicus,
  in mouth and saliva, 72, 78, 86, 87
Bacteroides oralis,
  in mouth and saliva, 72

Campylobacter,
  in mouth and saliva, 70, 72
Caries lesions,
  composition of bacterial plaques, 71
Cellular interactions,
  aggregation factor, 403, 404

Cellular interactions (*continued*)
egg–sperm interaction, 404, 407
embryonic chick retina, 416
fibronectin opsonic activity, 410,
411
marine sponge cell agglutination,
407–409
molecules involved, 431–433
role of galactolectins, 413–416
slime mold amoeba cells, 411, 412
slime mold discoidin, 412, 413
slime mold lectin receptors, 412
tissue-specific temporal regulation,
419, 420
yeast mating factors, 403, 404
Cervicitis,
bacterial adherence, 21
Cheek epithelial cells,
bacterial concentrations, 64
Chemotaxis,
see Bacterial motility and chemotaxis,
Clostridium spp.,
adherence to colon epithelia, 53
Columnar epithelium,
enterocytes, 35
goblet cells, 35
microbial adherence to, 52, 53
Coronal plaque,
bacterial concentration, 64
Corynebacterium renale,
fimbriae, 302, 304, 310, 334, 336
Corynebacterium spp.,
fimbriae, 302, 304, 310
Crypts of Lieberkuhn, 34, 35

Dental caries inhibition,
by dextranase, 127, 128
by immunization, 128, 129
Dental plaques, 65, 68
bacterial adsorption, 76, 82, 88,
96, 97
models for formation, 74–76
rate of formation, 71, 77, 88

Egg–sperm interaction.
egg sperm receptor, 405, 406

species specificity, 404–407
bindin-egg receptor on sperm, 406
407
Embryonic chick retina,
CAM protein, 416–420
cell-aggregating factor, 418
Endocarditis,
and bacterial adherence, 20
Endothelial cell,
streptococcal binding to, 14
Enterobacter cloacae,
adhesion to intestinal epithelia, 196
fimbriae, 298, 308
Enterobacteria,
haemagglutination, 188, 202
MRE adhesin activity, 190, 191, 195,
202–204, 206, 213
MR/K adhesin activity, 190, 191,
207, 214
MR/P adhesin activity, 190, 191,
214
MS adhesin activity, 190, 191, 195,
201, 213
MS adhesive fimbriae, 200, 201
Enterobacterial species,
types of adhesins, 192–194
Enterococci,
in mouth and saliva, 72
Enterocytes,
microvilli of, 35
Epithelial cell,
adherence, 5, 13, 19, 24, 25, 79,
445–447
Epithelial cell receptor, 172, 173
role of sugar residues, 8, 52
see also Receptors
Erythocytes,
lipoteichoic acid binding, 147–151
Escherichia coli,
adhesion to intestinal epithelia, 196,
205, 206
enterotoxin, 18, 19, 206
epithelial cell adherence, 13, 18, 23,
25, 201
fimbriae, 5, 189, 194, 298, 302, 308,
312, 334, 336
glucan synthesis, 111

Escherichia coli (*continued*)
glucosyltransferase binding, 109
haemagglutination, 188, 189, 202
hydrophobicity, 5, 199
K88 antigen, 19, 187, 206
MRE adhesins, 202–204
MRE adhesion and fimbriae, 205
MS adhesin, 194, 204
MS adhesin and fimbriae, 194, 195
MS haemagglutinin, 189
Eukaryotic cell interaction,
see Cellular interactions

Filamentous bacteria,
epithelial adherence, 54, 55
in mouth and saliva, 72
Fimbriae, 5, 10, 12, 15, 291
adhesion, 292
inhibition, 316, 317, 319,
324–328
inhibition by D-mannose, 199,
294, 296, 305, 306, 309, 311
313, 321, 322
amino acid composition of, 336–339
and bacterial adherence, 291, 396,
307, 309, 311–315,
317–320, 323
and haemagglutination, 294, 296, 297,
305–308, 310, 312, 320–322
and lipoteichoic acid, 154, 156
and M-protein, 154
and motility, 291
and pellicle formation, 291
antigenic structure of, 329–332
antisera against, 305–307, 314,
315, 330–332
binding site, 315–317, 319, 328
chemical structure of, 333–337,
339
classification of, 187, 211–213,
298–303
in phagocytosis, 23, 24
mannose receptor sites for, 199,
200, 321–323, 329
morphology of, 293–303
mutation occurrence of, 293

phase,
and cultural conditions, 197,
198, 202, 291, 293,
298–303
and environmental conditions,
198, 293
Fungal adherence to plants, 392, 393
Fusobacterium,
in mouth and saliva, 71, 72
Fusobacterium nucleatum,
glucosyltransferase binding, 109

GTase,
see Glucosyltransferase
Gastro-intestinal epithelia,
and adherence of,
Enterobacteriaceae, 196
lactobacilli, 41, 50
spirochetes, 41, 42, 44
streptococci, 41
Torulopsis, 41, 42, 47, 52
Vibrio cholerae, 235, 236
Gingival plaque,
bacterial concentration, 64, 67
Gingivitis,
composition of bacterial plaques,
71
Glucosyltransferase,
properties, 122, 123
Glycocalyx,
microbial adherence to, 40, 49, 54,
442
Gonococcal adhesion,
effect of gangliosides, 280–283
effect of pH, 276, 277, 284
human fallopian tube organ culture,
263, 274
Gonococci,
adherence to epithelial cells, 15,
254–257, 261–268, 272, 275,
276, 279
DLVO theory, 273–275, 282
electrostatic interactions, 270–272, 283
fimbriae, 16, 19, 258, 259, 263–267,
270, 276, 279, 284, 292, 295,
300, 310, 334

Gonococci (*continued*)
  fimbrial amino acids, 258, 336
  haemagglutination, 263, 265, 266,
    268, 273
  hydrophobicity, 278, 279, 283
  lipopolysaccharides, 261, 262, 265,
    267, 279, 280, 282
  outer membrane proteins, 259, 260,
    270, 283
  phagocytic attachment, 15, 257,
    263
  phagocytosis resistance, 269
  role of ionic bridging, 275

Hemophilus,
  in mouth and saliva, 72
Hydrophobicity,
  measurement, 5, 278, 360—363
Hydroxyapatite,
  bacterial adherence, 74, 75, 82—85,
    95

Infectious process, 3, 4. 18, 445—448
  predisposition to, 22
Intestinal epithelia,
  indigenous micro-organisms, 40,
    44, 45, 48, 49
  microbial adherence to, 36, 39,
    54, 196
  isotope scintillation counting, 37,
    38
Intestinal mucosa,
  microbial culture methods, 37, 38
  microscopy, 36—38

Klebsiella aerogenes,
  fimbriae, 298, 308
  MR/K adhesin and fimbriae,
    207—209
Klebsiella spp.,
  fimbriae, 298, 308
  MR/K adhesin, 208
  MS and MR/K adhesins, 208, 210

LTA,
  see Lipoteichoic acid

Lactobacilli,
  adherence to stomach epithelia,
    41, 46, 51
Lactobacillus,
  adherence to keratinized epithelia,
    50, 51
Lactobacillus spp.,
  in mouth and saliva, 72, 78
Lactobacillus casei,
  structure of lipoteichoic acid, 140
Lactobacillus fermentum,
  extracellular lipoteichoic acid, 145
  structure of lipoteichoic acid, 140
Lactobacillus helveticus,
  structure of lipoteichoic acid, 140
Lactobacillus plantarum,
  glucosyltransferase binding, 109
  glucan synthesis, 111
Lectins,
  see Cellular interactions; Platelet
    interactions; Receptors;
    Ligand; *Rhizobium* spp.,
Leptotrichia buccalis,
  glucosyltransferase binding, 109
Ligand, 6, 7, 9, 11
  antibodies to, 25
  binding sites, 7, 10
  genetic control, 12
  phenotypic variants, 12
  receptor interactions, 6, 7, 8
  see also individual organisms;
    Receptors; Bacterial adherence
Lipoteichoic acid, 10, 16, 139
  antibodies, 25, 155
  bacterial binding to teeth, 82, 147
  binding to erythrocytes, 147—151
  binding to lymphocytes, 152, 153
  binding to platelets, 151, 152
  erythrocytes sensitization, 147, 148
  excretion, 145, 155
  hydrophobic interaction, 155, 156
  location in bacteria, 143—145, 154
  phenol extraction, 141, 142, 155
  purification, 142, 143
  relationship to cell wall, 143, 144
  structure, 140

Marine bacteria,
  attachment in natural environment,
    347, 370
  attachment mechanisms, 350, 358,
    371
  attachment theory, 349, 350
  extracellular polymeric adhesives,
    348–350, 355, 358, 360,
    367, 371
  stages of attachment, 348, 349
Marine bacterial attachment,
  adhesive polymer analysis, 355, 356
  adsorption of adhesive, 360
  chemical treatment of adhesive,
    356–358
  cultural and environmental factors,
    364–367, 370
  electron microscopy, 350–354, 369
  hydrophobicity of substrate, 360–
    364
  inhibition by adsorbed polymers,
    363, 364
  inhibition of sequestering agents,
    367, 368
  physico-chemical properties of
    substrata, 358–360, 367–
    369
Meningococci,
  adherence to epithelial cells, 254,
    259
  capsules, 259, 269
  fimbriae, 258, 259, 300
  lipopolysaccharides, 261, 262
  outer membrane proteins, 261
  phagocytosis resistance, 269
Microvilli,
  glycocalyx, 35
  motility, 35
Motility,
  see Bacterial motility and chemotaxis
Moraxella spp.,
  fimbriae, 300, 310, 334, 336
Mycoplasma,
  cell membrane, 161, 163–166
  definition, 161
  ligands, 8
  morphology, 163

reproduction, 167
taxonomy, 162
Mycoplasma gallisepticum,
  morphology, 163
Mycoplasma infections,
  clinical features, 178, 179
  pathogenesis, 180
  pathology, 179
Mycoplasma pneumoniae,
  adherence mechanism, 169, 170
  anionic sites, 166
  attachment,
    to alveolar macrophages, 171
    to ciliated respiratory epithelium,
      173, 174, 176, 177, 180
    to platelets, 170
    to tissue cultures, 170, 171
    to white blood cells, 170
  hemadsorption, 168–170
  metabolism, 166, 167
  morphology, 163–165
  motility, 166
  organ culture, 173–176
  organ culture human fetal tracheal,
    176, 177
  phagocytosis, 170
  role of attachment, 167
  spermadsorption, 171
  tracheal epithelial cell adsorption,
    172, 173, 175
  virulence, 166, 173
Mycoplasma pulmonis,
  morphology, 163

Neisseria gonorrhoeae,
  fimbriae, *see* Gonococci, fimbriae,
    phagocytic resistance, 15
Neisseria infections,
  clinical aspects, 253, 254
Neisseria meningitidis,
  fimbriae, *see* Meningococci, fimbriae
Neisseria perflava,
  glucosyltransferase binding, 109
Neisseria spp.,
  fimbriae, 300, 310
  in mouth and saliva,
    72, 78

Oral epithelial cells,
    bacterial concentration, 79, 88

Phagocytic cells,
    bacterial attachment, 3, 14
    LTA receptors, 15
    surface hydrophobicity, 14
Phagocytosis,
    role of fimbriae, 23, 24
    resistance, 15, 16, 24
Pharyngitis,
    adherence *in vitro*, 20
Pili, *see* Fimbriae
Plaque matrix,
    bacterial adherence, 65, 67, 73, 99
Plasmids,
    adherence determination, 12
Platelet interactions, 420
    Glanzmann's thrombasthenia,
        427–430
    Glanzmann's thrombasthenia membrane
        defect, 428, 429
    plasma membrane composition and
        structure, 421, 422, 427
    platelet aggregation, 422–431
    platelet lectin activity, 422–425,
        429, 430
    platelet lectin receptor, 426–430
    thromblasthenic platelet lectins,
        428, 429
Proteus mirabilis,
    adherence, 23, 24
    epithelial attachment, 15
    fimbriae, 15, 19, 23, 298, 310
    glucosyltransferase binding, 109
    phagocytic attachment, 15
Proteus spp.,
    fimbriae, 298, 310
    MR/P adhesin and fimbriae, 210,
        211
Proteus vulgaris,
    fimbriae, 298
Pseudomonas aeruginosa,
    fimbriae, 300, 334, 336
Pseudomonas echinoides,
    fimbriae, 298, 310
Pseudomonas multivorans,

fimbriae, 300, 310
Pseudomonas spp.,
    fimbriae, 298
Pyelonephritis,
    and bacterial adherence *in vitro*, 21
Pyoderma,
    and bacterial adherence *in vitro*, 20

Receptors,
    accessibility, 10, 11
    for bacterial attachment, 6–11,
        442, 446, 447, 449
    ligand interaction, 6–10
    see also individual micro-organisms;
        Platelet interaction;
        lectin receptors; Cellular
        interactions; Egg–sperm
        interaction
Receptor analogues, 150, 151
Receptor substances in diet, 448–450
Rhizobium spp.,
    adherence to plant cell wall,
        385–389
    induction of root nodules, 385, 386
    and plant lectins, 387–389
    surface polysaccharides, 386–389

Saliva,
    adherence inhibitors, 85, 86
    bacterial agglutinins, 84, 85
    bacterial concentration, 64, 80, 86
Salmonella spp.,
    fimbriae, 19, 298, 308
Serratia marcescens,
    fimbriae, 298, 308
    glycan synthesis, 111
    glucosyltransferase binding, 109
Shigella flexneri,
    adhesion to intestinal epithelia, 196
    fimbriae, 298, 308
Spirochetes,
    adherence to cecum epithelia, 42
    adherence to colon epithelia, 44
    adherence to stomach epithelia, 41
    in mouth and saliva, 72
Squamous epithelium, keratinized,
    lactic acid bacteria adherence, 50, 51

Squamous epithelium, keratinized, (*continued*)
  microbial adherence, 43, 46, 49
Streptococci,
  adherence to esophageal epithelia, 41
  adherence to stomach epithelia, 41
  dextran producer, 19
  LTA receptors, 8, 16
  M protein, 16
  phagocytic attachment, 15
Streptococcus bovis,
  glucan synthesis, 111
Streptococcus, group E,
  glucan synthesis, 111
  glucosyltransferase binding, 109
Streptococcus faecalis,
  extracellular lipoteichoic acid, 145
  ^{3}H-lipoteichoic acid binding, 149
  structure of lipoteichoic acid, 140, 141, 143
Streptococcus lactis,
  structure of lipoteichoic acid, 140
Streptococcus mitis,
  glucan synthesis, 111
  in mouth and saliva, 72, 78
Streptococcus mutans,
  accumulation on teeth, 89, 90, 91, 115
  adherence to glass surface, 109, 110, 113, 114
  agglutination, 115–117, 119
  agglutination, dextran/glucan-induced, 116, 117, 119
  antibodies, 129, 130
  antisera, 128, 129
  attachment to tooth enamel, 70, 71, 81, 93, 107, 115, 119, 153
  binding site, 116–119
  caries induction, 107
  extracellular lipoteichoic acid, 145
  fructan synthesis, 126, 127
  genetic groups, 93
  glucan synthesis, 89–94, 107, 109, 110, 113, 115, 120, 124, 125
  glucosyltransferase binding, 109, 112, 113, 119, 121
  in mouth and saliva, 72, 78, 88, 119

  in supragingival plaques, 70, 71, 74
  plaque formation, 108, 115
  sucrose-dependent adherence, 107, 109, 110, 113, 117
Streptococcus pyogenes,
  binding to buccal mucosa, 154, 155
  binding to skin epithelia, 154
  fimbriae, 155
  glycosyltransferae binding, 109
  lipoteichoic acid, 154
  ^{3}H-lipoteichoic acid binding, 148, 149
  M protein, 154
Streptococcus salivarius,
  fructan synthesis, 126
  glucosyltransferase binding, 109
  in mouth and saliva, 72, 73, 78, 153
Streptococcus sanguis,
  glucan synthesis, 111, 124
  glucosyltransferase binding, 109
  in mouth and saliva, 72, 78, 153
Subgingival plaques, 67, 68
  associated with periodontitis, 67, 71
  bacterial composition, 67–69
Supragingival plasques,
  bacterial composition, 67, 68, 71

Teeth,
  acquired pellicles of, 73, 74
  bacterial adherence, 63, 67, 81
Teichoic acid, 139, 141
  see also Lipoteichoic acid
Torulopsis,
  adherence to stomach epithelia, 41, 42, 47, 52

Vaginitis,
  adherence *in vitro,* 21
Veillonella,
  in mouth and saliva, 72, 78
Vibrio cholerae,
  adhesins, 221, 222, 225, 226, 239–245
  adhesion inhibition, 241–243
  adhesion to brush-border membranes, 235–238, 241

Vibrio cholerae (*continued*)
  adhesion to intestinal mucosa,
        224–226, 235–239,
        242–245
  chemotaxis, 226, 236, 238, 241,
        243, 244
  fimbriae, 300, 310
  habitats, 221, 222, 224, 225, 235
  haemagglutinating activity,
        227–234, 240, 244
  haemagglutinins, 226, 233–235, 241
     D-mannose inhibition, 234, 241
  motility, 226, 236, 238, 241, 243,
        244
  taxonomy, 222, 223
Vibrio eltor,
  fimbriae, 300, 310
Vibrio spp.,
  fimbriae, 239, 240
  habitats, 223–225, 244